U0387611

教育部高等学校电子信息类专业教学指导委员会规划教材
高等学校电子信息类专业系列教材

Fundamentals
of Engineering Electromagnetics

工程电磁场基础

王薪　刘冰　陆明宇　主编

清华大学出版社
北京

内容简介

本书基于工科电类专业开设“电磁场理论”双语课的需求，紧扣国内高校电磁场理论教学的基本要求，并考虑与后续专业课程的有效衔接，详细阐述了电磁场与电磁波的基本概念、基本规律和基本分析方法。本书主要内容包括矢量、静电场、静电场边值问题、恒定电流场、恒定磁场、时变电磁场、平面电磁波及导行电磁波与谐振腔。

本书主要采用英文编写，对重要的概念、定理、定律及方法，提供中英文对照，让读者在掌握专业英文词汇和表达方法的同时，能够从中英文两种语言的角度对本课程的基本概念和基本理论有更为深入的理解，从而起到提高教学效果的目的。

图书在版编目(CIP)数据

工程电磁场基础=Fundamentals of Engineering Electromagnetics/王薪，刘冰，陆明宇主编. —北京：清华大学出版社，2022. 9

高等学校电子信息类专业系列教材

ISBN 978-7-302-61207-0

Ⅰ. ①工… Ⅱ. ①王… ②刘… ③陆… Ⅲ. ①电磁场-高等学校-教材 Ⅳ. ①O441. 4

中国版本图书馆 CIP 数据核字(2022)第 110201 号

责任编辑：王　芳
封面设计：李召霞
责任校对：李建庄
责任印制：刘海龙

出版发行：清华大学出版社
　　网　　址：http://www. tup. com. cn，http://www. wqbook. com
　　地　　址：北京清华大学学研大厦 A 座　　**邮　　编**：100084
　　社 总 机：010-83470000　　**邮　　购**：010-62786544
　　投稿与读者服务：010-62776969，c-service@tup. tsinghua. edu. cn
　　质量反馈：010-62772015，zhiliang@tup. tsinghua. edu. cn
　　课件下载：http://www. tup. com. cn，010-83470236
印 装 者：三河市龙大印装有限公司
经　　销：全国新华书店
开　　本：185mm×260mm　　**印　张**：20. 5　　**字　　数**：499 千字
版　　次：2022 年 11 月第 1 版　　**印　　次**：2022 年 11 月第1 次印刷
印　　数：1～1500
定　　价：79. 00 元

产品编号：093014-01

前言

PREFACE

双语教学是我国高等教育适应全球化趋势,增强学生国际交流与竞争能力的一项重要措施。"电磁场理论"作为电子信息与电气学科最重要的专业基础课程之一,包含的概念、原理和方法在后续多个课程中都有重要的体现。开设"电磁场理论"双语课,让学生在学习专业知识的过程中同时积累和掌握中英文相关专业词汇和表达方式,对于学生将来从事相应的科研和技术工作十分有益。而据编者所知,国内尚没有专门针对电磁场双语教学的教材。国内已开设相关双语课程的高校多采用英美原版英文教材,存在与国内电磁场理论教学要求不一致等问题。本书是在笔者多年从事"电磁场理论"双语教学经验的基础上编写而成,在内容安排、章节分配、知识侧重点和讲解顺序上更适合于国内双语教学。

由于理论性强,对学生的数学基础和空间想象力要求较高,"电磁场理论"普遍被认为是学习难度最大的电类基础课程之一。这主要表现在课程的概念公式多、理论抽象、数学方法较为复杂,学生在学习过程中容易产生迷失感。针对这一问题,笔者尝试在教材内容编排上突出知识点之间的逻辑联系,围绕"场"和"波"这两个大的概念体系,分别以矢量场的亥姆霍兹定理和亥姆霍兹方程在不同条件下的典型解作为逻辑主线,形成各自的核心问题,通过核心问题将各个部分的知识点串联形成易于掌握的逻辑链条,以帮助学生理解每个知识点在整个电磁场基础理论中的位置与作用,便于在学习过程中形成课程知识体系。

前5章的矢量分析与静态场着重建立"场"的概念体系。从矢量分析的亥姆霍兹定理出发,将矢量场的散度和旋度作为"场"的概念体系的核心,突出场源关系这一核心问题,即电、磁场与其通量源和漩涡源的关系。亥姆霍兹定理表明任何一个矢量场均能通过其散度和旋度确定,因此,本书组织"场"的概念体系的逻辑链条为:首先由实验定律出发确定场的通量源和漩涡源,即微分形式的场源关系,然后通过矢量分析工具得出相应物理定律,即积分形式的场源关系,在此基础上再衍生出其他物理概念和方法。具体而言,介绍静电场时,从库仑定律和电场强度的概念出发,推导得出静电场的通量源与涡旋源,进而推导出高斯定律和静电场的无旋特性,引申出利用高斯定律和标量电位求解电场分布的方法。介绍恒定磁场时,从磁感应强度的概念和比奥萨法定律出发,推导得出恒定磁场的通量源与漩涡源,进而推导出安培环路定律和恒定磁场的无散特性,形成利用安培环路定律或者矢量磁位求解磁场分布的方法。静态和时变场的边界条件是场源关系在不连续媒质分界面上的特殊体现。静电场边值问题的求解方法则给出了实际情况下电荷分布未知时分析场源关系的数学工具及其理论依据。介质的极化和磁化现象以及导体在电场中的特性都反应了不同媒质的存在对场源关系的影响。

第6章引入时变场,从法拉第电磁感应定律和时变条件下的电荷守恒定律出发,分别推导出时变电场和时变磁场的旋度,从而得出时变电场和时变磁场互为各自的漩涡源的结论,在此基础上推导得到波动方程，形成电磁波的概念，为后续章节过渡到波的概念体系打下基础。

后3章着重建立“波”的概念体系，突出无源区域的电磁波作为亥姆霍兹方程的解这一关键点，突出不同边界条件下亥姆霍兹方程的求解这一核心问题，以利于学生理解数学推导的目的性，将复杂的数学表达式与物理意义相结合。具体而言，在自由空间中，均匀平面波的概念作为亥姆霍兹方程的最简单的基本解被引入，在此基础上定义行波电磁波的波长、相速、群速、极化、波阻抗等概念。在无限大媒质分界面上，需要引入反射和折射的概念以形成同时满足亥姆霍兹方程与分界面边界条件的解，在此基础上定义反射系数、透射系数以及驻波的概念。在波导结构中，各个波导模式则代表了满足波导壁边界条件的亥姆霍兹方程的离散的特征解。在此基础上阐述不同波导模式对应的不同的截止频率、波导波长、相速度、波阻抗等概念。

针对双语教学要求，本书主要采用英文编写，对重要的概念、定理、定律以及方法，提供中英文对照，并配备了复习思考题和习题。本书的主要读者对象为电类工程专业的本科学生，特别适用于“电磁场理论”双语课程，以及针对国际留学生的“电磁场理论”本科课程教学，也可供一般的电子和电气工程专业技术人员参考。

因笔者水平所限，书中难免有不足之处，衷心欢迎读者和同行批评指正。

编著者
2022年10月

目 录

CONTENTS

Chapter 1 Vector Analysis(矢量分析)

1.1 Introduction(引言)

The physical quantities studied in electromagnetics can be classified into two categories: **scalars**(标量) and **vectors**(矢量). A scalar quantity, such as charge, current and energy, can be completely characterized by one real number along with a unit(单位). The real number together with its unit represents the magnitude of the scalar quantity. It should be noted that the real number can be either positive or negative. On the other hand, a vector quantity, such as force and speed intensity, requires three real numbers; the three-number representation is necessary because a vector quantity involves not only a magnitude but also a direction. ①

If a scalar or a vector quantity varies with respect to the spatial location, the space distribution of the scalar or vector is called a **scalar field**(标量场) or a **vector field**(矢量场). In a three-dimensional space, the three-number representation of a vector depends on the choice of a coordinate system. However, the choice of coordinate system has no impact on the physical meaning of the vector or the related physical laws and theorems. ②

The physical laws of electromagnetics are described by the differentiation and/or integration of scalar and vector fields. In this chapter, we define and discuss various kinds of differential and integral operators involving scalars and vectors. These operators are fundamental in formulating the electromagnetic theories. ③

1.2 Vector Representation(矢量的表示方式)

It is very important to distinguish a vector from a scalar in terms of notations. In this text, a vector is represented by an italic boldfaced letter(斜粗体字母), and a scalar is represented by

---

① 电磁场理论中涉及的物理量分为矢量和标量。标量是单个数加上单位就能确定的物理量(例如电荷、电流、电位、能量、功率等),通常是实数,可正可负,代表了该物理量的大小。矢量是既有大小又有方向的物理量(例如电场强度矢量和磁场强度矢量),在三维空间中,需要有三个数来表示。

② 电磁场理论中研究的标量和矢量都有可能是空间的函数,即空间中每一点对应一个唯一的标量或者矢量,这样形成的标量或者矢量空间分布被称为标量场或者矢量场。

③ 本章在引入矢量的概念及其代数运算的基础上,重点阐述标量场和矢量场的微分、积分运算,相关的概念和定理是电磁场理论建立的基础。

a regular italic letter. It is also a common practice to use an arrow or a bar over a letter as the notation of a vector, though it is not adopted in this book. ①

A vector $\boldsymbol{A}$ can be written as

$$\boldsymbol{A} = \boldsymbol{e}_A A \tag{1.1}$$

where $\boldsymbol{e}_A$ is a dimensionless(无量纲的) **unit vector(单位矢量)** with a unity magnitude(单位幅度) and the same direction of $\boldsymbol{A}$. A is the magnitude(with the unit) of $\boldsymbol{A}$, i. e. , ②

$$A = |\boldsymbol{A}| \tag{1.2}$$

which is always nonnegative for real vectors(实矢量). For complex vectors(复矢量), their magnitude may be complex. Both the magnitude and direction of a complex vector need to be understood differently. The complex vectors are usually used as the phasor form of a time-harmonic vector, which will be introduced and discussed in subsequent chapters.

From(1.1) and (1.2), we have

$$\boldsymbol{e}_A = \frac{\boldsymbol{A}}{|\boldsymbol{A}|} = \frac{\boldsymbol{A}}{A} \tag{1.3}$$

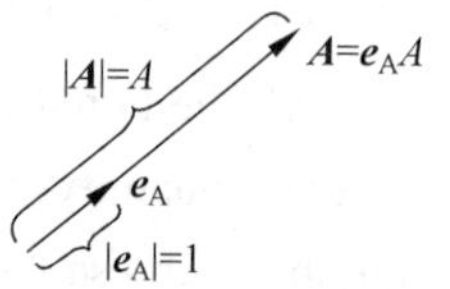

Figure 1-1 Graphical representation of vector $\boldsymbol{A}$

A vector $\boldsymbol{A}$ can be represented graphically by a directed line segment(有向线段) with a magnitude A and an arrowhead pointing in the direction of $\boldsymbol{e}_A$, as shown in Figure 1-1. It must be noted that A in Figure 1-1 does not have to represent a length. For example, if $\boldsymbol{A}$ is a speed vector, the length of A in Figure 1-1 indicates "how fast $\boldsymbol{A}$ is" rather than "how long $\boldsymbol{A}$ is."

A vector is the **zero vector** (or **null vector**, 零矢量), denoted by $\boldsymbol{0}$, if and only if its magnitude is zero. The direction of the zero vector is of no significance, and therefore, is not defined. ③

Two vectors are equal to each other if and only if they have the same magnitude and the same direction, even if they appear in different locations in space.

Two vectors are negative to each other if and only if they have the same magnitude but the opposite direction. The negative of a vector $\boldsymbol{A}$ is denoted as $-\boldsymbol{A}$. Therefore,

$$-\boldsymbol{A} = (-\boldsymbol{e}_A)A \tag{1.4}$$

where $-\boldsymbol{e}_A$ is a unit vector pointing toward the opposite direction of $\boldsymbol{e}_A$.

1.3 Addition and Subtraction(矢量的加减法)

Addition of two vectors $\boldsymbol{A}$ and $\boldsymbol{B}$ is defined by the parallelogram rule(平行四边形法则) or the head-to-tail rule(头尾连接法则) as shown in Figure 1-2.

---

① 本书中以粗体或者带箭头的斜体字母代表矢量(例如 $\boldsymbol{A}$),以常规的斜体字母代表标量(例如 u)。

② 矢量的幅度 A,作为物理量的大小,是有量纲的。而单位矢量 $\boldsymbol{e}_A$ 则没有量纲,其幅度就是实数 1。

③ 零矢量是幅度为零的特殊矢量,由于幅度为零,其方向就没有什么意义,可以认为其方向没有被定义,也可以认为其方向为任意方向。

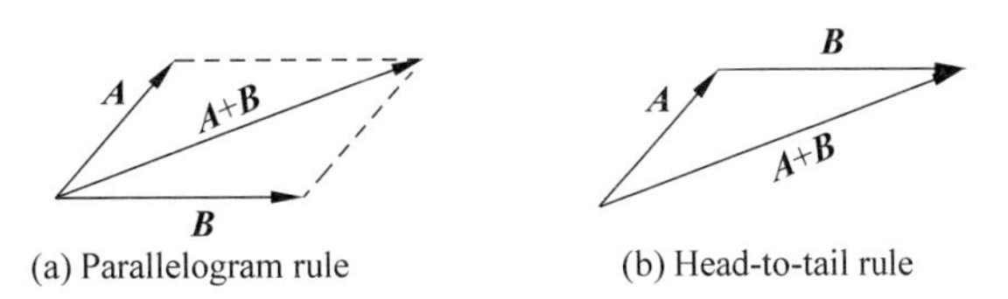

Figure 1-2 Vector addition: $\boldsymbol{A}+\boldsymbol{B}$

It is obvious that vector addition obeys the **commutative** and **associative laws**(矢量相加满足交换律与结合律),i. e.

$$\boldsymbol{A}+\boldsymbol{B}=\boldsymbol{B}+\boldsymbol{A} \tag{1.5}$$

$$\boldsymbol{A}+(\boldsymbol{B}+\boldsymbol{C})=(\boldsymbol{A}+\boldsymbol{B})+\boldsymbol{C} \tag{1.6}$$

Subtraction of a vector $\boldsymbol{B}$ from another vector $\boldsymbol{A}$ can be defined as the vector resulted from adding the negative of the vector $\boldsymbol{B}$ to $\boldsymbol{A}$,i. e. ,

$$\boldsymbol{A}-\boldsymbol{B}=\boldsymbol{A}+(-\boldsymbol{B}) \tag{1.7}$$

Figure 1-3 gives the graphical representation of vector subtraction. Obviously,for any two vectors $\boldsymbol{A}$ and $\boldsymbol{B}$, $\boldsymbol{A}-\boldsymbol{B}=\boldsymbol{0}$ if and only if $\boldsymbol{A}=\boldsymbol{B}$, which is useful for proving two vectors are identical.

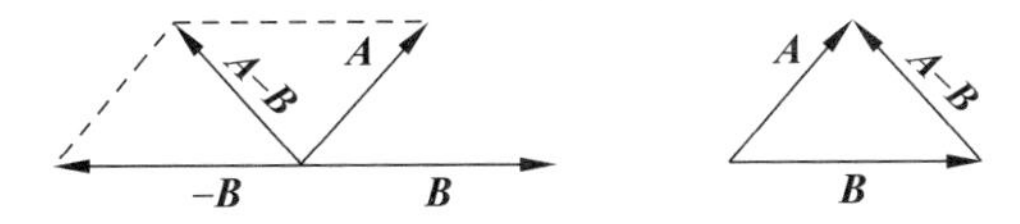

Figure 1-3 Vector subtraction: $\boldsymbol{A}-\boldsymbol{B}$

1.4 Products of Vectors(矢量乘积)

1.4.1 Multiplication by scalars(数乘)

Multiplying a vector $\boldsymbol{A}$ by a scalar k is equivalent to scaling the vector $\boldsymbol{A}$ by k times,i. e. ,

$$k\boldsymbol{A}=\boldsymbol{e}_{\mathrm{A}}(kA) \tag{1.8}$$

Notice that,if k is a positive number,the resulting vector has the same direction as $\boldsymbol{A}$. But if k is a negative number, the resulting vector is in the opposite direction of $\boldsymbol{A}$ and its unit vector becomes $-\boldsymbol{e}_{\mathrm{A}}$.

1.4.2 Dot Product/Scalar Product(点积/标量积)

The **dot product**(点积),or **scalar product**(标量积) of two vectors $\boldsymbol{A}$ and $\boldsymbol{B}$(denoted by $\boldsymbol{A}\cdot\boldsymbol{B}$) is defined to be a scalar that equals the product of the magnitude of $\boldsymbol{A}$,the magnitude of $\boldsymbol{B}$,and the cosine of the angle between them,i. e. ,

$$\boldsymbol{A}\cdot\boldsymbol{B}\triangleq AB\cos\theta_{\mathrm{AB}} \tag{1.9}$$

where θ_{AB} is the angle between $\boldsymbol{A}$ and $\boldsymbol{B}$ as illustrated in Figure 1-4.

Based on the above definition,the dot product of two non-zero vectors,$\boldsymbol{A}$ and $\boldsymbol{B}$,satisfies the following properties:

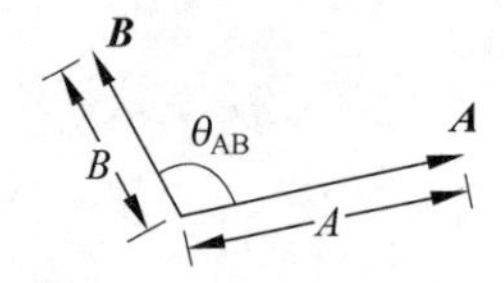

Figure 1-4 Graphical illustration of the dot product of two vectors **A** and **B**

(1) $|\boldsymbol{A}\cdot\boldsymbol{B}|\leqslant AB$, where the equal sign is valid only when **A** and **B** have the same direction;

(2) $\boldsymbol{A}\cdot\boldsymbol{B}>0$, if $\theta_{AB}<90°$; and $\boldsymbol{A}\cdot\boldsymbol{B}<0$ if $\theta_{AB}>90°$;

(3) $\boldsymbol{A}\cdot\boldsymbol{B}=0$, if **A** and **B** are **perpendicular**(垂直) to each other, i. e. , $\theta_{AB}=90°$.

The dot product is directly related to the vector **projection**(投影). As is shown in Figure 1-5, the projection of **B** upon the direction of **A** is $B\cos\theta_{AB}$. According to (1.9), the dot product $\boldsymbol{A}\cdot\boldsymbol{B}$ is equal to the product of the magnitude of **A** and the projection of **B** onto the direction of **A**. The projection $B\cos\theta_{AB}$ is obviously equal to the dot product $\boldsymbol{B}\cdot\boldsymbol{e}_A$, which is also called the **scalar component**(标量分量), or simply the **component**(分量) of **B** along the direction of **A**. The scalar component together with the unit vector $\boldsymbol{e}_A$ constitutes the **vector component** (**矢量分量**) along the direction of **A**. In other words, the vector component of **B** along the direction of **A** can be written as $(\boldsymbol{B}\cdot\boldsymbol{e}_A)\boldsymbol{e}_A$. [①]

Figure 1-5 Graphical illustration of projection of a vector **B** upon the direction of a vector **A**

From (1.9), it is evident that

$$\boldsymbol{A}\cdot\boldsymbol{A}=A^2 \text{ or } A=\sqrt{\boldsymbol{A}\cdot\boldsymbol{A}} \tag{1.10}$$

which is useful for finding the magnitude of a vector.

The dot product is commutative and distributive(可分配的,即满足分配律), i. e. ,

$$\boldsymbol{A}\cdot\boldsymbol{B}=\boldsymbol{B}\cdot\boldsymbol{A} \tag{1.11}$$

$$\boldsymbol{A}\cdot(\boldsymbol{B}+\boldsymbol{C})=\boldsymbol{A}\cdot\boldsymbol{B}+\boldsymbol{A}\cdot\boldsymbol{C} \tag{1.12}$$

It is not difficult to prove the above two identities from the definition of the dot product (1.9).

Example 1-1 Prove the law of cosines for a triangle(三角形的余弦定律) by using dot products of vectors.

Solution: Referring to Figure 1-6, the law of cosines states that

$$C=\sqrt{A^2+B^2-2AB\cos\theta_{AB}}$$

Obviously, C is the length of the vector $\boldsymbol{C}=\boldsymbol{A}-\boldsymbol{B}$. Therefore,

$$C^2=\boldsymbol{C}\cdot\boldsymbol{C}=(\boldsymbol{A}-\boldsymbol{B})\cdot(\boldsymbol{A}-\boldsymbol{B})=\boldsymbol{A}\cdot\boldsymbol{A}+\boldsymbol{B}\cdot\boldsymbol{B}-2\boldsymbol{A}\cdot\boldsymbol{B}=A^2+B^2-2AB\cos\theta_{AB}$$

---

① 一个矢量在某个方向上的投影可以表示为该矢量与代表该方向的单位矢量的点积。该投影也被称为矢量在该方向上的标量分量,或者简称分量。标量分量与其所在方向上的单位矢量的结合被称为该方向上的矢量分量。

which proves the law of cosines.

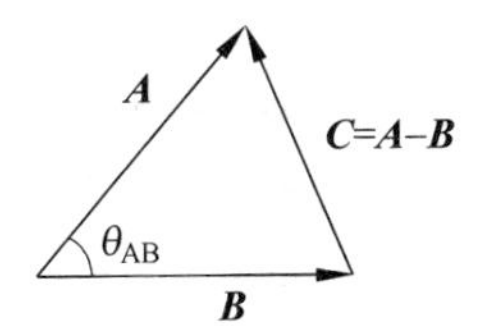

Figure 1-6 Illustration of Example 1-1

1.4.3 Cross Product/Vector Product(叉积/矢量积)

The cross or vector product of two vectors $\boldsymbol{A}$ and $\boldsymbol{B}$, denoted by $\boldsymbol{A}\times\boldsymbol{B}$, is defined as

$$\boldsymbol{A}\times\boldsymbol{B} \triangleq \boldsymbol{e}_{\mathrm{n}} AB\sin\theta_{\mathrm{AB}} \tag{1.13}$$

where θ_{AB} is the smaller angle between $\boldsymbol{A}$ and $\boldsymbol{B}$, and $\boldsymbol{e}_{\mathrm{n}}$ is the unit vector perpendicular to both $\boldsymbol{A}$ and $\boldsymbol{B}$. Obviously, a vector perpendicular to both $\boldsymbol{A}$ and $\boldsymbol{B}$ have two possible directions that are opposite to each other. The direction of $\boldsymbol{e}_{\mathrm{n}}$ in(1.13) is uniquely determined by the right-hand rule(右手法则) as is illustrated in Figure 1-7: when the right-hand fingers rotate from $\boldsymbol{A}$ to $\boldsymbol{B}$ through the angle θ_{AB}, the thumb points along the direction of $\boldsymbol{e}_{\mathrm{n}}$. Notice that, ($B\sin\theta_{\mathrm{AB}}$) is the height of the parallelogram(平行四边形) formed by the vectors $\boldsymbol{A}$ and $\boldsymbol{B}$, and hence, the magnitude of $\boldsymbol{A}\times\boldsymbol{B}$ equals to the area of the parallelogram.

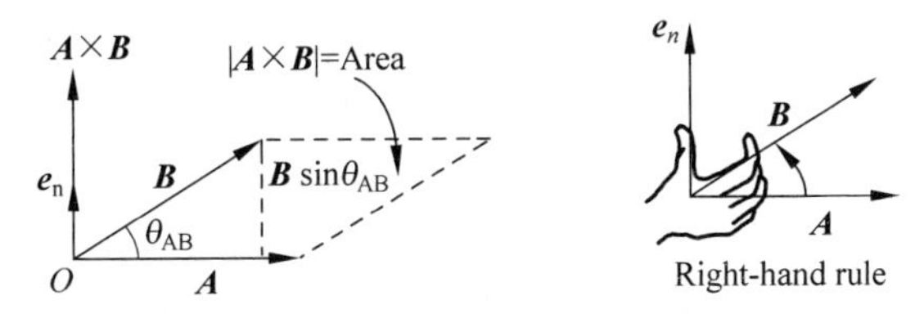

Figure 1-7 Cross product of vectors $\boldsymbol{A}$ and $\boldsymbol{B}$

Using the definition in(1.13) and following the right-hand rule, we find that

$$\boldsymbol{B}\times\boldsymbol{A} = -\boldsymbol{A}\times\boldsymbol{B} \tag{1.14}$$

Hence the cross product is NOT commutative(不满足交换律). However, we can prove that the cross product satisfies the distributive law(满足分配律),

$$\boldsymbol{A}\times(\boldsymbol{B}+\boldsymbol{C}) = \boldsymbol{A}\times\boldsymbol{B}+\boldsymbol{A}\times\boldsymbol{C} \tag{1.15}$$

It is easy to see that the vector product is NOT associative(不满足结合律), i.e.,

$$\boldsymbol{A}\times(\boldsymbol{B}\times\boldsymbol{C}) \neq (\boldsymbol{A}\times\boldsymbol{B})\times\boldsymbol{C}$$

Therefore, the order in which the two vector products are performed is vital, and the parentheses in the above expression should not be omitted.

1.4.4 Scalar and vector triple products(标量/矢量三重积)

There are two important vector identities(矢量恒等式) involving products of three vectors. One is about the **scalar triple product**(**标量三重积**) and the other is about the **vector triple product**(**矢量三重积**). The scalar triple product satisfies the following identity:

$$\boldsymbol{A}\cdot(\boldsymbol{B}\times\boldsymbol{C}) = \boldsymbol{B}\cdot(\boldsymbol{C}\times\boldsymbol{A}) = \boldsymbol{C}\cdot(\boldsymbol{A}\times\boldsymbol{B}) \tag{1.16}$$

Notice that the order of the three vectors in (1.16) follow the same the cyclic permutation **A**-**B**-**C**.①

(1.16) can be proved by noticing that the three vector triple products all have a magnitude equal to the volume of the parallelepiped(平行六面体) formed by the three vectors **A**, **B** and **C**. Take $\boldsymbol{A}\cdot(\boldsymbol{B}\times\boldsymbol{C})$ as an example. As is seen in Figure 1-8, the parallelepiped has a base with an area equal to $|\boldsymbol{B}\times\boldsymbol{C}|$. With the unit vector of $\boldsymbol{B}\times\boldsymbol{C}$ to be $\boldsymbol{e}_n$, the height of the parallelepiped is equal to $|\boldsymbol{A}\cdot\boldsymbol{e}_n|$. Hence the volume is $|\boldsymbol{A}\cdot\boldsymbol{e}_n||\boldsymbol{B}\times\boldsymbol{C}|=|\boldsymbol{A}\cdot(\boldsymbol{B}\times\boldsymbol{C})|$. Obviously, $\boldsymbol{A}\cdot(\boldsymbol{B}\times\boldsymbol{C})$ is equal to the positive or negative volume of the parallelepiped depending on whether the angle between the vector **A** and $(\boldsymbol{B}\times\boldsymbol{C})$ is smaller or larger than 90 degrees. Similarly, it is easy to prove that $\boldsymbol{B}\cdot(\boldsymbol{C}\times\boldsymbol{A})$ and $\boldsymbol{C}\cdot(\boldsymbol{A}\times\boldsymbol{B})$ are also equal to either the positive or negative volume of the same parallelepiped, and should all be equal to $\boldsymbol{A}\cdot(\boldsymbol{B}\times\boldsymbol{C})$.②

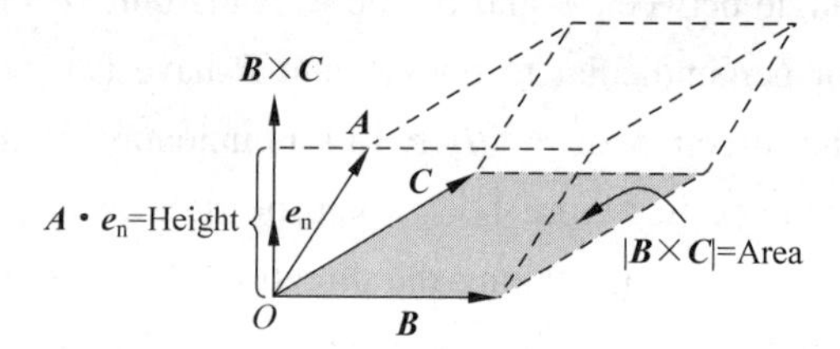

Figure 1-8 Illustrating the geometric meaning of $\boldsymbol{A}\cdot(\boldsymbol{B}\times\boldsymbol{C})$

Another vector identity involving the vector triple product $\boldsymbol{A}\times(\boldsymbol{B}\times\boldsymbol{C})$ is

$$\boldsymbol{A}\times(\boldsymbol{B}\times\boldsymbol{C})=\boldsymbol{B}(\boldsymbol{A}\cdot\boldsymbol{C})-\boldsymbol{C}(\boldsymbol{A}\cdot\boldsymbol{B}) \tag{1.17}$$

(1.17) is also known as the "BAC-CAB" rule, whose proof is left as an exercise.

1.5 Orthogonal Coordinate Systems(正交坐标系)

A coordinate system is needed to mathematically specify a point in the space. Although the laws of electromagnetism are invariant with **coordinate system**(坐标系), solutions to practical problems may be simplified when an appropriate coordinate system is selected.

In a three-dimensional space, a point can be located as the intersection of three surfaces, represented by u_i = constant (i=1, 2, or 3). These surfaces may be planar or curved surfaces. When these three surfaces are mutually perpendicular to one another, an **orthogonal coordinate system**(正交坐标系) is defined. u_i (i=1, 2, or 3) are the **coordinates**(坐标) of the orthogonal coordinate system. The **origin**(原点) of the coordinate system is the point with $u_1=u_2=u_3=0$. The origin is usually denoted by O in a coordinate system. Here we introduce three typical

---

① 注意式(1.16)中三项从左往右的循环排列都是**A**-**B**-**C**(与**B**-**C**-**A**、**C**-**A**-**B**是同一个循环排列)。交换点积顺序不会影响该循环排列，也不会影响最终乘积的结果。但交换叉积的顺序则会导致相反的循环排列**A**-**C**-**B**(与**B**-**A**-**C**、**C**-**B**-**A**是同一个循环排列)，并导致最终乘积相差一个负号。

② $\boldsymbol{A}\cdot(\boldsymbol{B}\times\boldsymbol{C})$是一个可正可负的标量，其绝对值等于**A**、**B**、**C**三个矢量构成的平行六面体的体积。从几何上可以看出，若**A**与$(\boldsymbol{B}\times\boldsymbol{C})$的夹角小于90°，**B**与$(\boldsymbol{C}\times\boldsymbol{A})$的夹角以及**C**与$(\boldsymbol{A}\times\boldsymbol{B})$的夹角都小于90°。式(1.16)中标量三重积均为正数。反之则均为负数。

orthogonal coordinate systems:

(1) **Cartesian**(or **rectangular**, or x-y-z) **coordinates**(笛卡儿坐标/直角坐标/x-y-z 坐标);

(2) **Cylindrical coordinates**(圆柱坐标);

(3) **Spherical coordinates**(圆球坐标).

In this section, expressions of vectors and vector algebra are given in each of the above three coordinate systems, with the following fundamental concepts.

(1) **Base vectors**(基矢量), denoted by $\boldsymbol{e}_{u1}$, $\boldsymbol{e}_{u2}$ and $\boldsymbol{e}_{u3}$, are the three unit vectors in the three coordinate directions. $\boldsymbol{e}_{ui}$ is everywhere perpendicular to the surface u_i = constant and pointing to the direction along which the coordinate u_i increases. Any vector can be written as the sum of its vector components in the three orthogonal directions.

(2) **Position vector**(位置矢量), denoted by $\boldsymbol{r}$, is a vector associated with a point in the space, pointing from the origin to the point(u_1, u_2, u_3) as shown in Figure 1-9. The position vector has different expressions in different coordinate systems. ①

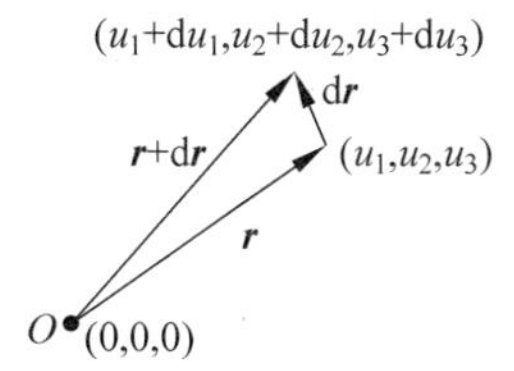

Figure 1-9 Illustration of position vector $\boldsymbol{r}$ and differential length vector d$\boldsymbol{r}$

(3) **Differential length vector**(微分长度矢量、微分线元矢量), denoted by d$\boldsymbol{l}$ or d$\boldsymbol{r}$, is a vector corresponding to a differential change of the position vector $\boldsymbol{r}$ as shown in Figure 1-9. d$\boldsymbol{r}$ will be used when line integrations are conducted, in which case is often written as d$\boldsymbol{l}$. ②

(4) **Differential area vector**(微分面元矢量), denoted by d$\boldsymbol{s}$, is a vector with its magnitude being the area of a differential element(ds) on a surface and its direction($\boldsymbol{e}_n$) normal to the surface(沿表面法向方向). In another word, d$\boldsymbol{s}$=$\boldsymbol{e}_n$ ds. Differential area vector is useful in finding the **flux**(通量、流量) of a vector field through a surface. ③

(5) **Differential volume**(微分体积元), dv is a scalar corresponding to the volume formed by differential coordinate changes du_1, du_2 and du_3 in directions $\boldsymbol{e}_{u1}$, $\boldsymbol{e}_{u2}$ and $\boldsymbol{e}_{u3}$ respectively. dv will be used when volume integrations are evaluated.

1.5.1 Cartesian Coordinates(笛卡儿坐标)

In a Cartesian coordinate,

$$(u_1, u_2, u_3) = (x, y, z)$$

---

① 位置矢量与空间中的位置一一对应,因此位置矢量本身也是空间位置的函数,是一个矢量场。

② 位置矢量 $\boldsymbol{r}$ 发生微小变化时,变化后的位置对应($\boldsymbol{r}$+d$\boldsymbol{r}$)。微分长度矢量 d$\boldsymbol{r}$ 也就是对位置矢量 $\boldsymbol{r}$ 的微分。d$\boldsymbol{r}$ 用于对矢量场作线积分时,通常将 d$\boldsymbol{r}$ 写成 d$\boldsymbol{l}$,代表积分路径上的一段线元矢量。

③ 微分面元矢量的方向是垂直于该面元的法向方向。显然,给定面元的法向有相反的两个方向,具体选取哪个方向作为微分面元矢量的方向将在 1.6 节中给出具体说明。

A point $P(x_0, y_0, z_0)$ in Cartesian coordinates is the intersection of three planes specified by $x=x_0$, $y=y_0$ and $z=z_0$ as shown in Figure 1-10.

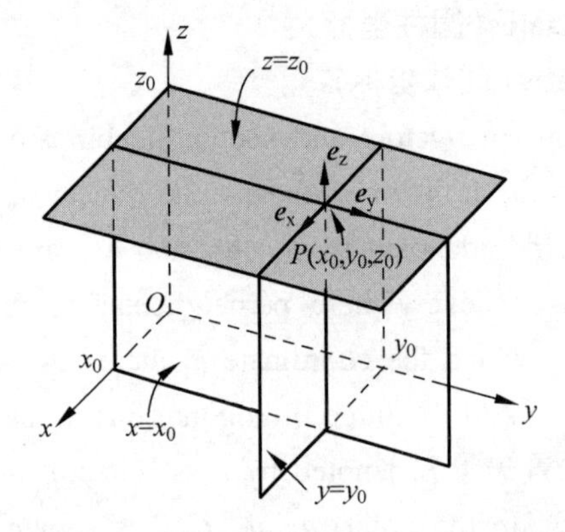

Figure 1-10 Illustration of Cartesian coordinates

It is a right-handed system with base vectors $\boldsymbol{e}_x$, $\boldsymbol{e}_y$ and $\boldsymbol{e}_z$ satisfying the following relations:

$$\begin{aligned} \boldsymbol{e}_x \times \boldsymbol{e}_y &= \boldsymbol{e}_z \\ \boldsymbol{e}_y \times \boldsymbol{e}_z &= \boldsymbol{e}_x \\ \boldsymbol{e}_z \times \boldsymbol{e}_x &= \boldsymbol{e}_y \end{aligned} \tag{1.18}$$

Apparently,

$$\begin{aligned} \boldsymbol{e}_x \cdot \boldsymbol{e}_y = \boldsymbol{e}_y \cdot \boldsymbol{e}_z = \boldsymbol{e}_x \cdot \boldsymbol{e}_z = 0 \\ \boldsymbol{e}_x \cdot \boldsymbol{e}_x = \boldsymbol{e}_y \cdot \boldsymbol{e}_y = \boldsymbol{e}_z \cdot \boldsymbol{e}_z = 1 \end{aligned} \tag{1.19}$$

A vector $\boldsymbol{A}$ in Cartesian coordinates can be written as

$$\boldsymbol{A} = \boldsymbol{e}_x A_x + \boldsymbol{e}_y A_y + \boldsymbol{e}_z A_z \tag{1.20}$$

where A_x, A_y and A_z are the components of $\boldsymbol{A}$ along x-, y- and z-directions respectively. In other words, A_x, A_y and A_z are the projections of $\boldsymbol{A}$ onto $\boldsymbol{e}_x$, $\boldsymbol{e}_y$ and $\boldsymbol{e}_z$:

$$\begin{aligned} A_x &= \boldsymbol{A} \cdot \boldsymbol{e}_x \\ A_y &= \boldsymbol{A} \cdot \boldsymbol{e}_y \\ A_z &= \boldsymbol{A} \cdot \boldsymbol{e}_z \end{aligned} \tag{1.21}$$

(1.21) can be readily derived by using the orthogonal relations in (1.19). The dot product of two vectors $\boldsymbol{A}$ and $\boldsymbol{B}$ is

$$\boldsymbol{A} \cdot \boldsymbol{B} = A_x B_x + A_y B_y + A_z B_z \tag{1.22}$$

which can be easily derived by using the distributive law (1.12) and the orthogonal relations in (1.19). Similarly, we can derive the expression for the cross product by using the distributive law (1.15) as

$$\begin{aligned} \boldsymbol{A} \times \boldsymbol{B} &= \boldsymbol{e}_x (A_y B_z - A_z B_y) + \boldsymbol{e}_y (A_z B_x - A_x B_z) + \boldsymbol{e}_z (A_x B_y - A_y B_x) \\ &= \begin{vmatrix} \boldsymbol{e}_x & \boldsymbol{e}_y & \boldsymbol{e}_z \\ A_x & A_y & A_z \\ B_x & B_y & B_z \end{vmatrix} \end{aligned} \tag{1.23}$$

The position vector associated with a point (x, y, z) is

$$\boldsymbol{r}=\boldsymbol{e}_x x+\boldsymbol{e}_y y+\boldsymbol{e}_z z \tag{1.24}$$

Since $\boldsymbol{e}_x$, $\boldsymbol{e}_y$ and $\boldsymbol{e}_z$ have unit magnitude and fixed direction independent of the position in the space, they are **constant vectors**(常矢量). Differential change of the vector $\boldsymbol{r}$ due to differential changes in the x, y and z coordinates can be obtained by taking differential on both sides of (1.24), which leads to

$$\mathrm{d}\boldsymbol{l}=\mathrm{d}\boldsymbol{r}=\boldsymbol{e}_x\mathrm{d}x+\boldsymbol{e}_y\mathrm{d}y+\boldsymbol{e}_z\mathrm{d}z \tag{1.25}$$

In Cartesian coordinates, three typical differential area vectors along the three coordinate directions are

$$\begin{aligned}\mathrm{d}\boldsymbol{s}_x&=\boldsymbol{e}_x\mathrm{d}y\mathrm{d}z=\boldsymbol{e}_x\mathrm{d}s_x\\ \mathrm{d}\boldsymbol{s}_y&=\boldsymbol{e}_y\mathrm{d}z\mathrm{d}x=\boldsymbol{e}_y\mathrm{d}s_y\\ \mathrm{d}\boldsymbol{s}_z&=\boldsymbol{e}_z\mathrm{d}x\mathrm{d}y=\boldsymbol{e}_z\mathrm{d}s_z\end{aligned} \tag{1.26}$$

as illustrated in Figure 1-11. Specifically, $\mathrm{d}\boldsymbol{s}_x$ is a differential area vector associated with a differential area element in the $y-z$ plane. The differential area element has a width $\mathrm{d}y$ and height $\mathrm{d}z$, and its **normal direction**(法向方向) is along $\boldsymbol{e}_x$. Similarly, $\mathrm{d}\boldsymbol{s}_y$ and $\mathrm{d}\boldsymbol{s}_z$ are the differential area vectors along $\boldsymbol{e}_y$ and $\boldsymbol{e}_z$ directions respectively. The expressions in(1.26) are particularly useful for surface integrals on surfaces parallel to the coordinate planes.

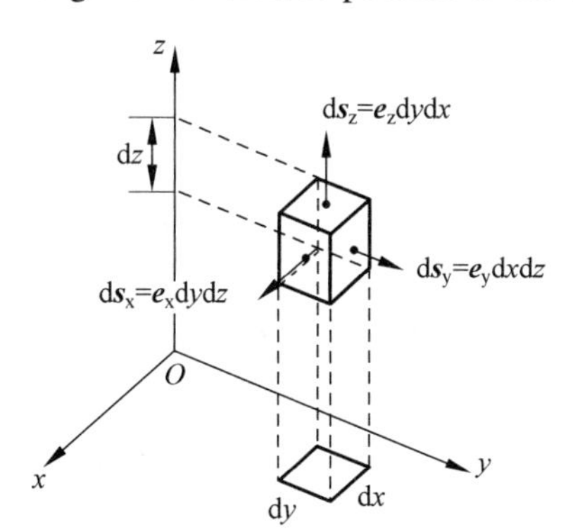

Figure 1-11 Differential area and volume in Cartesian coordinates

The differential volume associated with the differential changes $\mathrm{d}x$, $\mathrm{d}y$ and $\mathrm{d}z$ is expressed as

$$\mathrm{d}v=\mathrm{d}x\mathrm{d}y\mathrm{d}z \tag{1.27}$$

1.5.2 Cylindrical Coordinates(圆柱坐标)

In a cylindrical coordinate system,

$$(u_1,u_2,u_3)=(\rho,\phi,z)$$

A point $P(\rho_0,\phi_0,z_0)$ is determined by the intersection of a circular cylindrical surface $\rho=\rho_0$, a half-plane $\phi=\phi_0$, and a plane $z=z_0$ as is shown in Figure 1-12. The coordinate ρ is the distance from the z-axis, and the coordinate ϕ is the angle between the positive x-axis and the projection of the line segment OP on the $x-y$ plane(or the $z=0$ plane). The base vector $\boldsymbol{e}_\rho$ is everywhere perpendicular to the cylindrical surface, whereas the base vector $\boldsymbol{e}_\phi$ is everywhere tangential(切向的) to the cylindrical surface. The right-hand relations of $\boldsymbol{e}_\rho$, $\boldsymbol{e}_\phi$ and $\boldsymbol{e}_z$ are:

$$\boldsymbol{e}_\rho \times \boldsymbol{e}_\phi = \boldsymbol{e}_z$$
$$\boldsymbol{e}_\phi \times \boldsymbol{e}_z = \boldsymbol{e}_\rho \tag{1.28}$$
$$\boldsymbol{e}_z \times \boldsymbol{e}_\rho = \boldsymbol{e}_\phi$$

Apparently,

$$\begin{aligned} \boldsymbol{e}_\rho \cdot \boldsymbol{e}_\phi = \boldsymbol{e}_\phi \cdot \boldsymbol{e}_z = \boldsymbol{e}_\rho \cdot \boldsymbol{e}_z = 0 \\ \boldsymbol{e}_\rho \cdot \boldsymbol{e}_\rho = \boldsymbol{e}_\phi \cdot \boldsymbol{e}_\phi = \boldsymbol{e}_z \cdot \boldsymbol{e}_z = 1 \end{aligned} \tag{1.29}$$

Different from the base vectors in Cartesian coordinates, $\boldsymbol{e}_\rho$ and $\boldsymbol{e}_\phi$ are not constant vectors, because their directions may be different at different locations in the space. As shown in Figure 1-13,

$$\boldsymbol{e}_\rho \cdot \boldsymbol{e}_x = \cos\phi \tag{1.30a}$$

$$\boldsymbol{e}_\rho \cdot \boldsymbol{e}_y = \sin\phi \tag{1.30b}$$

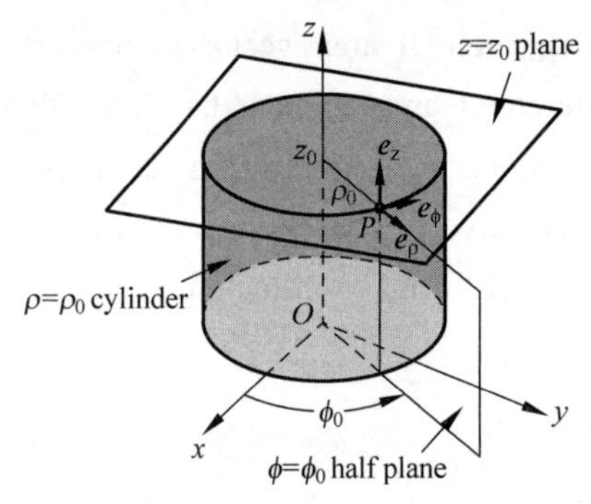

Figure 1-12 Illustration of cylindrical coordinates

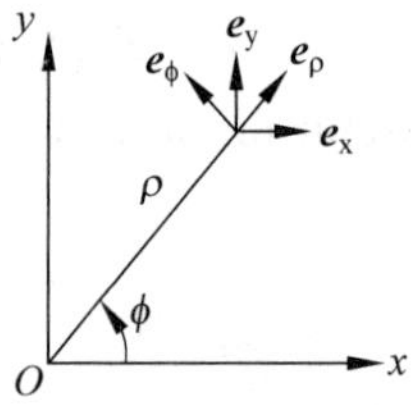

Figure 1-13 Relations among $\boldsymbol{e}_\rho$, $\boldsymbol{e}_\phi$, $\boldsymbol{e}_x$ and $\boldsymbol{e}_y$

Obviously, $\boldsymbol{e}_\rho$ only have components along the *x*-and *y*-direction. Therefore,

$$\boldsymbol{e}_\rho = \boldsymbol{e}_x\cos\phi + \boldsymbol{e}_y\sin\phi \tag{1.31}$$

Similarly, we have

$$\boldsymbol{e}_\phi = -\boldsymbol{e}_x\sin\phi + \boldsymbol{e}_y\cos\phi \tag{1.32}$$

Apparently, the directions of $\boldsymbol{e}_\rho$ and $\boldsymbol{e}_\phi$ is dependent on the ϕ coordinate.

A vector $\boldsymbol{A}$ in cylindrical coordinates can be written as

$$\boldsymbol{A} = \boldsymbol{e}_\rho A_\rho + \boldsymbol{e}_\phi A_\phi + \boldsymbol{e}_z A_z \tag{1.33}$$

The dot product of two vectors $\boldsymbol{A}$ and $\boldsymbol{B}$ is,

$$\boldsymbol{A} \cdot \boldsymbol{B} = A_\rho B_\rho + A_\phi B_\phi + A_z B_z \tag{1.34}$$

The expression for the cross product is

$$\begin{aligned} \boldsymbol{A} \times \boldsymbol{B} &= \boldsymbol{e}_\rho(A_\phi B_z - A_z B_\phi) + \boldsymbol{e}_\phi(A_z B_\rho - A_\rho B_z) + \boldsymbol{e}_z(A_\rho B_\phi - A_\phi B_\rho) \\ &= \begin{vmatrix} \boldsymbol{e}_\rho & \boldsymbol{e}_\phi & \boldsymbol{e}_z \\ A_\rho & A_\phi & A_z \\ B_\rho & B_\phi & B_z \end{vmatrix} \end{aligned} \tag{1.35}$$

Notice that the expressions (1.34) and (1.35) assume that the vectors $\boldsymbol{A}$ and $\boldsymbol{B}$ are expressed using the same base vectors. That is, the $\boldsymbol{e}_\rho$ and $\boldsymbol{e}_\phi$ used for vector $\boldsymbol{A}$ must be the same as those for $\boldsymbol{B}$. ①

In cylindrical coordinates, the position vector associated with a point(ρ,ϕ,z) is

$$\boldsymbol{r}=\boldsymbol{e}_\rho\rho+\boldsymbol{e}_z z \tag{1.36}$$

It is important to realize that the position vector $\boldsymbol{r}$ has no component along ϕ-direction. However, it does not mean that $\boldsymbol{r}$ is independent of the coordinate ϕ. Instead, as indicated by (1.31), the unit vector $\boldsymbol{e}_\rho$ is a function of ϕ, therefore $\boldsymbol{r}$ is also a function of ϕ. ②

Differential change of the position vector $\boldsymbol{r}$ in(1.36) leads to

$$\mathrm{d}\boldsymbol{l}=\mathrm{d}\boldsymbol{r}=\mathrm{d}(\boldsymbol{e}_\rho\rho+\boldsymbol{e}_z z)=\mathrm{d}\rho\boldsymbol{e}_\rho+\rho\mathrm{d}\boldsymbol{e}_\rho+\mathrm{d}z\boldsymbol{e}_z \tag{1.37}$$

With(1.31) and(1.32), the second term on the right hand side of(1.37) can be written as

$$\rho\mathrm{d}\boldsymbol{e}_\rho=\rho\mathrm{d}(\boldsymbol{e}_x\cos\phi+\boldsymbol{e}_y\sin\phi)=\rho(-\boldsymbol{e}_x\sin\phi+\boldsymbol{e}_y\cos\phi)\mathrm{d}\phi=\rho\boldsymbol{e}_\phi\mathrm{d}\phi \tag{1.38}$$

Substitute(1.38) into(1.37), we have

$$\mathrm{d}\boldsymbol{l}=\mathrm{d}\boldsymbol{r}=\mathrm{d}(\boldsymbol{e}_\rho\rho+\boldsymbol{e}_z z)=\mathrm{d}\rho\boldsymbol{e}_\rho+\rho\mathrm{d}\phi\boldsymbol{e}_\phi+\mathrm{d}z\boldsymbol{e}_z \tag{1.39}$$

By referring to Figure 1-14, three differential area vectors along the three coordinate directions are

$$\begin{aligned}\mathrm{d}\boldsymbol{s}_\rho&=\boldsymbol{e}_\rho\rho\mathrm{d}\phi\mathrm{d}z=\boldsymbol{e}_\rho\mathrm{d}s_\rho\\ \mathrm{d}\boldsymbol{s}_\phi&=\boldsymbol{e}_\phi\mathrm{d}\rho\mathrm{d}z=\boldsymbol{e}_\phi\mathrm{d}s_\phi\\ \mathrm{d}\boldsymbol{s}_z&=\boldsymbol{e}_z\rho\mathrm{d}\rho\mathrm{d}\phi=\boldsymbol{e}_z\mathrm{d}s_z\end{aligned} \tag{1.40}$$

The differential volume in a cylindrical coordinate system is expressed as

$$\mathrm{d}v=\rho\mathrm{d}\rho\mathrm{d}\phi\mathrm{d}z \tag{1.41}$$

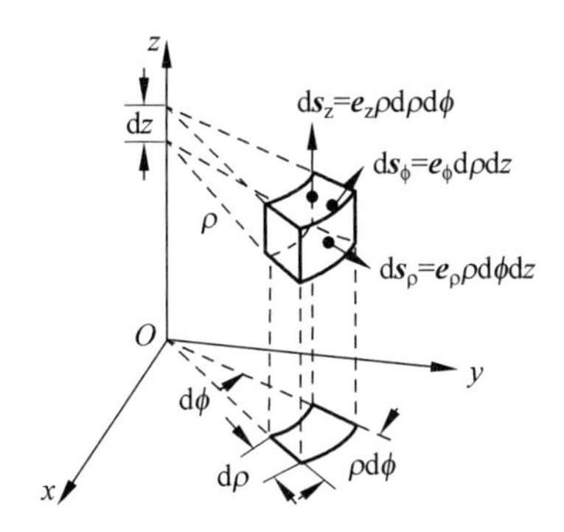

Figure 1-14 Differential area and volume in cylindrical coordinates

1.5.3 Spherical Coordinates(圆球坐标)

In a spherical coordinate system,

$$(u_1,u_2,u_3)=(r,\theta,\phi)$$

A point $P(r_0,\theta_0,\phi_0)$ is determined by the intersection of a spherical surface $r=r_0$, a circular

---

① 式(1.34)和式(1.35)成立的条件是：矢量$\boldsymbol{A}$和$\boldsymbol{B}$在柱坐标系展开时，采用同样的基矢量。由于$\boldsymbol{e}_\rho$和$\boldsymbol{e}_\phi$在不同的ϕ坐标下方向不同，因此采用式(1.34)和式(1.35)计算两个矢量的点积和叉积时，需要注意这两个矢量分解采用的单位矢量方向是否一致。

② 位置矢量$\boldsymbol{r}$是坐标ϕ的函数，但没有ϕ方向的分量。

conical surface(圆锥面) $\theta=\theta_0$, and a half-plane $\phi=\phi_0$ as shown in Figure 1-15. The coordinate r is the length of the position vector $\boldsymbol{r}$ of the point. The coordinate θ is the angle between the positive z-axis and the line segment OP. The angle ϕ and the base vector $\boldsymbol{e}_\phi$ are the same as those in cylindrical coordinates. The base vector $\boldsymbol{e}_\theta$ lies in the $\phi=\phi_0$ plane and is tangential to the spherical surface. The right-hand relations of $\boldsymbol{e}_r$, $\boldsymbol{e}_\theta$ and $\boldsymbol{e}_\phi$ are:

$$\begin{aligned} \boldsymbol{e}_r \times \boldsymbol{e}_\theta &= \boldsymbol{e}_\phi \\ \boldsymbol{e}_\theta \times \boldsymbol{e}_\phi &= \boldsymbol{e}_r \\ \boldsymbol{e}_\phi \times \boldsymbol{e}_r &= \boldsymbol{e}_\theta \end{aligned} \tag{1.42}$$

Further,

$$\begin{aligned} \boldsymbol{e}_r \cdot \boldsymbol{e}_\theta = \boldsymbol{e}_\theta \cdot \boldsymbol{e}_\phi = \boldsymbol{e}_\phi \cdot \boldsymbol{e}_r = 0 \\ \boldsymbol{e}_r \cdot \boldsymbol{e}_r = \boldsymbol{e}_\theta \cdot \boldsymbol{e}_\theta = \boldsymbol{e}_\phi \cdot \boldsymbol{e}_\phi = 1 \end{aligned} \tag{1.43}$$

As is shown in Figure 1-16,

$$\begin{aligned} \boldsymbol{e}_r \cdot \boldsymbol{e}_x &= \sin\theta\cos\phi \\ \boldsymbol{e}_r \cdot \boldsymbol{e}_y &= \sin\theta\sin\phi \\ \boldsymbol{e}_r \cdot \boldsymbol{e}_z &= \cos\theta \end{aligned} \tag{1.44}$$

Therefore,

$$\boldsymbol{e}_r = \boldsymbol{e}_x\sin\theta\cos\phi + \boldsymbol{e}_y\sin\theta\sin\phi + \boldsymbol{e}_z\cos\theta \tag{1.45}$$

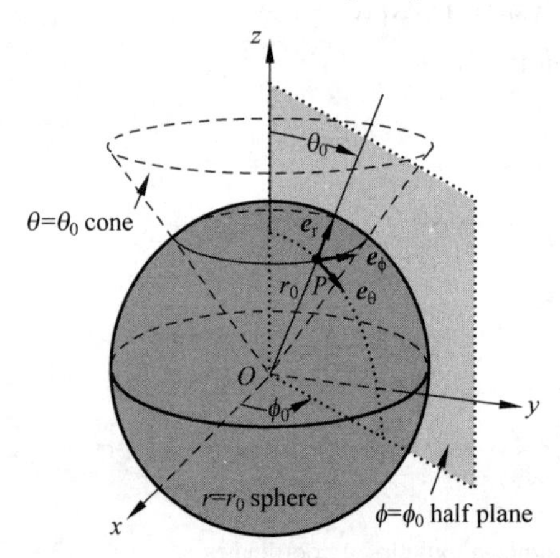

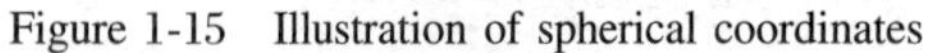
Figure 1-15 Illustration of spherical coordinates

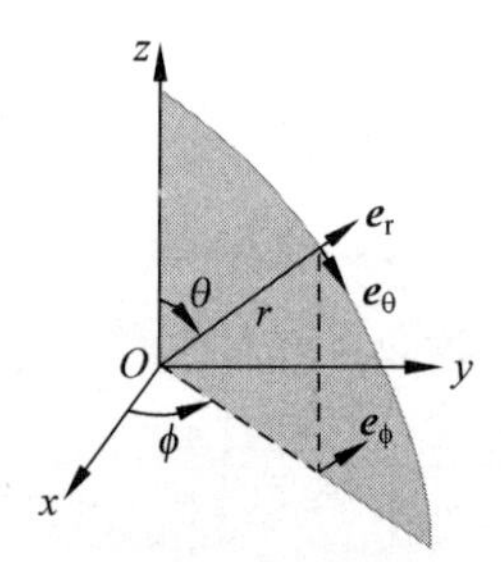

Figure 1-16 Relations among $\boldsymbol{e}_r$, $\boldsymbol{e}_\theta$, $\boldsymbol{e}_\phi$, $\boldsymbol{e}_x$, $\boldsymbol{e}_y$ and $\boldsymbol{e}_z$

Similarly, it is not difficult to find that

$$\boldsymbol{e}_\theta = \boldsymbol{e}_x\cos\theta\cos\phi + \boldsymbol{e}_y\cos\theta\sin\phi - \boldsymbol{e}_z\sin\theta \tag{1.46}$$

$$\boldsymbol{e}_\phi = -\boldsymbol{e}_x\sin\phi + \boldsymbol{e}_y\cos\phi \tag{1.47}$$

Obviously, none of the three base vectors $\boldsymbol{e}_r$, $\boldsymbol{e}_\theta$ and $\boldsymbol{e}_\phi$ is a constant vector.

A vector $\boldsymbol{A}$ in spherical coordinates can be written as

$$\boldsymbol{A} = \boldsymbol{e}_r A_r + \boldsymbol{e}_\theta A_\theta + \boldsymbol{e}_\phi A_\phi \tag{1.48}$$

The dot product of two vectors $\boldsymbol{A}$ and $\boldsymbol{B}$ is,

$$\boldsymbol{A} \cdot \boldsymbol{B} = A_r B_r + A_\theta B_\theta + A_\phi B_\phi \tag{1.49}$$

The expression of the cross product is

$$\boldsymbol{A}\times\boldsymbol{B}=\boldsymbol{e}_{\rm r}(A_\theta B_\phi - A_\phi B_\theta)+\boldsymbol{e}_\theta(A_\phi B_{\rm r} - A_{\rm r}B_\phi)+\boldsymbol{e}_\phi(A_{\rm r}B_\theta - A_\theta B_{\rm r})$$
$$=\begin{vmatrix}\boldsymbol{e}_{\rm r} & \boldsymbol{e}_\theta & \boldsymbol{e}_\phi \\ A_{\rm r} & A_\theta & A_\phi \\ B_{\rm r} & B_\theta & B_\phi\end{vmatrix} \tag{1.50}$$

Again, the expressions (1.49) and (1.50) assume that the vectors $\boldsymbol{A}$ and $\boldsymbol{B}$ are expressed by using the same unit vectors $\boldsymbol{e}_{\rm r}$, $\boldsymbol{e}_\theta$ and $\boldsymbol{e}_\phi$.

In spherical coordinates, the position vector associated with a point (r,θ,ϕ) is

$$\boldsymbol{r}=\boldsymbol{e}_{\rm r}r \tag{1.51}$$

Obviously, $\boldsymbol{r}$ has no component along θ-or ϕ-direction. However, according to (1.45), $\boldsymbol{r}$ is a function of both θ and ϕ. ①

Differential change of the vector $\boldsymbol{r}$ in (1.51) leads to

$$\mathrm{d}\boldsymbol{l}=\mathrm{d}\boldsymbol{r}=\mathrm{d}(\boldsymbol{e}_{\rm r}r)=\boldsymbol{e}_{\rm r}\mathrm{d}r+r\mathrm{d}\boldsymbol{e}_{\rm r} \tag{1.52}$$

With (1.45)-(1.47), the second term on the right hand side of (1.52) becomes

$$\begin{aligned} r\mathrm{d}\boldsymbol{e}_{\rm r} &= r\boldsymbol{e}_{\rm x}(\cos\theta\cos\phi\mathrm{d}\theta-\sin\theta\sin\phi\mathrm{d}\phi)+ \\ &\quad r\boldsymbol{e}_{\rm y}(\cos\theta\sin\phi\mathrm{d}\theta+\sin\theta\cos\phi\mathrm{d}\phi)-r\boldsymbol{e}_{\rm z}\sin\theta\mathrm{d}\theta \\ &=(\boldsymbol{e}_{\rm x}\cos\theta\cos\phi+\boldsymbol{e}_{\rm y}\cos\theta\sin\phi-\boldsymbol{e}_{\rm z}\sin\theta)r\mathrm{d}\theta+(-\boldsymbol{e}_{\rm x}\sin\phi+\boldsymbol{e}_{\rm y}\cos\phi)r\sin\theta\mathrm{d}\phi \\ &=\boldsymbol{e}_\theta r\mathrm{d}\theta+\boldsymbol{e}_\phi r\sin\theta\mathrm{d}\phi \end{aligned} \tag{1.53}$$

After substituting (1.53) into (1.52), we have

$$\mathrm{d}\boldsymbol{l}=\mathrm{d}\boldsymbol{r}=\boldsymbol{e}_{\rm r}\mathrm{d}r+\boldsymbol{e}_\theta r\mathrm{d}\theta+\boldsymbol{e}_\phi r\sin\theta\mathrm{d}\phi \tag{1.54}$$

By referring to Figure 1-17, three differential area vectors along the three spherical coordinate directions are

$$\begin{aligned} \mathrm{d}\boldsymbol{s}_{\rm r} &= \boldsymbol{e}_{\rm r}r^2\sin\theta\mathrm{d}\theta\mathrm{d}\phi=\boldsymbol{e}_{\rm r}\mathrm{d}s_{\rm r} \\ \mathrm{d}\boldsymbol{s}_\theta &= \boldsymbol{e}_\theta r\sin\theta\mathrm{d}r\mathrm{d}\phi=\boldsymbol{e}_\theta\mathrm{d}s_\theta \\ \mathrm{d}\boldsymbol{s}_\phi &= \boldsymbol{e}_\phi r\mathrm{d}r\mathrm{d}\theta=\boldsymbol{e}_\phi\mathrm{d}s_\phi \end{aligned} \tag{1.55}$$

The differential volume in a spherical coordinate system is expressed as

$$\mathrm{d}v=r^2\sin\theta\mathrm{d}r\mathrm{d}\theta\mathrm{d}\phi \tag{1.56}$$

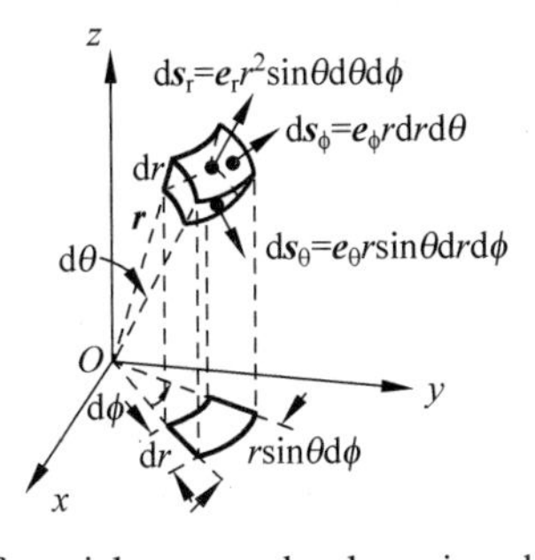

Figure 1-17 Differential areas and volume in spherical coordinates

① 位置矢量 $\boldsymbol{r}$ 只有 r 方向的分量,没有 θ 和 ϕ 方向的分量。但是 $\boldsymbol{e}_{\rm r}$ 是坐标 θ 和 ϕ 的函数,因此位置矢量 $\boldsymbol{r}$ 也是 θ 和 ϕ 的函数,当然同时也是 r 的函数。

1.6 Integrals of Vector Fields(矢量场的积分)

In electromagnetics, two types of integrals of vector fields are important and will be used extensively in this text: the **scalar line integral**(标量线积分) and the **scalar surface integral** (标量面积分). The scalar line integral, or simply **line integral** of a vector field $\boldsymbol{F}$ over a path C is defined as

$$\int_C \boldsymbol{F} \cdot \mathrm{d}\boldsymbol{l} = \lim_{N \to +\infty} \sum_{i=1}^{N} \boldsymbol{F}_i \cdot \Delta \boldsymbol{l}_i \tag{1.57}$$

where the integrand is the component of the vector $\boldsymbol{F}$ along the path C. According to (1.57), this line integral is evaluated by first dividing the path C into N tiny directed line segments $\Delta \boldsymbol{l}_i (i=1,2,\cdots,N)$ and assuming that $\boldsymbol{F}$ at the location of the segment $\Delta \boldsymbol{l}_i$ is a constant vector $\boldsymbol{F}_i$, as illustrated in Figure 1-18. Then we add up the dot products $\boldsymbol{F}_i \cdot \Delta \boldsymbol{l}_i$ for all the N line segments, and the limit of the summation when N approaches infinity and all the line segments $\Delta \boldsymbol{l}_i$ approach infinitely small is equal to the scalar line integral of the vector $\boldsymbol{F}$ over the path of integration C. In Cartesian coordinates, the line integral can be expressed as

$$\begin{aligned}\int_C \boldsymbol{F} \cdot \mathrm{d}\boldsymbol{l} &= \int_C [\boldsymbol{e}_x F_x + \boldsymbol{e}_y F_y + \boldsymbol{e}_z F_z] \cdot [\boldsymbol{e}_x \mathrm{d}x + \boldsymbol{e}_y \mathrm{d}y + \boldsymbol{e}_z \mathrm{d}z] \\ &= \int_C F_x \mathrm{d}x + \int_C F_y \mathrm{d}y + \int_C F_z \mathrm{d}z \end{aligned} \tag{1.58}$$

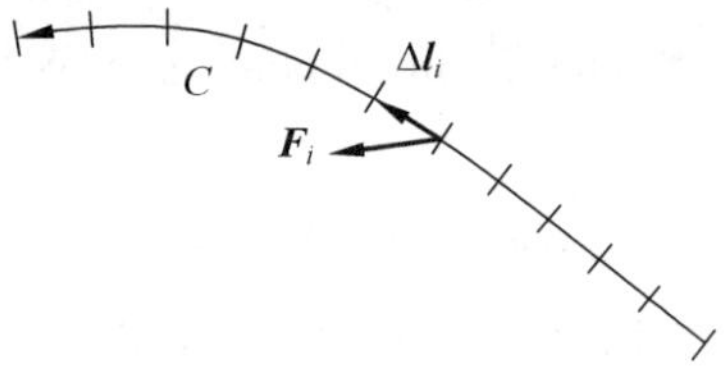

Figure 1-18 Illustration of the scalar line integral of the vector field $\boldsymbol{F}$ over a path C

The line integral is important because its physical meaning is work. If $\boldsymbol{F}$ is a force vector, the integral is the work done by the force in moving an object along the specified path C. If $\boldsymbol{F}$ is the electric field intensity, then the integral represents the work done by the electric field in moving a unit charge along the specified path C.[①]

The scalar surface integral, or simply the **surface integral** of a vector field $\boldsymbol{A}$ over a surface S is defined as

$$\int_S \boldsymbol{A} \cdot \mathrm{d}\boldsymbol{s} = \lim_{N \to +\infty} \sum_{i=1}^{N} \boldsymbol{A}_i \cdot \Delta \boldsymbol{s}_i \tag{1.59}$$

where the integrand is the component of the vector $\boldsymbol{A}$ along the normal direction of the surface S. To evaluate this surface integral, as illustrated in Figure 1-19, we first divide S into N small

① 式(1.57)定义的矢量线积分是矢量沿积分路径方向分量的积分,在物理学中,常用于计算机械力和电场力所做的功。

areas with each area represented by an area vector $\Delta \boldsymbol{s}_i (i=1,2,\cdots,N)$. The vector field $\boldsymbol{A}$ at the location of the ith area vector $\Delta \boldsymbol{s}_i$ is a constant vector A_i. Then we add up the dot products $\boldsymbol{A}_i \cdot \Delta \boldsymbol{s}_i$ for all the N small areas, and the limit of the summation when N approaches infinity and all the small area vectors $\Delta \boldsymbol{s}_i$ approach infinitely small is equal to the scalar surface integral of the vector $\boldsymbol{A}$ over the surface S. In Cartesian coordinates, the surface integral can be expressed as

$$\int_S \boldsymbol{A} \cdot \mathrm{d}\boldsymbol{s} = \int_S [\boldsymbol{e}_x A_x + \boldsymbol{e}_y A_y + \boldsymbol{e}_z A_z] \cdot \boldsymbol{e}_n \mathrm{d}s \qquad (1.60)$$

where $\boldsymbol{e}_n$ is the unit vector along the normal direction of the surface S. ①

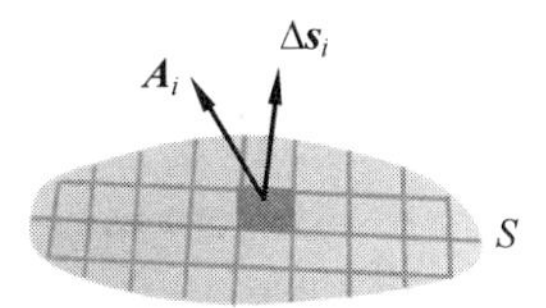

Figure 1-19 Illustration of the scalar surface integral of a vector field $\boldsymbol{A}$ over a surface S

The surface integral (1.60) embodies the flux, i.e., it measures how much the vector field $\boldsymbol{A}$ flows through the area S. In the integral, the differential area vector $\mathrm{d}\boldsymbol{s}=\boldsymbol{e}_\mathrm{n}\mathrm{d}s$ has a magnitude $\mathrm{d}s$ and a direction $\boldsymbol{e}_\mathrm{n}$ everywhere normal to the surface S. To determine the direction of $\boldsymbol{e}_\mathrm{n}$, we usually follow the following conventions: ②

(1) If S is an open surface, we first define the direction of the perimeter of the open surface, and then determine the direction of $\boldsymbol{e}_\mathrm{n}$ by the right-hand rule. As illustrated in Figure 1-20(a), if the fingers of the right hand follow the direction of the perimeter, the thumb points in the direction of $\boldsymbol{e}_\mathrm{n}$. Notice that, the direction of $\boldsymbol{e}_\mathrm{n}$ generally depends on the locations on the surface S. A planar surface, such as the disk in Figure 1-20(b), is a special case of an open surface on which $\boldsymbol{e}_\mathrm{n}$ is a constant vector.

(2) If S is a closed surface enclosing a volume, $\boldsymbol{e}_\mathrm{n}$ is always in the normal outward direction from the volume as illustrated in Figure 1-20(c). A small circle is added over the integral sign of the integration when S is a closed surface:

$$\oint_S \boldsymbol{A} \cdot \mathrm{d}\boldsymbol{s} = \oint_S \boldsymbol{A} \cdot \boldsymbol{e}_\mathrm{n}\mathrm{d}s$$

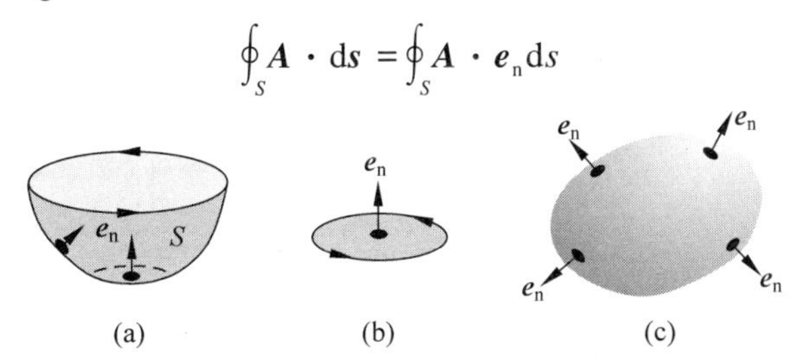

Figure 1-20 Determination of $\boldsymbol{e}_\mathrm{n}$ for various surfaces

① 式(1.59)和式(1.60)定义的矢量面积分计算的是矢量在某曲面的法向分量的积分,因此通常用于计算矢量物理量(如电流密度矢量)通过某个曲面的通量(如总电流)。

② 显然,给定曲面上任意一点垂直于该曲面的法向方向有两个。对矢量场做标量面积分时,为了唯一确定面元矢量的法向方向,采用的规则为:若积分曲面开放,即该曲面的边界为一个闭合曲线,则根据该闭合曲线的参考方向,按照右手定则确定面元矢量的法向;若积分曲面为闭合曲面,即该曲面包围形成一个内部空间,则默认取由内往外的法向方向为面元矢量的方向。

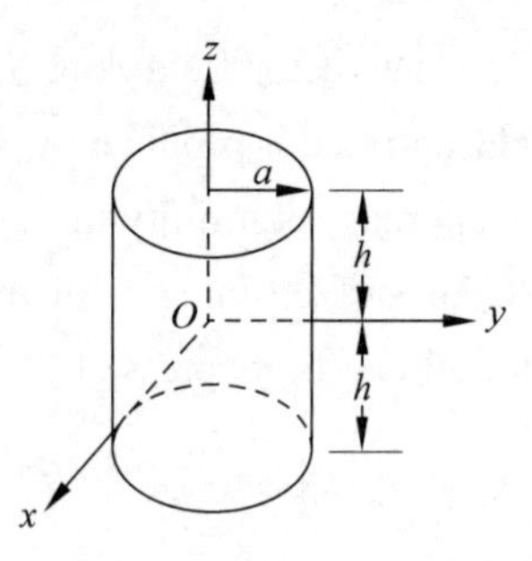

Figure 1-21 The cylinder in Example 1-2

Example 1-2 Given a vector field $\boldsymbol{F}(\boldsymbol{r})=\frac{\boldsymbol{e}_{r}}{r^2}$, evaluate the surface integral $\oint_S \boldsymbol{F}\cdot d\boldsymbol{s}$ over the surface of a cylinder with radius a and height $2h$ as is shown in Figure 1-21.

Solution: The cylinder has three surfaces: the top face, the bottom face, and the side wall. Therefore, the surface integral can be written as:

$$\oint_S \boldsymbol{F}\cdot \mathrm{d}\boldsymbol{s}=\int_{\substack{\text{top}\\ \text{face}}}\boldsymbol{F}\cdot \boldsymbol{e}_{\mathrm{n}}\mathrm{d}s+\int_{\substack{\text{bottom}\\ \text{face}}}\boldsymbol{F}\cdot \boldsymbol{e}_{\mathrm{n}}\mathrm{d}s+\int_{\substack{\text{side}\\ \text{wall}}}\boldsymbol{F}\cdot \boldsymbol{e}_{\mathrm{n}}\mathrm{d}s \tag{1.61}$$

(1) On the top face, $\boldsymbol{e}_{\mathrm{n}}=\boldsymbol{e}_z$, $\mathrm{d}s=\rho\mathrm{d}\rho\mathrm{d}\phi$, and

$$\boldsymbol{F}\cdot \boldsymbol{e}_{\mathrm{n}}=\frac{1}{r^2}\boldsymbol{e}_{\mathrm{r}}\cdot \boldsymbol{e}_z=\frac{1}{r^2}\cos\theta=\frac{1}{r^2}\frac{h}{r}=\frac{h}{(\rho^2+h^2)^{3/2}}$$

Therefore,

$$\int_{\substack{\text{top}\\ \text{face}}}\boldsymbol{F}\cdot \boldsymbol{e}_{\mathrm{n}}\mathrm{d}s=\int_0^{2\pi}\int_0^a\frac{h}{(\rho^2+h^2)^{3/2}}\rho\mathrm{d}\rho\mathrm{d}\phi=2\pi\left[1-\frac{h}{\sqrt{h^2+a^2}}\right] \tag{1.62}$$

(2) On the bottom face, $\boldsymbol{e}_{\mathrm{n}}=-\boldsymbol{e}_z$, $\mathrm{d}s=\rho\mathrm{d}\rho\mathrm{d}\phi$, and

$$\boldsymbol{F}\cdot \boldsymbol{e}_{\mathrm{n}}=\frac{1}{r^2}\boldsymbol{e}_{\mathrm{r}}\cdot(-\boldsymbol{e}_z)=\frac{1}{r^2}(-\cos\theta)=\frac{1}{r^2}\frac{h}{r}=\frac{h}{(\rho^2+h^2)^{3/2}}$$

Therefore, we have the same integral result

$$\int_{\substack{\text{bottom}\\ \text{face}}}\boldsymbol{F}\cdot \boldsymbol{e}_{\mathrm{n}}\mathrm{d}s=\int_0^{2\pi}\int_0^a\frac{h}{(\rho^2+h^2)^{3/2}}\rho\mathrm{d}\rho\mathrm{d}\phi=2\pi\left[1-\frac{h}{\sqrt{h^2+a^2}}\right] \tag{1.63}$$

(3) On the side wall, $\boldsymbol{e}_{\mathrm{n}}=\boldsymbol{e}_{\rho}$, $\mathrm{d}s=\rho\mathrm{d}\phi\mathrm{d}z=a\mathrm{d}z\mathrm{d}\phi$, and

$$\boldsymbol{F}\cdot \boldsymbol{e}_{\mathrm{n}}=\frac{1}{r^2}\boldsymbol{e}_{\mathrm{r}}\cdot \boldsymbol{e}_{\rho}=\frac{1}{r^2}\frac{a}{r}=\frac{a}{(z^2+a^2)^{3/2}}$$

Therefore,

$$\begin{aligned}\int_{\substack{\text{side}\\ \text{wall}}}\boldsymbol{F}\cdot \boldsymbol{e}_{\mathrm{n}}\mathrm{d}s&=\int_0^{2\pi}\int_{-h}^{h}\frac{a}{(z^2+a^2)^{3/2}}a\mathrm{d}z\mathrm{d}\phi=2\pi\int_{-h}^{h}\frac{a^2}{(z^2+a^2)^{3/2}}\mathrm{d}z\\&=2\pi\frac{z}{\sqrt{z^2+a^2}}\Bigg|_{-h}^{h}=4\pi\frac{h}{\sqrt{h^2+a^2}}\end{aligned} \tag{1.64}$$

Substituting (1.62) ~ (1.64) into (1.61), we have

$$\oint_S \boldsymbol{F}\cdot \mathrm{d}\boldsymbol{s}=4\pi$$

This result suggests that the net outward flux of this specific field $\boldsymbol{F}$ through the closed cylindrical surface is independent of the geometrical parameters of the cylinder (as long as the spatial origin is inside the cylinder). This fact can be easily proved by the divergence theorem (散度定理), which will be discussed later in this chapter.

1.7 Gradient of a Scalar Field(标量场的梯度)

For a scalar field φ, what interests us is the space rate of change(空间变化率) of the field. Given a point P and a direction l in the space, as shown in Figure 1-22, we define the **directional derivative**(方向导数) along l as

$$\left.\frac{\partial \varphi}{\partial l}\right|_P = \lim_{\Delta l \to 0} \frac{\varphi(P') - \varphi(P)}{\Delta l} \tag{1.65}$$

Apparently, the direction derivative $\partial \varphi / \partial l$ is dependent on the direction l. For each point P in the space, there exists a direction along which $\partial \varphi / \partial l$ reaches the maximum value, which is the maximum space rate of increase of the field φ at the point P. We define the **gradient**(梯度) of a scalar field as the vector that represents both the magnitude and the direction of the maximum space rate of increase of the scalar field. Specifically, the gradient of scalar field φ is written as

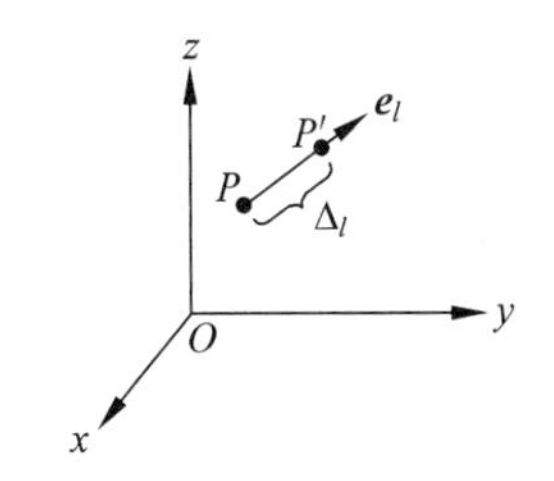

Figure 1-22 Illustrating directional derivative along the direction l

$$\mathbf{grad}\varphi \triangleq \boldsymbol{e}_{1,\max} \left.\frac{\partial \varphi}{\partial l}\right|_{\max} \tag{1.66}$$

where $\boldsymbol{e}_{1,\max}$ is the direction along which the directional derivative reaches the maximum value. ①

To derive an expression for the gradient, we notice that the numerator in the right-hand side of (1.65) becomes a total differential along the direction l as $\Delta l \to 0$, i. e.

$$\varphi(P') - \varphi(P) \to \frac{\partial \varphi}{\partial x}\mathrm{d}x + \frac{\partial \varphi}{\partial y}\mathrm{d}y + \frac{\partial \varphi}{\partial z}\mathrm{d}z \quad \text{as} \quad \Delta l \to 0$$

In the meantime, Δl becomes $\mathrm{d}l$ along the direction l. Therefore, (1.65) can be rewritten as

$$\left.\frac{\partial \varphi}{\partial l}\right|_P = \frac{\frac{\partial \varphi}{\partial x}\mathrm{d}x + \frac{\partial \varphi}{\partial y}\mathrm{d}y + \frac{\partial \varphi}{\partial z}\mathrm{d}z}{\mathrm{d}l} = \left(\boldsymbol{e}_x \frac{\partial \varphi}{\partial x} + \boldsymbol{e}_y \frac{\partial \varphi}{\partial y} + \boldsymbol{e}_z \frac{\partial \varphi}{\partial z}\right) \cdot \left(\frac{\boldsymbol{e}_x \mathrm{d}x + \boldsymbol{e}_y \mathrm{d}y + \boldsymbol{e}_z \mathrm{d}z}{\mathrm{d}l}\right) \tag{1.67}$$

From (1.25), we recognize that $\boldsymbol{e}_x \mathrm{d}x + \boldsymbol{e}_y \mathrm{d}y + \boldsymbol{e}_z \mathrm{d}z$ is the differential change of position $\mathrm{d}\boldsymbol{l}$. Therefore,

$$\frac{\boldsymbol{e}_x \mathrm{d}x + \boldsymbol{e}_y \mathrm{d}y + \boldsymbol{e}_z \mathrm{d}z}{\mathrm{d}l} = \frac{\mathrm{d}\boldsymbol{l}}{\mathrm{d}l} = \boldsymbol{e}_1 \tag{1.68}$$

is the unit vector along the direction l. For the sake of brevity, we introduce symbol ∇ as a differential operator that maps a scalar field φ to a vector field

$$\nabla \varphi \triangleq \boldsymbol{e}_x \frac{\partial \varphi}{\partial x} + \boldsymbol{e}_y \frac{\partial \varphi}{\partial y} + \boldsymbol{e}_z \frac{\partial \varphi}{\partial z} \tag{1.69}$$

---

① 给定标量场在某一点的梯度被定义为这样一个矢量,其幅度为该标量场在该点上空间变化率(即方向导数)的最大值,其方向为该标量场在该点上空间变化率最大的方向。

∇ is called **del operator** (**del 算子**), **vector differential operator** (**矢量微分算子**), or **Hamilton operator**(**哈密尔顿算子**). Substituting(1.68) and(1.69) into(1.67), we have

$$\left.\frac{\partial \varphi}{\partial l}\right|_P = \nabla \varphi \cdot \boldsymbol{e}_l \tag{1.70}$$

which states that $\partial\varphi/\partial l$ is the projection of the vector $\nabla\varphi$ onto the direction l. Since the vector $\nabla\varphi$ is independent of the direction l, $\partial\varphi/\partial l$ reaches maximum when $\boldsymbol{e}_l$ is in the same direction of $\nabla\varphi$. This means that the magnitude and direction of the maximum $\partial\varphi/\partial l$ is identical to the magnitude and direction of $\nabla\varphi$. Therefore, $\nabla\varphi$ is exactly the gradient of the field φ. ①

(1.70) gives an important relation between the directional derivative and the gradient of a scalar field. This relation can also be interpreted graphically by considering two **constant-value surfaces**(**等值面**) as illustrated in Figure 1-23. A constant-value surface is a surface formed by all the points at which the scalar field has the same values. Given a point P on the surface φ_0, let d$\boldsymbol{n}$ be the normal vector(法向矢量) pointing from P to P_1, the corresponding point on surface φ_0+dφ. Let d$\boldsymbol{l}$($\neq$d$\boldsymbol{n}$) be another vector pointing from P to P_2, another point close to P_1 on surface φ_0+dφ. When dφ is infinitely small, the three points P, P_1 and P_2 are all within a very small region in which the two constant-value surfaces are two planes parallel to each other. In this case, dn is obviously the shortest distance between the two surfaces, and dn=d$l$cosα, where α is the angle between d$\boldsymbol{l}$ and d$\boldsymbol{n}$. Therefore, we have

$$\left.\frac{\partial \varphi}{\partial l}\right|_P = \frac{\mathrm{d}\varphi}{\mathrm{d}l} = \frac{\mathrm{d}\varphi}{\mathrm{d}n}\frac{\mathrm{d}n}{\mathrm{d}l} = \frac{\mathrm{d}\varphi}{\mathrm{d}n}\cos\alpha = \frac{\mathrm{d}\varphi}{\mathrm{d}n}\boldsymbol{e}_\mathrm{n} \cdot \boldsymbol{e}_l \tag{1.71}$$

where $\boldsymbol{e}_l$ and $\boldsymbol{e}_\mathrm{n}$ are respectively the unit vectors of d$\boldsymbol{l}$ and d$\boldsymbol{n}$ respectively. By comparing(1.71) and(1.70), we have

$$\nabla \varphi = \frac{\mathrm{d}\varphi}{\mathrm{d}n}\boldsymbol{e}_\mathrm{n} \tag{1.72}$$

which means that the direction of the gradient at any point is normal to the constant-value

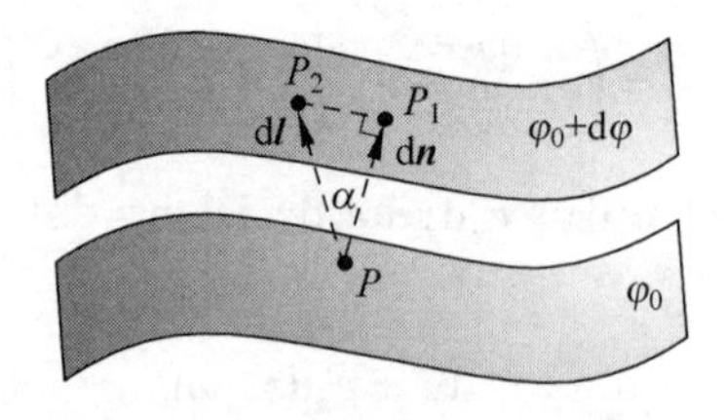

Figure 1-23 Relation between directional derivative and constant value surfaces

① 式(1.70)是由多元函数的全微分公式推导得出的。该式表明,标量场 φ 在某个方向上的变化率等于矢量 $\nabla\varphi$ 在该方向上的投影。显然,任何矢量在自身方向上的投影达到最大,且投影最大值就是该矢量的幅度。因此,矢量 $\nabla\varphi$ 的幅度就是 φ 的方向导数的最大值, $\nabla\varphi$ 的方向就是 φ 的方向导数达到最大值的方向。这意味着矢量 $\nabla\varphi$ 正好符合式(1.66)给出的梯度的定义,所以矢量 $\nabla\varphi$ 就是 φ 的梯度。换言之,式(1.70)给出了标量场梯度的一个非常重要的性质:标量场梯度在某个方向的投影就等于该标量场在该方向上的方向导数(空间变化率)。

surface containing the point. ①

Equation(1.70) can also be written as

$$\mathrm{d}\varphi = \nabla\varphi \cdot \mathrm{d}\boldsymbol{l} \tag{1.73}$$

which is a useful expression of the total differential of φ due to a change of position represented by the differential length vector $\mathrm{d}\boldsymbol{l}$.

Equation(1.69) gives the expression of the gradient in Cartesian coordinates. The gradient expressions in cylindrical and spherical coordinates can be found in a similar fashion, which leads to

$$\nabla\varphi = \left(\boldsymbol{e}_\rho \frac{\partial}{\partial\rho} + \boldsymbol{e}_\phi \frac{\partial}{\rho\partial\phi} + \boldsymbol{e}_z \frac{\partial}{\partial z}\right)\varphi \tag{1.74}$$

$$\nabla\varphi = \left(\boldsymbol{e}_r \frac{\partial}{\partial r} + \boldsymbol{e}_\theta \frac{\partial}{r\partial\theta} + \boldsymbol{e}_\phi \frac{\partial}{r\sin\theta\partial\phi}\right)\varphi \tag{1.75}$$

Apparently, gradient is a linear differential operator. For any two scalar fields φ and ψ, we have

$$\nabla(\varphi + \psi) = \nabla\varphi + \nabla\psi \tag{1.76}$$

Example 1-3 Given $\boldsymbol{r}'=\boldsymbol{e}_x x'+\boldsymbol{e}_y y'+\boldsymbol{e}_z z'$ to be the position vector of a fixed point(x', y', z') in the space. Then $\boldsymbol{R}=\boldsymbol{r}-\boldsymbol{r}'$ is a function of $\boldsymbol{r}$ and can be considered as a vector field. $R=|\boldsymbol{R}|$ is also a function of $\boldsymbol{r}$ and can be considered as a scalar field. Prove that

$$\nabla R = \frac{\boldsymbol{R}}{R} = \boldsymbol{e}_R \tag{1.77}$$

$$\nabla\left(\frac{1}{R}\right) = -\frac{\boldsymbol{R}}{R^3} = -\frac{\boldsymbol{e}_R}{R^2} \tag{1.78}$$

Solution: The above equations can be proved directly by applying the expressions of gradient in Cartesian coordinates. Here, to reduce mathematical complexity, we first let $\boldsymbol{r}'$ be the zero vector so that the following relations can be readily proved by using(1.75) in the spherical coordinate system:

$$\nabla r = \frac{\boldsymbol{r}}{r} = \boldsymbol{e}_r \tag{1.79}$$

$$\nabla\left(\frac{1}{r}\right) = -\frac{\boldsymbol{r}}{r^3} = -\frac{\boldsymbol{e}_r}{r^2} \tag{1.80}$$

With(1.79) and(1.80), (1.77) and(1.78) are self-evident due to the translational-invariance of the gradient operator. That is, for any scalar field $\varphi(\boldsymbol{r})$, if $\nabla\varphi(\boldsymbol{r})=\boldsymbol{F}(\boldsymbol{r})$, then $\nabla\varphi(\boldsymbol{r}-\boldsymbol{r}')=\boldsymbol{F}(\boldsymbol{r}-\boldsymbol{r}')$.

Since the scalar R defined in Example 1-3 is the distance between two positions $\boldsymbol{r}$ and $\boldsymbol{r}'$, the vector $\boldsymbol{R}$ can be called the **distance vector**(**距离矢量**) pointing from $\boldsymbol{r}'$ to $\boldsymbol{r}$. By default, the

① 式(1.72)表明标量场中任一点的梯度矢量均垂直于通过该点的等值面。该结论也可以从式(1.70)直接得到：在式(1.70)中取 l 的方向为 P 点等值面的任意一个切线方向，则$(\partial\varphi/\partial l)_P=0$。因此，只要 $\boldsymbol{e}_l$ 的方向为等值面的切线方向，$\nabla\varphi$ 必然垂直于 $\boldsymbol{e}_l$。换言之，标量场在任一点的梯度垂直于过该点的等值面的所有切线方向，因而垂直于该等值面。

gradient ∇ is a differential operator with respect to $\boldsymbol{r}$. For a scalar field that is only a function of R, it is useful to define a gradient operator ∇' with respect to $\boldsymbol{r}'$. Specifically, in the Cartesian coordinates, we can define

$$\nabla'\varphi \triangleq \boldsymbol{e}_x \frac{\partial\varphi}{\partial x'} + \boldsymbol{e}_y \frac{\partial\varphi}{\partial y'} + \boldsymbol{e}_z \frac{\partial\varphi}{\partial z'} \tag{1.81}$$

By following the same procedure in Example 1-3, we can easily prove that

$$\nabla'\left(\frac{1}{R}\right) = \frac{\boldsymbol{R}}{R^3} = \frac{\boldsymbol{e}_R}{R^2} \tag{1.82}$$

In fact, for any scalar field f that depends only on R, we can prove that

$$\nabla' f(R) = -\nabla f(R) \tag{1.83}$$

1.8 Divergence of a Vector Field(矢量场的散度)

In a vector field, every point in the space is associated with a vector with a certain magnitude and a certain direction. To illustrate the variations of the vector field graphically, we introduce the **flux lines**(通量线) or **streamline**(流线). As illustrated in Figure 1-24, flux lines are a set of directed lines or curves. At any point on a flux line, the direction of the vector field is the same as the tangential direction of the flux line. The magnitude of the vector field is indicated by the density of the flux lines near the point. Therefore, in Figure 1-24(a), the vector field in region A is stronger than that in region B, whereas the vector field in Figure 1-24(b) has a uniform magnitude.

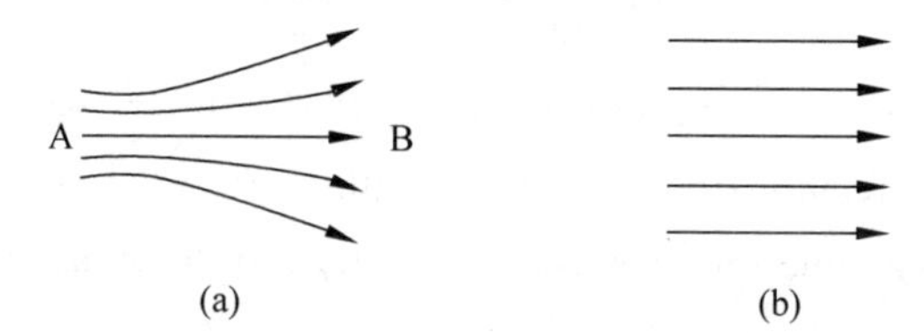

Figure 1-24 Vector fields represented by flux lines

With the aid of flux lines, we can estimate the **flux**(通量) of a vector field through a surface. Mathematically, the flux of a vector field is calculated as the surface integral of the vector field as defined in(1.60). Graphically, the flux through a surface is represented by the number of flux lines through the surface. Particularly of interest to us is the total flux through a closed surface. Figure 1-25 illustrates three typical cases in which a vector field flows through the bounding surface of a volume.

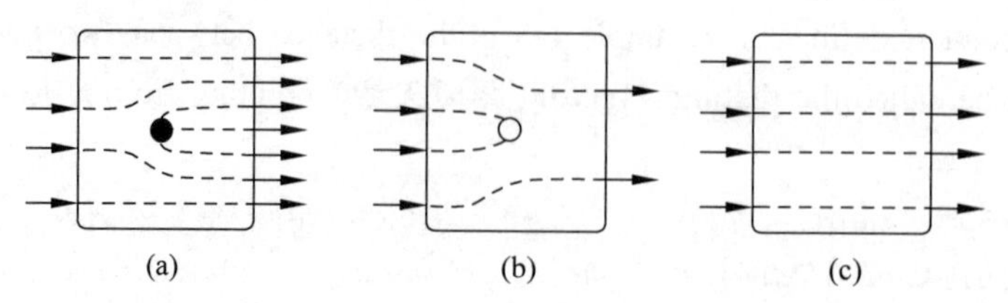

Figure 1-25 Relation between flow sources and flux of a vector field through an enclosed surface

In Figure 1-25(a), there are 4 flux lines passing through the left side of the closed surface into the volume, whereas there are 7 flux lines coming out from the volume through the right side. Therefore, the inward flux through the left side is less than the outward flux through the right side, and the net outward flux is positive. Clearly, the positive outward flux is due to the fact that 3 additional flux lines originate somewhere within the volume. The origins of the flux lines are tantamount to the positive **flow sources**(**通量源、流量源**) within the volume. In Figure 1-25(b), on the contrary, there is a net inward flux of the vector field through the closed surface. In other words, the net outward flux is negative, which is obviously due to the fact that some of the flux lines terminate somewhere within the volume. The terminal points of the flux lines are tantamount to the negative flow sources. In Figure 1-25(c), there is an equal amount of inward and outward flux going through the closed surface. That means the volume contains no sources, or there is equal amount of positive and negative sources. Therefore, the net outward flux of a vector field through an enclosed surface is a measure of the total flow sources within the volume bounded by the surface, and further, the flux per unit volume is a measure of the flow source density of the vector field. ①

Now, we define the **divergence**(**散度**) of a vector field $\boldsymbol{A}$ at a point, denoted by $\nabla \cdot \boldsymbol{A}$, as the net outward flux of $\boldsymbol{A}$ per unit volume as the volume containing the point approaches zero: ②

$$\nabla \cdot \boldsymbol{A} \triangleq \lim_{\Delta v \to 0} \frac{\oint_S \boldsymbol{A} \cdot \mathrm{d}\boldsymbol{s}}{\Delta v} \tag{1.84}$$

The numerator in(1.84), representing the net outward flux, is an integral over the entire surface S that bounds the volume. Obviously, the divergence of a vector field is a scalar field.

The definition of divergence(1.83) holds for any coordinate system. In the following, we derive the expression for the divergence in Cartesian coordinates.

Consider a differential volume of sides Δx, Δy and Δz centered around a point $P(x_0, y_0, z_0)$ in a vector field $\boldsymbol{A}$, as shown in Figure 1-26. Since the differential volume has six faces, the surface integral in the numerator of(1.84) can be decomposed into six parts:

$$\oint_S \boldsymbol{A} \cdot \mathrm{d}\boldsymbol{s} = \left[\int_{\substack{\text{front}\\\text{face}}} + \int_{\substack{\text{back}\\\text{face}}} + \int_{\substack{\text{right}\\\text{face}}} + \int_{\substack{\text{left}\\\text{face}}} + \int_{\substack{\text{top}\\\text{face}}} + \int_{\substack{\text{bottom}\\\text{face}}}\right] \boldsymbol{A} \cdot \mathrm{d}\boldsymbol{s} \tag{1.85}$$

On the front face, when the differential volume is infinitely small,

$$\int_{\substack{\text{front}\\\text{face}}} \boldsymbol{A} \cdot \mathrm{d}\boldsymbol{s} \approx \boldsymbol{A}_{\substack{\text{front}\\\text{face}}} \cdot \Delta \boldsymbol{s}_{\substack{\text{front}\\\text{face}}} = \boldsymbol{A}_{\substack{\text{front}\\\text{face}}} \cdot \boldsymbol{e}_x(\Delta y \Delta z) = A_x\left(x_0 + \frac{\Delta x}{2}, y_0, z_0\right) \Delta y \Delta z \tag{1.86}$$

---

① 矢量场通过闭合曲面向外的净通量代表了该闭合曲面包围区域内通量源的大小。如果该通量为正,意味着闭合曲面内存在通量线的起始点,即正的通量源;如果该通量为负,说明闭合曲面内存在通量线的终止点,即负的通量源。然而通量仅能表示给定区域内通量源的总量,不能描述源的分布特性。为此,需要引入矢量场的散度概念,以描述通量源密度的分布。

② 矢量场在空间某一点的散度的几何意义是该点的通量密度,即包含该点的趋于无限小的体积元表面的通量与该体积元体积之比的极限。由于体积元表面通量代表体积元内部总的通量源,这样得到的通量密度也就代表了该点的通量源密度。

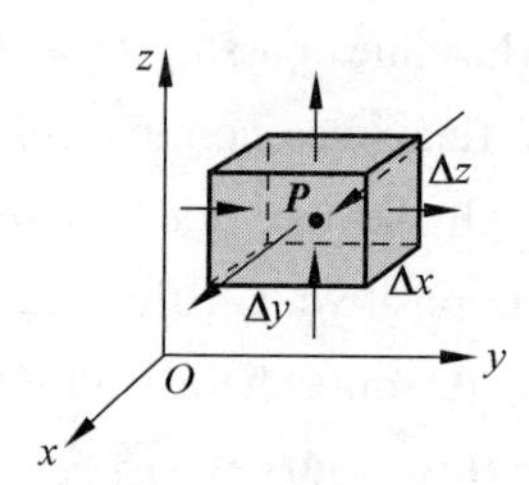

Figure 1-26 A differential volume in Cartesian coordinates

where $A_x(x_0+\Delta x/2, y_0, z_0)$ is the x-component of the $\boldsymbol{A}$ vector at the center point of the front face. Since Δx is infinitely small, it can be approximated by the first two terms of the Taylor series expansion of A_x at (x_0, y_0, z_0) as

$$A_x\left(x_0+\frac{\Delta x}{2}, y_0, z_0\right) \approx A_x(x_0, y_0, z_0) + \frac{\Delta x}{2}\frac{\partial A_x}{\partial x}\bigg|_{(x_0,y_0,z_0)} \tag{1.87}$$

in which high-order terms of Δx are omitted. Similarly, on the back face

$$\int_{\substack{\text{back}\\\text{face}}}\boldsymbol{A}\cdot \mathrm{d}\boldsymbol{s} \approx \boldsymbol{A}_{\substack{\text{back}\\\text{face}}}\cdot \Delta\boldsymbol{s}_{\substack{\text{back}\\\text{face}}} = \boldsymbol{A}_{\substack{\text{back}\\\text{face}}}\cdot(-\boldsymbol{e}_x\Delta y\Delta z) = -A_x\left(x_0-\frac{\Delta x}{2}, y_0, z_0\right)\Delta y\Delta z \tag{1.88}$$

The Taylor-series expansion of $A_x(x_0-\Delta x/2, y_0, z_0)$ gives the following approximation:

$$A_x\left(x_0-\frac{\Delta x}{2}, y_0, z_0\right) \approx A_x(x_0, y_0, z_0) - \frac{\Delta x}{2}\frac{\partial A_x}{\partial x}\bigg|_{(x_0,y_0,z_0)} \tag{1.89}$$

Substituting (1.87) into (1.86) and (1.89) into (1.88) and adding the two contributions, we have

$$\left[\int_{\substack{\text{front}\\\text{face}}} + \int_{\substack{\text{back}\\\text{face}}}\right]\boldsymbol{A}\cdot \mathrm{d}\boldsymbol{s} = \frac{\partial A_x}{\partial x}\bigg|_{(x_0,y_0,z_0)}\Delta x\Delta y\Delta z \tag{1.90}$$

Following the same procedure for the left/right and top/bottom faces, we find

$$\left[\int_{\substack{\text{right}\\\text{face}}} + \int_{\substack{\text{left}\\\text{face}}}\right]\boldsymbol{A}\cdot \mathrm{d}\boldsymbol{s} = \frac{\partial A_y}{\partial y}\bigg|_{(x_0,y_0,z_0)}\Delta x\Delta y\Delta z \tag{1.91}$$

$$\left[\int_{\substack{\text{top}\\\text{face}}} + \int_{\substack{\text{bottom}\\\text{face}}}\right]\boldsymbol{A}\cdot \mathrm{d}\boldsymbol{s} = \frac{\partial A_z}{\partial z}\bigg|_{(x_0,y_0,z_0)}\Delta x\Delta y\Delta z \tag{1.92}$$

Substituting (1.90), (1.91) and (1.92) into (1.85), we have

$$\oint_s \boldsymbol{A}\cdot \mathrm{d}\boldsymbol{s} = \left(\frac{\partial A_x}{\partial x}+\frac{\partial A_y}{\partial y}+\frac{\partial A_z}{\partial z}\right)\bigg|_{(x_0,y_0,z_0)}\Delta x\Delta y\Delta z \tag{1.93}$$

Substituting (1.93) into (1.84) yields the following expression of the divergence in Cartesian coordinates:

$$\nabla\cdot\boldsymbol{A} \triangleq \lim_{\Delta v\to 0}\frac{\oint_S \boldsymbol{A}\cdot \mathrm{d}\boldsymbol{s}}{\Delta v} = \frac{\partial A_x}{\partial x}+\frac{\partial A_y}{\partial y}+\frac{\partial A_z}{\partial z} \tag{1.94}$$

Expressions of the divergence in cylindrical and spherical coordinates can be derived in a similar fashion, which gives:

$$\nabla\cdot\boldsymbol{A} = \frac{1}{\rho}\frac{\partial}{\partial\rho}(\rho A_\rho) + \frac{1}{r}\frac{\partial A_\phi}{\partial\phi} + \frac{\partial A_z}{\partial z} \tag{1.95}$$

$$\nabla \cdot \boldsymbol{A} = \frac{1}{r^2}\frac{\partial}{\partial r}(r^2 A_r) + \frac{1}{r\sin\theta}\frac{\partial}{\partial\theta}(A_\theta \sin\theta) + \frac{1}{r\sin\theta}\frac{\partial A_\phi}{\partial\phi} \tag{1.96}$$

1.9 Divergence Theorem(散度定理)

With the concept of divergence, we have the following identity, which is called the **divergence theorem**(散度定理):

$$\int_V \nabla \cdot \boldsymbol{A} \mathrm{d}v = \oint_S \boldsymbol{A} \cdot \mathrm{d}\boldsymbol{s} \tag{1.97}$$

It states that the volume integral of the divergence of a vector field equals the total outward flux of the vector through the surface that bounds the volume.[①] The divergence theorem (1.97) applies to any volume V that is bounded by the surface S. The direction of $\mathrm{d}\boldsymbol{s}$ is always the outward normal direction, perpendicular to the surface and directed away from the volume.

To prove the divergence theorem, we subdivide an arbitrary volume V into infinite number of infinitely small differential volumes as shown Figure 1-27. For the jth volume element Δv_j bounded by a surface S_j, the definition of divergence in (1.84) gives

$$(\nabla \cdot \boldsymbol{A})_j \Delta v_j = \oint_{S_j} \boldsymbol{A} \cdot \mathrm{d}\boldsymbol{s} \tag{1.98}$$

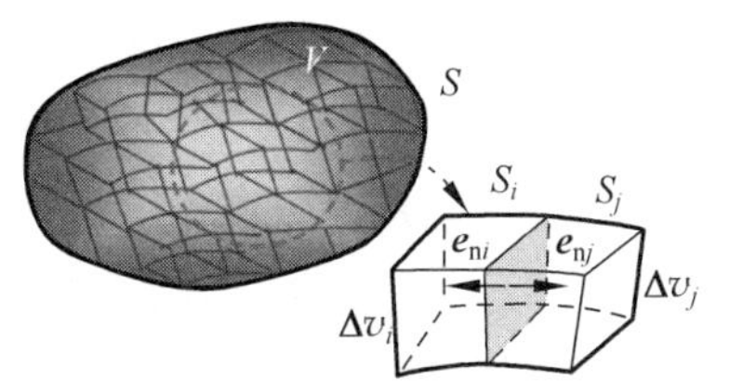

Figure 1-27 Proving divergence theorem by subdividing a volume

Adding up the contributions of all the differential volumes to both sides of (1.98), we have

$$\lim_{\Delta v_j \to 0}\left[\sum_j (\nabla \cdot \boldsymbol{A})_j \Delta v_j\right] = \lim_{\Delta v_j \to 0}\left[\sum_j \oint_{S_j} \boldsymbol{A} \cdot \mathrm{d}\boldsymbol{s}\right] \tag{1.99}$$

The left side of (1.99) is, by definition, the volume integral of $\nabla \cdot \boldsymbol{A}$, i.e.,

$$\lim_{\Delta v_j \to 0}\left[\sum_{j=1} (\nabla \cdot \boldsymbol{A})_j \Delta v_j\right] = \int_V (\nabla \cdot \boldsymbol{A}) \mathrm{d}v \tag{1.100}$$

The surface integrals on the right side of (1.99) are summed over all the faces of all the differential volume elements. However, the contributions from the internal surfaces of adjacent elements (e.g. Δv_i and Δv_j in Figure 1-27) will cancel each other, because on the shared surfaces, the outward normals ($\boldsymbol{e}_{ni}$ and $\boldsymbol{e}_{nj}$) of the adjacent elements point in opposite directions. Hence the surface integral of the right side of (1.99) is contributed only by the external surface S bounding the volume V, i.e.,

---

① 散度定理：任意区域内矢量场散度的体积分等于该区域表面矢量场向外的通量。从数学角度看，高斯定理建立了面积分和体积分的关系。从物理角度看，由于矢量场通过闭合曲面向外的通量等于该闭合曲面内通量源的总量，高斯定理说明矢量场的散度代表了该矢量场的通量源密度。

$$\lim_{\Delta v_j \to 0}\left[\sum_{j=1}\oint_{S_j} \boldsymbol{A}\cdot \mathrm{d}\boldsymbol{s}\right]=\oint_S \boldsymbol{A}\cdot \mathrm{d}\boldsymbol{s} \tag{1.101}$$

Substitution of (1.100) and (1.101) in (1.99) yields the divergence theorem in (1.97). Notice that this proof requires that the vector field $\boldsymbol{A}$, as well as its first derivatives, exist and be continuous both in V and on S, which we will assume to be true in most practical electromagnetic problems except at locations where spatial discontinuities exist (e. g, at the interface between different media).

The divergence theorem is an important identity in vector analysis. Mathematically, it converts a volume integral to a closed surface integral, and vice versa. Physically, as discussed previously, the net outward flux of a vector field through a closed surface is a measure of the total flow sources enclosed by the surface. The divergence theorem indicates that the volume integral of the divergence of the vector field is also a measure of the total flow sources of a vector field within a given volume. Therefore, we can conclude that the divergence of a vector field is a measure of the density of the flow source.

Example 1-4 Following Example 1-3, in which $\boldsymbol{R}(\boldsymbol{r})=\boldsymbol{r}-\boldsymbol{r}'=\boldsymbol{e}_{\mathrm{R}} R$ and $R=|\boldsymbol{R}|$, show that

$$\nabla\cdot\left(\frac{\boldsymbol{e}_{\mathrm{R}}}{R^2}\right)=4\pi\delta(\boldsymbol{r}-\boldsymbol{r}') \tag{1.102}$$

where $\delta(\boldsymbol{r}-\boldsymbol{r}')$ is the Dirac delta function that satisfies:

$$\delta(\boldsymbol{r}-\boldsymbol{r}')=\begin{cases}\infty & \text{for } \boldsymbol{r}=\boldsymbol{r}'\\ 0 & \text{for } \boldsymbol{r}\neq\boldsymbol{r}'\end{cases} \tag{1.103}$$

$$\int_V \delta(\boldsymbol{r}-\boldsymbol{r}')\,\mathrm{d}v=\begin{cases}1 & \text{if } \boldsymbol{r}'\in V\\ 0 & \text{if } \boldsymbol{r}'\notin V\end{cases} \tag{1.104}$$

It is evident from (1.104) that, given any function $f(\boldsymbol{r})$ which is continuous at $\boldsymbol{r}=\boldsymbol{r}'$, we have

$$\int_V f(\boldsymbol{r})\delta(\boldsymbol{r}-\boldsymbol{r}')\,\mathrm{d}v=\begin{cases}f(\boldsymbol{r}') & \text{if } \boldsymbol{r}'\in V\\ 0 & \text{if } \boldsymbol{r}'\notin V\end{cases} \tag{1.105}$$

Solution: Since the divergence operator is translational-invariant, we follow the same strategy in Example 1-3 by first letting $\boldsymbol{r}'=0$ and prove

$$\nabla\cdot\left(\frac{\boldsymbol{e}_{\mathrm{r}}}{r^2}\right)=4\pi\delta(\boldsymbol{r}) \tag{1.106}$$

First, by using (1.96), it is trivial to prove that, for any $\boldsymbol{r}\neq\boldsymbol{0}$,

$$\nabla\cdot\left(\frac{\boldsymbol{e}_{\mathrm{r}}}{r^2}\right)=\frac{1}{r^2}\frac{\partial}{\partial r}\left(r^2\frac{1}{r^2}\right)=0 \quad \text{for} \quad \boldsymbol{r}\neq\boldsymbol{0} \tag{1.107}$$

As a result of (1.107), for any volume V that does not contain the origin, we have

$$\int_V \nabla\cdot\left(\frac{\boldsymbol{e}_{\mathrm{r}}}{r^2}\right)\mathrm{d}v=0 \quad \text{for any } V \text{ not containing the origin} \tag{1.108}$$

Now consider the above integral with V containing the origin. In this case, we can always find a small sphere centered at the origin and fully contained by V. Let the sphere have a radius r_0, by using(1.107) and the divergence theorem,

$$\int_V \nabla \cdot \left(\frac{\boldsymbol{e}_r}{r^2}\right) dv = \int_{r<r_0} \nabla \cdot \left(\frac{\boldsymbol{e}_r}{r^2}\right) dv = \int_{r=r_0} \left(\frac{\boldsymbol{e}_r}{r^2}\right) \cdot \boldsymbol{e}_r ds = \frac{1}{r_0^2}\int_{r=r_0} ds = 4\pi \tag{1.109}$$

(1.107) and(1.109) also imply that

$$\nabla \cdot \left(\frac{\boldsymbol{e}_r}{r^2}\right) = \infty \quad \text{for} \quad \boldsymbol{r} = \boldsymbol{0} \tag{1.110}$$

Equations(1.107)-(1.110) indicate that $\left(\frac{1}{4\pi}\right)\nabla \cdot \left(\frac{\boldsymbol{e}_r}{r^2}\right)$ satisfies the definitional equations of the Dirac delta function(1.103) and(1.104) with $\boldsymbol{r}'=0$. Therefore, (1.106) is proved. Due to the translation-invariance of the divergence operator, the equation(1.102) is also proved.

Example 1-5 Repeat solving Example 1-2 by using divergence theorem.

Solution: According to(1.106), we have $\nabla \cdot \boldsymbol{F}(\boldsymbol{r}) = 4\pi\delta(\boldsymbol{r})$. Therefore, using divergence theorem and equation(1.104), we have

$$\oint_S \boldsymbol{F} \cdot d\boldsymbol{s} = \oint_V \nabla \cdot \boldsymbol{F} dv = \oint_V 4\pi\delta(\boldsymbol{r}) dv = 4\pi$$

as long as the origin is contained by the volume V.

1.10 Curl of a Vector Field(矢量场的旋度)

The flow source discussed in the previous section produces net flux through enclosed surfaces. There is another kind of source, called **vortex source**(**漩涡源**) that produces vector fields with streamlines closing upon themselves. That means the streamlines due to vortex sources do not have a start or end point, and therefore, cannot create a nonzero outward flux through any closed surface. To study the relation between a vortex source and the field it produces, we first introduce the concept of **circulation**(**环量**). The circulation of a vector field around a closed path(闭合路径) is defined as the scalar line integral of the vector over the path:

$$\text{Circulation of } \boldsymbol{A} \text{ around closed path } C \triangleq \oint_C \boldsymbol{A} \cdot d\boldsymbol{l} \tag{1.111}$$

A typical example of the field produced by vortex source is the magnetic field due to a current. The magnetic field lines run around the current with a circulation proportional to the current. Therefore, the circulation of a vector field around a closed path is a measure of the vortex source enclosed by the path.①

① 漩涡源产生的矢量线是闭合的,没有起点和终点,因此在任一闭合曲面上生成的净通量必然为零。闭合的矢量线如果在闭合曲面的某个位置穿入,必然会在另一个地方穿出,否则将无法形成闭合曲线。换言之,闭合的矢量线不可能是由通量源产生的。例如,磁场不是由通量源产生的(自然界不存在磁荷),电流产生的磁力线是闭合的。闭合磁力线的环量正比于其包围的电流大小。因此,矢量场在闭合路径上的环量代表了该闭合路径包围的漩涡源的大小。进一步,为了描述漩涡源的空间分布特性,需要引入环量面密度和旋度的概念。

In order to define a point function as a measure of the vortex source density, we take a small surface Δs containing a given point, calculate the ratio between the circulation around the contour(闭合围线) C of the surface and the surface area Δs. If the limit of the ratio exists as Δs approaches zero, we obtain a **circulation density**(**环量密度**) at the given point as the limit of the ratio.

The circulation defined in(1.111) is obviously dependent on the closed path C, including the path's shape, length and orientation(朝向). As a result, the circulation density at a point P is not unique. It is dependent on the orientation of the small surface Δs in the above definition. Naturally, the orientation of Δs is prescribed by its normal direction at the point P. There must be a normal direction of Δs with which the circulation density reaches its maximum. We define the **curl**(**旋度**) of a vector field $\boldsymbol{A}$ at a given point, denoted by $\nabla\times\boldsymbol{A}$, as a vector whose magnitude is the maximum circulation density at this point, and whose direction is along the normal direction of the surface Δs when its orientation maximizes the circulation density. That is,

$$\nabla\times\boldsymbol{A} \triangleq \lim_{\Delta s\to 0}\frac{1}{\Delta s}\left[\boldsymbol{e}_{\mathrm{n}}\oint_C \boldsymbol{A}\cdot \mathrm{d}\boldsymbol{l}\right]_{\max} \tag{1.112}$$

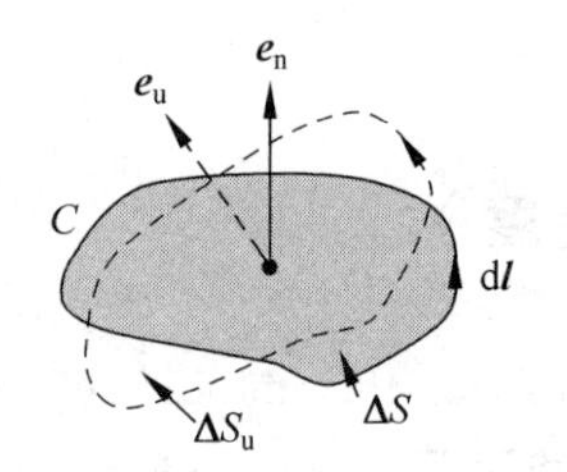

Figure 1-28 Relation between the direction of $\boldsymbol{e}_{\mathrm{n}}$ and d$\boldsymbol{l}$ in defining curl

where Δs is the area of the surface bounded by the contour C, the direction $\boldsymbol{e}_{\mathrm{n}}$ is normal to the area Δs and is determined by the right-hand rule: when the fingers of the right hand follow the direction of d$\boldsymbol{l}$, the thumb points to the $\boldsymbol{e}_{\mathrm{n}}$ direction. This is illustrated in Figure 1-28. ①

From the definition, $\nabla\times\boldsymbol{A}$ is also a vector field. One important property of the curl is that, the component of $\nabla\times\boldsymbol{A}$ in any other direction $\boldsymbol{e}_{\mathrm{u}}$ is the circulation density of the vector $\boldsymbol{A}$ with the normal direction of the defining surface Δs pointing along $\boldsymbol{e}_{\mathrm{u}}$, i. e. ,

$$(\nabla\times\boldsymbol{A})_{\mathrm{u}} = \boldsymbol{e}_{\mathrm{u}}\cdot(\nabla\times\boldsymbol{A}) = \lim_{\Delta s_{\mathrm{u}}\to 0}\frac{1}{\Delta s_{\mathrm{u}}}\left(\oint_{C_{\mathrm{u}}}\boldsymbol{A}\cdot\mathrm{d}\boldsymbol{l}\right) \tag{1.113}$$

where the direction $\boldsymbol{e}_{\mathrm{u}}$ is normal to the surface Δs_{u}, Δs_{u} is bounded by the contour C_{u}, and the direction of line integration around C_{u} is related to $\boldsymbol{e}_{\mathrm{u}}$ via the right-hand rule. ②

We now use(1.113) to find the expression of $\nabla\times\boldsymbol{A}$ in Cartesian coordinates. For an arbitrary point $P(x_0, y_0, z_0)$ in space, we first take a rectangular differential area centered at the point P, as is shown in Figure 1-29. The differential area is parallel to the $y-z$ plane and has sides Δy and Δz. Then we have $\boldsymbol{e}_{\mathrm{u}}=\boldsymbol{e}_{\mathrm{x}}$ and $\Delta s_{\mathrm{u}}=\Delta y\Delta z$, and the contour C_{u} consists of four sides

---

① 旋度的定义：矢量场在空间某一点的旋度仍然是一个矢量，其幅度为该矢量场在该点上环流面密度的最大值，其方向为该矢量场在该点上取得最大环流面密度时，环流所在面元的法向方向。

② 式(1.113)既是矢量场旋度的重要性质，也可以作为旋度的定义：对于给定的矢量场 $\boldsymbol{A}$，如果在空间任意一点存在一个矢量，记为 $\nabla\times\boldsymbol{A}$，其在任意方向 u 上的投影等于 $\boldsymbol{A}$ 在该点沿 u 方向的微分面元上的环流密度，那么该矢量 $\nabla\times\boldsymbol{A}$ 就定义为矢量场 $\boldsymbol{A}$ 在该点的旋度。

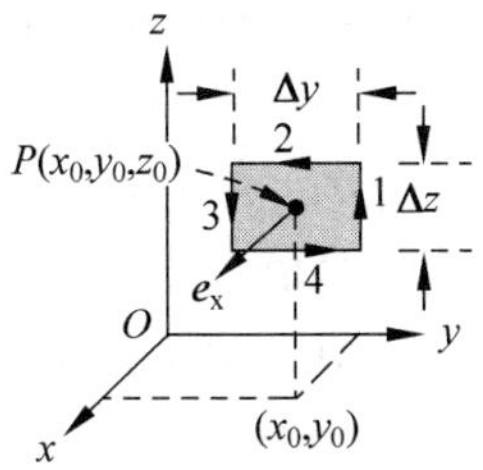

Figure 1-29 Derivation of$(\nabla\times\boldsymbol{A})_x$

1,2,3 and 4. From(1.113), we have

$$(\nabla\times\boldsymbol{A})_x=\lim_{\Delta y\Delta z\to 0}\frac{1}{\Delta y\Delta z}\left(\oint_{\substack{\text{sides}\\1,2,3,4}}\boldsymbol{A}\cdot d\boldsymbol{l}\right)\tag{1.114}$$

The line integral contains contributions from four sides. For side 1, $d\boldsymbol{l}=\boldsymbol{e}_z\Delta z$. As $\Delta z\to 0$, we have

$$\int_{\text{side1}}\boldsymbol{A}\cdot d\boldsymbol{l}\approx A_z\left(x_0,y_0+\frac{\Delta y}{2},z_0\right)\Delta z\approx\left\{A_z(x_0,y_0,z_0)+\frac{\Delta y}{2}\frac{\partial A_z}{\partial y}\bigg|_{(x_0,y_0,z_0)}\right\}\Delta z\tag{1.115}$$

For side 3, $d\boldsymbol{l}=-\boldsymbol{e}_z\Delta z$, and

$$\int_{\text{side3}}\boldsymbol{A}\cdot d\boldsymbol{l}\approx A_z\left(x_0,y_0-\frac{\Delta y}{2},z_0\right)(-\Delta z)\approx-\left\{A_z(x_0,y_0,z_0)-\frac{\Delta y}{2}\frac{\partial A_z}{\partial y}\bigg|_{(x_0,y_0,z_0)}\right\}\Delta z\tag{1.116}$$

Combining(1.115) and(1.116), we have

$$\int_{\text{side1}}\boldsymbol{A}\cdot d\boldsymbol{l}+\int_{\text{side3}}\boldsymbol{A}\cdot d\boldsymbol{l}=\frac{\partial A_z}{\partial y}\bigg|_{(x_0,y_0,z_0)}\Delta y\Delta z\tag{1.117}$$

Similarly, we can derive

$$\int_{\text{side2}}\boldsymbol{A}\cdot d\boldsymbol{l}+\int_{\text{side4}}\boldsymbol{A}\cdot d\boldsymbol{l}=-\frac{\partial A_y}{\partial z}\bigg|_{(x_0,y_0,z_0)}\Delta y\Delta z\tag{1.118}$$

Substituting(1.117) and(1.118) into(1.114), we obtain

$$(\nabla\times\boldsymbol{A})_x=\frac{\partial A_z}{\partial y}-\frac{\partial A_y}{\partial z}\tag{1.119}$$

Similar derivations can be performed to obtain the y and z component of $\nabla\times\boldsymbol{A}$ in Cartesian coordinates, which leads to an expression of the vector $\nabla\times\boldsymbol{A}$ as follows:

$$\nabla\times\boldsymbol{A}=\boldsymbol{e}_x\left(\frac{\partial A_z}{\partial y}-\frac{\partial A_y}{\partial z}\right)+\boldsymbol{e}_y\left(\frac{\partial A_x}{\partial z}-\frac{\partial A_z}{\partial x}\right)+\boldsymbol{e}_z\left(\frac{\partial A_y}{\partial x}-\frac{\partial A_x}{\partial y}\right)\tag{1.120}$$

Notice the cyclic permutation order in x, y and z in the above expression. An easier way to remember the expression of $\nabla\times\boldsymbol{A}$ is by arranging it in a determinant form similar to the cross-product expression in(1.23):

$$\nabla\times\boldsymbol{A}=\begin{vmatrix}\boldsymbol{e}_x & \boldsymbol{e}_y & \boldsymbol{e}_z\\ \dfrac{\partial}{\partial x} & \dfrac{\partial}{\partial y} & \dfrac{\partial}{\partial z}\\ A_x & A_y & A_z\end{vmatrix}\tag{1.121}$$

Putting together(1.69),(1.94) and(1.121), we find it is convenient to consider the del operator ∇ in Cartesian coordinates as an ostensible vector:

$$\nabla \equiv \boldsymbol{e}_x \frac{\partial}{\partial x} + \boldsymbol{e}_y \frac{\partial}{\partial y} + \boldsymbol{e}_z \frac{\partial}{\partial z} \tag{1.122}$$

Then the expression of the three differential operations, gradient, divergence and curl of a field, can be easily remembered as the multiplication, dot product and cross product of the vector of (1.122) with the corresponding scalar or vector fields. ①

Expressions of the curl in cylindrical and spherical coordinates can be found by following the similar procedure, which gives:

$$\nabla \times \boldsymbol{A} = \frac{1}{\rho} \begin{vmatrix} \boldsymbol{e}_\rho & \boldsymbol{e}_\phi \rho & \boldsymbol{e}_z \\ \frac{\partial}{\partial \rho} & \frac{\partial}{\partial \phi} & \frac{\partial}{\partial z} \\ A_\rho & \rho A_\phi & A_z \end{vmatrix} \tag{1.123}$$

$$\nabla \times \boldsymbol{A} = \frac{1}{r^2 \sin\theta} \begin{vmatrix} \boldsymbol{e}_r & \boldsymbol{e}_\theta r & \boldsymbol{e}_\phi r\sin\theta \\ \frac{\partial}{\partial r} & \frac{\partial}{\partial \theta} & \frac{\partial}{\partial \phi} \\ A_r & rA_\theta & r\sin\theta A_\phi \end{vmatrix} \tag{1.124}$$

1.11 Stokes's Theorem(斯托克斯定理)

Stokes's theorem states that the surface integral of the curl of a vector field over an open surface is equal to the closed line integral of the vector along the contour bounding the surface, which can be expressed as

$$\int_S (\nabla \times \boldsymbol{A}) \cdot \mathrm{d}\boldsymbol{s} = \oint_C \boldsymbol{A} \cdot \mathrm{d}\boldsymbol{l} \tag{1.125}$$

where S is an arbitrary surface bounded by the contour C. The directions of $\mathrm{d}\boldsymbol{s}$ and $\mathrm{d}\boldsymbol{l}$ are determined by following the right-hand rule. ②

To prove the Stokes's theorem, we subdivide an arbitrary surface S into infinite number of infinitely small differential areas as shown Figure 1-30. For the jth differential area $\Delta \boldsymbol{s}_j$ bounded by the contour C_j, the property of curl in (1.113) leads to

$$(\nabla \times \boldsymbol{A})_j \cdot (\Delta \boldsymbol{s}_j) = \oint_{C_j} \boldsymbol{A} \cdot \mathrm{d}\boldsymbol{l} \tag{1.126}$$

Applying (1.126) to all the differential areas and adding them up, we have

$$\lim_{\Delta s_j \to 0} \sum_{j=1}^{N} (\nabla \times \boldsymbol{A})_j \cdot (\Delta \boldsymbol{s}_j) = \lim_{\Delta s_j \to 0} \sum_{j=1}^{N} \left(\oint_{C_j} \boldsymbol{A} \cdot \mathrm{d}\boldsymbol{l} \right) \tag{1.127}$$

The left side of (1.127) becomes the flux of $\nabla \times \boldsymbol{A}$ through the surface S, i.e.

---

① 形式上将 del 算子(哈密尔顿算子)看成式(1.122)所示的一个矢量,该矢量与标量场作乘积,就得到直角坐标系下梯度的表达式,与矢量作点积和叉积,则会分别得到散度和旋度的表达式。

② 斯托克斯定理:矢量场的旋度在任意一个曲面上的通量面积分等于该矢量场在该曲面的闭合围线上的线积分(即环量)。

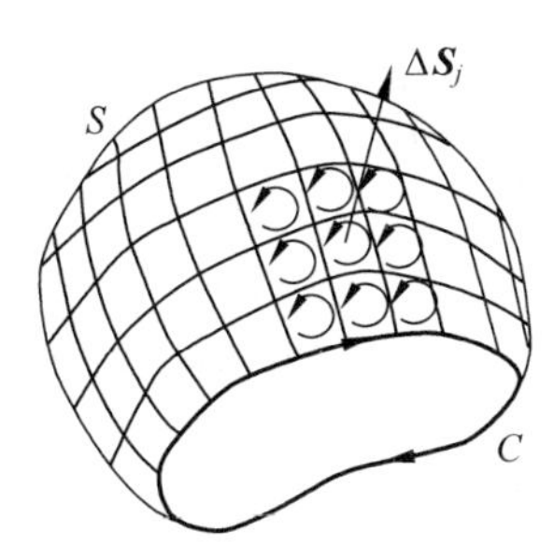

Figure 1-30 Proving Stokes's theorem by subdividing a surface

$$\lim_{\Delta s_j \to 0} \sum_{j=1}^{N} (\nabla \times \boldsymbol{A})_j \cdot (\Delta \boldsymbol{S}_j) = \int_S (\nabla \times \boldsymbol{A}) \cdot \mathrm{d}\boldsymbol{s} \tag{1.128}$$

In the right side of (1.127), the shared segment of the contours of two adjacent elements is in opposite directions along the integration paths. The net contribution of all the internal shared segments to the total line integral is zero, and only the contribution from the external contour C remains after the summation:

$$\lim_{\Delta s_j \to 0} \sum_{j=1}^{N} \left(\oint_{C_j} \boldsymbol{A} \cdot \mathrm{d}\boldsymbol{l} \right) = \oint_C \boldsymbol{A} \cdot \mathrm{d}\boldsymbol{l} \tag{1.129}$$

By substituting (1.128) and (1.129) into (1.127), we obtain Stokes's theorem expressed by (1.125).

Stokes's theorem converts a surface integral of the curl of a vector to a line integral of the vector, and vice versa. Like the divergence theorem, Stokes's theorem is an important identity in vector analysis, and we will use it frequently in establishing other theorems and relations in electromagnetics.

Notice that, if the surface integral of $\nabla \times \boldsymbol{A}$ is carried over a closed surface, there will be no surface-bounding external contour, and Stokes's theorem (1.125) tells us that

$$\oint_S (\nabla \times \boldsymbol{A}) \cdot \mathrm{d}\boldsymbol{s} = 0 \tag{1.130}$$

for any closed surface S.

1.12 Laplacian Operator(拉普拉斯算子)

The **Laplacian operator** (拉普拉斯算子), denoted by ∇^2 (del square), is a second derivative operator that can be applied on both scalar and vector fields.

For a scalar φ, $\nabla^2\varphi$ is defined as "the divergence of the gradient of" the scalar field φ, i.e.

$$\nabla^2\varphi = \nabla \cdot (\nabla\varphi) \tag{1.131}$$

In Cartesian coordinates, we have

$$\nabla^2\varphi = \nabla \cdot \nabla\varphi = \left(\boldsymbol{e}_x \frac{\partial}{\partial x} + \boldsymbol{e}_y \frac{\partial}{\partial y} + \boldsymbol{e}_z \frac{\partial}{\partial z} \right) \cdot \left(\boldsymbol{e}_x \frac{\partial \varphi}{\partial x} + \boldsymbol{e}_y \frac{\partial \varphi}{\partial y} + \boldsymbol{e}_z \frac{\partial \varphi}{\partial z} \right) = \frac{\partial^2 \varphi}{\partial x^2} + \frac{\partial^2 \varphi}{\partial y^2} + \frac{\partial^2 \varphi}{\partial z^2} \tag{1.132a}$$

It is straightforward to derive the expressions for $\nabla^2\varphi$ in cylindrical and spherical coordinates as

$$\nabla^2\varphi = \frac{1}{\rho}\frac{\partial}{\partial\rho}\left(\rho\frac{\partial\varphi}{\partial\rho}\right) + \frac{1}{\rho^2}\frac{\partial^2\varphi}{\partial\phi^2} + \frac{\partial^2\varphi}{\partial z^2} \tag{1.132b}$$

$$\nabla^2\varphi = \frac{1}{r^2}\frac{\partial}{\partial r}\left(r^2\frac{\partial\varphi}{\partial r}\right) + \frac{1}{r^2\sin\theta}\frac{\partial}{\partial\theta}(\sin\theta\frac{\partial\varphi}{\partial\theta}) + \frac{1}{r^2\sin^2\theta}\frac{\partial^2\varphi}{\partial\phi^2} \tag{1.132c}$$

Example 1-6 Following Example 1-3 and Example 1-4, show that

$$\nabla^2\left(\frac{1}{R}\right) = -4\pi\delta(\boldsymbol{r}-\boldsymbol{r}') \tag{1.133}$$

Solution: By using(1.78) in Example 1-3 and(1.102) in Example 1-4, we have

$$\nabla^2\left(\frac{1}{R}\right) = \nabla\cdot\left[\nabla\left(\frac{1}{R}\right)\right] = \nabla\cdot\left(-\frac{\boldsymbol{e}_R}{R^2}\right) = -4\pi\delta(\boldsymbol{r}-\boldsymbol{r}')$$

For a vector field $\boldsymbol{A}$, the Laplacian operator is defined by the following identity

$$\nabla\times\nabla\times\boldsymbol{A} = \nabla(\nabla\cdot\boldsymbol{A}) - \nabla^2\boldsymbol{A} \tag{1.134}$$

or

$$\nabla^2\boldsymbol{A} = \nabla(\nabla\cdot\boldsymbol{A}) - \nabla\times\nabla\times\boldsymbol{A} \tag{1.135}$$

For Cartesian coordinates, it can be readily verified by direct substitution that

$$\nabla^2\boldsymbol{A} = \boldsymbol{e}_x\nabla^2 A_x + \boldsymbol{e}_y\nabla^2 A_y + \boldsymbol{e}_z\nabla^2 A_y \tag{1.136}$$

That is, for Cartesian coordinates, the Laplacian of a vector field $\boldsymbol{A}$ is another vector field whose components are the Laplacian of the corresponding components of $\boldsymbol{A}$. However, note that it is not true for cylindrical and spherical coordinate systems.

Example 1-7 Given a scalar field $f(\boldsymbol{r})$ and a constant vector $\boldsymbol{C}$, prove that

$$\boldsymbol{C}[\nabla^2 f(\boldsymbol{r})] = \nabla^2[\boldsymbol{C}f(\boldsymbol{r})] \tag{1.137}$$

Notice that the two ∇^2 operators in(1.137) have different definitions. On the left side, ∇^2 operating on the scalar field $f(\boldsymbol{r})$ is defined by(1.131), whereas on the right side, ∇^2 operating on a vector field $\boldsymbol{C}f(\boldsymbol{r})$ is defined by(1.135).

Solution: Let $\boldsymbol{F}(\boldsymbol{r})=\boldsymbol{C}f(\boldsymbol{r})$, which is a vector field. Under Cartesian coordinates, $\boldsymbol{C}=\boldsymbol{e}_x C_x+\boldsymbol{e}_y C_y+\boldsymbol{e}_z C_z$

Then, $\boldsymbol{F}(\boldsymbol{r}) = \boldsymbol{e}_x C_x f(\boldsymbol{r}) + \boldsymbol{e}_y C_y f(\boldsymbol{r}) + \boldsymbol{e}_z C_z f(\boldsymbol{r})$. With (1.136), the right side of (1.137) becomes

$$\nabla^2[\boldsymbol{C}f(\boldsymbol{r})] = \nabla^2\boldsymbol{F}(\boldsymbol{r}) = \boldsymbol{e}_x\nabla^2[C_x f(\boldsymbol{r})] + \boldsymbol{e}_y\nabla^2[C_y f(\boldsymbol{r})] + \boldsymbol{e}_z\nabla^2[C_z f(\boldsymbol{r})]$$

With(1.132a), the left side of(1.137) becomes

$$\begin{aligned}\boldsymbol{C}[\nabla^2 f(\boldsymbol{r})] &= (\boldsymbol{e}_x C_x + \boldsymbol{e}_y C_y + \boldsymbol{e}_z C_z)\nabla^2 f(\boldsymbol{r}) \\ &= \boldsymbol{e}_x\nabla^2[C_x f(\boldsymbol{r})] + \boldsymbol{e}_y\nabla^2[C_y f(\boldsymbol{r})] + \boldsymbol{e}_z\nabla^2[C_z f(\boldsymbol{r})]\end{aligned}$$

in which we use the fact that C_x, C_y and C_z are all constant scalars, and that ∇^2 is a linear operator. Since both sides of(1.137) end up with the same expressions, and therefore, (1.137) is proved.

1.13 Curl-free and Divergence-free Fields (无旋场与无散场)

A **curl-free field**(无旋场) in a particular region is a vector field that has zero curl throughout the region. Apparently, there is no vortex sources within the particular region on which the curl-free field is defined. An important null identity related to curl-free fields is

$$\nabla\times(\nabla\varphi)\equiv 0 \tag{1.138}$$

which states that the gradient of any scalar field is a curl-free field.[①] It can be proved readily by using the expressions of gradient and curl in Cartesian coordinates. In general, it can be proved by using Stokes's theorem. Specifically, for any surface S bounded by contour C, replace $\boldsymbol{A}$ with $\nabla\varphi$ in (1.125), we have

$$\int_S [\nabla\times(\nabla\varphi)]\cdot d\boldsymbol{s}=\oint_C (\nabla\varphi)\cdot d\boldsymbol{l} \tag{1.139}$$

From (1.73),

$$\oint_C (\nabla\varphi)\cdot d\boldsymbol{l}=\oint_C d\varphi=0 \tag{1.140}$$

(1.139) and (1.140) leads to the fact that the surface integral of $\nabla\times(\nabla\varphi)$ over any surface is zero. Therefore, the integrand $\nabla\times(\nabla\varphi)$ must be zero everywhere, which proves the identity in (1.138).

The converse of the identity (1.138) is also true, i. e., any curl-free vector field can be expressed as the gradient of a scalar field. According to (1.140), any closed line integral of the gradient of a scalar field is zero, therefore, a curl-free field is also called **irrotational**(无旋的) vector field. (1.140) also means that the line integral of a curl-free field is path independent. Physically, in a system with conserved energy, the line integral of forces is also path independent, therefore a curl-free field is also a **conservative vector field** (保守矢量场). As will be studied later, the static electric field $\boldsymbol{E}$ is curl-free/conservative and can be expressed as the negative of the gradient of another scalar field called electric potential φ.[②]

Example 1-8 Following Example 1-3, where $\boldsymbol{R}=\boldsymbol{r}-\boldsymbol{r}'$ and $R=|\boldsymbol{R}|$, show that

$$\nabla\times\left(\frac{\boldsymbol{e}_R}{R^2}\right)=0 \tag{1.141}$$

Solution: This equation can be proved directly by applying the expressions of curl in Cartesian coordinates (1.121). Or it can be proved under spherical coordinates by first letting $\boldsymbol{r}'=0$ and then using the translation-invariant property of the curl operator. Another way of proving (1.141) is by noticing that $(\boldsymbol{e}_R/R^2)$ can be expressed as the gradient of $(-1/R^2)$ according to (1.78). Then it is evident from the identity (1.138) that (1.141) must hold.

A **divergence-free** (or **divergenceless**) **field**(无散场) in a particular region is a vector field that has zero divergence throughout the region. Apparently, there is no flow sources within the region on which the divergence-free field is defined. An important null identity related to divergence-free fields is

---

① 式(1.138)表明任意标量场的梯度都是无旋场。

② 无旋场的几个性质：无旋场可以表示为一个标量场的梯度；无旋场沿任意闭合回路的环量为零；任意两点之间无旋场的线积分与积分路径无关；重力场和静电场是两种典型的无旋场，其物理意义体现了能量守恒定律，因此无旋场也被称为保守场。

$$\nabla \cdot (\nabla \times \boldsymbol{A}) \equiv 0 \tag{1.142}$$

This identity states that the curl of any vector field is a divergence-free field. [①] It can be proved by using the expressions of divergence and curl in Cartesian coordinates. It can also be proved generally by using divergence theorem. Specifically, for any volume V bounded by surface S, applying(1.97) gives us

$$\int_V \nabla \cdot (\nabla \times \boldsymbol{A}) \mathrm{d}v = \oint_S (\nabla \times \boldsymbol{A}) \cdot \mathrm{d}\boldsymbol{s} = 0 \tag{1.143}$$

In the above derivation, the equation(1.130) is used. Since(1.143) is true for any arbitrary volume V, the integrand itself must be zero, which proves the identity(1.142).

The converse of the identity(1.142) states that, any divergence-free field can be expressed as the curl of another vector field. It is seen from(1.143) that the net out-ward flux of a divergence-free field through any closed surface within the particular region is zero. This also means that the flux lines starting from any point within the particular region must end at the same point, forming closed paths. Therefore, a divergence-free field is also called a **solenoidal field**(**螺旋场**). [②] Strictly speaking, if the region on which the divergence-free field is defined is finite, the flux lines may also start and end at the boundary; if the region extends to infinity, the flux lines may also start and end at infinity. In both cases, the field can be seen as solenoidal if the boundary of the region or the infinity is viewed as one point. As will be studied later, the magnetic flux density $\boldsymbol{B}$ is solenoidal and can be expressed as the curl of another vector field called magnetic vector potential $\boldsymbol{A}$.

1.14 Helmholtz's Theorem(亥姆霍兹定理)

In previous sections, we introduced the concepts of divergence-free(solenoidal) field and curl-free(irrotational) field. Since the divergence of a vector field is a measure of its flow source whereas the curl of a vector field is a measure of its vortex source, divergence-free fields must only have vortex source whereas curl-free field must only have flow source in the region they are defined. Generally, a vector field may have both flow and vortex sources. The flow source determines the curl-free component of the field, whereas the vortex source determines the divergence-free component, which leads to the following important theorem.

Helmholtz's theorem (**亥姆霍兹定理**) states that a vector field $\boldsymbol{F}$ can be written as the sum of a divergence-free field and a curl-free field as follows:

$$\boldsymbol{F} = -\nabla\varphi + \nabla \times \boldsymbol{A} \tag{1.144}$$

where $-\nabla\varphi$ is the curl-free part and $\nabla \times \boldsymbol{A}$ is the divergence-free part. φ and $\boldsymbol{A}$ are determined by the flow and vortex sources of the field respectively as

---

① 式(1.142)表明,任意矢量场的旋度是一个无散场。

② 无散场的几个性质:无散场可以表示为某个矢量场的旋度;无散场在任意闭合曲面上的通量为零;无散场中任意位置发出的矢量线必然终止于该位置,因而无散场也被称为螺旋场。

$$\varphi(\boldsymbol{r})=\frac{1}{4\pi}\int_{V'}\frac{\nabla'\cdot\boldsymbol{F}(\boldsymbol{r}')}{|\boldsymbol{r}-\boldsymbol{r}'|}\mathrm{d}v' \tag{1.145}$$

$$\boldsymbol{A}(\boldsymbol{r})=\frac{1}{4\pi}\int_{V'}\frac{\nabla'\times\boldsymbol{F}(\boldsymbol{r}')}{|\boldsymbol{r}-\boldsymbol{r}'|}\mathrm{d}v' \tag{1.146}$$

where V' is all space or the region that contains all the flow and vortex sources of the field $\boldsymbol{F}$.

According to the Helmholtz's theorem, a vector field is determined if both its divergence and its curl are specified everywhere. ①

Summary

Concepts

Scalar(标量)

Vector(矢量)

Scalar field(标量场)

Vector field(矢量场)

Dot product(点积)

Cross product(叉积)

Directional derivative(方向导数)

Gradient(梯度)

Divergence(散度)

Curl(旋度)

Divergence-free(solenoid) field(无散场)

Curl-free(irrotational) field(无旋场)

Laws & Theorems

Divergence theorem(散度定理)

Stokes's theorem(旋度定理)

Helmholtz's theorem(亥姆霍兹定理)

Two null identities(两个零恒等式)

Methods

Finding directional derivatives by using projection of the gradient;

Conversion between volume and surface integrations using divergence theorem;

Conversion between surface and line integrals using Stokes's theorem.

Review Questions

1.1 分别在什么条件下 $\boldsymbol{A}\cdot\boldsymbol{B}$ 的结果为正数、负数、零?

1.2 如何通过点积运算判断两个矢量是否相互平行,以及是否相互垂直?

1.3 若 $\boldsymbol{A}\cdot\boldsymbol{B}=\boldsymbol{A}\cdot\boldsymbol{C}$,是否可以得到结论 $\boldsymbol{B}=\boldsymbol{C}$? 为什么?

1.4 给定两个矢量 $\boldsymbol{A}$ 和 $\boldsymbol{B}$,写出①$\boldsymbol{A}$ 沿 $\boldsymbol{B}$ 所在方向的分量,②$\boldsymbol{B}$ 沿 $\boldsymbol{A}$ 所在方向的分量。

1.5 如何通过叉积运算判断两个矢量是否相互平行,以及是否相互垂直?

1.6 若 $\boldsymbol{A}\times\boldsymbol{B}=\boldsymbol{A}\times\boldsymbol{C}$,是否可以得到结论 $\boldsymbol{B}=\boldsymbol{C}$? 为什么?

1.7 什么条件下 $\boldsymbol{A}\times\boldsymbol{B}=\boldsymbol{B}\times\boldsymbol{A}$?

---

① 亥姆霍兹定理表明,任意矢量场可以分解为一个无旋场和一个无散场的叠加,其中无旋场部分由该矢量场的散度代表的通量源产生,无散场部分由该矢量场的旋度代表的漩涡源产生。因此,矢量场可由其散度和旋度确定,而矢量场的散度及旋度特性是矢量场研究的首要问题。

1.8 写出直角坐标系下$\boldsymbol{A}\cdot\boldsymbol{B}$和$\boldsymbol{A}\times\boldsymbol{B}$的数学表达式。

1.9 简述标量场的梯度的物理意义。

1.10 已知标量场的梯度,如何计算该标量场在给定方向上的方向导数?

1.11 直角坐标系下∇算子代表什么?

1.12 简述矢量场的散度的定义及其物理意义。

1.13 什么是散度定理?其物理意义是什么?

1.14 简述矢量场的旋度的定义及其物理意义。

1.15 什么是斯托克斯定理?其物理意义是什么?

1.16 什么是亥姆霍兹定理?其物理意义是什么?

Problems

1.1 Prove the distributive law (1.12) for the dot product and (1.15) for the cross product. ①

1.2 Prove that, if $\boldsymbol{A}\cdot\boldsymbol{B}=\boldsymbol{A}\cdot\boldsymbol{C}$, $\boldsymbol{A}\times\boldsymbol{B}=\boldsymbol{A}\times\boldsymbol{C}$, and $\boldsymbol{A}$ is not a zero vector, then $\boldsymbol{B}=\boldsymbol{C}$.

1.3 (1) Given the **unit normal vector** (**单位法向矢量**) $\boldsymbol{e}_{\mathrm{n}}$ of a surface, find the expressions for the **normal and tangential vector components** (i. e. the vector components along normal and tangential directions of the surface, **法向与切向矢量分量**) of a vector $\boldsymbol{A}$ using cross products only.

(2) Repeat (1), but using dot products only.

(3) Given a unit vector $\boldsymbol{e}_{\mathrm{k}}$, find an equation (in terms of $\boldsymbol{e}_{\mathrm{k}}$ and position vector $\boldsymbol{r}$) for the plane that satisfies the following: (i) the plane is perpendicular to $\boldsymbol{e}_{\mathrm{k}}$, (ii) the plane contains the point (0,0,1) in rectangular coordinates. (Hint: the two conditions mean that the vector pointing from any point on the plane to the point (0,0,1) is perpendicular to $\boldsymbol{e}_{\mathrm{k}}$).

1.4 As shown in the Figure 1-31, $\boldsymbol{r}_{\mathrm{P}}$, $\boldsymbol{r}_{\mathrm{Q}_1}$, $\boldsymbol{r}_{\mathrm{Q}_2}$, and $\boldsymbol{r}_{\mathrm{Q}_3}$ are respectively the 4 position vectors of the 4 points P, Q_1, Q_2, Q_3 in the space. P_0 is the projection of P onto the plane determined by Q_1, Q_2 and Q_3. Find an expression for the position vector $\boldsymbol{r}_{P_0}$ of the point P_0 (in terms of $\boldsymbol{r}_{\mathrm{P}}$, $\boldsymbol{r}_{\mathrm{Q}_1}$, $\boldsymbol{r}_{\mathrm{Q}_2}$ and $\boldsymbol{r}_{\mathrm{Q}_3}$).

Figure 1-31 Illustration of Problem 1.4

1.5 Given two vectors $\boldsymbol{A}=5\boldsymbol{e}_{\mathrm{x}}-3\boldsymbol{e}_{\mathrm{y}}+2\boldsymbol{e}_{\mathrm{z}}$, $\boldsymbol{B}=\boldsymbol{e}_{\mathrm{x}}-6\boldsymbol{e}_{\mathrm{y}}-5\boldsymbol{e}_{\mathrm{z}}$, find

(1) $\boldsymbol{A}+\boldsymbol{B}$, $\boldsymbol{A}-\boldsymbol{B}$.

(2) $\boldsymbol{A}\cdot\boldsymbol{B}$, $\boldsymbol{A}\times\boldsymbol{B}$.

(3) The angle between $\boldsymbol{A}$ and $\boldsymbol{B}$.

1.6 Given the following four vectors: $\boldsymbol{A}=2\boldsymbol{e}_{\mathrm{x}}-\alpha\boldsymbol{e}_{\mathrm{y}}+\boldsymbol{e}_{\mathrm{z}}$, $\boldsymbol{B}=\alpha\boldsymbol{e}_{\mathrm{x}}+\boldsymbol{e}_{\mathrm{y}}-2\boldsymbol{e}_{\mathrm{z}}$, $\boldsymbol{C}=\boldsymbol{e}_{\mathrm{x}}+\alpha\boldsymbol{e}_{\mathrm{y}}+2\boldsymbol{e}_{\mathrm{z}}$, and $\boldsymbol{D}=\alpha^2\boldsymbol{e}_{\mathrm{x}}+\alpha\boldsymbol{e}_{\mathrm{y}}+\boldsymbol{e}_{\mathrm{z}}$, find the value of α for each of the following cases:

---

① 矢量点积和叉积的分配律的证明必须从点积和叉积的几何定义出发,而不能利用坐标系下的表达式(1.22)或式(1.23)。因为式(1.22)或式(1.23)本身就是由分配律得到的。

(1) $\boldsymbol{A}$ is perpendicular to $\boldsymbol{B}$.

(2) $\boldsymbol{B}$ is parallel to $\boldsymbol{C}$.

(3) $\boldsymbol{D}$ is perpendicular to both $\boldsymbol{A}$ and $\boldsymbol{B}$.

1.7 Calculate the volume of a parallelepiped formed by vectors $\boldsymbol{A}=2\boldsymbol{e}_x+2\boldsymbol{e}_y-2\boldsymbol{e}_z$, $\boldsymbol{B}=3\boldsymbol{e}_x+5\boldsymbol{e}_y-4\boldsymbol{e}_z$, and $\boldsymbol{C}=5\boldsymbol{e}_x-2\boldsymbol{e}_y+2\boldsymbol{e}_z$.

1.8 Given vector $\boldsymbol{A}=3\boldsymbol{e}_\rho+2\boldsymbol{e}_\phi+2\boldsymbol{e}_z$ at point $P\left(3,\frac{\pi}{6},5\right)$, and $\boldsymbol{B}=-2\boldsymbol{e}_\rho+3\boldsymbol{e}_\phi-5\boldsymbol{e}_z$ at point $Q\left(4,\frac{\pi}{3},3\right)$, evaluate $\boldsymbol{C}=\boldsymbol{A}-\boldsymbol{B}$ in rectangular coordinates.

1.9 Show that the relation between the unit vector in Cartesian coordinates and the unit vectors in spherical coordinates $\boldsymbol{e}_r, \boldsymbol{e}_\theta, \boldsymbol{e}_\phi$ is $\boldsymbol{e}_x=\boldsymbol{e}_r\sin\theta\cos\phi+\boldsymbol{e}_\theta\cos\theta\cos\phi-\boldsymbol{e}_\phi\sin\phi$.

1.10 Given a scalar field

$$\varphi=\left(\sin\frac{\pi}{2}x\right)\left(\sin\frac{\pi}{3}y\right)\boldsymbol{e}^{-z}$$

calculate

(1) the magnitude and the direction of the maximum rate of increase of φ at point $P(1,2,3)$.

(2) the rate of increase of φ at P in the direction toward the origin$(0,0,0)$.

1.11 Given a complex scalar field $f(\boldsymbol{r})=\boldsymbol{e}^{-jk\boldsymbol{e}_k\cdot\boldsymbol{r}}$ and a vector field $\boldsymbol{E}(\boldsymbol{r})=\boldsymbol{E}_0 f(\boldsymbol{r})$, where k is a constant scalar, $\boldsymbol{e}_k$ is a constant unit vector, and $\boldsymbol{E}_0$ is a constant vector, prove that

(1) $\nabla f=-jkf\boldsymbol{e}_k$.

(2) $\nabla\cdot\boldsymbol{E}=-jk\boldsymbol{e}_k\cdot\boldsymbol{E}$.

(3) $\nabla\times\boldsymbol{E}=-jk\boldsymbol{e}_k\times\boldsymbol{E}$.

1.12 Find the divergence of the vector field $\boldsymbol{F}(r)=\boldsymbol{e}_r r^n$.

1.13 For two differentiable vector fields $\boldsymbol{E}$ and $\boldsymbol{H}$, prove that $\nabla\cdot(\boldsymbol{E}\times\boldsymbol{H})=\boldsymbol{H}\cdot(\nabla\times\boldsymbol{E})-\boldsymbol{E}\cdot(\nabla\times\boldsymbol{H})$.

1.14 Given a vector field $\boldsymbol{F}=\boldsymbol{e}_x(x+c_1z)+\boldsymbol{e}_y(c_2x-3z)+\boldsymbol{e}_z(x+c_3y+c_4z)$, determine

(1) the constants c_1, c_2 and c_3 if $\boldsymbol{F}$ is irrotational,

(2) the constant c_4, if in addition to(1), $\boldsymbol{F}$ is also solenoidal,

(3) a scalar potential function φ whose negative gradient equals $\boldsymbol{F}$.

1.15 Evaluate $\oint \boldsymbol{r}\cdot d\boldsymbol{s}$ over the closed surface of a cube with $0\leqslant x\leqslant 2$, $0\leqslant y\leqslant 2$, and $0\leqslant z\leqslant 2$.

1.16 Given two arbitrary scalar fields ψ and φ, and two arbitrary vector fields $\boldsymbol{A}$ and $\boldsymbol{B}$, prove that

$$\nabla(\psi\varphi)=\psi\nabla\varphi+\varphi\nabla\psi \tag{1.147}$$

$$\nabla\cdot(\psi\boldsymbol{A})=\psi\nabla\cdot\boldsymbol{A}+\boldsymbol{A}\cdot\nabla\psi \tag{1.148}$$

$$\nabla\times(\psi\boldsymbol{A})=\psi\nabla\times\boldsymbol{A}+\nabla\psi\times\boldsymbol{A} \tag{1.149}$$

$$\nabla\cdot(\boldsymbol{A}\times\boldsymbol{B})=\boldsymbol{B}\cdot\nabla\times\boldsymbol{A}-\boldsymbol{A}\cdot\nabla\times\boldsymbol{B} \tag{1.150}$$

1.17 For a scalar field f, use the Stokes's Theorem to prove

$$\oint_C f\mathrm{d}\boldsymbol{l} = -\int_S \nabla f \times \mathrm{d}\boldsymbol{s} \tag{1.151}$$

where S is a surface bounded by contour C. (Hint: let $\boldsymbol{A} = \boldsymbol{C}f$, where $\boldsymbol{C}$ is an arbitrary constant vector.)

1.18 For a vector field $\boldsymbol{F}$ with continuous first derivatives, prove that

$$\int_V (\nabla \times \boldsymbol{F})\mathrm{d}v = -\oint_S \boldsymbol{F} \times \mathrm{d}\boldsymbol{s} \tag{1.152}$$

where S is the surface enclosing the volume V. (Hint: Apply the divergence theorem to $(\boldsymbol{F} \times \boldsymbol{C})$, where $\boldsymbol{C}$ is a constant vector.)

1.19 Derive the expressions of $\nabla \cdot \nabla\varphi$ in Cartesian, cylindrical and spherical coordinates.

Chapter 2 Static Electric Fields(静电场)

2.1 Introduction(引言)

Electric fields due to stationary(static) electric charges in free space is the simplest and special case of electromagnetics. In this case, electric fields do not change with time, and are often called static electric fields or electrostatic fields. In electrostatics, there are no currents, and therefore, no magnetic fields.

In this chapter, we start with the definition of electric field and derivation of the source-field relation based on Coulomb's law. We study the divergence of electrostatic fields and formulate Gauss's law of electrostatics, which shows that static charges are the flow sources of the electrostatic fields. We study the curl of electrostatic fields, which shows that electrostatic fields are curl-free fields. Then we introduce the concept of electric potential. One general problem of electrostatics is that of finding the distribution of fields produced by given sources in a specified region with given media(媒质) and boundaries. The media can be conductors or dielectrics. Polarization in dielectrics is analyzed, based on which the displacement vector is introduced. Boundary conditions involving dielectrics and conductors are derived. Then we show how to calculate capacitances. Finally, electric energy is defined and expressed in terms of electric fields.①

2.2 Electric Fields and Charges(电场与电荷)

The vector quantity of static electric fields is called **electric field intensity**(电场强度), or simply electric field, which is defined as the force per unit charge experienced by a very small stationary test charge when it is placed in a region where an electric field exists, i. e.

---

① 静电场是静止的电荷产生的电场。在静电场中,场和源都不随时间变化,电荷的位置不随时间变化,因此也没有电流,不涉及磁场。静电场是典型的无旋场,电荷是其通量源。由场的保守特性可以引入标量电位。本章从真空中静电场的场源关系出发,讨论导体和电介质在电场中的特性并引入电通密度的概念,介绍由电荷分布求解电场分布的典型方法,并应用于求解电容问题,最后讨论了电场储能的定义和应用。

$$\boldsymbol{E} = \lim_{\Delta q \to 0} \frac{\boldsymbol{F}}{\Delta q} \tag{2.1}$$

If the force $\boldsymbol{F}$ is measured in newton(N) and charge Δq in coulombs(C), then $\boldsymbol{E}$ is in newton per coulomb(N/C), which is the same as volts per meter(V/m). The limit in the definition expression of(2.1) means that the test charge Δq should be small enough so that the source of $\boldsymbol{E}$ as well as the field $\boldsymbol{E}$ itself is not disturbed by the existence of Δq. ①

2.2.1 Electric Fields Due to Discrete Charges (离散电荷产生的电场)

The simplest form of static electric field is the one produced by a single **point charge**(**点电荷**). According to the experimental **Coulomb's law**(**库仑定律**) formulated in 1785, the force between two point charges, q_1 and q_2 as shown in Figure 2-1, can be written mathematically as

$$\boldsymbol{F}_{12} = \boldsymbol{e}_{12} \frac{q_1 q_2}{4\pi\varepsilon_0 R_{12}^2} \tag{2.2}$$

where $\boldsymbol{F}_{12}$ is the vector force exerted by q_1 on q_2, $\boldsymbol{e}_{12}$ is a unit vector in the direction from q_1 to q_2, and R_{12} is the distance between the two charges. ε_0 is the **permittivity of the vacuum**(**真空介电常数**), or **free space**(**自由空间**) which is found from experiment to be

$$\varepsilon_0 = 8.854 \times 10^{-12} \approx \frac{1}{36\pi} \times 10^{-9} \left(\frac{\mathrm{F}}{\mathrm{m}}\right) \tag{2.3}$$

In(2.2), if we take q_2 to be a test charge Δq located at a position $\boldsymbol{r}$, and q_1 to be a source charge q at a position $\boldsymbol{r}'$ as shown in Figure 2-2, then the electric field produced by the point charge q can be found by substitute(2.2) into(2.1), which leads to

$$\boldsymbol{E}(\boldsymbol{r}) = \boldsymbol{e}_{\mathrm{R}} \frac{q}{4\pi\varepsilon_0 R^2} = \frac{q(\boldsymbol{r} - \boldsymbol{r}')}{4\pi\varepsilon_0 |\boldsymbol{r} - \boldsymbol{r}'|^3} \tag{2.4}$$

where $\boldsymbol{e}_{\mathrm{R}}$ is the unit vector of the distance vector $\boldsymbol{R}=\boldsymbol{r}-\boldsymbol{r}'$ as is introduced in Chapter 1. That is,

$$\boldsymbol{e}_{\mathrm{R}} = \frac{\boldsymbol{r} - \boldsymbol{r}'}{|\boldsymbol{r} - \boldsymbol{r}'|} \tag{2.5}$$

and

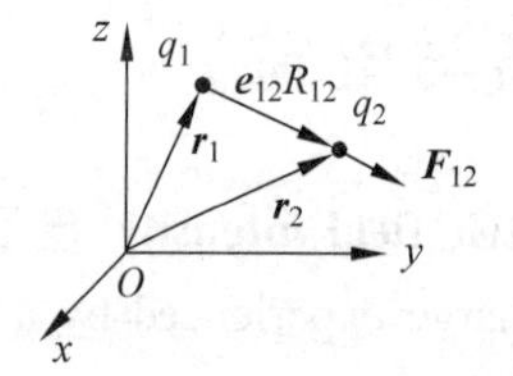

Figure 2-1 Illustration of Coulomb's law

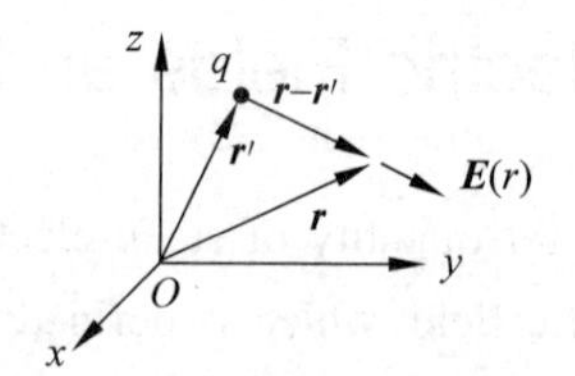

Figure 2-2 Electric field observed at $\boldsymbol{r}$ produced by point source q at $\boldsymbol{r}'$

① 电场强度的概念可以描述为"单位电荷在电场中受到的电场力"。这个概念是从实验角度给出的，即将一个测试电荷放置于电场中，测量该测试电荷受到的电场力，然后计算电场力与该电荷大小的比值。需要注意这个实验定义的前提是该测试电荷的电荷量和体积都足够小，以至于引入该电荷对原有的电场分布不产生任何影响。

$$R = |\boldsymbol{r} - \boldsymbol{r}'| \tag{2.6}$$

is the distance between the locations of $\boldsymbol{r}$ and $\boldsymbol{r}'$. (2.4) shows that the electric field due to a point charge obeys the **inverse square law**(**平方反比定律**). ①

Since electric field intensity is a linear function (线性函数) of q, the principle of superposition(叠加原理) can be used for calculating electric fields produced by multiple charges. For N discrete(离散的) point charges $q_1, q_2, \cdots, q_N$ located at different positions, the total $\boldsymbol{E}$ field is the vector sum of the fields caused by all the individual charges. From(2.4), the total $\boldsymbol{E}$ field at an **observation point**(**观察点**) $\boldsymbol{r}$ is

$$\boldsymbol{E}(\boldsymbol{r}) = \frac{1}{4\pi\varepsilon_0}\sum_{k=1}^{n}\frac{q_k(\boldsymbol{r} - \boldsymbol{r}'_k)}{|\boldsymbol{r} - \boldsymbol{r}'_k|^3} \tag{2.7}$$

where $\boldsymbol{r}'_k$ represents the location of the charge q_k. ②

Example 2-1 Find the $\boldsymbol{E}$ field of an **electric dipole**(**电偶极子**), which consists of a pair of equal and opposite charges $+q$ and $-q$ with a small separation d.

Solution: Let $\boldsymbol{d}$ be the vector pointing from $-q$ to $+q$. Let the center of the dipole coincide with the origin and the two charges be located on the z-axis as shown in Figure 2-3. By using (2.7), the $\boldsymbol{E}$ field of the dipole at an observation point $\boldsymbol{r}$ can be written as

$$\boldsymbol{E}(\boldsymbol{r}) = \frac{q}{4\pi\varepsilon_0}\left\{\frac{\boldsymbol{r} - \frac{\boldsymbol{d}}{2}}{\left|\boldsymbol{r} - \frac{\boldsymbol{d}}{2}\right|^3} - \frac{\boldsymbol{r} + \frac{\boldsymbol{d}}{2}}{\left|\boldsymbol{r} + \frac{\boldsymbol{d}}{2}\right|^3}\right\} \tag{2.8}$$

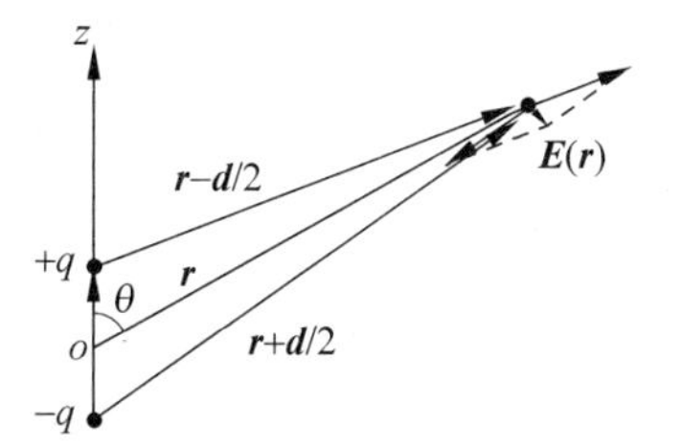

Figure 2-3 Electric field produced by an electric dipole

If $d \ll r$, i. e. , if the observation point is in the **far zone**(**远区**), the first term on the right side of(2.8) can be simplified by using

$$\begin{aligned}\left|\boldsymbol{r} - \frac{\boldsymbol{d}}{2}\right|^{-3} &= \left[\left(\boldsymbol{r} - \frac{\boldsymbol{d}}{2}\right)\cdot\left(\boldsymbol{r} - \frac{\boldsymbol{d}}{2}\right)\right]^{\frac{-3}{2}} = \left[r^2 - \boldsymbol{r}\cdot\boldsymbol{d} + \frac{d^2}{4}\right]^{\frac{-3}{2}} \\ &\approx r^{-3}\left[1 - \frac{\boldsymbol{r}\cdot\boldsymbol{d}}{r^2}\right]^{\frac{-3}{2}} \approx r^{-3}\left[1 + \frac{3}{2}\frac{\boldsymbol{r}\cdot\boldsymbol{d}}{r^2}\right]\end{aligned} \tag{2.9}$$

where the Taylor series expansion has been used and all terms containing the second and

① 平方反比定律即点源产生的全向场如果作用至无穷远,其场强将与距离的平方成反比关系。

② 从点电荷电场的表达式(2.4) 推导出式(2.7)利用了电场强度与电荷的线性叠加关系,本质上也是库仑定律的线性叠加关系,即存在多个源电荷时,测试点电荷受到的电场力等于各个源电荷单独存在时对测试点电荷施加的电场力的矢量和。

higher powers of(d/r) have been neglected. Similarly, for the second term on the right side of (2.8), we have

$$\left|\boldsymbol{r}+\frac{\boldsymbol{d}}{2}\right|^{-3} \approx r^{-3}\left[1-\frac{3}{2}\frac{\boldsymbol{r}\cdot\boldsymbol{d}}{r^{2}}\right] \tag{2.10}$$

Substitution of(2.9) and(2.10) into(2.8) leads to

$$\boldsymbol{E}(\boldsymbol{r}) \approx \frac{q}{4\pi\varepsilon_0 r^3}\left[3\frac{\boldsymbol{r}\cdot\boldsymbol{d}}{r^2}\boldsymbol{r}-\boldsymbol{d}\right] \tag{2.11}$$

The electric dipole is an important entity in the study of the electric field in dielectric media. We define the **electric dipole moment**(**电偶极矩**) $\boldsymbol{p}$ as

$$\boldsymbol{p}=q\boldsymbol{d} \tag{2.12}$$

Then(2.11) can be rewritten as

$$\boldsymbol{E}(\boldsymbol{r})=\frac{1}{4\pi\varepsilon_0 r^3}\left[3\frac{\boldsymbol{r}\cdot\boldsymbol{p}}{r^2}\boldsymbol{r}-\boldsymbol{p}\right] \tag{2.13}$$

For the dipole lying along $+z$-direction, $\boldsymbol{d}=\boldsymbol{e}_z d$ and we have

$$\boldsymbol{p}=\boldsymbol{e}_z qd=\boldsymbol{e}_z p=p(\boldsymbol{e}_r\cos\theta-\boldsymbol{e}_\theta\sin\theta) \tag{2.14}$$

where $p=qd$ is the magnitude of the electric dipole moment. Hence

$$\boldsymbol{r}\cdot\boldsymbol{p}=rp\cos\theta \tag{2.15}$$

The $\boldsymbol{E}$ field expression of(2.13) becomes

$$\boldsymbol{E}(\boldsymbol{r})=\frac{p}{4\pi\varepsilon_0 r^3}(\boldsymbol{e}_r 2\cos\theta+\boldsymbol{e}_\theta\sin\theta) \tag{2.16}$$

(2.16) shows that the $\boldsymbol{E}$ field due to an electric dipole is inversely proportional to r^3 in the far zone. This is reasonable because as r increases, the fields due to the closely spaced $+q$ and $-q$ tend to be equal in magnitude and opposite in direction. Thus, their summation decreases more rapidly than the $\boldsymbol{E}$ field of a single point charge. ①

Generally, for any orientation of the dipole moment $\boldsymbol{p}$ centered at an arbitrary position $\boldsymbol{r}'$, the $\boldsymbol{E}$ field at an observation point $\boldsymbol{r}$ far away from the dipole (i. e., $|\boldsymbol{r}-\boldsymbol{r}'|\gg d$) can be expressed as the spatial translation of(2.13), i. e.,

$$\boldsymbol{E}(\boldsymbol{r})=\frac{1}{4\pi\varepsilon_0|\boldsymbol{r}-\boldsymbol{r}'|^3}\left[3\frac{(\boldsymbol{r}-\boldsymbol{r}')\cdot\boldsymbol{p}}{|\boldsymbol{r}-\boldsymbol{r}'|^2}(\boldsymbol{r}-\boldsymbol{r}')-\boldsymbol{p}\right]=\frac{1}{4\pi\varepsilon_0 R^3}[3(\boldsymbol{e}_R\cdot\boldsymbol{p})\,\boldsymbol{e}_R-\boldsymbol{p}] \tag{2.17}$$

where R and $\boldsymbol{e}_R$ are the magnitude and unit vector of the distance vector $\boldsymbol{R}=\boldsymbol{r}-\boldsymbol{r}'$ as defined by (2.5) and(2.6).

2.2.2 Electric Fields Due to Continuous Charge Distributions (连续分布电荷产生的电场)

In practice, a macroscopic(宏观的) charge cannot be located in a single point. Therefore,

---

① 式(2.16)表明电偶极子产生的远区电场与距离的3次方成反比。回到式(2.7),可以证明,若静电荷且分布于有限区域,且总电荷量不为零,其产生的远区电场应当与距离平方成反比,即满足平方反比定律。典型的例子就是单个点电荷产生的电场。然而,若多个点电荷的总电荷量为零,则该远区电场的大小随距离衰减的速度可能比 $1/r^2$ 更快。电偶极子就是这样的一个典型例子。

the ideal point charge does not exist. Macroscopic charges exist in the form of a distribution in the space. Referring to Figure 2-4(a), we define a **volume charge density**(**体电荷密度**) ρ_v as

$$\rho_v = \lim_{\Delta v \to 0} \frac{\Delta q}{\Delta v} (\text{C/m}^3) \tag{2.18}$$

where Δq is the amount of charge in a very small volume Δv. In certain cases, as shown in Figure 2-4(b) and Figure 2-4(c), the charge Δq may be distributed within a very thin sheet or line, and its variation along thickness direction of the sheet or over the cross-section of the line is negligible or unimportant. Then it is more convenient to define a **surface charge density** (**面电荷密度**) ρ_s and a **line charge density** (**线电荷密度**) ρ_l respectively as

$$\rho_s = \lim_{\Delta s \to 0} \frac{\Delta q}{\Delta s} (\text{C/m}^2) \tag{2.19}$$

$$\rho_l = \lim_{\Delta l \to 0} \frac{\Delta q}{\Delta l} (\text{C/m}) \tag{2.20}$$

(a) volume charge (b) surface charge (c) line charge

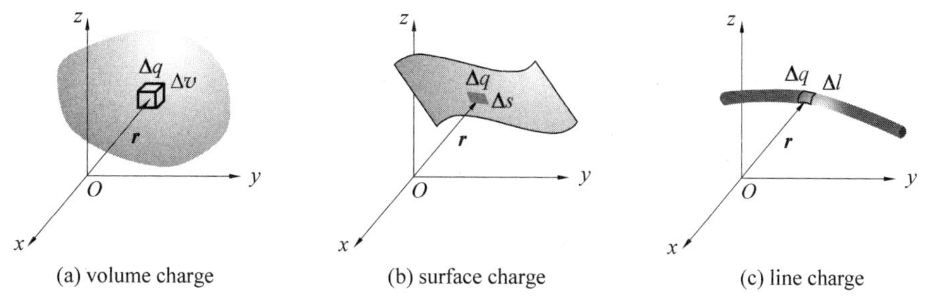

Figure 2-4 Definition of continuous charge densities

To find the electric field caused by a continuous distribution of charge density ρ_v, we first divide the volume charge into infinite number of differential elements, and treat each differential element as a point charge. As shown in Figure 2-5, from (2.4), the contribution of the point charge $\rho_v \, dv'$ (located at $\boldsymbol{r}'$) to the electric field (at point $\boldsymbol{r}$) is

$$d\boldsymbol{E}(\boldsymbol{r}) = \boldsymbol{e}_R \frac{\rho_v(\boldsymbol{r}')dv'}{4\pi\varepsilon_0 R^2} = \frac{\rho_v(\boldsymbol{r}')dv'}{4\pi\varepsilon_0} \frac{\boldsymbol{r}-\boldsymbol{r}'}{|\boldsymbol{r}-\boldsymbol{r}'|^3} \tag{2.21}$$

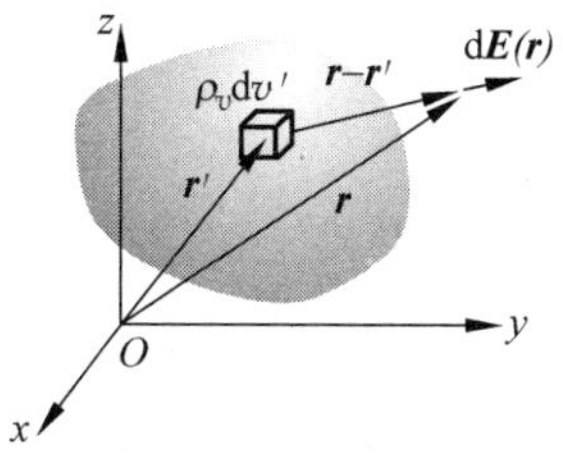

Figure 2-5 Electric field by a continuous distribution of charge

By integrating both sides of the above equation, we have the total electric field produced by the volume charge density ρ_v to be

$$\boldsymbol{E}(\boldsymbol{r}) = \frac{1}{4\pi\varepsilon_0}\int_{V'} \boldsymbol{e}_R \frac{\rho_v(\boldsymbol{r}')}{R^2} dv' = \frac{1}{4\pi\varepsilon_0}\int_{V'} \rho_v(\boldsymbol{r}') \frac{\boldsymbol{r}-\boldsymbol{r}'}{|\boldsymbol{r}-\boldsymbol{r}'|^3} dv' \tag{2.22}$$

(2.22) is a more general formulation of the relation between electric charges and electric fields. The expression of electric fields due to point charges (2.7) can also be written in the form of (2.22) with ρ_v of the point charges Q_k at position $\boldsymbol{r}'_k$ expressed as $Q_k\delta(\boldsymbol{r}-\boldsymbol{r}'_k)$.

We have seen quite a few times using the vector $\boldsymbol{r}$ as the observation point of a field, and using the vector $\boldsymbol{r}'$ as the source location. In electromagnetic theory, it is a convention to use $\boldsymbol{r}$ to represent a **field point** (**场点**) at which the field is observed, and use $\boldsymbol{r}'$ to represent a **source point** (**源点**) at which the source that produces the field is located. Then, the vector $\boldsymbol{R}=\boldsymbol{r}-\boldsymbol{r}'$ represents the distance vector pointing from the field point $\boldsymbol{r}$ to the source point $\boldsymbol{r}'$. ①

If the source is a surface charge density ρ_s distributed on a surface S' or a line charge ρ_l along a line (not necessarily straight) L', then the above equation can be respectively written as

$$\boldsymbol{E}(\boldsymbol{r}) = \frac{1}{4\pi\varepsilon_0}\int_{S'} \boldsymbol{e}_R \frac{\rho_s(\boldsymbol{r}')}{R^2}\mathrm{d}s' = \frac{1}{4\pi\varepsilon_0}\int_{S'} \rho_s(\boldsymbol{r}') \frac{\boldsymbol{r}-\boldsymbol{r}'}{|\boldsymbol{r}-\boldsymbol{r}'|^3}\mathrm{d}s' \tag{2.23}$$

$$\boldsymbol{E}(\boldsymbol{r}) = \frac{1}{4\pi\varepsilon_0}\int_{L'} \boldsymbol{e}_R \frac{\rho_l(\boldsymbol{r}')}{R^2}\mathrm{d}l' = \frac{1}{4\pi\varepsilon_0}\int_{L'} \rho_l(\boldsymbol{r}') \frac{\boldsymbol{r}-\boldsymbol{r}'}{|\boldsymbol{r}-\boldsymbol{r}'|^3}\mathrm{d}l' \tag{2.24}$$

Example 2-2 Find the electric field on the axis of a circular ring disk with outer radius a, inner radius b, and uniform surface charge density ρ_s.

Solution: We use (2.23) to evaluate the $\boldsymbol{E}$ field. Referring to Figure 2-6, we have

$$\mathrm{d}s' = \rho'\mathrm{d}\rho'\mathrm{d}\phi', \quad \boldsymbol{r}-\boldsymbol{r}' = \boldsymbol{e}_z z - \boldsymbol{e}'_\rho \rho', \quad |\boldsymbol{r}-\boldsymbol{r}'| = R = \sqrt{z^2+\rho'^2}$$

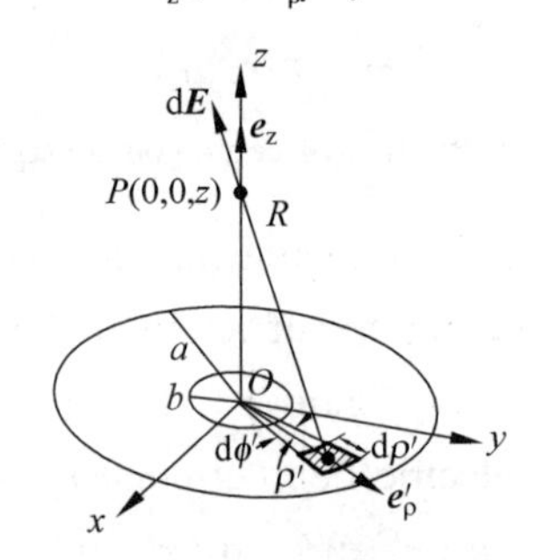

Figure 2-6 Solving the electric field of a surface charge (Example 2-2)

The symbol $\boldsymbol{e}'_\rho$ is used to represent the ρ-direction base vector at the source point. Then the electric field at the point $P(0,0,z)$ is

$$\begin{aligned}\boldsymbol{E} &= \frac{\rho_s}{4\pi\varepsilon_0}\int_0^{2\pi}\int_b^a \frac{\boldsymbol{e}_z z - \boldsymbol{e}'_\rho \rho'}{(z^2+\rho'^2)^{3/2}}\rho'\mathrm{d}\rho'\mathrm{d}\phi' \\ &= \boldsymbol{e}_z\frac{\rho_s}{4\pi\varepsilon_0}\int_b^a \frac{z\rho'}{(z^2+\rho'^2)^{3/2}}\mathrm{d}\rho'\int_0^{2\pi}\mathrm{d}\phi' - \frac{\rho_s}{4\pi\varepsilon_0}\int_b^a \frac{\rho'^2}{(z^2+\rho'^2)^{3/2}}\mathrm{d}\rho'\int_0^{2\pi}\boldsymbol{e}'_\rho\mathrm{d}\phi'\end{aligned} \tag{2.25}$$

① 式(2.22)是一个三维卷积表达式,其中体电荷密度ρ_v是源点$\boldsymbol{r}'$的函数,被积函数既是源点$\boldsymbol{r}'$的函数,又是场点$\boldsymbol{r}$的函数。卷积积分变量是$\boldsymbol{r}'$的坐标(x',y',z'),因此卷积结果得到的电场就只是$\boldsymbol{r}$的函数,而不是$\boldsymbol{r}'$的函数。在此类积分表达式中,通常将源(包括电荷源和后续章节中涉及的电流源)所在位置以$\boldsymbol{r}'$或者带撇号的坐标(例如x',y',z'或者ρ',ϕ',θ'等)表示,而电场的观察位置则以$\boldsymbol{r}$或者不带撇号的坐标表示。需要强调的是,标量R和单位矢量$\boldsymbol{e}_R$都既是场点$\boldsymbol{r}$的函数,也是源点$\boldsymbol{r}'$的函数。场点$\boldsymbol{r}$也可能位于电荷源分布区域,即$\boldsymbol{r}$和$\boldsymbol{r}'$有可能重合。

Notice that $\boldsymbol{e}_z$ is a constant vector, and hence, is moved out of the surface integral. However, $\boldsymbol{e}'_\rho$ is a function of the coordinate ϕ', and by using (1.31), we have

$$\int_0^{2\pi} \boldsymbol{e}'_\rho \mathrm{d}\phi' = \int_0^{2\pi} (\boldsymbol{e}_x \cos\phi' + \boldsymbol{e}_y \sin\phi') \mathrm{d}\phi' = 0$$

Therefore, (2.25) becomes

$$\boldsymbol{E} = \boldsymbol{e}_z \frac{\rho_s z}{2\varepsilon_0} \int_b^a \frac{\rho'}{(z^2 + \rho'^2)^{3/2}} \mathrm{d}\rho' = \boldsymbol{e}_z \frac{\rho_s z}{2\varepsilon_0} [(z^2 + b^2)^{-1/2} - (z^2 + a^2)^{-1/2}] \tag{2.26}$$

For very large z in the positive z axis, we can have the approximation

$$z(z^2 + b^2)^{-1/2} = \left(1 + \frac{b^2}{z^2}\right)^{-1/2} \approx 1 - \frac{b^2}{2z^2}, \quad \text{and similarly,} \quad z(z^2 + a^2)^{-1/2} \approx 1 - \frac{a^2}{2z^2} \tag{2.27}$$

Substitute (2.27) into (2.26), we have

$$\boldsymbol{E} \approx \boldsymbol{e}_z \frac{\rho_s}{2\varepsilon_0}\left[\frac{a^2}{2z^2} - \frac{b^2}{2z^2}\right] = \boldsymbol{e}_z \frac{\pi(a^2 - b^2)\rho_s}{4\pi\varepsilon_0 z^2} = \boldsymbol{e}_z \frac{Q}{4\pi\varepsilon_0 z^2}, \quad \text{as } z \gg b \tag{2.28}$$

where $Q = \pi(a^2 - b^2)\rho_s$ is the total charge on the disk. We see that, when the observation point is very far away from the charged disk, the $\boldsymbol{E}$ field approximates that of a point charge with the same amount of charge Q. ①

Example 2-3 Find the $\boldsymbol{E}$ field of an infinitely long, straight, line charge of a uniform density ρ_l.

Solution: Assume that the line charge lies along the z axis as shown in Figure 2-7. Since the problem has a cylindrical symmetry (that is, the $\boldsymbol{E}$ field is independent of the azimuth angle ϕ), it would be more convenient to use the cylindrical coordinate. As the charge is infinitely long, the $\boldsymbol{E}$ field is independent of the z coordinate. Therefore, we only need to solve for the field in the $z=0$ plane. Consider the field point P located at $\boldsymbol{r} = \boldsymbol{e}_\rho \rho$ and a source point located at $\boldsymbol{r}' = \boldsymbol{e}_z z'$. Then we have

$$\boldsymbol{r} - \boldsymbol{r}' = \boldsymbol{e}_\rho \rho - \boldsymbol{e}_z z' \tag{2.29}$$

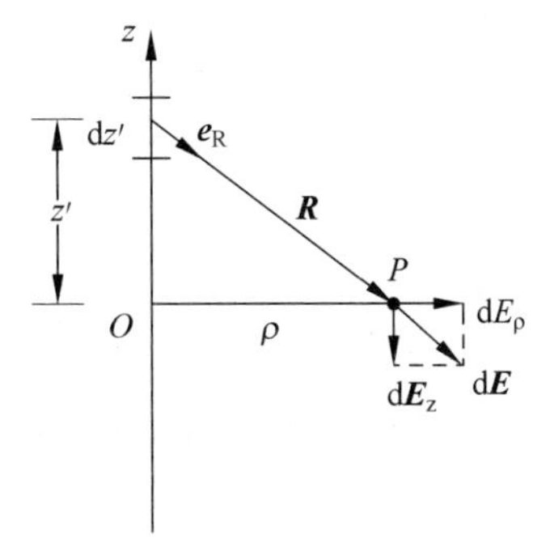

Figure 2-7 Electric field produced by an infinitely long, straight line charge

① 例 2-2 的结果表明,如果电荷源分布在有限区域内,且总电荷量不为零,则其产生的电场在足够远的区域近似于一个同等电荷量的点电荷产生的电场。

Substitute(2.29) into(2.24) and notice that $dl'=dz'$. We have

$$E=\frac{\rho_l}{4\pi\varepsilon_0}\int_{L'}\frac{e_\rho\rho-e_z z'}{(\rho^2+z'^2)^{3/2}}dz'=e_\rho\int_{L'}dE_\rho+e_z\int_{L'}dE_z \tag{2.30}$$

where

$$dE_\rho=\frac{\rho_l}{4\pi\varepsilon_0}\frac{\rho}{(\rho^2+z'^2)^{3/2}}dz' \tag{2.31}$$

is the ρ component of the E field due to the differential line charge element $\rho_l dl'=\rho_l dz'$, and

$$dE_z=-\frac{\rho_l}{4\pi\varepsilon_0}\frac{z'}{(\rho^2+z'^2)^{3/2}}dz' \tag{2.32}$$

is the z component. Since dE_z is an odd function of z', the integral of dE_z over the entire z axis is zero. In fact, for every $\rho_l dz'$ at the point $z=+z'$, there is a charge element $\rho_l dz'$ at the point $z=-z'$, both producing a dE_z of the same magnitude but opposite direction. Hence the z components cancel in the integration process. We only need to integrate the ρ component in (2.30), which leads to

$$E=e_\rho\frac{\rho_l}{4\pi\varepsilon_0}\int_{-\infty}^{\infty}\frac{\rho}{(\rho^2+z'^2)^{3/2}}dz'=e_\rho\frac{\rho_l}{2\pi\varepsilon_0\rho} \tag{2.33}$$

Equation(2.33) is an important result for an infinite line charge. Of course, there is no such infinitely long line charge in reality. Nevertheless, (2.33) gives a good approximation of the E field at a point close to a long straight line charge. ①

2.3 Divergence of Electrostatic Fields and Gauss's Law (静电场的散度与高斯定律)

According to the Helmholtz's theorem, a vector field is determined by the divergence and curl of the field itself. Therefore, it is important to find out the divergence and curl of static electric fields. Here, we first derive the divergence of electrostatic fields directly by applying the differential operators on both sides of the equation(2.22).

As the divergence is taken with respect to the field point r, and the integral in(2.22) is with respect to the source point r', we can interchange the sequence of integration and divergence, and obtain

$$\nabla\cdot E(r)=\nabla\cdot\left(\frac{1}{4\pi\varepsilon_0}\int_{V'}e_R\frac{\rho_v}{R^2}dv'\right)=\frac{1}{4\pi\varepsilon_0}\int_{V'}\rho_v\,\nabla\cdot\left(\frac{e_R}{R^2}\right)dv' \tag{2.34}$$

By noticing(1.102) and utilizing the property of Dirac function given by(1.105), the last integral in(2.34) can be evaluated as

$$\int_{V'}\rho_v\,\nabla\cdot\left(\frac{e_R}{R^2}\right)dv'=\int_{V'}\rho_v[4\pi\delta(r-r')]dv'=4\pi\rho_v(r) \tag{2.35}$$

---

① 根据式(2.33),无限长均匀线电荷产生的电场与距离 ρ 成反比,这是因为无限长的线电荷不是分布在有限区域。当电荷分布至无穷远处时,电场强度大小与距离平方成反比的关系就有可能不再成立。

Combining(2.34) and(2.35), we obtain the very important property of electrostatic fields,

$$\nabla \cdot \boldsymbol{E} = \frac{\rho_{\mathrm{v}}}{\varepsilon_0} \tag{2.36}$$

which is also called the **differential form of Gauss's Law**(**微分形式的高斯定律**) in free space.

By performing a volume integration on both sides of (2.36) and then applying the divergence theorem, we have

$$\oint_S \boldsymbol{E} \cdot \mathrm{d}\boldsymbol{s} = \frac{Q}{\varepsilon_0} \tag{2.37}$$

where Q is the total charge contained in the region bounded by the closed surface S. (2.37) is the **integral form of Gauss's Law**(**积分形式的高斯定律**), which states that the total outward flux of the electric field intensity over any closed surface in free space is equal to the total charge enclosed in the surface divided by ε_0.①

Gauss's law is particularly useful in determining the electrostatic field of charge distributions with certain symmetry conditions. Under these symmetry conditions, the normal component(法向分量) of the electric field intensity is constant over a particular surface. As a result, the surface integral on the left side of(2.37) would be very easy to evaluate, and using Gauss's law would be a much more efficient way for finding the electric field intensity than using(2.22), (2.23) or(2.24) directly. On the other hand, when symmetry conditions do not exist, Gauss's law cannot be used directly. The key of applying Gauss's law is to take advantage of the symmetry condition of the problem and choose a surface over which the normal component of the $\boldsymbol{E}$ field is a constant. Such a surface is referred to as a **Gaussian surface**(**高斯面**).②

Example 2-4 Use Gauss's law to determine the electric field intensity of a point charge q at the origin.

Solution: Due the spherical symmetry of this problem, we use a spherical surface S of a radius r centered at q as the Gaussian surface as shown in Figure 2-8. The $\boldsymbol{E}$ field on the Gaussian surface is everywhere radial and has a constant magnitude. That is, $\boldsymbol{E} = \boldsymbol{e}_{\mathrm{r}} E_{\mathrm{r}}(\boldsymbol{r})$. The total outward flux of the $\boldsymbol{E}$ field over the Gaussian surface is

$$\oint_S \boldsymbol{E} \cdot \mathrm{d}\boldsymbol{s} = \oint_S \boldsymbol{e}_{\mathrm{r}} E_{\mathrm{r}} \cdot (\boldsymbol{e}_{\mathrm{r}} \mathrm{d}s) = E_{\mathrm{r}} \oint_S \mathrm{d}s = 4\pi r^2 E_{\mathrm{r}}$$

The total charge Q enclosed in the spherical Gaussian surface is q. Then according to Gauss's law(2.37), we have

---

① 高斯定律表明电荷是电场的通量源,电荷产生的电场强度矢量线在包围该电荷的闭合曲面上形成的通量等于该曲面包围的总电荷数除以 ε_0.

② 在电场分布具有一定对称性的情况下,利用高斯定律计算电场强度较为方便,其关键是选择符合条件的高斯面,可以判断在该高斯面上电场强度的法向分量均匀,从而将式(2.37)中的电场强度通量积分转换为电场强度法向分量大小与高斯面面积的乘积,继而直接求解得到电场强度的大小。

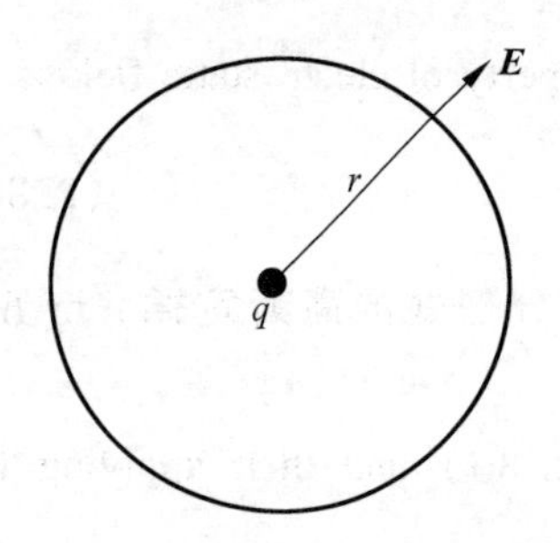

Figure 2-8 Application of Gauss's law to a point charge

$$4\pi r^2 E_r = \frac{q}{\varepsilon_0}$$

Therefore,

$$\boldsymbol{E} = \boldsymbol{e}_r E_r = \boldsymbol{e}_r \frac{q}{4\pi\varepsilon_0 r^2} \tag{2.38}$$

Example 2-5 Use Gauss's law to determine the electric field intensity of an infinitely long, straight line charge of a uniform density ρ_l in air.

Solution: This is the same problem as Example 2-3, which can be solved more easily here by using Gauss's law. Since the line charge is infinitely long, there is no variation along the line (z-direction as shown in Figure 2-9). Also due to the cylindrical symmetry of the problem, we can infer that the magnitude of the field has no variation along the ϕ-direction. The $\boldsymbol{E}$ field lines originate from the line charge, and must be radial and perpendicular to the line charge. Therefore, the $\boldsymbol{E}$ field can only have component along the ρ-direction. That is, $\boldsymbol{E} = \boldsymbol{e}_\rho E_\rho(\rho)$. We construct a cylindrical Gaussian surface of a radius ρ and an arbitrary length L with the line charge as its axis. On the side wall of the cylinder, E_ρ is constant, and $d\boldsymbol{s} = \boldsymbol{e}_\rho \rho d\phi dz$, and the total outward flux of the $\boldsymbol{E}$ field is

$$\oint_S \boldsymbol{E} \cdot d\boldsymbol{s} = \int_0^L \int_0^{2\pi} E_\rho \rho d\phi dz = 2\pi\rho L E_\rho$$

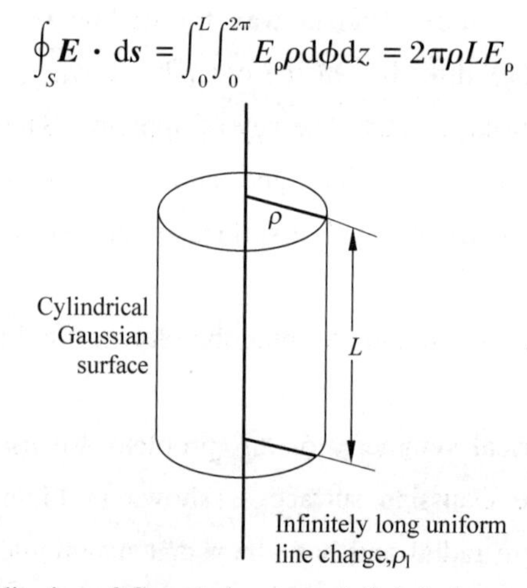

Figure 2-9 Application of Gauss's law to an infinitely long, straight line charge

There is no contribution from the top or the bottom face of the cylinder because on both faces the $\boldsymbol{E}$ field is perpendicular to the differential area vector $d\boldsymbol{s}$. The total charge enclosed within the cylinder is $Q=\rho_l L$. Then according to Gauss's law, we have

$$2\pi\rho L E_\rho = \frac{\rho_l L}{\varepsilon_0}$$

which means that

$$\boldsymbol{E} = \boldsymbol{e}_\rho E_\rho = \boldsymbol{e}_\rho \frac{\rho_l}{2\pi\varepsilon_0 \rho} \tag{2.39}$$

Example 2-6 Determine the electric field intensity of an infinite planar charge with a uniform surface charge density ρ_s.

Solution: Assume the charge surface lies on the xy-plane. As shown in Figure 2-10, we choose the Gaussian surface to be a rectangular box with top and bottom faces of an arbitrary area A equidistant from the planar charge. Due to the symmetry of the problem, the $\boldsymbol{E}$ field at any location cannot have x or y component. The field on the top face of the box must have the same magnitude but opposite direction as that on the bottom face. On the top face,

$$\boldsymbol{E}\cdot \mathrm{d}\boldsymbol{s}=(\boldsymbol{e}_z E_z)\cdot(\boldsymbol{e}_z \mathrm{d}s)=E_z \mathrm{d}s$$

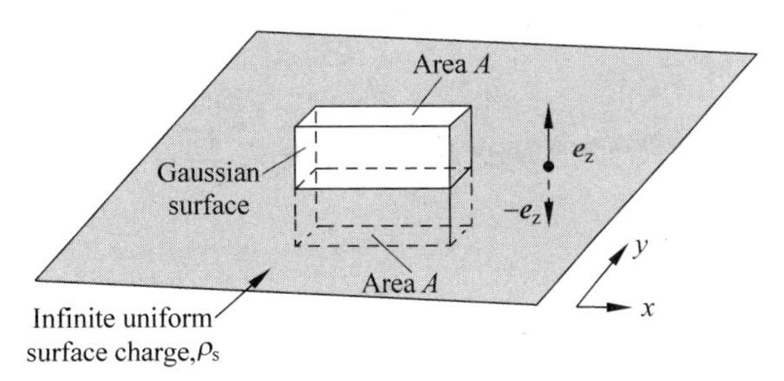

Figure 2-10 Application of Gauss's law to an infinitely large plane surface charge

Then on the bottom face,

$$\boldsymbol{E}\cdot \mathrm{d}\boldsymbol{s}=(-\boldsymbol{e}_z E_z)\cdot(-\boldsymbol{e}_z \mathrm{d}s)=E_z \mathrm{d}s$$

Since there is no contribution from the side faces, we have

$$\oint_S \boldsymbol{E}\cdot \mathrm{d}\boldsymbol{s}=2E_z\int_A \mathrm{d}s=2E_z A$$

The total charge enclosed within the box is $Q=\rho_s A$. Therefore,

$$2E_z A=\frac{\rho_s A}{\varepsilon_0}$$

which leads to

$$E_z=\frac{\rho_s}{2\varepsilon_0}$$

And we have

$$\boldsymbol{E}=\begin{cases}\boldsymbol{e}_z\dfrac{\rho_s}{2\varepsilon_0} & \text{for } z>0\\[2ex] -\boldsymbol{e}_z\dfrac{\rho_s}{2\varepsilon_0} & \text{for } z<0\end{cases} \tag{2.40}$$

which is constant above and below the plane of the surface charge respectively. ①

① 比较式(2.38)、式(2.39)和式(2.40)可以看出,不同性质的电荷源产生的电场随距离的增加表现出不同的变化特性。点电荷产生的场随距离增加呈平方衰减;无限长线电荷产生的场随着距离的增加呈线性衰减;无限大面电荷随着距离的增加保持不变。无限长线电荷与无线大面电荷生成的电场均不满足平方反比定律,这是因为这两种电荷不是分布在有限区域内的,总的电荷数量也不是有限的。虽然无限长的线电荷和无限大的面电荷在现实中并不存在,但在足够长的线电荷和足够大的面电荷附近,仍然能够观察到类似的变化规律。

Example 2-7 Determine the $\boldsymbol{E}$ field caused by a uniform charge distribution in a sphere with a volume density $\rho_v = \rho_{v0}$ for $0 \leqslant r \leqslant a$ and $\rho_v = 0$ for $r > a$.

Solution: Due to the spherical symmetry of the source distribution, we choose concentric spherical surfaces as the Gaussian surfaces. $\boldsymbol{E}$ has only a radial component with constant magnitude over the Gaussian surfaces. Hence the total outward flux of $\boldsymbol{E}$ over a Gaussian surface S with radius r is

$$\oint_S \boldsymbol{E} \cdot \mathrm{d}\boldsymbol{s} = E_r \int_S \mathrm{d}s = 4\pi r^2 E_r$$

(1) For region $0 \leqslant r \leqslant a$, as shown in Figure 2-11(a), the total charge enclosed within the surface S is

$$Q = \int_V \rho_v \mathrm{d}v = \rho_{v0} \int_V \mathrm{d}v = \rho_{v0} \frac{4\pi}{3} r^3$$

By Gauss's law, we have

$$4\pi r^2 E_r = \rho_{v0} \frac{4\pi}{3\varepsilon_0} r^3$$

which leads to

$$\boldsymbol{E} = \boldsymbol{e}_r \frac{\rho_{v0}}{3\varepsilon_0} r$$

(2) For region $r > a$, as shown in Figure 2-11(b), the charge enclosed within a Gaussian surface S is

$$Q = \rho_{v0} \frac{4\pi}{3} a^3$$

By Gauss's law, we have

$$4\pi r^2 E_r = \rho_{v0} \frac{4\pi}{3\varepsilon_0} a^3$$

which leads to

$$\boldsymbol{E} = \boldsymbol{e}_r \frac{\rho_{v0} a^3}{3\varepsilon_0 r^2} \tag{2.41}$$

Since the total charge contained in the entire sphere is

$$Q_{\text{total}} = \rho_{v0} \frac{4\pi}{3} a^3$$

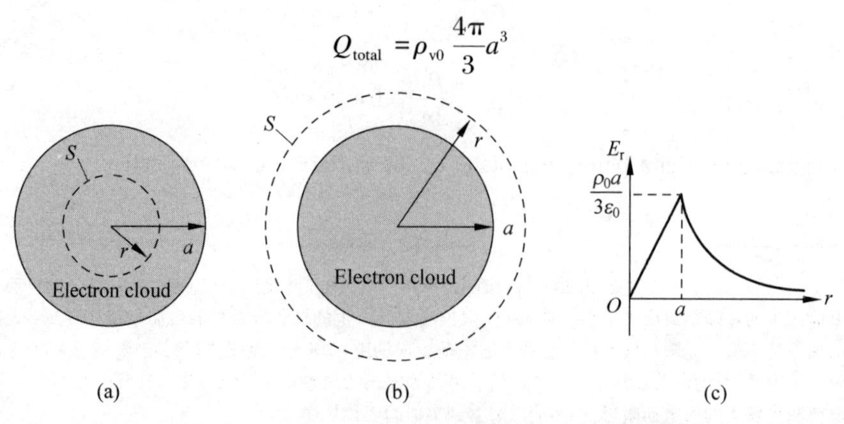

Figure 2-11 Solving electric field due to a spherical volume charge using Gauss's law (Example 2-7)

(2.41) can be written as

$$E = e_r \frac{Q_{\text{total}}}{4\pi\varepsilon_0 r^2}$$

We observe that outside the volume charge, the E field follows the inverse square law and is the same as if the total charge is concentrated on a single point at the center. This is true, in general, for any charge distribution with spherical symmetry. The variation of E_r versus r is plotted in Figure 2-11(c).

2.4 Curl of Electrostatic Fields and Electric Potential (静电场的旋度与电位)

By taking the curl of the electric field expressed in (2.22), we have

$$\nabla \times E(r) = \nabla \times \left(\frac{1}{4\pi\varepsilon_0} \int_{V'} e_R \frac{\rho_v}{R^2} dv' \right) = \frac{1}{4\pi\varepsilon_0} \int_{V'} \rho_v \nabla \times \left(\frac{e_R}{R^2} \right) dv' \tag{2.42}$$

By (1.141), the integrand in the rightmost side of (2.42) vanishes everywhere, and therefore we have

$$\nabla \times E = 0 \tag{2.43}$$

Equation (2.43) asserts that static electric fields are curl-free (or irrotational). An integral form can be obtained by integrating both sides of (2.43) over an open surface, and then converting the surface integral to a closed-path line integral by using Stokes's theorem, which leads to

$$\oint_C E \cdot dl = 0 \tag{2.44}$$

Here, the line integral is performed over a closed path C bounding an arbitrary surface, and therefore, C is also arbitrary.

(2.44) states that the scalar line integral of the static electric field intensity around any closed path vanishes. In circuit theory, the scalar product $E \cdot dl$ integrated over any path is the voltage drop along that path. Thus (2.44) is an expression of **Kirchhoff's voltage law(基尔霍夫电压定律)** which states that the algebraic sum of voltage drops around any closed circuit (outside the sources) is zero. ①

Let the closed path C be a contour made of two paths C_1 and C_2 as is shown in Figure 2-12. The line integral of a static E field from point P_1 to P_2 along C_1 together with the line integral from point P_2 to P_1 along C_2 constitutes a closed path line integral of the static E field, and according to (2.44), must be zero. That is, we have

$$\oint_{C_1C_2} E \cdot dl = \int_{P_1 \atop \text{Along}C_1}^{P_2} E \cdot dl + \int_{P_2 \atop \text{Along}C_2}^{P_1} E \cdot dl = 0$$

And as a result,

---

① 式(2.43)和式(2.44)给出了静电场的一个重要的基本性质：静电场是无旋场，其沿任意闭合路径的线积分为零。式(2.44)对应直流电路中的基尔霍夫电压定律，即沿着闭合回路所有器件两端电势差(电压)的代数和等于零。

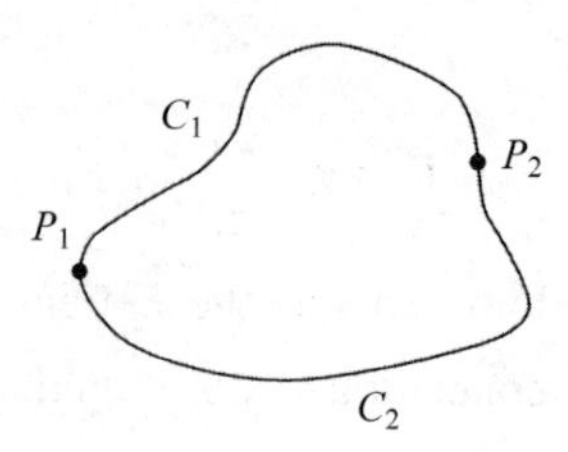

Figure 2-12 Illustration of line integral over a closed path that is divided into two parts: C_1 and C_2

$$\int_{P_1 \atop \text{Along}C_1}^{P_2} \boldsymbol{E} \cdot \mathrm{d}\boldsymbol{l} = -\int_{P_2 \atop \text{Along}C_2}^{P_1} \boldsymbol{E} \cdot \mathrm{d}\boldsymbol{l} = \int_{P_1 \atop \text{Along}C_2}^{P_2} \boldsymbol{E} \cdot \mathrm{d}\boldsymbol{l} \tag{2.45}$$

This mcans that the line integral of the static $\boldsymbol{E}$ field from P_1 to P_2 remains the same no matter which path to take. Since the electric field is defined as the force acting on a unit test charge, the line integral of $\boldsymbol{E}$ can be interpreted as the work done by the electric field in moving a unit charge from point P_1 to point P_2. (2.45) shows that the work done by a static electric field does not depend on the path, but only depends on the positions of the start and end points. In other words, static electric fields are **conservative fields**(**保守场**). In fact, we will soon see that the integral in(2.45) equals to the potential difference between the two positions P_1 and P_2. ①

Since static electric fields are curl-free, and according to(1.138), a curl-free field can be expressed as the gradient of a scalar field. We define the **electric potential**(**电位**), denoted by φ, such that

$$\boldsymbol{E} = -\nabla\varphi \tag{2.46}$$

where the negative sign means that $\boldsymbol{E}$ is along the direction of maximum space rate of **decrease** of φ. This conforms to the convention that, along the direction of the $\boldsymbol{E}$ field, the electric potential decreases. ②

Substituting(2.46) into(2.45), we have

$$\int_{P_1}^{P_2} \boldsymbol{E} \cdot \mathrm{d}\boldsymbol{l} = -\int_{P_1}^{P_2} (\nabla\varphi) \cdot \mathrm{d}\boldsymbol{l} = -\int_{P_1}^{P_2} \mathrm{d}\varphi = \varphi_1 - \varphi_2 \tag{2.47}$$

In physics, (2.47) states that the work done by $\boldsymbol{E}$ in moving a unit charge from point P_1 to point P_2 equals the electric potential energy decrease of the charge. The unit of the potential is J/C or V.

The definition of φ in(2.46) only specifies the space rate of φ. For a given $\boldsymbol{E}$ field, (2.46) can only determine the relative potential difference between two points as is given in(2.47). To determine the absolute value of φ, a reference zero-potential point needs to be specified first. In most cases, especially when the source charges are distributed within a finite region, the zero-potential point is taken at infinity. Otherwise, the reference zero-potential point should be

① 从静电场是无旋场出发,直接推导出静电场在闭合曲线上的线积分恒等于零,从而证明静电场的线积分只与积分线的起点和终点位置有关,与路径无关。进而得到结论:静电场力对电荷做的功也只与电荷移动的起点和终点位置有关,与路径无关。这也就是保守场的意义。

② 式(2.46)表明电场强度矢量的方向即为电位下降最快的方向,电场强度矢量的大小就是电位最快的下降变化率;而沿着电场反方向,则电位上升最快。

explicitly specified. For example, in determining the field between two conductors, we usually specify the potential of one of the conductors as zero. ①

By letting P_2 in (2.47) be the reference point and P_1 be an arbitrary point P, i.e., $\varphi_1=\varphi_P$ and $\varphi_2=0$, we have an expression for the potential field:

$$\varphi_P=\int_P^{\text{Reference Point}} \boldsymbol{E}\cdot d\boldsymbol{l}=-\int_{\text{Reference Point}}^{P} \boldsymbol{E}\cdot d\boldsymbol{l} \tag{2.48}$$

in which the integral path can be selected arbitrarily as long as the start and end points are fixed.

We know from Section 1.7 that the direction of the gradient of the scalar field φ is perpendicular to the surfaces of constant φ. Therefore, the streamlines of an $\boldsymbol{E}$ field are everywhere perpendicular to **equipotential lines**(等位线) or **equipotential surfaces**(等位面).

2.4.1 Electric Potential Due to Discrete Charges (离散电荷产生的电位)

Let's first find the electric potential due to a point charge q at position $\boldsymbol{r}'$. We specify the reference zero-potential point to be infinity, then φ at a field point $\boldsymbol{r}$ at a distance R from the point charge can be obtained readily from (2.48):

$$\varphi(\boldsymbol{r})=-\int_{\infty}^{R}\left(\boldsymbol{e}_R\frac{q}{4\pi\varepsilon_0 R^2}\right)\cdot d\boldsymbol{l}' \tag{2.49}$$

To make evaluation of (2.49) easier, we select the integral path to be a straight radial line from infinity to the field point along the direction $-\boldsymbol{e}_R$. In this case, $d\boldsymbol{l}'=\boldsymbol{e}_R\,dR$ (notice that $d\boldsymbol{l}'$ is in the opposite direction of $\boldsymbol{e}_R$ because dR is negative) and we have

$$\varphi=-\int_{\infty}^{R}\left(\boldsymbol{e}_R\frac{q}{4\pi\varepsilon_0 R^2}\right)\cdot(\boldsymbol{e}_R dR)=\frac{q}{4\pi\varepsilon_0 R}=\frac{q}{4\pi\varepsilon_0|\boldsymbol{r}-\boldsymbol{r}'|} \tag{2.50}$$

Based on (2.50) and the linear relation between the charge and potential, the φ at a field point $\boldsymbol{r}$ due to N discrete point charges $q_1, q_2, \cdots, q_N$ located at source points $\boldsymbol{r}_1', \boldsymbol{r}_2', \cdots, \boldsymbol{r}_N'$ is, by superposition, the sum of the potentials due to the individual charges. That is,

$$\varphi(\boldsymbol{r})=\frac{1}{4\pi\varepsilon_0}\sum_{k=1}^{n}\frac{q_k}{|\boldsymbol{r}-\boldsymbol{r}_k'|} \tag{2.51}$$

Since this is a scalar sum, it is generally easier to determine $\boldsymbol{E}$ by taking the negative gradient of φ than from the vector sum in (2.7) directly. ②

Example 2-8 Find the electric potential of an electric dipole, and derive the expression of the electric field produced by the dipole from the its electric potential.

Solution: By using (2.51) and referring to Figure 2-13, the electric potential of an electric dipole consisting of charges $+q$ and $-q$ with a small separation d can be written down as

---

① 唯一确定电位值需要指定零电位参考点。当电荷分布于有限区域时，默认选取无限远为零电位参考点，否则就需要指定零点位参考点。

② 式(2.50)和式(2.51)说明，若电荷分布于有限区域，且总电荷量不为零，则其远区电位应当与距离成反比。典型的例子就是单个点电荷产生的电场。然而，若多个点电荷的总电荷量为零，则该远区电场的大小随距离衰减的速度可能比 1/r 更快，典型的例子就是例 2-8 分析的电偶极子。

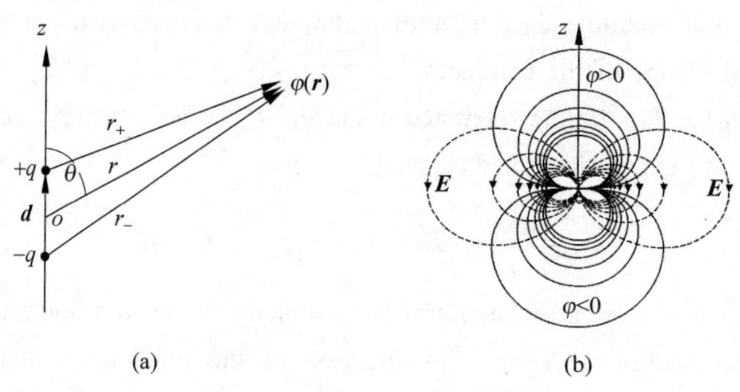

Figure 2-13 Solving the electric potential and field of an electric dipole(Example 2-8)

$$\varphi(\boldsymbol{r})=\frac{q}{4\pi\varepsilon_0}\left(\frac{1}{r_+}-\frac{1}{r_-}\right) \tag{2.52}$$

where r_+ and r_- are respectively the distances from the positive and negative charges to a field point $\boldsymbol{r}$. If $d \ll r$, we have

$$\frac{1}{r_+}\approx\left(r-\frac{d}{2}\cos\theta\right)^{-1}\approx r^{-1}\left(1+\frac{d}{2r}\cos\theta\right) \tag{2.53}$$

and

$$\frac{1}{r_-}\approx\left(r+\frac{d}{2}\cos\theta\right)^{-1}\approx r^{-1}\left(1-\frac{d}{2r}\cos\theta\right) \tag{2.54}$$

Substitution of(2.53) and(2.54) into(2.52) gives

$$\varphi(\boldsymbol{r})=\frac{qd\cos\theta}{4\pi\varepsilon_0 r^2}=\frac{\boldsymbol{p}\cdot\boldsymbol{e}_r}{4\pi\varepsilon_0 r^2} \tag{2.55}$$

where $\boldsymbol{p}=q\boldsymbol{d}$. The $\boldsymbol{E}$ field can be obtained as

$$\boldsymbol{E}=-\nabla\varphi=-\boldsymbol{e}_r\frac{\partial\varphi}{\partial r}-\boldsymbol{e}_\theta\frac{\partial\varphi}{r\partial\theta}=\frac{p}{4\pi\varepsilon_0 r^3}(\boldsymbol{e}_r 2\cos\theta+\boldsymbol{e}_\theta\sin\theta) \tag{2.56}$$

which is the same as(2.16). A two-dimensional sketch of the equipotential lines and the electric field streamlines are given in Figure 2-13(b). Notice that the electric field vectors are everywhere perpendicular to the equipotential lines.

Generally, for an arbitrary orientation of the dipole moment $\boldsymbol{p}$, centered at an arbitrary position $\boldsymbol{r}'$, the potential at an observation point $\boldsymbol{r}$ far away from the dipole can be expressed as

$$\varphi(\boldsymbol{r})=\frac{\boldsymbol{p}\cdot\boldsymbol{e}_R}{4\pi\varepsilon_0 R^2} \tag{2.57}$$

where $\boldsymbol{e}_R$ and R are the same as defined by(2.5) and(2.6).①

2.4.2 Electric Potential Due to a Continuous Charge Distribution(连续分布电荷产生的电位)

The electric potential due to a continuous distribution of charge confined in a given region

① 式(2.55)和式(2.57)表明电偶极子的远区电位与距离的平方成反比。这是由于电偶极子所带的电荷总量为零,在距离足够远时,正负电荷对观察点的贡献趋近于相互抵消。

is obtained by first dividing the charge into infinite number of differential elements, treating each differential element as a point charge, and then integrating the contribution from all the point charges. For a volume charge distribution ρ_v, we have ①

$$\varphi(\boldsymbol{r}) = \frac{1}{4\pi\varepsilon_0}\int_{V'} \frac{\rho_v(\boldsymbol{r}')}{|\boldsymbol{r} - \boldsymbol{r}'|}\mathrm{d}v' \tag{2.58}$$

For a surface charge distribution ρ_s,

$$\varphi(\boldsymbol{r}) = \frac{1}{4\pi\varepsilon_0}\int_{S'} \frac{\rho_s(\boldsymbol{r}')}{|\boldsymbol{r} - \boldsymbol{r}'|}\mathrm{d}s' \tag{2.59}$$

and for a line charge distribution ρ_l,

$$\varphi(\boldsymbol{r}) = \frac{1}{4\pi\varepsilon_0}\int_{L'} \frac{\rho_l(\boldsymbol{r}')}{|\boldsymbol{r} - \boldsymbol{r}'|}\mathrm{d}l' \tag{2.60}$$

Example 2-9 Find the electric field on the axis of a circular ring disk with outer radius a, inner radius b, and uniform surface charge density ρ_s. (The same example as Example 2-2)

Solution: We first find the potential distribution along the axis, and then solve for $\boldsymbol{E}$ field by using (2.46). (We cannot use Gauss's law directly here. Why?) Referring to Figure 2-14 and using (2.59),

$$\mathrm{d}s' = \rho'\mathrm{d}\rho'\mathrm{d}\phi', \quad R = \sqrt{z^2 + \rho'^2}$$

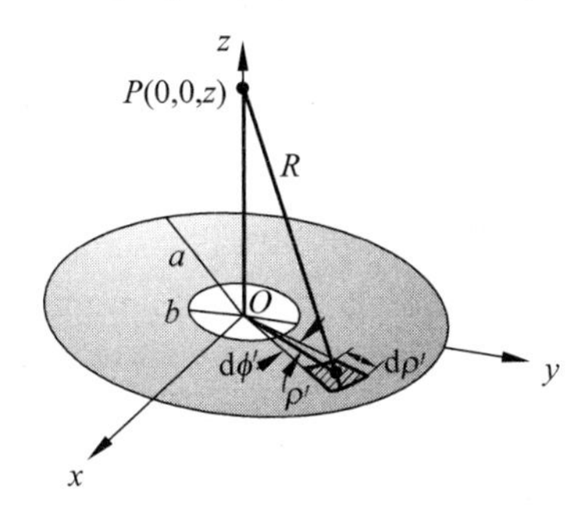

Figure 2-14 Solving the electric potential and field of a surface charge (Example 2-9)

The electric potential at the point $P(0,0,z)$ is

$$\varphi = \frac{\rho_s}{4\pi\varepsilon_0}\int_0^{2\pi}\int_b^a \frac{\rho'}{(z^2 + \rho'^2)^{1/2}}\mathrm{d}\rho'\mathrm{d}\phi' = \frac{\rho_s}{2\varepsilon_0}[(z^2 + a^2)^{1/2} - (z^2 + b^2)^{1/2}] \tag{2.61}$$

Due to the cylindrical symmetry, we expect the $\boldsymbol{E}$ field only have z component. Therefore,

$$\boldsymbol{E} = -\nabla\varphi = -\boldsymbol{e}_z\frac{\partial\varphi}{\partial z} = \boldsymbol{e}_z\frac{\rho_s z}{2\varepsilon_0}[(z^2 + b^2)^{-1/2} - (z^2 + a^2)^{-1/2}] \tag{2.62}$$

which is the same as (2.26). Notice that determination of $\boldsymbol{E}$ field at an off-axis point will be more complicated as it will involve expression of φ off-axis and partial derivatives of φ along ρ-direction. For very large z, we can have the approximation

① 电位计算表达式(2.58)也可以由1.14节的亥姆霍兹定理直接得到(见习题2.10)。

$$(z^2+a^2)^{1/2}=|z|\left(1+\frac{a^2}{z^2}\right)^{1/2}\approx|z|\left(1-\frac{a^2}{2z^2}\right),\quad \text{and similarly,}$$

$$(z^2+b^2)^{1/2}\approx|z|\left(1-\frac{b^2}{2z^2}\right) \tag{2.63}$$

Substitute(2.63) into(2.61), we have

$$\varphi\approx\frac{\rho_s}{2\varepsilon_0}\left[|z|\left(1+\frac{a^2}{2z^2}\right)-|z|\left(1+\frac{b^2}{2z^2}\right)\right]=\frac{\pi(a^2-b^2)\rho_s}{4\pi\varepsilon_0|z|}=\frac{Q}{4\pi\varepsilon_0|z|} \tag{2.64}$$

where $Q=\pi(a^2-b^2)\rho_s$ is the total charge on the disk. Therefore, the electric field for very large z, can be approximated by taking negative gradient of(2.64),

$$\boldsymbol{E}=-\nabla\varphi\approx\boldsymbol{e}_z\frac{Q}{4\pi\varepsilon_0 z^2} \tag{2.65}$$

which is the same as(2.28). We see that, when the observation point is very far away from the charged disk, both the potential and the $\boldsymbol{E}$ field approximates those of a point charge with the same amount of total charge Q.

It needs to be noted that the expressions for electric potentials(2.58) ~ (2.60) are based on the assumption that the charge is distributed within a finite region. With this assumption, it is also implied by(2.58) ~ (2.60) that the potential at infinity is zero. However, if the charge is not confined in a finite region, equations(2.58) ~ (2.60) may lead to an infinitely large result. In that case, we cannot use infinity as the reference zero-potential point. Nevertheless, we might still define a potential with the reference zero-potential point selected somewhere other than infinity, which is shown in the examples that follow. ①

Example 2-10 Determine the electric potential of an infinitely long, straight line charge of a uniform density ρ_l in air.

Solution: In Example 2-5, we solved the problem of the electric field due to an infinitely long line charge. Based on the $\boldsymbol{E}$ field solution, we can use(2.48) to obtain the electric potential directly. Again, we assume the line charge is placed along the z axis. The solution as given by (2.39) shows that the $\boldsymbol{E}$ field is along the ρ-direction and symmetric about the z axis. Therefore, we expect the solution of the potential φ is only a function of ρ, and it is most convenient to choose the integration path in(2.48) to be along the ρ-direction. By substituting (2.39) into(2.48), and letting $\mathrm{d}\boldsymbol{l}=\boldsymbol{e}_\rho\mathrm{d}\rho$, we have an expression for the electric potential at a point P with coordinate $\rho=\rho_P$:

$$\varphi(\rho_P)=-\int_{\text{Reference point}}^{\rho_P}\left(\boldsymbol{e}_\rho\frac{\rho_l}{2\pi\varepsilon_0\rho}\right)\cdot\boldsymbol{e}_\rho\mathrm{d}\rho=\frac{\rho_l}{2\pi\varepsilon_0}\ln\left(\frac{\rho_{\text{Reference point}}}{\rho_P}\right) \tag{2.66}$$

where $\rho_{\text{Reference point}}$ is the ρ coordinate of the zero-potential reference point. Obviously, we cannot

---

① 分布于有限区域的电荷产生的电位可以由式(2.58)~式(2.60)直接计算,其计算结果隐含了无穷远处电位为零的假设。但是如果电荷分布不是有限的,就有可能无法由式(2.58)~式(2.60)计算电位。这时,可能的解决办法是根据需要指定某个位置为参考零电位点。例如例题 2-10 和例题 2-11,从这两个例子还可以看出,当电荷分布区域不是有限区域时,远区电位随距离的衰减可能比 1/r 更慢。

select the reference zero potential point to be infinity, because that will make $\rho_{\text{Reference point}} \to +\infty$, and consequently, make $\varphi(\rho_P)$ infinitely large everywhere. This is not surprising because the charge distribution of the field extends to infinity. However, we have the freedom to select the reference zero potential point to be any location. Therefore, we can simply select $\rho_{\text{Reference point}}$ to be any nonzero constant ρ_0 as shown in Figure 2-15. And then, (2.66) can be rewritten as

$$\varphi(\rho) = \frac{\rho_l}{2\pi\varepsilon_0}\ln\left(\frac{\rho_0}{\rho}\right) \tag{2.67}$$

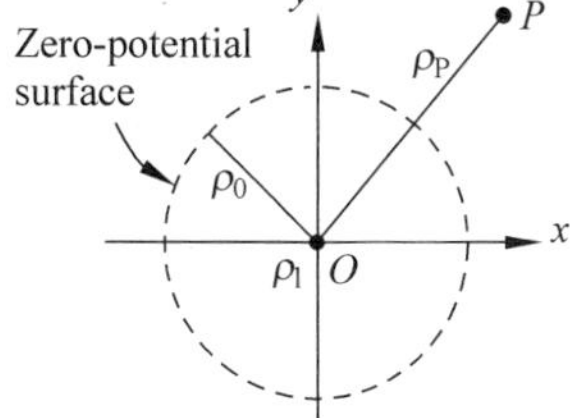

Figure 2-15 Solving the electric potential of an infinitely long, straight line charge

where the subscript P is omitted. Notice that, if $\rho_l > 0$, the electric potential approaches $-\infty$ as ρ increases to infinity, reaches zero at $\rho = \rho_0$, and approaches $+\infty$ as ρ approaches zero.

Example 2-11 Two infinitely long, straight line charges with uniform density of $+\rho_l$ and $-\rho_l$ respectively, are parallel to each other and separated by a distance d_0. Determine the electric potential produced by the two line charges.

Solution: The electric potential produced by a single infinitely long line charge can be expressed by (2.67) as derived in Example 2-10. Based on it, we can write down the electric potential due to the two line charges as

$$\varphi = \frac{\rho_l}{2\pi\varepsilon_0}\ln\left(\frac{\rho_0}{\rho_+}\right) - \frac{\rho_l}{2\pi\varepsilon_0}\ln\left(\frac{\rho_0}{\rho_-}\right) = \frac{\rho_l}{2\pi\varepsilon_0}\ln\left(\frac{\rho_-}{\rho_+}\right) \tag{2.68}$$

where ρ_+ and ρ_- are respectively the distance from the field point to the line charge $+\rho_l$ and $-\rho_l$ as is illustrated in Figure 2-16. (2.68) implies that the reference zero-potential point is selected to be equidistant (ρ_0) from both line charges, which is reasonable as any location with $\rho_+ = \rho_-$ is naturally a zero-potential point. From (2.68), we notice that equipotential surfaces can be specified by letting the ratio between ρ_- and ρ_+ be a constant, i. e., by letting

$$\frac{\rho_-}{\rho_+} = C \tag{2.69}$$

where C is a constant. Particularly, the plane $x = 0$ is a zero potential surface with $C = 1$. At infinity, C also approaches 1, and therefore, infinity can be taken as a reference zero potential point. Now we write down the expression of ρ_- and ρ_+ in the Cartesian coordinates

$$\rho_- = \sqrt{\left(x + \frac{d_0}{2}\right)^2 + y^2}, \quad \rho_+ = \sqrt{\left(x - \frac{d_0}{2}\right)^2 + y^2}$$

and substitute them into (2.69). With some mathematical manipulation, we have an equation for the equipotential surface that is determined by d_0 and C:

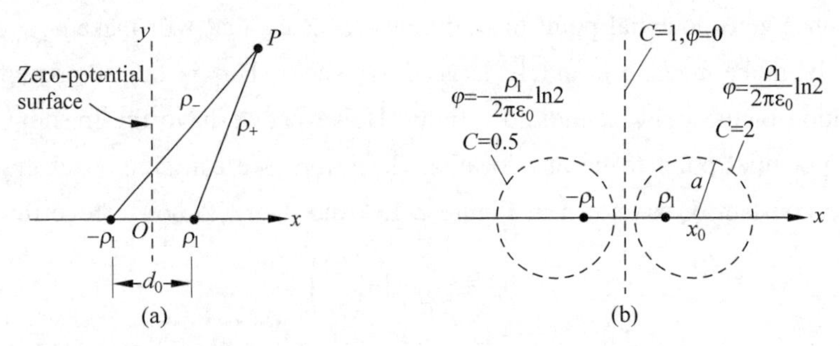

Figure 2-16 Solving the electric potential of two infinitely long, straight line charges

$$(x - x_0)^2 + y^2 = a^2 \tag{2.70}$$

where

$$x_0 = \frac{d_0}{2}\left(\frac{C^2 + 1}{C^2 - 1}\right) \quad \text{and} \quad a = d_0\left|\frac{C}{C^2 - 1}\right| \tag{2.71}$$

Apparently, the equipotential surface described by (2.70) is an infinitely long cylinder wall with the axis along the z-direction. The axis of the cylinder is located at ($x=x_0$, $y=0$), the radius is a, and the potential on the cylinder wall is $(\rho_l/2\pi\varepsilon_0)\ln(C)$. For $C>1$, $x_0>0$ and the entire cylinder wall is in the $x>0$ region with a constant positive potential. For $C<1$, $x_0<0$ and the entire cylinder wall is in the $x<0$ region with a constant negative potential. Two typical equipotential surfaces, with $C=2$ and $C=0.5$ respectively, are illustrated in Figure 2-16(b). It is seen that the two equipotential surfaces are of the same radius and located symmetrically with respective to the y axis. And the potentials on the two surfaces are opposite to each other. This is generally true for any two equipotential surfaces with the ratio between ρ_- and ρ_+ being reciprocal of each other.

2.5 Conductors in Static Electric Field (静电场中的导体)

All the discussions in previous sections are about static electric fields in free space. In reality, the presence of **material media** (物质媒质) will affect the field distribution. Material media can be generally classified into three types: **conductors** (导体), **semiconductors** (半导体), and **dielectrics/insulators** (介质/绝缘体) according to their electrical properties. Conductors have plenty of electrons loosely held around nucleus. Here, "plenty" means there will always be more than enough movable electrons. These electrons can migrate easily from one atom to another. Most metals are conductors. The electrons in the atoms of dielectrics (or insulators), however, are confined to their orbits around the positively charged nuclei. In other words, there are no freely movable charges or conducting currents in dielectric media. Semiconductors, as suggested by the name, have a relatively small number (not plenty) of freely movable charges. Analysis of the electrical properties of semiconductors involves solid state physics, which is not covered in this text. In this section, we discuss the behavior of static electric fields with the presence of conductors. ①

Since conductors are assumed to always have plenty of movable electrons, any electric field within a conductor will make the electrons move, causing a nonzero current. Static current fields will be discussed in Chapter 4. For the electric fields discussed in this chapter, we assume **electrostatic balance**(静电平衡) is reached, and therefore, no current exists in the conductor. This means that there is no electric field within conductors. And therefore, according to(2.36), there is no net volume charge within conductors. In conclusion, under the static condition, we have

$$\rho_v = 0 \quad \text{and} \quad \boldsymbol{E} = 0 \tag{2.72}$$

inside a conductor. ②

Since $\boldsymbol{E}=0$ inside a conductor, the entire conductor is an **equipotential body**(等位体), and the conductor surface is an equipotential surface under static conditions. Therefore, the $\boldsymbol{E}$ field on a conductor surface has no **tangential component**(切向分量). Otherwise, the electrons on the conductor surface will move under the influence of the tangential $\boldsymbol{E}$ field. In other words, the $\boldsymbol{E}$ field on a conductor surface is everywhere normal to the surface. Detailed derivation of the boundary conditions on the conductor surface is given as follows by using Figure 2-17.

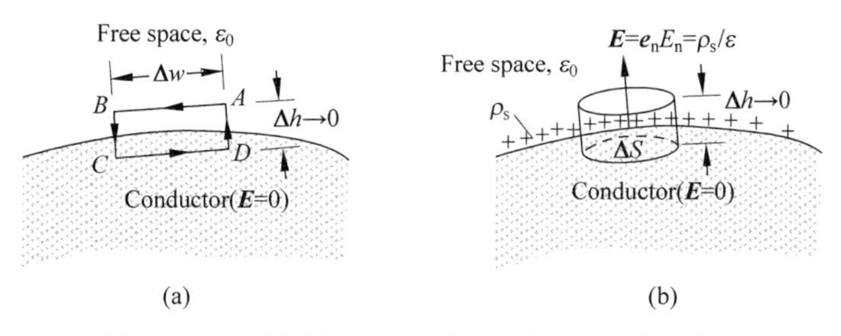

Figure 2-17 Fields on a conductor-free space interface

Consider a very small rectangular contour $ABCDA$ across the interface between a conductor and free space as shown in Figure 2-17(a). Sides $AB=CD=\Delta w$ are parallel to the interface. Sides $BC=DA=\Delta h$ are normal to the interface. Applying(2.44) to the rectangle, letting $\Delta h \to 0$, and noting that $\boldsymbol{E}$ in the conductor is zero, we have

$$\oint_{ABCDA} \boldsymbol{E} \cdot \mathrm{d}\boldsymbol{l} = E_{AB}\Delta w = 0$$

Therefore, E_{AB}, the component of the $\boldsymbol{E}$ field along the direction $A \to B$, must be zero. Since $A \to B$ can be taken as any direction parallel to the interface, it means that

$$E_t = 0 \quad \text{or} \quad \boldsymbol{e}_n \times \boldsymbol{E} = \boldsymbol{0} \tag{2.73}$$

① 导体被定义为含有大量可自由移动的电荷的媒质。这里,“大量”的意思是指永远不会缺乏。换言之,我们假定导体内部永远存在取之不尽的可移动的自由电荷(电子)。

② 静电场是静电平衡条件下的场,其中没有定向移动的电荷。因此,静电场中导体内部没有电流,这也就意味着导体内电场为零。否则,由于导体内存在大量能够自由移动的电荷,必然会在电场力的作用下形成电流,即定向移动的电荷。导体内电场为零,根据微分形式的高斯定律,电荷密度也必然为零(注意:这里指的是包含能够自由移动和不能自由移动的电荷的总电荷密度为零)。

which states that the tangential component of the $\boldsymbol{E}$ field on a conductor surface is zero. ①

Another boundary condition is about E_n, the normal component of $\boldsymbol{E}$ at the surface of the conductor. For that, we construct a thin box with the top face outside the conductor, and the bottom face inside the conductor where $\boldsymbol{E}=0$. The area of the top and bottom faces is ΔS, and the height of the box is Δh. Applying Gauss's law(2.37) to the region enclosed by the box and letting $\Delta h \to 0$, we have

$$\oint_S \boldsymbol{E} \cdot \mathrm{d}\boldsymbol{s} = E_n \Delta S = \frac{\rho_s \Delta S}{\varepsilon_0}$$

where E_n is the normal component of the $\boldsymbol{E}$ field pointing outwardly from the conductor surface, and ρ_s is the surfacc charge density on the conductor. Notice that the total charge enclosed by the box remains $\rho_s\ \Delta S$ even when $\Delta h \to 0$. Therefore, we have

$$E_n = \frac{\rho_s}{\varepsilon_0} \quad \text{or} \quad \boldsymbol{e}_n \cdot \boldsymbol{E} = \frac{\rho_s}{\varepsilon_0} \tag{2.74}$$

which states that the normal component of the static $\boldsymbol{E}$ field at a conductor/free space boundary is equal to the surface charge density on the conductor divided by the permittivity of free space. ②

When a conductor is placed in a static electric field, the **external field**(**外加电场**) will cause electrons in the conductor to move along the opposite direction of the field, leading to a nonzero surface charge distribution. Partial surface will have a negative charge distribution due to accumulation of the electrons. Partial surface will have a positive charge distribution due to loss of the electrons. These induced surface charges create an induced field that completely counteracts the external field both inside the conductor and tangential to its surface.

Example 2-12 A positive point charge Q is at the center of a spherical conducting shell with an inner radius r_i and an outer radius r_o. Determine the distribution of $\boldsymbol{E}$ and φ field. Find the surface charge distribution on the inner and outer surfaces of the conducting shell.

Solution: Due to spherical symmetry, $\boldsymbol{E}$ has only radial component, and we can use Gauss's law to determine E_r and then find φ by integration. The surface charge distribution can be found by using the boundary conditions. As is shown in Figure 2-18, we consider three distinct regions: ① $r>r_o$, ② $r_i<r<r_o$, and ③ $r<r_i$. In each of these regions, we construct a spherical Gaussian surface centered at the origin (location of the charge Q) and apply the Gauss's law as shown in Figure 2-18(a) ~ Figure 2-18(c). Apparently, over the Gaussian surface with radius r, the outward flux of $\boldsymbol{E}$ is $4\pi r^2 E_r$.

(1) $r>r_o$: As shown in Figure 2-18(a), the total charges contained by the Gaussian surface is Q(since the net charge contained in the conducting shell must be zero). Therefore, according to Gauss's law, $4\pi r^2 E_r = Q/\varepsilon_0$, and

① 式(2.73)给出的静电场中导体基本性质：导体表面电场强度的切向分量为零。

② 式(2.74)给出的静电场中导体基本性质：导体与自由空间分界面上导体外侧电场的法向分量大小等于导体表面的面电荷密度除以真空中的介电常数。

$$\boldsymbol{E}=\boldsymbol{e}_r E_r=\boldsymbol{e}_r \frac{Q}{4\pi\varepsilon_0 r^2} \quad \text{for} \quad r \geqslant r_o$$

The $\boldsymbol{E}$ field is the same as that of a point charge Q without the presence of the shell. Since all the charges are confined in a finite region, the reference zero-potential point is taken to be at infinity, and the potential outside the conducting shell is

$$\varphi(\boldsymbol{r})=-\int_{\infty}^{r} \boldsymbol{E}\cdot\boldsymbol{e}_r \mathrm{d}r=-\int_{\infty}^{r} \frac{Q}{4\pi\varepsilon_0 r^2}\mathrm{d}r=\frac{Q}{4\pi\varepsilon_0 r} \quad \text{for} \quad r \geqslant r_o$$

where the integral path is selected to be a straight radial line from infinity to the field point along the direction $(-\boldsymbol{e}_r)$. The surface charge on the outer surface of the shell is, by (2.74),

$$\rho_s\big|_{r=r_o}=\varepsilon_0\,\boldsymbol{e}_n\cdot\boldsymbol{E}(r_o^+)=\varepsilon_0\,\boldsymbol{e}_r\cdot\boldsymbol{e}_r\frac{Q}{4\pi\varepsilon_0 r_o^2}=\frac{Q}{4\pi r_o^2} \tag{2.75}$$

Here, the superscript "+" represents the free-space side ($r>r_o$) of the outer surface of the shell.

(2) $r_i<r<r_o$: We know that the static electric field inside a conductor is always zero. Therefore,

$$\boldsymbol{E}=\mathbf{0}$$

The conducting shell is an equipotential body. Hence,

$$\varphi(\boldsymbol{r})=\varphi(r_o)=\frac{Q}{4\pi\varepsilon_0 r_o} \quad \text{for} \quad r_i \leqslant r < r_o$$

(3) $r<r_i$: As shown in Figure 2-18(c), application of Gauss's law yields the same expression for $\boldsymbol{E}$ as in the region(1), i.e.,

$$\boldsymbol{E}=\boldsymbol{e}_r E_r=\boldsymbol{e}_r \frac{Q}{4\pi\varepsilon_0 r^2} \quad \text{for} \quad r < r_i$$

The potential in this region is

$$\varphi(r)=\varphi(r_i)-\int_{r_i}^{r} \boldsymbol{E}\cdot\boldsymbol{e}_r \mathrm{d}r=\frac{Q}{4\pi\varepsilon_0 r_o}-\int_{r_i}^{r}\frac{Q}{4\pi\varepsilon_0 r^2}\mathrm{d}r=\frac{Q}{4\pi\varepsilon_0}\left(\frac{1}{r}+\frac{1}{r_o}-\frac{1}{r_i}\right) \quad \text{for} \quad r < r_i$$

The surface charge on the inner surface of the shell is

$$\rho_s\big|_{r=r_i}=\varepsilon_0\,\boldsymbol{e}_n\cdot\boldsymbol{E}(r_i^-)=\varepsilon_0(-\boldsymbol{e}_r)\cdot\left(\boldsymbol{e}_r\frac{Q}{4\pi\varepsilon_0 r_i^2}\right)=-\frac{Q}{4\pi r_i^2} \tag{2.76}$$

Here, the superscript "−" represents the free-space side ($r<r_i$) of the inner surface of the shell. The variations of E_r and φ versus r in all three regions are plotted in Figure 2-18(d) and(e). Notice that the electric intensity is discontinuous across the interface between air and the conductor, while the potential remains continuous. This is because the electric potential is related to the line integral of the electric field intensity and therefore, it cannot have discontinuity unless the electric field intensity goes to infinity at some point.

The surface charge densities(2.75) and(2.76) can also be found by applying Gauss's law to the region inside the conducting shell, as shown in Figure 2-18(b). Since the total charge enclosed in the Gaussian surface S must be zero, an amount of negative charge equal to Q must be induced on the inner shell surface at $r=r_i$. The charge $-Q$ must be uniformly distributed on the inner surface, leading to the surface charge density given by(2.76). As the conducting shell is neutral, this also means that an amount of positive charge equal to $+Q$ is induced on the outer shell surface at $r=r_o$, leading to the surface charge given by(2.75).

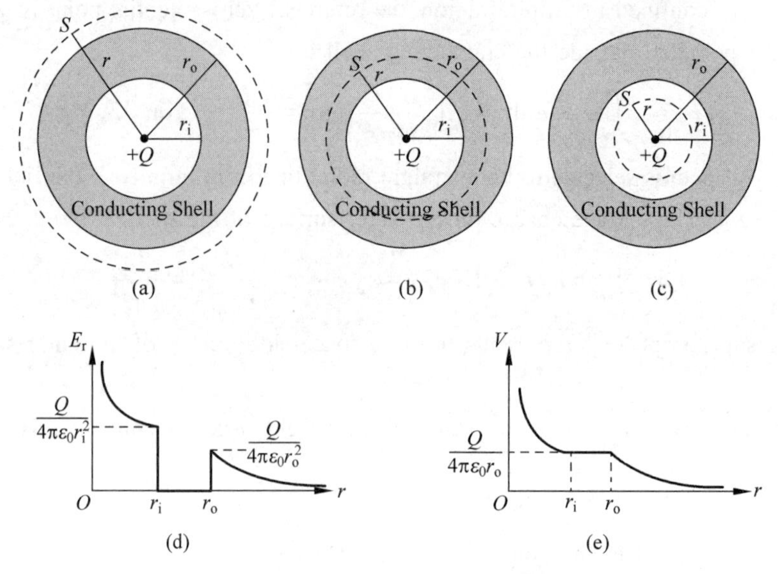

Figure 2-18 A point charge $+Q$ at the center of a conducting shell(Example 2-12)

2.6 Dielectrics in Static Electric Fields(静电场中的介质)

All materials are composed of atoms or molecules with a positively charged nucleus and negatively charged electrons. In ideal dielectrics, these charges are bound charges that are not freely movable. If the positive and negative charges are evenly distributed over the molecule, the molecule does not possess a dipole moment, and is called **nonpolar molecule**(非极化分子). However, when a dielectric body containing nonpolar molecules is placed in an **external electric field**(外加电场), a force is applied on the charges, causing small displacements of positive and negative charges in opposite directions. As a result, the molecules are polarized with a nonzero electric dipole moment, and the dielectric body becomes polarized with induced electric dipoles, as shown in Figure 2-19. The induced electric dipoles will create additional electric field both inside and outside the dielectric material. ①

The molecules of some dielectrics possess permanent dipole moments, even in the absence of an external polarizing field. Such molecules are called **polar molecules**(极化分子). When there is no external field, the individual dipoles are randomly oriented, producing no net dipole

---

① 组成介质的原子或分子结构中,正负电荷紧密结合而不能自由移动。若正负电荷的中心重合,则所有的分子对外都不显电性,这样的分子称为非极性分子。在外加电场的作用下,正电荷沿电场方向移动,负电荷沿电场反方向移动,导致正负电荷重心不再重合,每个分子都等效形成具有一定电偶极矩的电偶极子,这种现象称为介质的极化。极化后的介质包含大量与外加电场方向一致的电偶极子,这些感应形成的电偶极子产生的电场将改变原来的电场分布。

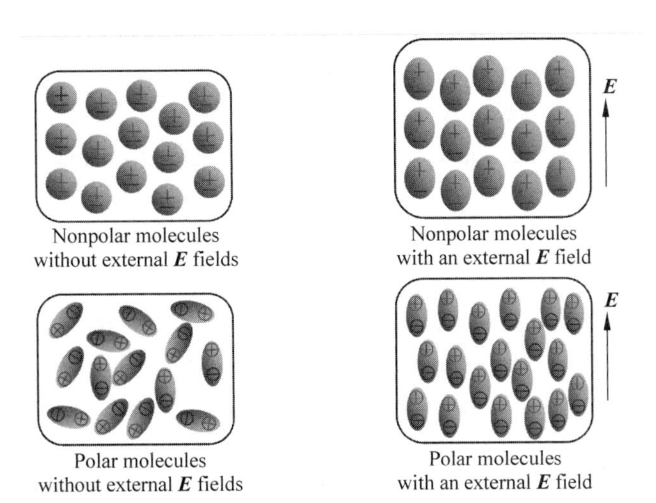

Figure 2-19 Polarization of a dielectric medium due to external electric fields

moment macroscopically. However, when the dielectric body with polar molecules is placed in an external electric field, a **torque**(**力矩**) will be exerted on the dipoles, and all the dipoles tend to be aligned with the same orientation, creating a nonzero dipole moment macroscopically. ①

To analyze the macroscopic effect of induced dipoles we define a **polarization vector**(**极化强度矢量**) $\boldsymbol{P}$ as

$$\boldsymbol{P}=\lim_{\Delta v\to 0}\frac{\sum_{k=1}^{n\Delta v}\boldsymbol{p}_k}{\Delta v}(\mathrm{C/m^2}) \tag{2.77}$$

where n is the number of molecules per unit volume and the numerator represents the vector sum of the induced dipole moments contained in a very small volume Δv. According to the definition, $\boldsymbol{P}$ is the volume density of the induced **molecular dipole moment**(**分子偶极矩**). Then, the differential dipole moment $\mathrm{d}\boldsymbol{p}$ within a differential volume $\mathrm{d}v'$ is $\mathrm{d}\boldsymbol{p}=\boldsymbol{P}\mathrm{d}v'$. According to(2.55), $\mathrm{d}\boldsymbol{p}$ produces an electrostatic potential

$$\mathrm{d}\varphi=\frac{\boldsymbol{P}\cdot\boldsymbol{e}_R}{4\pi\varepsilon_0 R^2}\mathrm{d}v' \tag{2.78}$$

where R is the distance from the elemental volume $\mathrm{d}v'$ to a fixed field point. Integrating over the volume V' of the dielectric, we obtain the potential due to the polarized dielectric to be

$$\varphi=\frac{1}{4\pi\varepsilon_0}\int_{V'}\frac{\boldsymbol{P}\cdot\boldsymbol{e}_R}{R^2}\mathrm{d}v' \tag{2.79}$$

According to(1.82), the above equation can be rewritten as

$$\varphi=\frac{1}{4\pi\varepsilon_0}\int_{V'}\boldsymbol{P}\cdot\nabla'\left(\frac{1}{R}\right)\mathrm{d}v' \tag{2.80}$$

---

① 若分子的正负电荷中心不重合,即便没有外加电场,也等效于一个电偶极子。但由于大量的电偶极子排列杂乱无章,合成的电偶极矩相互抵消,宏观上对外不显电性。而在外加电场的作用下,电场力作用于极性分子形成力矩,使得每个电偶极子的排列倾向于与电场方向一致,从而对外产生的电场不再为零。

with the identity(1.148), in which let $\boldsymbol{A}=\boldsymbol{P}$ and $\psi=1/R$, the potential φ further becomes

$$\varphi=\frac{1}{4\pi\varepsilon_0}\left[\int_{V'}\nabla'\cdot\left(\frac{\boldsymbol{P}}{R}\right)\mathrm{d}v'-\int_{V'}\frac{\nabla'\cdot\boldsymbol{P}}{R}\mathrm{d}v'\right] \tag{2.81}$$

By applying the divergence theorem to the first volume integral on the right side of(2.81), we have

$$\varphi=\frac{1}{4\pi\varepsilon_0}\oint_{S'}\frac{\boldsymbol{P}\cdot\boldsymbol{e}_\mathrm{n}'}{R}\mathrm{d}s'+\frac{1}{4\pi\varepsilon_0}\int_{V'}\frac{(-\nabla'\cdot\boldsymbol{P})}{R}\mathrm{d}v' \tag{2.82}$$

where $\boldsymbol{e}_\mathrm{n}'$ is the outward normal from the surface element $\mathrm{d}s'$ of the dielectric. Now we see that, the first integral in(2.82) is the same as that in(2.59) if $\boldsymbol{P}\cdot\boldsymbol{e}_\mathrm{n}'$ is replaced by ρ_s, and the second integral in(2.82) is the same as that in(2.58) if $(-\nabla'\cdot\boldsymbol{P})$ is replaced by ρ_v. Therefore, the electric potential due to a polarized dielectric may be considered as the contribution of an **equivalent polarization surface charge density**(**等效极化面电荷密度**) ρ_ps and an **equivalent polarization volume charge density** (**等效极化体电荷密度**)ρ_pv. That is

$$\varphi=\frac{1}{4\pi\varepsilon_0}\oint_{S'}\frac{\rho_\mathrm{ps}}{R}\mathrm{d}s'+\frac{1}{4\pi\varepsilon_0}\int_{V'}\frac{\rho_\mathrm{pv}}{R}\mathrm{d}v' \tag{2.83}$$

where

$$\rho_\mathrm{ps}=\boldsymbol{P}\cdot\boldsymbol{e}_\mathrm{n} \tag{2.84}$$

$$\rho_\mathrm{pv}=-\nabla\cdot\boldsymbol{P} \tag{2.85}$$

These polarization charge densities are also called **bound-charge densities** (**束缚电荷密度**) as they are bounded to the respective molecules and cannot freely move within the dielectric.①

(2.84) and(2.85) were derived mathematically with the aid of a vector identity. They can also be derived physically with the aid of Figure 2-20. Consider an arbitrary volume bounded by a closed surface S, as shown in Figure 2-20(a). The total polarized volume charge within S is contributed only by polarized molecules that cross the interface, with the positive charge of the molecules on one side of S and the negative charge on the other side. For a differential area element $\mathrm{d}\boldsymbol{s}$ on S, the molecules crossing S through $\mathrm{d}\boldsymbol{s}$ must have their center located within an **oblique**(**倾斜的**) cylinder with volume ΔV. Assume each polarized molecule has the same dipole moment $q\boldsymbol{d}$, where $\boldsymbol{d}$ is the vector pointing from the negative charge $-q$ to the positive charge $+q$. Then $\Delta V=\mathrm{d}\boldsymbol{s}\cdot\boldsymbol{d}$. The number of molecules within this cylinder is $n\Delta V$, where n is the number of molecules per unit volume. The net charge left inside the volume is $\mathrm{d}Q=-qn\Delta V=-qn\mathrm{d}\boldsymbol{s}\cdot\boldsymbol{d}$. By noticing that $qn\boldsymbol{d}=\boldsymbol{P}$, we have

$$\mathrm{d}Q=-qn\boldsymbol{d}\cdot\mathrm{d}\boldsymbol{s}=-\boldsymbol{P}\cdot\mathrm{d}\boldsymbol{s} \tag{2.86}$$

Therefore, the total charge remaining within the volume bounded by S is

$$Q=\oint_S\mathrm{d}Q=-\oint_S\boldsymbol{P}\cdot\mathrm{d}\boldsymbol{s}=\int_V(-\nabla\cdot\boldsymbol{P})\,\mathrm{d}v=\int_V\rho_\mathrm{pv}\mathrm{d}v$$

which leads to the expression for the volume charge density in(2.85). If $\mathrm{d}\boldsymbol{s}$ is on the surface of

① 根据介质极化的分子模型,介质极化后对外呈现的电特性表现为大量分子电偶极矩产生的电场。从数学上证明,该电场可以被认为是由两种等效电荷共同产生的,即介质表面的等效极化面电荷和介质内部区域的等效极化体电荷。以下从物理和几何的角度说明,极化形成的大量分子偶极矩确实等效于在介质表面和内部感应出的束缚电荷。

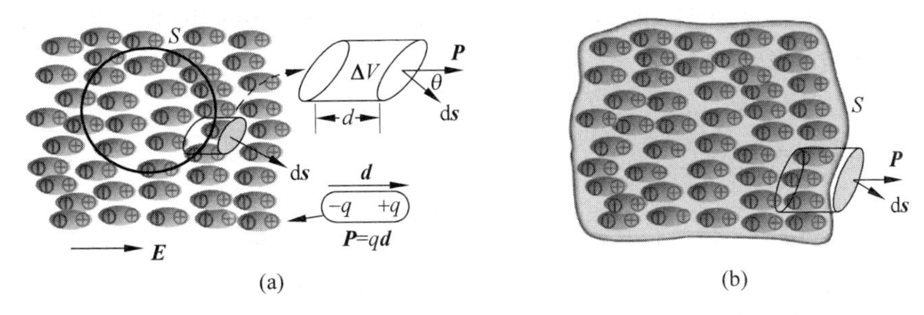

Figure 2-20 Physical interpretation of polarized volume and surface charges

the dielectric, as shown in Figure 2-20(b), then the net charge on the dielectric surface has a density of

$$\rho_{\mathrm{ps}} = \frac{\boldsymbol{P} \cdot \mathrm{d}\boldsymbol{s}}{\mathrm{d}s}$$

which is the same as (2.84).

Notice that, if a dielectric body is electrically neutral before polarization, the total charge of the dielectric after polarization must remain zero. This can be readily verified by noting that

$$\begin{aligned}\text{Total charge} &= \oint_S \rho_{\mathrm{ps}} \mathrm{d}s + \int_V \rho_{\mathrm{pv}} \mathrm{d}v \\ &= \oint_S \boldsymbol{P} \cdot \boldsymbol{e}_{\mathrm{n}} \mathrm{d}s - \int_V \nabla \cdot \boldsymbol{P} \mathrm{d}v = 0 \end{aligned} \qquad (2.87)$$

which is the direct result from the divergence theorem. ①

2.7 Electric Flux Density and Gauss's Law (电通密度与高斯定律)

As a result of the polarization discussed in the previous section, the divergence equation (2.36) needs to be modified to include the effect of the equivalent polarization charges, that is,

$$\nabla \cdot \boldsymbol{E} = \frac{1}{\varepsilon_0}(\rho_{\mathrm{v}} + \rho_{\mathrm{pv}}) \qquad (2.88)$$

where ρ_{v} is the volume density of the **free charge** (自由电荷), and ρ_{pv} is the equivalent polarization volume charge density. Notice that ρ_{v} is called "free charge" simply means that ρ_{v} is independent of polarization, but not necessarily freely movable as the electrons in conductors. On the contrary, ρ_{v} is completely due to the polarization of the dielectric. ②

Using (2.85), we have

$$\nabla \cdot (\varepsilon_0 \boldsymbol{E} + \boldsymbol{P}) = \rho_{\mathrm{v}} \qquad (2.89)$$

---

① 如前所述,极化电荷是由分子内正负电荷位移产生的,不能自由移动。因此介质极化过程中总的电荷量不变。若发生极化之前介质为中性,则极化之后极化电荷的总量必然也为零。

② 所谓“自由电荷”指的是与介质极化无关的电荷,也可以理解为产生外加电场的源。自由电荷可能存在于介质内,也可能存在于介质外的导体中,因此并不一定是能够自由移动的。但自由电荷一定不是介质极化的结果,相反,它是介质极化的原因。而极化电荷则是介质极化的结果。

Now, we define a new field quantity, the **electric flux density** (**电通密度**), or **electric displacement** (**电位移**), $\boldsymbol{D}$ as

$$\boldsymbol{D} = \varepsilon_0 \boldsymbol{E} + \boldsymbol{P} (\mathrm{C/m^2}) \tag{2.90}$$

Then from (2.89), we have

$$\nabla \cdot \boldsymbol{D} = \rho_v \tag{2.91}$$

which gives a divergence relation between the electric displacement field and free charge. Notice that the polarization vector $\boldsymbol{P}$ or the polarization charge density ρ_{pv} are not involved in (2.91).

In fact, equations (2.91) and (2.43) are considered as two fundamental governing differential equations for electrostatics in any medium. The corresponding integral form of (2.91) is obtained by taking the volume integral of both sides. We have

$$\int_V \nabla \cdot \boldsymbol{D} \mathrm{d}v = \int_V \rho_v \mathrm{d}v$$

which leads to

$$\oint_S \boldsymbol{D} \cdot \mathrm{d}\boldsymbol{s} = \int_V \rho_v \mathrm{d}v = Q_{\text{total}} \tag{2.92}$$

Equation (2.92) is another form of **Gauss's law**, which states that the total outward flux of the electric displacement (or the total outward electric flux) over any closed surface is equal to the total free charge enclosed in the surface. ①

As discussed in Section 2.3, Gauss's law is useful in determining the electric field due to charge distributions under symmetry conditions. Using (2.92) we can easily obtain a solution of electric flux density on a Gaussian surface over which the normal component of the $\boldsymbol{D}$ field is a constant. Now the question is how to find the electric field density $\boldsymbol{E}$ from the electric flux density $\boldsymbol{D}$. Fortunately, for most media, there exists a simple relation between $\boldsymbol{E}$ and $\boldsymbol{D}$ if the medium is **linear** (**线性**) and **isotropic** (**各向同性的**,即沿不同的方向没有不同,或者说与方向无关).

In a linear isotropic medium, the polarization is directly proportional to the electric field intensity, and the proportionality coefficient is independent of the magnitude and direction of the field. So, we can write

$$\boldsymbol{P} = \varepsilon_0 \chi_e \boldsymbol{E} \tag{2.93}$$

where χ_e is a dimensionless quantity called **electric susceptibility** (**电极化率**). If χ_e is independent of space coordinates, the dielectric medium is called **homogeneous** (**均匀的**). Substituting (2.93) into (2.90) yields the following **constitutive relation** (**本构关系**) between $\boldsymbol{D}$ and $\boldsymbol{E}$ fields:

$$\boldsymbol{D} = \varepsilon_0 (1 + \chi_e) \boldsymbol{E} = \varepsilon_0 \varepsilon_r \boldsymbol{E} = \varepsilon \boldsymbol{E} \tag{2.94}$$

where

$$\varepsilon_r = 1 + \chi_e = \frac{\varepsilon}{\varepsilon_0} \tag{2.95}$$

---

① 式(2.92)为电位移矢量表示的高斯定律：穿过任意闭合曲面向外的电位移矢量的通量等于该曲面包围的总电荷数。

is known as the **relative permittivity**(相对介电常数) or the dielectric constant of the medium. The coefficient $\varepsilon=\varepsilon_0\varepsilon_r$ is the **absolute permittivity** (绝对介电常数)(or simply **permittivity**, 介电常数) of the medium. The unit of ε is farad per meter(F/m), which is the same as that of ε_0. Air has a dielectric constant of 1.00059, which is usually considered equal to that of free space. Compare(2.90) and(2.94), we have

$$\boldsymbol{P}=(\varepsilon-\varepsilon_0)\boldsymbol{E}=\varepsilon_0(\varepsilon_r-1)\boldsymbol{E} \tag{2.96}$$

A linear, homogeneous, and isotropic medium is called a **simple medium**(简单媒质). The relative permittivity of a simple medium is a constant.

For **anisotropic** (**各向异性的**,即沿不同的方向呈现出不同的性质) materials, the dielectric constant is different for different directions of the electric field. As a result, $\boldsymbol{D}$ and $\boldsymbol{E}$ vectors generally have different directions in an anisotropic medium, and their relation has to be expressed with a **tensor**(张量). In matrix form, the constitutive relation becomes

$$\begin{bmatrix} D_x \\ D_y \\ D_z \end{bmatrix}=\begin{bmatrix} \varepsilon_{11} & \varepsilon_{12} & \varepsilon_{13} \\ \varepsilon_{21} & \varepsilon_{22} & \varepsilon_{23} \\ \varepsilon_{31} & \varepsilon_{32} & \varepsilon_{33} \end{bmatrix}\begin{bmatrix} E_x \\ E_y \\ E_z \end{bmatrix} \tag{2.97}$$

For crystals, when the coordinate axes are selected to be along the principal axes of the crystal, the off-diagonal terms of the permittivity matrix in(2.97) become zero. Then we have

$$\begin{bmatrix} D_x \\ D_y \\ D_z \end{bmatrix}=\begin{bmatrix} \varepsilon_1 & 0 & 0 \\ 0 & \varepsilon_2 & 0 \\ 0 & 0 & \varepsilon_3 \end{bmatrix}\begin{bmatrix} E_x \\ E_y \\ E_z \end{bmatrix} \tag{2.98}$$

Media having the property represented by(2.98) are said to be **biaxial** (双轴的). If $\varepsilon_1=\varepsilon_2$ in (2.98), the medium is said to be **uniaxial**(单轴的).

Example 2-13 A positive point charge Q is at the center of a spherical dielectric shell of an inner radius r_i and an outer radius r_o. The dielectric constant of the shell is ε_r. Determine the distribution $\boldsymbol{D}$, $\boldsymbol{E}$, φ, and $\boldsymbol{P}$; find the polarization charges inside and on the surface of the dielectric shell.

Solution: We take the advantage of the spherical symmetry of this problem, which means that the vector $\boldsymbol{D}$ and $\boldsymbol{E}$ have only radial components. We first apply Gauss's law to find $\boldsymbol{D}$, and then find $\boldsymbol{E}$ in three regions: ① $r>r_o$; ② $r_i<r<r_o$; and ③ $r<r_i$. After that, $\boldsymbol{P}$ is determined by the relation(2.96), and potential φ is found from the negative line integral of $\boldsymbol{E}$.

Over a Gaussian surface(sphere surface with center at the point charge) with radius r, the outward flux of $\boldsymbol{D}$ is $4\pi r^2 D_r$, and the total charge enclosed by the Gaussian surface is Q. According to Gauss's law(2.92), we have $4\pi r^2 D_r=Q$, therefore

$$\boldsymbol{D}=\boldsymbol{e}_r D_r=\boldsymbol{e}_r\frac{Q}{4\pi r^2}$$

This expression for $\boldsymbol{D}$ vector is the same in all the three regions. Now we can find $\boldsymbol{E}$, $\boldsymbol{P}$ and φ for the three regions separately.

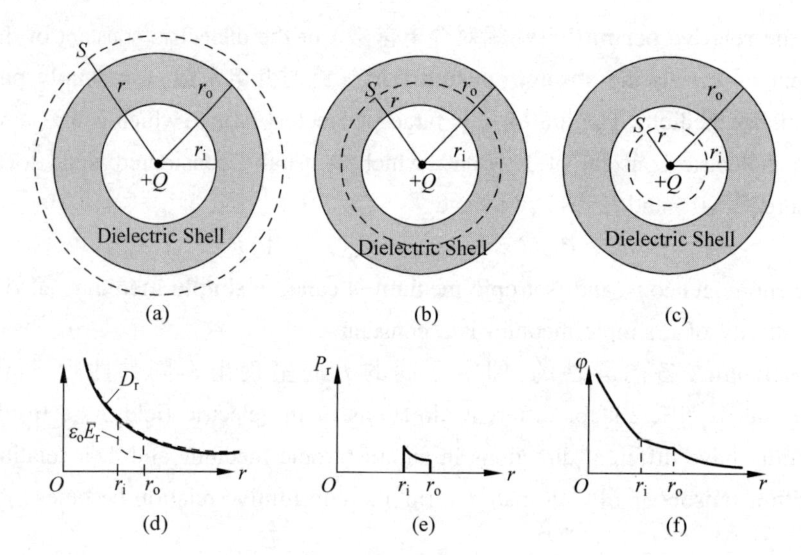

Figure 2-21 A point charge $+Q$ at the center of a dielectric shell (Example 2-13)

(1) $r>r_o$: As shown in Figure 2-21(a), the medium is free space, therefore,

$$\boldsymbol{E} = \frac{\boldsymbol{D}}{\varepsilon_0} = \boldsymbol{e}_r \frac{Q}{4\pi\varepsilon_0 r^2}$$

$$\boldsymbol{P} = \boldsymbol{D} - \varepsilon_0 \boldsymbol{E} = 0$$

$$\varphi(r) = -\int_\infty^r \boldsymbol{E}\cdot\boldsymbol{e}_r \mathrm{d}r = -\int_\infty^r \frac{Q}{4\pi\varepsilon_0 r^2}\mathrm{d}r = \frac{Q}{4\pi\varepsilon_0 r}$$

(2) $r_i<r<r_o$: As shown in Figure 2-21(b), the medium is dielectric ε_r, therefore,

$$\boldsymbol{E} = \frac{\boldsymbol{D}}{\varepsilon_0\varepsilon_r} = \boldsymbol{e}_r \frac{Q}{4\pi\varepsilon_0\varepsilon_r r^2}$$

$$\boldsymbol{P} = \boldsymbol{D} - \varepsilon_0 \boldsymbol{E} = \boldsymbol{e}_r\left(1-\frac{1}{\varepsilon_r}\right)\frac{Q}{4\pi r^2}$$

$$\varphi(r) = \varphi(r_o) - \int_{r_o}^r \boldsymbol{E}\cdot\boldsymbol{e}_r \mathrm{d}r = \frac{Q}{4\pi\varepsilon_0 r_o} - \int_{r_o}^r \frac{Q}{4\pi\varepsilon_0\varepsilon_r r^2}\mathrm{d}r = \frac{Q}{4\pi\varepsilon_0}\left[\frac{1}{r_o} + \frac{1}{\varepsilon_r}\left(\frac{1}{r} - \frac{1}{r_o}\right)\right]$$

On the inner surface of the shell, from (2.84), we find

$$\rho_{ps}\Big|_{r=r_i} = \boldsymbol{P}\cdot(-\boldsymbol{e}_r)\Big|_{r=r_i} = -\left(1-\frac{1}{\varepsilon_r}\right)\frac{Q}{4\pi r_i^2} \tag{2.99}$$

Similarly, on the outer surface of the shell, we find

$$\rho_{ps}\Big|_{r=r_o} = \boldsymbol{P}\cdot\boldsymbol{e}_r\Big|_{r=r_o} = \left(1-\frac{1}{\varepsilon_r}\right)\frac{Q}{4\pi r_o^2} \tag{2.100}$$

Inside the dielectric shell, from (2.85),

$$\rho_p = -\nabla\cdot\boldsymbol{P} = -\frac{1}{r^2}\frac{\partial}{\partial r}(r^2 P_r) = 0 \tag{2.101}$$

Equations (2.99), (2.100) and (2.101) indicate that there is no net polarization volume charge inside the dielectric shell. However, negative polarization surface charges exist on the inner surface. Positive polarization surface charges exist on the outer surface. And the total polarization charges of the dielectric shell are zero, which is as expected since the shell is apparently neutral.

(3) $r<r_i$: As shown in Figure 2-21(c), the medium is free space, and we have

$$\boldsymbol{E}=\frac{\boldsymbol{D}}{\varepsilon_0}=\boldsymbol{e}_r\frac{Q}{4\pi\varepsilon_0 r^2}$$

$$\boldsymbol{P}=\boldsymbol{D}-\varepsilon_0\boldsymbol{E}=0$$

$$\varphi(r)=\varphi(r_i)-\int_{r_i}^{r}\boldsymbol{E}\cdot\boldsymbol{e}_r\mathrm{d}r=\underbrace{\frac{Q}{4\pi\varepsilon_0 r_o}}_{\text{I}}+\underbrace{\frac{Q}{4\pi\varepsilon_0\varepsilon_r}\left(\frac{1}{r_i}-\frac{1}{r_o}\right)}_{\text{II}}+\underbrace{\frac{Q}{4\pi\varepsilon_0}\left(\frac{1}{r}-\frac{1}{r_i}\right)}_{\text{III}} \tag{2.102}$$

Notice the three terms in the right side of $\varphi(r)$ expression(2.102). Term Ⅰ represents the potential at the outer shell surface, which is contributed by the integral of $\boldsymbol{E}$ field in the region $r>r_o$. Term Ⅱ represents the contribution of the $\boldsymbol{E}$ field integration inside the shell. Term Ⅲ represents the contribution of the $\boldsymbol{E}$ field integration in the region $r<r_i$.

The variations of D_r and $\varepsilon_0 E_r$ versus r are plotted in Figure 2-21(d). The difference($D_r-\varepsilon_0 E_r$) is P_r and is shown in Figure 2-21(e). The plot for φ is shown in Figure 2-21(f). Notice that D_r is a continuous curve exhibiting no sudden changes in going from one medium to another and that P_r exists only in the dielectric region.

Compare Figure 2-18 of Example 2-12 with Figure 2-21 of Example 2-13. In both examples, surface charges are induced on the inner and outer surface of the shells, producing an electric field intensity that is in the opposite direction of the field produced by the free charge Q alone. For the conducting shell, negative induced charges and positive induced charges have the same amount as the point charge Q, which cancels out the $\boldsymbol{E}$ field inside the shell. For the dielectric shell, the amount of the induced polarization surface charges is less than Q, which reduces(but cannot cancel out) the $\boldsymbol{E}$ field in the shell.

2.8 Boundary Conditions for Electrostatic Fields (静电场的边界条件)

When electromagnetic problems involve multiple media with different electric properties (for example, different dielectric constants), the $\boldsymbol{E}$ and $\boldsymbol{D}$ vectors may exhibit discontinuity across the interfaces, and there may be surface charges induced on the interfaces. The relations between the fields on different sides of the interfaces are called boundary conditions. The boundary conditions cannot be derived by using differential equations due to the discontinuities at the boundaries. Nevertheless, integral equations(2.44) and(2.92) can be used to find the boundary conditions. We already derived the boundary conditions (2.73) and (2.74) at a conductor/free space interface. In this section, we derive the general boundary conditions of electrostatic fields between two general media shown in Figure 2-22. The two media can be conductors or dielectrics with permittivity of ε_1 and ε_2 respectively. On the interface, there may

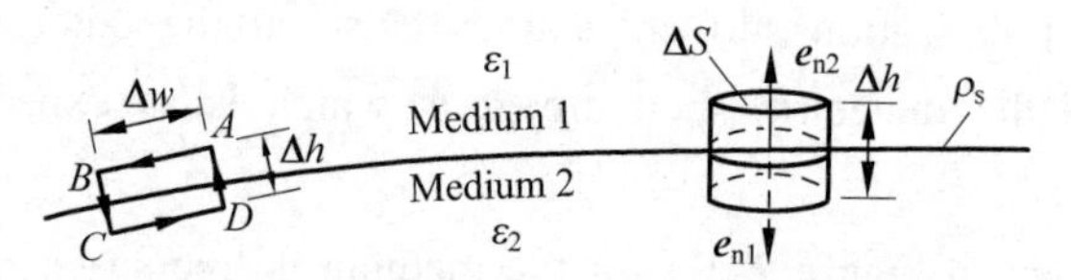

Figure 2-22 An interface between two media

also exist a surface charge distribution ρ_s.

As is illustrated in Figure 2-22, we first construct a small rectangular contour *ABCD* with sides *AB* in medium 1 and *CD* in medium 2, both being parallel to the interface and equal to Δw in length. Let the vector pointing from *A* to *B* be $\Delta \boldsymbol{w}$. Then the vector pointing from *C* to *D* is $(-\Delta \boldsymbol{w})$. Let sides $BC = DA = \Delta h$ approach zero, and apply (2.44) to the contour *ABCD*. Then we have

$$\oint_{ABCDA} \boldsymbol{E} \cdot \mathrm{d}\boldsymbol{l} = \boldsymbol{E}_1 \cdot \Delta \boldsymbol{w} + \boldsymbol{E}_2 \cdot (-\Delta \boldsymbol{w}) = E_{\mathrm{AB}} \Delta w - E_{\mathrm{CD}} \Delta w = 0$$

where $\boldsymbol{E}_1$ and $\boldsymbol{E}_2$ are respectively the electric field intensity in medium 1 and medium 2. E_{AB} and E_{CD} are respectively the component of $\boldsymbol{E}_1$ and $\boldsymbol{E}_2$ along the direction $A \to B$. Since $A \to B$ can be taken as any direction parallel to the interface, Therefore,

$$E_{1\mathrm{t}} = E_{2\mathrm{t}} \quad \text{or} \quad \boldsymbol{e}_{\mathrm{n}} \times \boldsymbol{E}_1 = \boldsymbol{e}_{\mathrm{n}} \times \boldsymbol{E}_2 \tag{2.103}$$

where $E_{1\mathrm{t}}$ and $E_{2\mathrm{t}}$ are the tangential component of $\boldsymbol{E}_1$ and $\boldsymbol{E}_2$ respectively. (2.103) gives the tangential boundary conditions of electrostatic fields, which states that the tangential component of an $\boldsymbol{E}$ field is continuous across an interface.① (2.103) simplifies to (2.73) if one of the media is a conductor. By using the constitutive relation (2.94), we also have

$$\frac{D_{1\mathrm{t}}}{\varepsilon_1} = \frac{D_{2\mathrm{t}}}{\varepsilon_2} \tag{2.104}$$

For the boundary conditions of the normal components of the fields, we construct a small box with its top face in medium 1 and bottom face in medium 2, as illustrated in Figure 2-22. The top and bottom faces both have an area ΔS. Let the height of the box Δh approaches zero and apply Gauss's law (2.92) to the box. Then we have

$$\begin{aligned} \oint_S \boldsymbol{D} \cdot \mathrm{d}\boldsymbol{s} &= (\boldsymbol{D}_1 \cdot \boldsymbol{e}_{\mathrm{n2}} + \boldsymbol{D}_2 \cdot \boldsymbol{e}_{\mathrm{n1}}) \Delta S = \boldsymbol{e}_{\mathrm{n2}} \cdot (\boldsymbol{D}_1 - \boldsymbol{D}_2) \Delta S \\ &= \rho_s \Delta S \end{aligned} \tag{2.105}$$

where $\boldsymbol{D}_1$ and $\boldsymbol{D}_2$ are respectively the electric displacement vectors in medium 1 and medium 2, $\boldsymbol{e}_{\mathrm{n1}}$ and $\boldsymbol{e}_{\mathrm{n2}}$ are respectively the outward unit normals from medium 1 and medium 2, and apparently $\boldsymbol{e}_{\mathrm{n2}} = -\boldsymbol{e}_{\mathrm{n1}}$.

From (2.105) we obtain

$$\boldsymbol{e}_{\mathrm{n2}} \cdot (\boldsymbol{D}_1 - \boldsymbol{D}_2) = \rho_s \quad \text{or} \quad D_{1\mathrm{n}} - D_{2\mathrm{n}} = \rho_s \tag{2.106}$$

where $D_{1\mathrm{n}}$ and $D_{2\mathrm{n}}$ are the tangential component of $\boldsymbol{D}_1$ and $\boldsymbol{D}_2$ respectively. Notice that the subscript n represents the normal direction pointing from medium 2 to medium 1. (2.106) states

① 静电场切向边界条件：任意分界面上电场强度的切向分量连续。

that the normal component of $\boldsymbol{D}$ field is discontinuous across an interface if a surface charge exists on the interface. The amount of discontinuity is equal to the surface charge density. ①

Equations(2.103) and(2.106) are the general forms of the boundary conditions for static electric fields. If medium 2 is a conductor, $\boldsymbol{D}_2=0$ and(2.106) becomes

$$D_{1n}=\varepsilon_1 E_{1n}=\rho_s \tag{2.107}$$

which is the same as(2.74) when medium 1 is free space.

On the interface between two dielectrics, there are usually no free charges. That is, $\rho_s=0$ in (2.106), and hence

$$D_{1n}=D_{2n} \quad \text{or} \quad \varepsilon_1 E_{1n}=\varepsilon_2 E_{2n} \tag{2.108}$$

Example 2-14 As is shown in Figure 2-23, an FR4 sheet ($\varepsilon_r=4.4$) is introduced perpendicularly in a uniform electric field in free space. Assume that the electric field outside the FR4 sheet is measured to be $\boldsymbol{E}_o=\boldsymbol{e}_x E_o$. Determine $\boldsymbol{E}_i$, $\boldsymbol{D}_i$, and $\boldsymbol{P}_i$ inside the sheet.

Solution: FR4 is a dielectric material widely used for printed circuit boards(PCB). On the boundaries between free space and the dielectric sheet, the electric field only has normal component and no free charge exists. According to boundary condition(2.106), we have

$$\boldsymbol{D}_i=\boldsymbol{e}_x D_i=\boldsymbol{e}_x D_o=\boldsymbol{e}_x \varepsilon_0 E_o$$

Then we have

$$\boldsymbol{E}_i=\frac{1}{\varepsilon}\boldsymbol{D}_i=\frac{1}{\varepsilon_0\varepsilon_r}\boldsymbol{e}_x\varepsilon_0 E_o=\boldsymbol{e}_x\frac{E_o}{4.4}=\boldsymbol{e}_x 0.23E_o$$

$$\boldsymbol{P}_i=\boldsymbol{D}_i-\varepsilon_0\boldsymbol{E}_i=\boldsymbol{e}_x\left(\varepsilon_0 E_o-\varepsilon_0\frac{E_o}{4.4}\right)=\boldsymbol{e}_x 0.77\varepsilon_0 E_o$$

Example 2-15 Two dielectric media with permittivities ε_1 and ε_2 are separated by a charge-free boundary as shown in Figure 2-24. At a point on the boundry, the electric field intensity on the medium 1 side is denoted by $\boldsymbol{E}_1$, and the electric field intensity on the medium 2 side is denoted by $\boldsymbol{E}_2$. Assume that the magnitude of $\boldsymbol{E}_1$ is known to be E_1. The angle between the direction of $\boldsymbol{E}_1$ and the normal is known to be α_1. Determine the magnitude and direction of $\boldsymbol{E}_2$ in terms of E_1 and α_1.

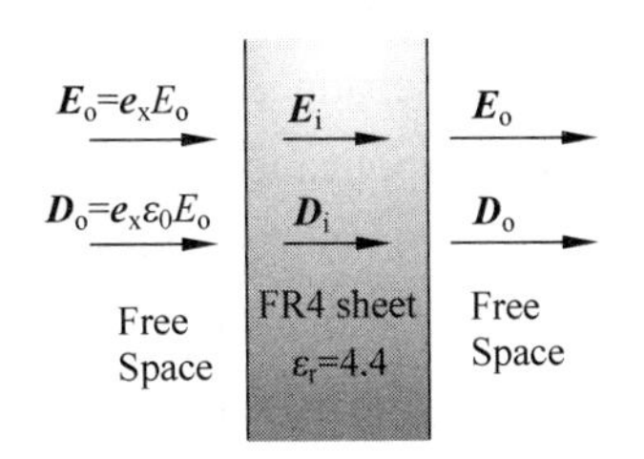

Figure 2-23 An FR4 sheet in a uniform electric field(Example 2-14)

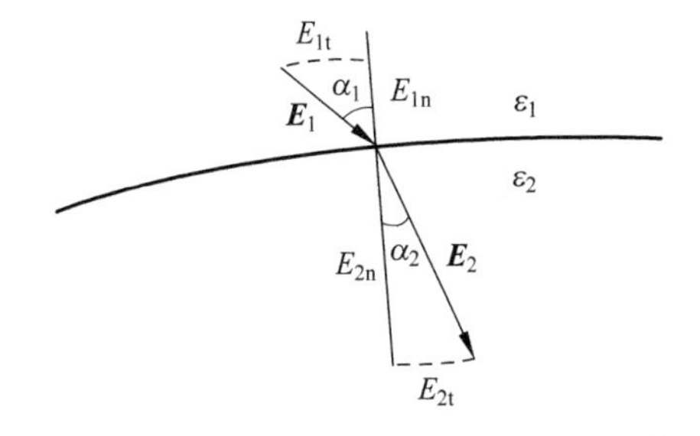

Figure 2-24 Boundary conditions at the interface between two dielectric media

① 静电场法向边界条件：任意分界面上电位移矢量法向分量的突变量等于该分界面上的面电荷密度。

Solution Using(2.103) and(2.108)

$$E_2\sin\alpha_2 = E_1\sin\alpha_1 \quad (2.109)$$

and

$$\varepsilon_2 E_2\cos\alpha_2 = \varepsilon_1 E_1\cos\alpha_1 \quad (2.110)$$

Division of(2.109) by(2.110) gives

$$\frac{\tan\alpha_2}{\tan\alpha_1} = \frac{\varepsilon_2}{\varepsilon_1} \quad (2.111)$$

The magnitude of E_2 is

$$E_2 = \sqrt{E_{2t}^2 + E_{2n}^2} = \sqrt{(E_2\sin\alpha_2)^2 + (E_2\cos\alpha_2)^2}$$

$$= \sqrt{(E_1\sin\alpha_1)^2 + \left(\frac{\varepsilon_1}{\varepsilon_2}E_1\cos\alpha_1\right)^2} = E_1\sqrt{\sin^2\alpha_1 + \left(\frac{\varepsilon_1}{\varepsilon_2}\cos\alpha_1\right)^2}$$

2.9 Capacitance and Capacitors(电容)

From Section 2.5, we know that a conductor in a static electric field is an equipotential body and the charges carried by the conductor are distributed on its surface. For an isolated conductor, suppose no free charges exist except those on the surface of the conductor. Then the constant potential of the conductor can be expressed by(2.59) with S' to be the conductor surface and observation point $\boldsymbol{r}$ to be any location within or on the surface of the conductor. Due to linearity of the source-field relation in(2.59), increasing the surface charge density ρ_s everywhere by a factor k would increase the potential φ of the conductor by the same factor, and obviously would also increase the total charge Q by the same factor. Therefore, we conclude that the potential of an isolated conductor is directly proportional to the total charge on it. The proportional constant, denoted by C, is defined as the **capacitance**(电容) of the isolated conducting body, i.e.,

$$C = \frac{Q}{\varphi}(\mathrm{F}) \quad (2.112)$$

The unit of capacitance is coulomb per volt, or farad(F).

The concept of capacitance is obviously related to one of the basic circuit elements, the **capacitor**(电容器). A capacitor usually consists of two conductors separated by free space or a dielectric medium, which provides a specific capacitance for circuit applications. The conductors may be of arbitrary shapes as illustrated in Figure 2-25. When a **DC**(**direct-current**, **直流**) voltage source with output voltage U is connected between the conductors, the two conductors will be charged respectively with $+Q$ and $-Q$, producing electric fields originating from the positively charged conductor and terminating on the negatively charged conductor. Notice that the field lines are perpendicular to the conductor surfaces, which are equipotential surfaces. In this case, the capacitance is defined as

$$C = \frac{Q}{U} = \frac{Q}{\varphi_+ - \varphi_-} \quad (2.113)$$

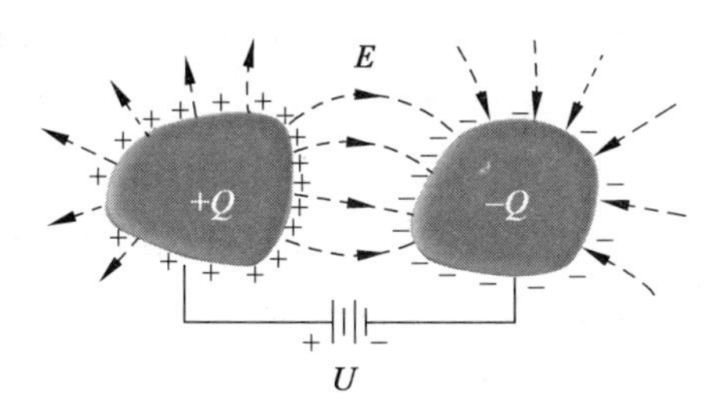

Figure 2-25 A two-conductor capacitor with applied potential difference φ

where φ_+ and φ_- are the potential of the conductors carrying $+Q$ and $-Q$ respectively.

The capacitance of a capacitor is a physical property of the two-conductor system. It depends on the geometry of the conductors and on the permittivity of the medium between them, but it does NOT depend on the actual charge Q or the potential difference U. A capacitor has a capacitance even when no voltage is applied or no free charge exists on its conductors. Nevertheless, capacitance C are usually calculated by using the its definition (2.113). Two methods can be employed to find the capacitance between two isolated conductors. The first method assumes a potential difference U across the two conductors, and then determines the total charge Q in terms of U. The second method assumes a total charge Q carried by the conductors and then determines the potential difference U in terms of Q. The first method usually involves solving boundary-value problems, which will be studied in the next chapter. Here we find C by using the second method in the following steps:

(1) Assume charges $+Q$ and $-Q$ on the conductors.

(2) Find $\boldsymbol{E}$ from Q by Gauss's law, boundary conditions, or other relations.

(3) Find the potential difference U by integrating the $\boldsymbol{E}$ field from the conductor with $-Q$ to $+Q$.

$$U = \varphi_+ - \varphi_- = -\int_-^+ \boldsymbol{E} \cdot \mathrm{d}\boldsymbol{l}$$

(4) Find C by taking the ratio Q/U.

The capacitance of a single isolated conductor can also be found by following the above procedure, but with the conductor carrying $-Q$ selected to be infinity. The potential φ of the conductor is calculated as the integral of the $\boldsymbol{E}$ field from infinity to any point on the surface of the conductor. And then, (2.112) can be used to determine the capacitance.

Example 2-16 Neglecting **fringing effect** (边缘效应), determine the capacitance of a parallel-plate capacitor filled with a dielectric of ε. As shown in Figure 2-26, the plates have an area of S and are separated by a distance d.

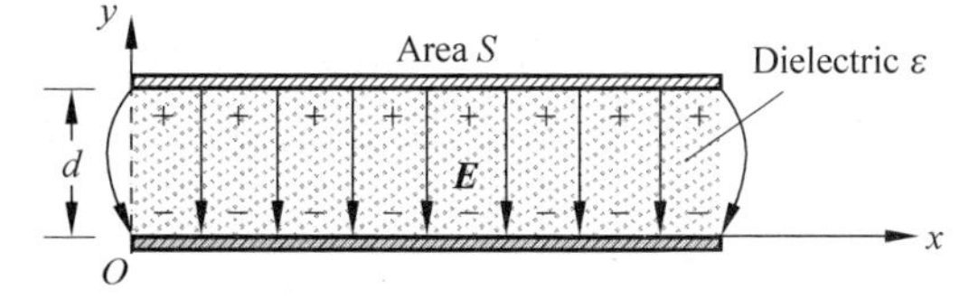

Figure 2-26 Parallel-plate capacitor (Example 2-16)

Solution: Assume the upper and lower conducting plates carry $+Q$ and $-Q$ charges respectively. Neglecting fringing effect at the edges of the plates means that the fields in the capacitor is the same as if the area S is infinite. This also means that the charges are assumed to be uniformly distributed, and the $\boldsymbol{E}$ field is constant everywhere between the two plates. Then, the surface densities on the upper and lower plates are uniform $+\rho_s$ and $-\rho_s$ where

$$\rho_s = \frac{Q}{S}$$

From(2.107), we have

$$\boldsymbol{E} = -\boldsymbol{e}_z \frac{\rho_s}{\varepsilon} = -\boldsymbol{e}_z \frac{Q}{\varepsilon S}$$

which is constant within the capacitor.

$$U = -\int_{z=0}^{z=d} \boldsymbol{E} \cdot d\boldsymbol{l} = -\int_0^d \left(-\boldsymbol{e}_z \frac{Q}{\varepsilon S}\right) \cdot (\boldsymbol{e}_z dz) = \frac{Q}{\varepsilon S} d$$

Therefore, for a parallel-plate capacitor,

$$C = \frac{Q}{U} = \varepsilon \frac{S}{d} \tag{2.114}$$

which is obviously independent of Q or U.

For this problem we could also start by assuming a potential difference U between the upper and lower plates. The electric field intensity between the plates is uniform and equals

$$\boldsymbol{E} = -\boldsymbol{e}_z \frac{U}{d}$$

which is a result of the solution of a boundary value problem (see Example 3-1). The surface charge densities at the upper and lower conducting plates are $+\rho_s$ and $-\rho_s$ respectively, with

$$\rho_s = \varepsilon E_z = \varepsilon \frac{U}{d}$$

Therefore, $Q = \rho_s S = (\varepsilon S/d) U$, and $C = Q/U = \varepsilon S/d$.

Example 2-17 A cylindrical capacitor consists of an inner conductor of radius a and an outer conductor whose inner radius is b. The space between the conductors is filled with a dielectric of permittivity ε, and the length of the capacitor is L. Determine the capacitance of this capacitor.

Solution: We use cylindrical coordinates for this problem. Again, we neglect the fringing effect of the field near the edges of the conductors. First, assume charges $+Q$ and $-Q$ on the surface of the inner conductor and the inner surface of the outer conductor, respectively. The $\boldsymbol{E}$ field in the dielectric can be obtained by applying Gauss's law to a cylindrical Gaussian surface within the dielectric $a<\rho<b$. Referring to Figure 2-27 and applying Gauss's law, we have

$$\boldsymbol{E} = \boldsymbol{e}_\rho E_\rho = \boldsymbol{e}_\rho \frac{Q}{2\pi\varepsilon L \rho} \tag{2.115}$$

The potential difference between the inner and outer conductors is

$$U = -\int_{\rho=b}^{\rho=a} \boldsymbol{E} \cdot \mathrm{d}\boldsymbol{l} = -\int_b^a \left(\boldsymbol{e}_\rho \frac{Q}{2\pi\varepsilon L\rho}\right) \cdot (\boldsymbol{e}_\rho \mathrm{d}\rho) = \frac{Q}{2\pi\varepsilon L}\ln\left(\frac{b}{a}\right) \tag{2.116}$$

Therefore, for a cylindrical capacitor,

$$C = \frac{Q}{U} = \frac{2\pi\varepsilon L}{\ln\left(\frac{b}{a}\right)} \tag{2.117}$$

This problem can also be solved by assuming a potential difference U between the inner and outer conductors. Then we obtain the potential distribution between the two conductors by solving a boundary value problem, which will be demonstrated in Example 3-2.

Example 2-18 As is shown in Figure 2-28, a spherical capacitor consists of an inner conducting sphere of radius a and an outer conductor with a spherical inner surface of radius b. The space in between is filled with a dielectric of permittivity ε. Determine the capacitance.

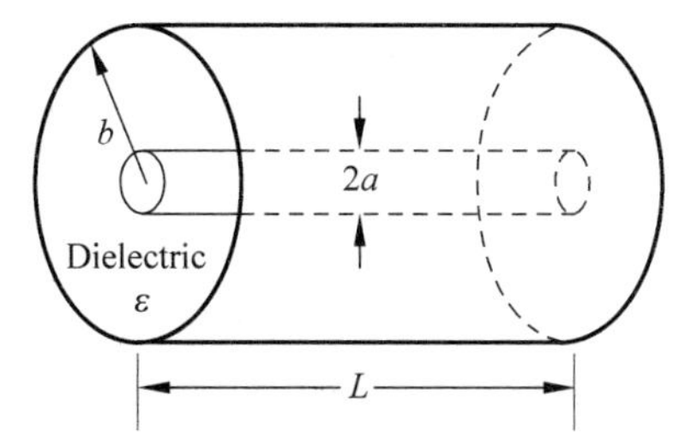

Figure 2-27 Cylindrical capacitor (Example 2-17)

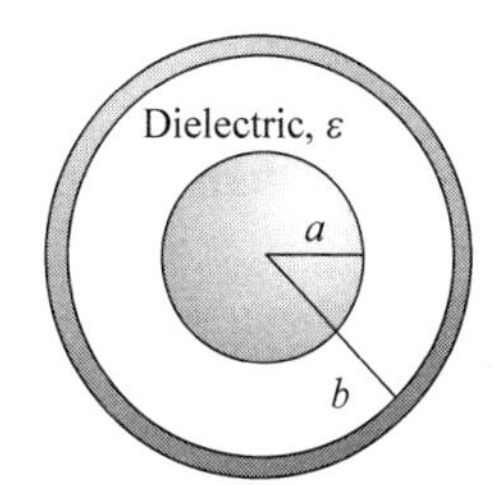

Figure 2-28 Spherical capacitor (Example 2-18)

Solution: Assume the inner and outer conductors of the spherical capacitor are respectively charged with $+Q$ and $-Q$. Applying Gauss's law to a spherical Gaussian surface with radius $r(a<r<b)$, we have

$$\boldsymbol{E} = \boldsymbol{e}_r E_r = \boldsymbol{e}_r \frac{Q}{4\pi\varepsilon r^2}$$

then

$$U = -\int_b^a \boldsymbol{E} \cdot (\boldsymbol{e}_r \mathrm{d}r) = -\int_b^a \frac{Q}{4\pi\varepsilon r^2}\mathrm{d}r = \frac{Q}{4\pi\varepsilon}\left(\frac{1}{a} - \frac{1}{b}\right)$$

Therefore, the capacitance of a spherical capacitor is

$$C = \frac{Q}{U} = \frac{4\pi\varepsilon}{\frac{1}{a} - \frac{1}{b}} = \frac{4\pi\varepsilon ab}{b - a} \tag{2.118}$$

For an isolated conducting sphere of a radius a, the capacitance can be directly obtained from (2.118) by letting $b \to \infty$, that is,

$$C = 4\pi\varepsilon a \tag{2.119}$$

Again, this problem can be solved by assuming a potential difference U between the inner and outer conductors, as will be shown in Example 3-3.

2.10 Electrostatic Energy(静电能)

2.10.1 Electrostatic Energy in Terms of Charge and Potential (电荷与电位表示的静电能)

It has been discussed in Section 2.4 that, the physical significance of the electric potential at a point is the work required to bring a unit positive charge to the point from the reference zero-potential point. Therefore, work is needed for bringing charges together to form a charge distribution. The work is done by external forces and eventually converted to **electrostatic energy**(静电能) when a static charge distribution is formed.

To derive the formula for calculating the stored energy from a given charge distribution, we first consider the work needed for bringing two point charges Q_1 and Q_2 together. In this case, the reference zero-potential point is infinity. Let Q_1 be the first charge brought from infinity. Assume that Q_1 is moved slow enough so that the kinetic energy and radiation effects can be neglected. Then no work is needed because Q_1 experiences no electric forces. Now when the second charge Q_2 is brought from infinity, it experiences an electrostatic force due to the field produced by Q_1, and work is required to overcome the electrostatic force. If the eventual distance between Q_1 and Q_2 is R_{12}, the amount of the required work is

$$W_2 = Q_2\varphi_2 = Q_2\frac{Q_1}{4\pi\varepsilon_0 R_{12}} \tag{2.120}$$

where φ_2 is the potential at the final location of Q_2 due to the charge Q_1. This work is stored as the **potential energy**(势能). Notice that(2.120) can also be written as

$$W_2 = Q_1\frac{Q_2}{4\pi\varepsilon_0 R_{12}} = Q_1\varphi_1 \tag{2.121}$$

where φ_1 is the potential at the location of Q_1 due to the charge Q_2. Combining (2.120) and(2.121), we have

$$W_2 = \frac{1}{2}(Q_1\varphi_1 + Q_2\varphi_2) \tag{2.122}$$

Now suppose another charge Q_3 is brought from infinity to a point that is R_{13} from Q_1 and R_{23} from Q_2. Then an additional amount of work ΔW is done during the process, which is calculated as

$$\Delta W = Q_3\varphi_3 = Q_3\left(\frac{Q_1}{4\pi\varepsilon_0 R_{13}} + \frac{Q_2}{4\pi\varepsilon_0 R_{23}}\right) \tag{2.123}$$

The sum of ΔW and W_2 is the potential energy, W_3, stored in the assembly of the three charges Q_1, Q_2 and Q_3. That is,

$$W_3 = W_2 + \Delta W = \frac{1}{4\pi\varepsilon_0}\left(\frac{Q_1Q_2}{R_{12}} + \frac{Q_1Q_3}{R_{13}} + \frac{Q_2Q_3}{R_{23}}\right) \tag{2.124}$$

We can rewrite W_3 in the following form:

$$W_3 = \frac{1}{2}\left[Q_1\left(\frac{Q_2}{4\pi\varepsilon_0 R_{12}} + \frac{Q_3}{4\pi\varepsilon_0 R_{13}}\right) + Q_2\left(\frac{Q_1}{4\pi\varepsilon_0 R_{12}} + \frac{Q_3}{4\pi\varepsilon_0 R_{23}}\right) + Q_3\left(\frac{Q_1}{4\pi\varepsilon_0 R_{13}} + \frac{Q_2}{4\pi\varepsilon_0 R_{23}}\right)\right]$$

$$= \frac{1}{2}(Q_1\varphi_1 + Q_2\varphi_2 + Q_3\varphi_3) \tag{2.125}$$

In(2.125), φ_1, the potential at the location of Q_1, is due to the charges Q_2 and Q_3. Similarly, φ_2 is the potential at the location of Q_2 due to Q_1 and Q_3, and φ_3 is the potential at the location of Q_3 due to Q_1 and Q_2. If the above procedure is continued with more charges being brought in, we arrive at the following general expression for the potential energy W_e of a group of N discrete point charges:

$$W_e = \frac{1}{2}\sum_{k=1}^{N} Q_k\varphi_k \tag{2.126}$$

where φ_k, is the electric potential at the location of Q_k due to all the other charges, i. e.,

$$\varphi_k = \frac{1}{4\pi\varepsilon_0}\sum_{\substack{j=1 \\ (j\neq k)}}^{N} \frac{Q_j}{R_{jk}} \tag{2.127}$$

Notice that, W_e in(2.126) represents only the **interaction energy**(互作用能) or **mutual energy**(互能). It does not include the work required to assemble the individual point charges themselves (self-energy). Assembly of discrete point charges may need positive work or negative work, therefore, W_e can be either positive or negative. For instance, W_2 in(2.121) will be negative if Q_1 and Q_2 are of opposite signs. In that case, the electric field does positive work whereas the external force does negative work while moving Q_2 from infinity.

For a continuous charge distribution of density ρ_v, the formula for W_e can be derived by treating the charge in a differential volume $\rho_v dv$ as a point charge, and hence(2.126) becomes

$$W_e = \frac{1}{2}\int_{V'} \rho_v \varphi dv \tag{2.128}$$

where φ is the potential at the point where the volume charge density is ρ_v, and V' represents the region where ρ_v exists.

Example 2-19 Find the electric energy of a uniform charge distribution ρ_v in a sphere with radius a.

Solution: Due to the spherical symmetry of the problem (notice that ρ_v is a constant), we have

$$W_e = \frac{\rho_v}{2}\int_{V'} \varphi dv = \frac{\rho_v}{2}\int_0^a \varphi 4\pi r^2 dr \tag{2.129}$$

To find the potential φ, we first find the $\boldsymbol{E}$ distribution using Gauss's law, which gives

$$\boldsymbol{E} = \begin{cases} \boldsymbol{e}_r \dfrac{Q\big|_{r>a}}{4\pi\varepsilon_0 r^2} = \boldsymbol{e}_r \dfrac{\rho_v a^3}{3\varepsilon_0 r^2} & r > a \\[2ex] \boldsymbol{e}_r \dfrac{Q\big|_{0<r\leqslant a}}{4\pi\varepsilon_0 r^2} = \boldsymbol{e}_r \dfrac{\rho_v r}{3\varepsilon_0} & 0 < r \leqslant a \end{cases}$$

Consequently, we obtain the potential φ for any point inside the sphere($r \leqslant a$) to be

$$\varphi(r)=-\int_{\infty}^{r} \boldsymbol{E} \cdot \mathrm{d}\boldsymbol{l}=-\left[\int_{\infty}^{a} \frac{\rho_{\mathrm{v}} a^{3}}{3 \varepsilon_{0} r^{2}} \mathrm{d}r+\int_{a}^{r} \frac{\rho_{\mathrm{v}} r}{3 \varepsilon_{0}} \mathrm{d}r\right]=\frac{\rho_{\mathrm{v}}}{3 \varepsilon_{0}}\left(a^{2}+\frac{a^{2}}{2}-\frac{r^{2}}{2}\right)=\frac{\rho_{\mathrm{v}}}{6 \varepsilon_{0}}(3 a^{2}-r^{2}) \tag{2.130}$$

Substituting(2.130) in(2.129), we have

$$W_{\mathrm{e}}=\frac{\rho_{\mathrm{v}}}{2} \int_{0}^{a} \frac{\rho_{\mathrm{v}}}{6 \varepsilon_{0}}(3 a^{2}-r^{2}) 4 \pi r^{2} \mathrm{d}r=\frac{4 \pi \rho_{\mathrm{v}}^{2} a^{5}}{15 \varepsilon_{0}} \tag{2.131}$$

In terms of the total charge $Q=4\pi a^3\rho_{\mathrm{v}}/3$, (2.131) can be rewritten as

$$W_{\mathrm{e}}=\frac{3 Q^{2}}{20 \pi \varepsilon_{0} a} \tag{2.132}$$

Notice that in (2.128), the potential φ is contributed by all the charges including the differential charge $\rho_{\mathrm{v}}\mathrm{d}v$ at the observation point of φ. That means W_{e} in (2.128) includes the self-energy required to assemble a distribution of volume charge, which is different from (2.126) that does not include the self-energies of point charges. In fact, the self-energy of a point charge Q is infinite, which can be seen by letting the radius a in(2.132) approach zero. Of course, there are no point charges in reality. In reality, for a volume charge distribution, every point has a finite charge density ρ_{v}. The self-energy of the charge in a differential volume $\rho_{\mathrm{v}}\mathrm{d}v$ is zero, as is seen by letting the radius a in(2.131) approach zero.

For a two-conductor capacitor, we can prove from(2.128) that the static energy stored in the capacitor is

$$W_{\mathrm{e}}=\frac{1}{2} Q U \tag{2.133a}$$

where U is the potential difference between the two conductors and Q is the amount of charges carried by the capacitor (one conductor carries $+Q$ and the other carries $-Q$). The proof of (2.133a) is left as an exercise(Problem 2.29). Since $Q=CU$, (2.133a) can be rewritten in two other forms:

$$W_{\mathrm{e}}=\frac{1}{2} C U^{2} \tag{2.133b}$$

and

$$W_{\mathrm{e}}=\frac{Q^{2}}{2 C} \tag{2.133c}$$

(2.133) can be used for finding the capacitance of two-conductor capacitors. Examples are shown in the next subsection.

2.10.2 Electrostatic Energy in Terms of Electric Field Quantities(电场表示的静电能)

The expression of electrostatic energy(2.128) is in terms of the source charge density and the electric potential, which does not involve electric field quantities. Here, we show that the electrostatic energy can also be expressed in terms of field quantities $\boldsymbol{E}$ and/or $\boldsymbol{D}$, which is equivalent to(2.128). To do that, we first substitute ρ_v in(2.128) with $\nabla \cdot \boldsymbol{D}$, which leads to

$$W_e = \frac{1}{2}\int_{V'} (\nabla \cdot \boldsymbol{D})\varphi \mathrm{d}v \tag{2.134}$$

From the vector identity(1.148), $\nabla \cdot (\varphi \boldsymbol{D}) = \varphi \nabla \cdot \boldsymbol{D} + \boldsymbol{D} \cdot \nabla\varphi$. Hence(2.134) can be written as

$$\begin{aligned} W_e &= \frac{1}{2}\int_{V'} \nabla \cdot (\varphi \boldsymbol{D})\mathrm{d}v - \frac{1}{2}\int_{V'} \boldsymbol{D} \cdot \nabla\varphi \mathrm{d}v \\ &= \frac{1}{2}\oint_{S'} \varphi \boldsymbol{D} \cdot \boldsymbol{e}_n \mathrm{d}s + \frac{1}{2}\int_{V'} \boldsymbol{D} \cdot \boldsymbol{E}\mathrm{d}v \end{aligned} \tag{2.135}$$

where the divergence theorem is used to convert the first volume integral into a closed surface integral, and $\nabla\varphi$ is substituted by $-\boldsymbol{E}$ in the second volume integral. Since V' can be any volume that includes all the charges, we may choose it to be a sphere large enough to contain all the charges. Let the radius of the sphere $r \to \infty$, the electric potential φ and the magnitude of the electric displacement $\boldsymbol{D}$ decreases at least as fast as $1/r$ and $1/r^2$, respectively. The area of the bounding surface S' increases as r^2. Hence the surface integral in(2.135) decreases at least as fast as $1/r$ and approaches zero as $r \to \infty$. Then(2.135) is reduced to

$$W_e = \frac{1}{2}\int_{V_\infty} \boldsymbol{D} \cdot \boldsymbol{E}\mathrm{d}v \tag{2.136a}$$

where V_∞ indicates that the limit of integration is the entire space(or at least covers all the region in which the $\boldsymbol{E}$ and $\boldsymbol{D}$ field is nonzero). Using the relation $\boldsymbol{D} = \varepsilon \boldsymbol{E}$ for a linear medium, (2.136a) can be written in two other forms:

$$W_e = \frac{1}{2}\int_{V_\infty} \varepsilon E^2 \mathrm{d}v \tag{2.136b}$$

$$W_e = \frac{1}{2}\int_{V_\infty} \frac{D^2}{\varepsilon}\mathrm{d}v \tag{2.136c}$$

Now we can define an **electrostatic energy density**(**静电能量密度**) w_e mathematically, such that its volume integral equals the total electrostatic energy:

$$W_e = \int w_e \mathrm{d}v \tag{2.137}$$

We can therefore write

$$w_e = \frac{1}{2}\boldsymbol{D} \cdot \boldsymbol{E} = \frac{1}{2}\varepsilon E^2 = \frac{D^2}{2\varepsilon} \tag{2.138}$$

Example 2-20 Use energy formulas(2.138) and(2.133) to find the capacitance of a cylindrical capacitor having a length L, an inner conductor of radius a, an outer conductor of inner radius b, and a dielectric of permittivity ε, as shown in Figure 2-27.

Solution: By applying Gauss's law, we know that

$$\boldsymbol{E} = \boldsymbol{e}_\rho E_\rho = \boldsymbol{e}_\rho \frac{Q}{2\pi\varepsilon L\rho}, \quad a < \rho < b$$

The electrostatic energy stored in the dielectric region is, from(2.138),

$$W_e = \frac{1}{2}\int_a^b \varepsilon\left(\frac{Q}{2\pi\varepsilon L\rho}\right)^2 (L2\pi\rho \mathrm{d}\rho) = \frac{Q^2}{4\pi\varepsilon L}\int_a^b \frac{\mathrm{d}\rho}{\rho} = \frac{Q^2}{4\pi\varepsilon L}\ln\frac{b}{a} \tag{2.139}$$

On the other hand, W_e can also be expressed in the form of(2. 133c). Compare(2. 133c) and(2. 139), we obtain

$$\frac{Q^2}{2C}=\frac{Q^2}{4\pi\varepsilon L}\ln\frac{b}{a}$$

or

$$C=\frac{2\pi\varepsilon L}{\ln\frac{b}{a}}$$

which is the same as that given in(2. 117).

Summary

Concepts

Electric field intensity(电场强度) Electric potential(电位)

Electric flux density/electric displacement(电通密度/电位移)

Electric dipole(电偶极子) Conductor(导体)

Dielectric(介质) Polarization(极化)

Polarization volume charge(极化体电荷)

Polarization surface charge(极化面电荷) Boundary condition(边界条件)

Capacitance(电容)

Electrostatic energy/Electrostatic energy density(静电能量/静电能量密度)

Laws & Theorems

Gauss's law(高斯定律)

Curl-free property of the electrostatic field(静电场的无旋性)

Field and potential distribution of typical charges (point charge, dipole, infinitely long straight line charge, etc.)(典型电荷源产生的电场和电位分布(点电荷、电偶极子、无限长直线电荷等))

Boundary conditions of electrostatic fields among two general media, two dielectric media, conductor and dielectric(静电场边界条件：两种一般媒质分界面；两种介质的分界面上；导体与介质分界面)

Methods

Find $\boldsymbol{E}$ field distribution given charge distributions using direct integration/superposition;

Find $\boldsymbol{E}$ field distribution given charge distributions using Gauss's law;

Find $\boldsymbol{E}$ field distribution given charge distributions using boundary conditions;

Find potential distributions given electric field distributions, and vice versa;

Find capacitance of typical capacitors(parallel-plates, cylindrical, spherical).

Review Questions

2.1 静电场的源是什么? 静电场的源是通量源还是漩涡源?

2.2 写出微分形式和积分形式的高斯定律,并阐明其物理意义。

2.3 高斯定律适用于求解什么条件下的静电场问题?

2.4 以下电荷源产生的电场强度随距离增加呈现何种变化? ①点电荷; ②电偶极子; ③无限长均匀直线线电荷; ④无限大均匀平面面电荷。

2.5 电位的定义是基于静电场的什么性质? 如何唯一确定电位分布?

2.6 画出以下两种电荷源产生的电力线和等位线: ①点电荷; ②电偶极子。

2.7 为什么说静电场的电场线是不可能闭合的,而且也不可能相交?

2.8 定义极化强度矢量,其单位是什么?

2.9 解释等效极化面电荷和等效极化面电荷的物理意义。

2.10 何为简单媒质? 简单媒质的介电常数有什么特征?

2.11 定义电极化率,其单位是什么?

2.12 定义电位移矢量,其单位是什么?

2.13 静电场在两种不同媒质分界面上的边界条件是什么?

2.14 静电场在导体和介质的分界面上的边界条件是什么?

2.15 静电场在两种不同介质分界面上的边界条件是什么?

2.16 给定静电场的电位分布,如何计算导体和介质分界面上的面电荷密度?

2.17 如何定义孤立导体的电容?

2.18 写出离散静止点电荷的互能表达式。

2.19 写出连续分布体静止体电荷的电场储能表达式。

2.20 写出以 $\boldsymbol{E}$ 和 $\boldsymbol{D}$ 表示的静电场储能表达式。

Problems

2.1 Two point charges, 2 C and −4 C, are located at (1,1,0) and (0,0,1), respectively.

(1) Determine the electric field intensity at (2,5,5).

(2) Find the location at which the electric field is zero.

2.2 An infinitely large plate with uniform surface charge density sits on the $x-y$ plane. Prove that, half of the electric field at a position with z coordinate to be z_0 is contributed by the charges within the circle right below the position and with a radius $\sqrt{3}z_0$.

2.3 Determine the charge density in each of the following electric fields.

(1) $\boldsymbol{E}=\boldsymbol{e}_x 8xy\varepsilon_0+\boldsymbol{e}_y 4x^2\varepsilon_0$

(2) $\boldsymbol{E}=\boldsymbol{e}_\rho \rho\sin\phi\varepsilon_0+\boldsymbol{e}_\phi 2\rho\cos\phi\varepsilon_0+\boldsymbol{e}_z 2z^2\varepsilon_0$

(3) $\boldsymbol{E}=\boldsymbol{e}_r \dfrac{2\cos\theta}{r^3}\varepsilon_0+\boldsymbol{e}_\theta \dfrac{\sin\theta}{r^3}\varepsilon_0$

2.4 Given an electric field intensity $\boldsymbol{E}=\boldsymbol{e}_x 100x$ (V/m) in free space, find the total electric charge contained inside

(1) a cube centered at the origin with side length of 100 mm;

(2) a cylinder centered at the origin with a radius of 50 mm and a height along the z-direction of 100 mm.

2.5 In free space, there exists a volume charge distribution with density

$$\rho_v = \begin{cases} \dfrac{10}{r^2}(\text{mC/m}^3) & 1 < r < 4 \\ 0 & \text{otherwise} \end{cases}$$

(1) Determine the $\boldsymbol{E}$ field everywhere.

(2) What is the total flux of the $\boldsymbol{E}$ field passing through a spherical surface with $r=3$?

2.6 A spherical cavity made of free space resides inside a uniformly charged sphere with charge density ρ_{v0}. As shown in Figure 2-29, the charged sphere is centered at the origin, and the position vector of the spherical cavity center is $\boldsymbol{r}_c$. Calculate the electric field inside the cavity.

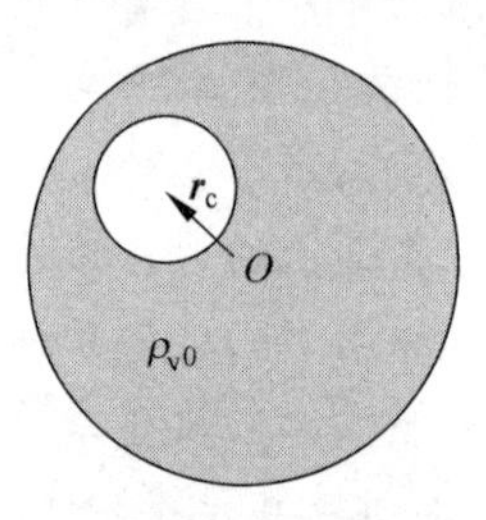

Figure 2-29 A small spherical cavity inside a uniformly charged sphere (Problem 2.6)

2.7 Two surface charges ρ_{sa} and ρ_{sb} are uniformly distributed on the infinitely long coaxial cylindrical surfaces, $\rho = a$ and $\rho = b$ $(b > a)$ respectively. Determine the electric field $\boldsymbol{E}$ everywhere.

2.8 A very long coaxial cable consists of an inner conductor with radius a and an outer conducting shell with its inner and outer radii of b and c respectively. The region between the two conductors carries a volume charge density of k/ρ, where k is a constant. Determine the electric field intensity and the electric flux density at any point in space when the outer conductor is ① grounded, and ② not grounded.

2.9 A discrete charge system with three point charges $+q$, $-2q$ and $+q$ located on the z-axis at $z=d/2$, $z=0$ and $z=-d/2$ respectively is called an electrostatic **quadrupole(四极子)**.

(1) Determine φ and $\boldsymbol{E}$ in the far zone.

(2) Find the equations for equipotential surfaces and streamlines.

2.10 Derive the expression of electric potential due to a continuous charge distribution (2.58) by using Helmholtz's theorem.

2.11 A volume charge distribution $\rho_v = \rho_0[1-(r^2/a^2)]$ exists in the region $0 \leqslant r \leqslant a$. The charge is surrounded by a concentric conductor shell with inner radius a and outer radius b. Determine the electric field $\boldsymbol{E}$ and the potential field φ everywhere.

2.12 The polarization vector in a dielectric cube is $\boldsymbol{P}=P_0(\boldsymbol{e}_x x+\boldsymbol{e}_y y+\boldsymbol{e}_z z)$. The dielectric cube is centered at the origin and its edge length is L.

(1) Determine the surface and volume polarization charge densities.

(2) Show that the total polarization charge is zero.

2.13 A small spherical cavity made of free space resides inside an infinitely large dielectric with constant polarization vector $\boldsymbol{P}$. Find the electric field $\boldsymbol{E}$ at the center of the cavity. (Hint: No polarization volume charge exists, and therefore, the $\boldsymbol{E}$ field is only due to the polarization surface charge.)

2.14 The $z=0$ plane is the interface between two dielectric regions with $\varepsilon_{r1}=2$ and $\varepsilon_{r2}=3$. If we know that $\boldsymbol{E}_1$ in region 1 is $\boldsymbol{e}_x 2y-\boldsymbol{e}_y 3x+\boldsymbol{e}_z(5+z)$.

(1) What do we know about $\boldsymbol{E}_2$ and $\boldsymbol{D}_2$ in region 2?

(2) What do we know about the surface charge density on the interface?

2.15 Derive the boundary conditions for the tangential and the normal components of $\boldsymbol{P}$ on the interface between two dielectrics with dielectric constants ε_{r1} and ε_{r2} respectively.

2.16 Derive the boundary conditions for the electric potential on the interface between two dielectrics with dielectric constants ε_{r1} and ε_{r2} respectively.

2.17 The dielectric material inside a parallel-plate capacitor has a nonuniform permittivity that varies linearly from ε_1 at the bottom plate (in the $z=0$ plane) to ε_2 at the top plate (in the $z=d$ plane). Assume the plates have surface area of S and the fringing effect can be neglected, find the capacitance.

2.18 Assume that the outer conductor of the cylindrical capacitor in Example 2-17 is grounded and that the inner conductor has a potential U_0.

(1) Find the electric field intensity, $E(a)$, at the surface of the inner conductor.

(2) With the radius b fixed, find the optimum a so that $E(a)$ is minimized.

(3) Find this minimum $E(a)$.

(4) Determine the capacitance under the conditions of part (b).

2.19 A parallel-plate capacitor is filled with three dielectric media as shown in Figure 2-30. Calculate its capacitance.

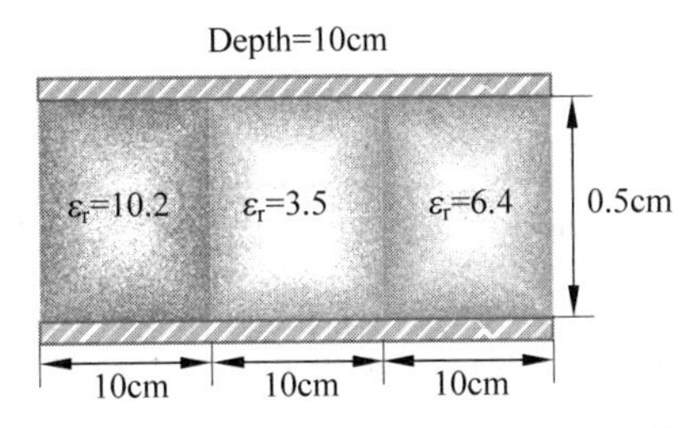

Figure 2-30 A parallel-plate capacitor filled with three dielectric media (Problem 2.19)

2.20 A parallel-plate capacitor is filled with a two-layer dielectric medium as shown in Figure 2-31. Calculate its capacitance.

2.21 A cylindrical capacitor of length L consists of an inner conductor of radius a and an outer conductor whose inner radius is b. The space between the two conductors is filled with two coaxial layers of dielectrics. The dielectric constant of the layer for $a<\rho<d$ is ε_{r1} and the dielectric constant of the layer for $d<\rho<b$ is ε_{r2}. Determine the capacitance of this capacitor.

2.22 A cylindrical capacitor of length L consists of coaxial conducting surfaces of radii a and b respectively. Two dielectric media of different dielectric constants ε_{r1} and ε_{r2} fill the space between the conducting surfaces as shown in Figure 2-32. Determine its capacitance.

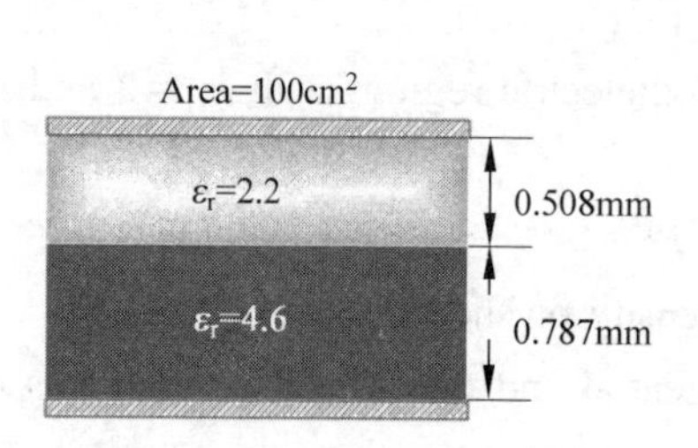

Figure 2-31 A parallel-plate capacitor filled with a two-layer dielectric medium(Problem 2.20)

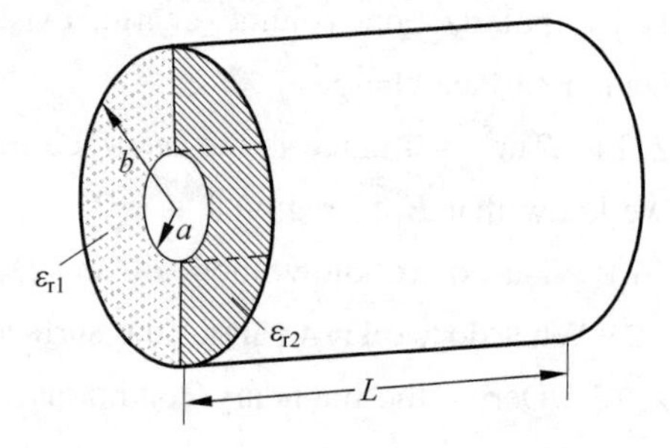

Figure 2-32 A cylindrical capacitor with two dielectric media(Problem 2.22)

2.23 A capacitor consists of two coaxial metallic cylindrical surfaces of a length 30mm and radii 5mm and 7mm respectively. The dielectric material between the surfaces has a relative permittivity $\varepsilon_r = 2 + (4/\rho)$, where ρ is the cylindrical coordinate in mm. Determine the capacitance of the capacitor.

2.24 A conductor sphere with radius a is coated with a dielectric layer with dielectric constant ε_r and thickness h. Determine its capacitance.

2.25 A spherical capacitor consists of an inner conductor of radius a and an outer conductor of radius b. The space between them is filled with two concentric dielectric layers. The dielectric constant of the layer for $a<\rho<d$ is ε_r, and the dielectric constant of the layer for $d<\rho<b$ is $2\varepsilon_r$.

(1) Determine $\boldsymbol{E}$ and $\boldsymbol{D}$ inside the capacitor in terms of an applied voltage U.

(2) Determine the capacitance.

2.26 For a uniform sphere of charge with radius a and volume charge density ρ_v, find the electrostatic energy stored in the following regions: ① inside the sphere, ② outside the sphere.

2.27 According to Einstein's theory of relativity, the work required to assemble a charge (the self-energy) is stored as energy given by the famous formula $E=mc^2$, where m is the mass and c is the velocity of light. For an electron with mass of 9.1×10^{-31}kg and charge of -1.6×10^{-19}C, find its radius.

2.28 Find the electrostatic energy of an electric dipole of moment $\boldsymbol{p}$ stored in the region $r>a$ in the far zone.

2.29 Prove that equation (2.133a) for stored electrostatic energy is true for any two-conductor capacitor.

Chapter 3 Solution of Electrostatic Boundary Value Problems (静电场边界值问题求解)

3.1 Introduction(引言)

Electrostatic problems are those to find electric potential and/or electric field intensity due to static electric charges. In Chapter 2, several methods have been developed to find the electric potential and the electric field intensity when the charge distribution is known. In practical problems, however, the exact charge distribution is usually unknown, and as a result, the formulas in Chapter 2 cannot be applied directly. Instead, practical electrostatic problems might involve conducting bodies with given potentials, which can be modeled as a boundary-value problem in terms of the electric potential. In these cases, the electric fields can be found by solving a partial differential equation subject to the known boundary conditions on the surfaces of conducting bodies. Analytical solutions of the partial differential equation may be obtained if the electrostatic problem can be reduced to one-dimensional. For two-dimensional or three-dimensional problems, analytical solutions generally do not exist. Nevertheless, if the boundaries are of certain simple geometries, the method of images or the method of separation of variables can be used to provide analytical or semi-analytical solutions. ①

3.2 Poisson's and Laplace's Equations (泊松方程、拉普拉斯方程)

In Chapter 2, two fundamental equations governing the electrostatic fields are formulated as

$$\nabla \cdot \boldsymbol{D} = \rho_{\mathrm{v}} \tag{3.1}$$

$$\nabla \times \boldsymbol{E} = \boldsymbol{0} \tag{3.2}$$

From (3.2), we introduced the electric potential φ that satisfies

$$\boldsymbol{E} = -\nabla\varphi \tag{3.3}$$

---

① 第2章给出了从已知的电荷分布出发求解电场的几种方法。然而在实际静电场问题中,电荷分布常常是未知的,而问题所在区域的边界上电位分布可能已知,这种情况下可以将静电场问题表述为关于电位的边界值问题,即给定边界条件,通过求解电位满足的二阶偏微分方程得到电位的解,然后对电位取负梯度,得到电场的解。本章从静电场电位满足的泊松/拉普拉斯方程出发,基于唯一性定理,介绍静电场边界值问题的几种典型的求解方法。

In a linear and isotropic medium, $\boldsymbol{D}=\varepsilon\boldsymbol{E}$. Therefore, (3.1) becomes

$$\nabla\cdot(\varepsilon\boldsymbol{E})=\rho_{v} \tag{3.4}$$

Substituting (3.3) into (3.4) leads to

$$\nabla\cdot(\varepsilon\nabla\varphi)=-\rho_{v} \tag{3.5}$$

where the permittivity ε can be a function of position. For a simple medium, ε is a constant and can be taken out of the divergence operation. Then we have

$$\nabla^{2}\varphi=-\frac{\rho_{v}}{\varepsilon} \tag{3.6}$$

where ∇^2 is the Laplacian operator as introduced in Section 1-12. (3.6) is known as **Poisson's equation**(泊松方程). It states that the Laplacian of φ equals $-\rho_v/\varepsilon$ for a simple medium where ρ_v is the volume density of free charges (which may be a function of space coordinates). If the charge distribution ρ_v is known everywhere in the entire free space, the solution of equation (3.6) is known as (2.58), which is rewritten as

$$\varphi(\boldsymbol{r})=\frac{1}{4\pi\varepsilon_{0}}\int_{V'}\frac{\rho_{v}(\boldsymbol{r}')}{|\boldsymbol{r}-\boldsymbol{r}'|}\mathrm{d}v' \tag{3.7}$$

However, in practical problems, the function ρ_v may not be known, or may be too complicated, which makes it difficult to evaluate the integration in (3.7). Then, instead of using (3.7), it is usually more practical to formulate the electrostatic problems as solving the Poisson's equation (3.6) subject to prescribed boundary conditions (e.g., given φ on certain conducting bodies).①

Poisson's equation (3.6) is a second-order partial differential equation, which, in Cartesian coordinates, becomes

$$\frac{\partial^{2}\varphi}{\partial x^{2}}+\frac{\partial^{2}\varphi}{\partial y^{2}}+\frac{\partial^{2}\varphi}{\partial z^{2}}=-\frac{\rho_{v}}{\varepsilon} \tag{3.8}$$

In cylindrical and spherical coordinates, the Poisson's equation becomes, respectively,

$$\frac{1}{\rho}\frac{\partial}{\partial\rho}\left(\rho\frac{\partial\varphi}{\partial\rho}\right)+\frac{1}{\rho^{2}}\frac{\partial^{2}\varphi}{\partial\phi^{2}}+\frac{\partial^{2}\varphi}{\partial z^{2}}=-\frac{\rho_{v}}{\varepsilon} \tag{3.9}$$

and

$$\frac{1}{r^{2}}\frac{\partial}{\partial r}\left(r^{2}\frac{\partial\varphi}{\partial r}\right)+\frac{1}{r^{2}\sin\theta}\frac{\partial}{\partial\theta}\left(\sin\theta\frac{\partial\varphi}{\partial\theta}\right)+\frac{1}{r^{2}\sin^{2}\theta}\frac{\partial^{2}\varphi}{\partial\phi^{2}}=-\frac{\rho_{v}}{\varepsilon} \tag{3.10}$$

At points in a simple medium where there is no free charge, $\rho_v=0$ and the Poisson's equation (3.6) reduces to

$$\nabla^{2}\varphi=0 \tag{3.11}$$

which is known as **Laplace's equation**(**拉普拉斯方程**). Laplace's equation is the governing

① 泊松方程(3.6)是根据静电场特性引入电位后,从静电场满足的基本方程直接推导得到的,也是静电场电位必须满足的基本方程。式(3.7)是泊松方程的解,然而,采用式(3.7)计算电位的前提条件是其中的电荷密度 ρ_v 在整个空间中已知,且积分区域 V' 包含所有的自由电荷。如果不知道电荷分布,无法直接采用式(3.7)得到静电场的解,但可以通过求解满足特定边界条件的泊松方程(3.6)得到静电场的解。

equation for many electrostatic problems involving a set of conductors maintained at given potentials. Once φ is found by solving the Laplace's equation, the electric field can be determined from $-\nabla\varphi$, and the charge distribution on the conductor surfaces can be determined from the boundary condition $\rho_s = \varepsilon E_n$. ①

Example 3-1 As shown in Figure 3-1, the potential difference across a parallel-plate capacitor is maintaned at U_0. The separation between the two plates of the capacitor is d. Assume the fringing effect can be neglected. Determine ① the potential distribution between the plates, and ② the surface charge densities on the plates.

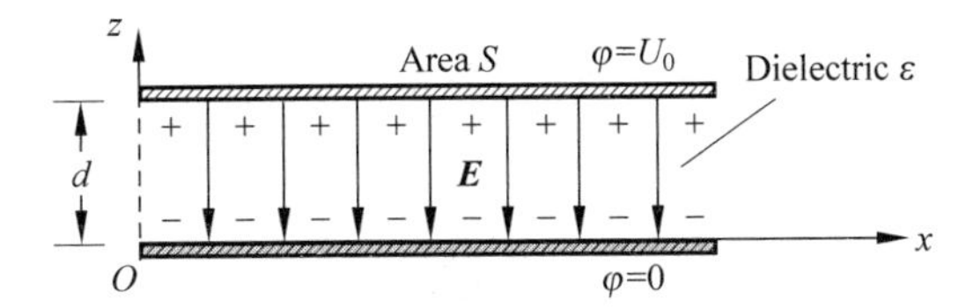

Figure 3-1 A parallel capacitor

Solution: This is essentially the same problem as Example 2-16. Now we solve it by solving the Laplace's equation satisfied by the electric potential since the charge density $\rho_v = 0$ between the plates.

(1) By ignoring the fringing effect of the electric field, we assume the field distribution is the same as if the plates were infinitely large. In other words, the potential φ has no variation in the x-and y-directions. Hence, Laplace's equation is then simplified to

$$\frac{d^2\varphi}{dz^2} = 0 \tag{3.12}$$

where d^2/dz^2 is used instead of $\partial^2/\partial z^2$ because z is the only variable in this problem. Integration of (3.12) with respect to z gives

$$\frac{d\varphi}{dz} = C_1$$

where C_1 is an unknown constant coefficient. Integrating again, we obtain

$$\varphi = C_1 z + C_2 \tag{3.13}$$

To determine the two unknown coefficients C_1 and C_2, we use the following two boundary conditions:

$$\text{At } z = 0, \quad \varphi = 0 \tag{3.14a}$$

$$\text{At } z = d, \quad \varphi = U_0 \tag{3.14b}$$

Substitution of (3.14a) and (3.14b) respectively into (3.13) yields two equations, from which the two unknown coefficients can be solved to obtain $C_1 = U_0/d$ and $C_2 = 0$. Hence the potential distribution between the plates is

① 在无源区域(即电荷密度为零的区域,或者说是电荷分布区域以外的空间),泊松方程变为拉普拉斯方程。在很多实际问题,包括以下几个例子中,感兴趣的电场都是分布在无源区域中,因此可以通过求解满足特定边界条件的拉普拉斯方程(3.11)得到静电场的解。

$$\varphi = \frac{U_0}{d}z \tag{3.15}$$

(2) The surface charge densities can be found by using the boundary condition of the $\boldsymbol{E}$ field on the surfaces of the conducting plates($z=0$ and $z=d$). We first find the $\boldsymbol{E}$ field by using (3.3):

$$\boldsymbol{E} = -\boldsymbol{e}_z \frac{\mathrm{d}\varphi}{\mathrm{d}z} = -\boldsymbol{e}_z \frac{U_0}{d}$$

Then the surface charge densities at the conducting plates are obtained as

$$\rho_s = \varepsilon \boldsymbol{e}_n \cdot \boldsymbol{E} = \varepsilon \boldsymbol{e}_n \cdot \left(-\boldsymbol{e}_z \frac{U_0}{d}\right)$$

On the surface of the lower plate,

$$\boldsymbol{e}_n = \boldsymbol{e}_z, \quad \rho_s = -\frac{\varepsilon U_0}{d}$$

On the surface of the upper plate,

$$\boldsymbol{e}_n = -\boldsymbol{e}_z, \quad \rho_s = \frac{\varepsilon U_0}{d}$$

This agrees with the fact that electric field lines in an electrostatic field originate from positive charges and terminate in negative charges.

Example 3-2 A cylindrical capacitor consists of an inner conductor of radius a and an outer conductor whose inner radius is b. The space between the conductors is filled with a dielectric of permittivity ε, and the length of the capacitor is L. The outer conductor is grounded, and the inner conductor is maintained at potential U_0. Determine ① the potential distribution between the two conductors, and ② the capacitance of this capacitor.

Solution: This is the same problem as Example 2-17, which is solved by applying Gauss's law. Here we solve it by solving the one-dimensional Laplace's equation under the cylindrical coordinate system.

(1) Due to cylindrical symmetry, φ has no variation along the ϕ-and z-directions(assuming no fringing effect). Laplace's equation(3.9) is then simplified to

$$\frac{1}{\rho}\frac{\mathrm{d}}{\mathrm{d}\rho}\left(\rho\frac{\mathrm{d}\varphi}{\mathrm{d}\rho}\right) = 0 \tag{3.16}$$

Integration of(3.16) with respect to ρ gives

$$\frac{\mathrm{d}\varphi}{\mathrm{d}\rho} = \frac{C_1}{\rho}$$

where C_1 is an unknown constant coefficient. Integrating again, we obtain

$$\varphi = C_1 \ln\rho + C_2 \tag{3.17}$$

To determine the two unknown coefficients C_1 and C_2, two boundary conditions are used:

$$\text{At } \rho = b, \quad \varphi = 0 \tag{3.18a}$$

$$\text{At } \rho = a, \quad \varphi = U_0 \tag{3.18b}$$

Substitution of(3.18a) and(3.18b) into(3.17) yields two equations, from which the two unknowns are solved to be $C_1 = U_0/\ln(a/b)$ and $C_2 = -U_0 \ln(b)/\ln(a/b)$. Hence the potential distribution between the conductors is

$$\varphi = \frac{U_0}{\ln\left(\frac{a}{b}\right)} \ln\left(\frac{\rho}{b}\right) \tag{3.19}$$

(2) In order to find the capacitance, we first find the distribution of $\boldsymbol{E}$ within the capacitor. From (3.3) and (3.19) we have

$$\boldsymbol{E}(\rho) = -\boldsymbol{e}_\rho \frac{\mathrm{d}\varphi}{\mathrm{d}\rho} = -\boldsymbol{e}_\rho \frac{U_0}{\ln\left(\frac{a}{b}\right)} \frac{1}{\rho} \tag{3.20}$$

At the surface of the inner conductor ($\rho = a$), we have

$$E_n(a) = \boldsymbol{e}_n \cdot \boldsymbol{E}(a) = \boldsymbol{e}_\rho \cdot (-\boldsymbol{e}_\rho) \frac{U_0}{\ln\left(\frac{a}{b}\right)} \frac{1}{a} = \frac{U_0}{\ln\left(\frac{b}{a}\right)} \frac{1}{a}$$

which is a constant. The surface charge densities at the conducting plates are obtained by using the boundary condition, i. e.,

$$\rho_s = \varepsilon E_n = \frac{\varepsilon U_0}{\ln\left(\frac{b}{a}\right)} \frac{1}{a}$$

The total charge on the inner conductor is

$$Q = \int_S \rho_s \mathrm{d}s = 2\pi a L \rho_s = \frac{2\pi\varepsilon L U_0}{\ln\left(\frac{b}{a}\right)} \tag{3.21}$$

We can verify easily that the charge carried by the outer conductor is $-Q$. Therefore, the capacitance is calculated as

$$C = \frac{Q}{U_0} = \frac{2\pi\varepsilon L}{\ln\left(\frac{b}{a}\right)}$$

which is the same as the result of Example 2-17.

Example 3-3 A spherical capacitor consists of an inner conducting sphere of radius a and an outer conductor with inner radius b. The space in between is filled with a dielectric of permittivity ε. The outer conductor is grounded, and the inner conductor is maintained at a potential U_0. Determine ① the potential distribution between the two conductors, and ② the capacitance of this capacitor.

Solution: This is essentially the same problem as Example 2-18. Here we solve it based on the Laplace's equation in spherical coordinates.

(1) Due to symmetry, φ has no variation along the ϕ- and θ- directions. Hence φ between the two conductors satisfies the one-dimensional Laplace's equation

$$\frac{1}{r^2} \frac{\mathrm{d}}{\mathrm{d}r}\left(r^2 \frac{\mathrm{d}\varphi}{\mathrm{d}r}\right) = 0 \tag{3.22}$$

Integration of(3.22) with respect to r gives

$$\frac{\mathrm{d}\varphi}{\mathrm{d}r}=\frac{C_1}{r^2}$$

where C_1 is an unknown constant coefficient. Integrating again, we obtain

$$\varphi=-\frac{C_1}{r}+C_2 \tag{3.23}$$

To determine the two unknown coefficients C_1 and C_2, two boundary conditions are used:

$$\text{At } r=b,\quad \varphi=0 \tag{3.24a}$$

$$\text{At } r=a,\quad \varphi=U_0 \tag{3.24b}$$

which leads to the solution of φ as

$$\varphi=\frac{U_0}{\frac{1}{a}-\frac{1}{b}}\left(\frac{1}{r}-\frac{1}{b}\right) \tag{3.25}$$

(2) From(3.3) and(3.25) we have

$$\boldsymbol{E}=-\boldsymbol{e}_{\mathrm{r}}\frac{\mathrm{d}\varphi}{\mathrm{d}r}=\boldsymbol{e}_{\mathrm{r}}\frac{U_0}{\frac{1}{a}-\frac{1}{b}}\left(\frac{1}{r^2}\right) \tag{3.26}$$

At the surface of the inner conductor($r=a$), we have

$$E_{\mathrm{n}}(a)=\boldsymbol{e}_{\mathrm{n}}\cdot\boldsymbol{E}(a)=\boldsymbol{e}_{\mathrm{r}}\cdot\boldsymbol{e}_{\mathrm{r}}\frac{U_0}{\frac{1}{a}-\frac{1}{b}}\left(\frac{1}{a^2}\right)=\frac{U_0}{\frac{1}{a}-\frac{1}{b}}\left(\frac{1}{a^2}\right)$$

The surface charge density at the inner conductor is obtained by using the boundary condition, i.e.,

$$\rho_{\mathrm{s}}=\varepsilon E_{\mathrm{n}}=\frac{\varepsilon U_0}{\frac{1}{a}-\frac{1}{b}}\left(\frac{1}{a^2}\right)$$

The total charge on the inner conductor is

$$Q=\int_S\rho_{\mathrm{s}}\mathrm{d}s=4\pi a^2\rho_{\mathrm{s}}=\frac{4\pi\varepsilon U_0}{\frac{1}{a}-\frac{1}{b}} \tag{3.27}$$

We can verify easily that the charge carried by the outer conductor is $-Q$. Therefore, the capacitance is calculated as

$$C=\frac{Q}{U_0}=\frac{4\pi\varepsilon}{\frac{1}{a}-\frac{1}{b}}$$

which is the same as the result of Example 2-18.

Example 3-4 Determine the $\boldsymbol{E}$ field caused by a uniform charge distribution in a sphere with a volume density $\rho_{\mathrm{v}}=\rho_0$ for $0\leqslant r\leqslant a$ and $\rho_{\mathrm{v}}=0$ for $r>a$.

Solution: This is the same problem as Example 2-7, which is solved by applying Gauss's law. Here we solve it by direct solving the one-dimensional Poisson's and Laplace's equations. By the spherical symmetry, there are no variations in θ- and ϕ- direction. Therefore, the fields including $\boldsymbol{E}$ and φ are functions of the r coordinates only. Both Poisson's and Laplace's equations are reduced to one-dimensional.

(1) For region $0\leqslant r\leqslant a, \rho_v=\rho_0$. The potential must satisfy 1-D Poisson's equation

$$\frac{1}{r^2}\frac{\mathrm{d}}{\mathrm{d}r}\left(r^2\frac{\mathrm{d}\varphi}{\mathrm{d}r}\right)=-\frac{\rho_0}{\varepsilon_0}$$

Integration of the above equation gives

$$\frac{\mathrm{d}\varphi}{\mathrm{d}r}=-\frac{\rho_0}{3\varepsilon_0}r+\frac{C_1}{r^2} \tag{3.28}$$

Therefore, the electric field intensity inside the region is

$$\boldsymbol{E}=-\nabla\varphi=-\boldsymbol{e}_r\left(\frac{\mathrm{d}\varphi}{\mathrm{d}r}\right)=\boldsymbol{e}_r\frac{\rho_0}{3\varepsilon_0}r \quad (0\leqslant r\leqslant a) \tag{3.29}$$

Here, we have used the fact that C_1 in (3.28) must be zero because otherwise, $\boldsymbol{E}$ will become infinite at $r=0$.

(2) For region $r>a, \rho_v=0$. The potential must satisfy 1-D Laplace's equation

$$\frac{1}{r^2}\frac{\partial}{\partial r}\left(r^2\frac{\mathrm{d}\varphi}{\mathrm{d}r}\right)=0 \tag{3.30}$$

Integrating of the above equation gives

$$\frac{\mathrm{d}\varphi}{\mathrm{d}r}=\frac{C_2}{r^2} \tag{3.31}$$

and therefore,

$$\boldsymbol{E}=-\nabla\varphi=-\boldsymbol{e}_r\frac{\mathrm{d}\varphi}{\mathrm{d}r}=-\boldsymbol{e}_r\frac{C_2}{r^2} \quad (r>a) \tag{3.32}$$

The integration constant C_2 can be found by equating $\boldsymbol{E}$ at $r=a$, which is the boundary condition of normal continuity of $\boldsymbol{D}$ vector (the permittivity is the same ε_0 inside and outside the source region). Therefore, from (3.29) and (3.32), we have

$$\frac{\rho_0}{3\varepsilon_0}a=-\frac{C_2}{a^2}$$

which gives

$$C_2=-\frac{\rho_0 a^3}{3\varepsilon_0} \tag{3.33}$$

Substitution of (3.33) into (3.32) gives

$$\boldsymbol{E}=\boldsymbol{e}_r\frac{\rho_0 a^3}{3\varepsilon_0 r^2} \quad (r>a) \tag{3.34}$$

which is the same as the results obtained in Example 2-7. We can continue to find the potential distribution as a function of r. For the region $0\leqslant r\leqslant a$, integrating (3.28) in which C_1 is already determined to be zero, we have

$$\varphi = -\frac{\rho_0 r^2}{6\varepsilon_0} + C_1' \quad (0 \leqslant r \leqslant a) \tag{3.35}$$

where C'_1 is a new integration constant that will be determined later. For the region $r>a$, substituting(3.33) into(3.31) and integrating both sides of the resulted equation, we obtain

$$\varphi = \frac{\rho_0 a^3}{3\varepsilon_0 r} \quad (r > a) \tag{3.36}$$

Here, we do not include an additional unknown constant in the integration result because the potential φ is zero at infinity($r \to \infty$). The only unknown left is C_1' in(3.35), which can be determined by the continuity condition of φ across the boundary. Let φ in(3.35) and(3.36) be equal at the boundary $r=a$, we have

$$-\frac{\rho_0 a^2}{6\varepsilon_0} + C_1' = \frac{\rho_0 a^2}{3\varepsilon_0}$$

then

$$C_1' = \frac{\rho_0 a^2}{2\varepsilon_0} \tag{3.37}$$

Substitute(3.37) into(3.35), we have

$$\varphi = \frac{\rho_0}{3\varepsilon_0}\left(\frac{3a^2}{2} - \frac{r^2}{2}\right) \quad (0 \leqslant r \leqslant a) \tag{3.38}$$

3.3 Uniqueness of Electrostatic Solutions (静电场解的唯一性)

In the examples in the last section, we obtained the solutions by direct integration. However, direct integration can be used only if Poisson's(or Laplace's) equation is reduced to one dimensional due to the symmetry. In more complicated situations involving two- or three-dimensional partial differential equations, the solution usually cannot be obtained by direct integration. Nevertheless, in some special cases, analytical or semi-analytical solutions can still be obtained by using special methods such as the method of images and the method of separation of variables that will be introduced later in this chapter. These two methods are both based on the important **uniqueness theorem**(唯一性定理).

The uniqueness theorem states that the solution of Poisson's (or Laplace's) equation satisfying the given boundary conditions is a unique solution. ① This means that, no matter what method we use to obtain a solution of the Poisson's (or Laplace's), it must be the correct solution as long as the boundary conditions are satisfied.

To prove the uniqueness theorem, we take an arbitrary volume V bounded by a closed surface S_o which may be a surface at infinity. Inside the closed surface S_o, the volume V may also be bounded by some interior surfaces $S_1, S_2, \cdots, S_N$ as depicted in Figure 3-2. Now assume

① 静电场解的唯一性定理可以表述为：满足给定边界条件的泊松方程或拉普拉斯方程的解是唯一存在的。

that there are two solutions, φ_1 and φ_2, to the same Poisson's equation in V, i. e. ,

$$\nabla^2 \varphi_1 = -\frac{\rho_v}{\varepsilon} \tag{3.39a}$$

$$\nabla^2 \varphi_2 = -\frac{\rho_v}{\varepsilon} \tag{3.39b}$$

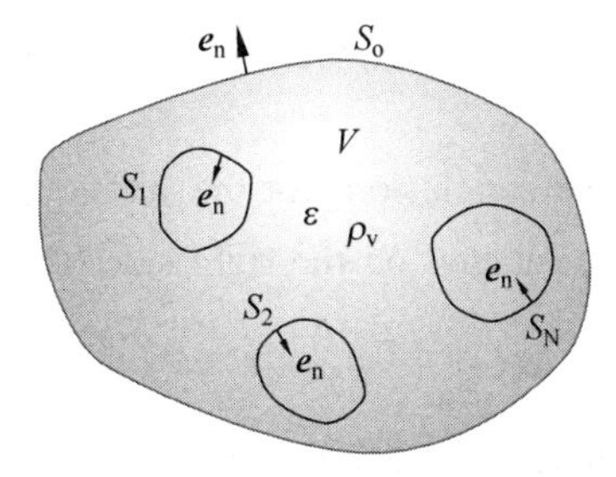

Figure 3-2 A region V bounded by an external surface S_o and possible internal surfaces $S_1, S_2, \cdots, S_N$

where ρ_v is the charge density within the volume V. Then we only need to prove that the difference between φ_1 and φ_2 in the volume V must be zero if φ_1 and φ_2 satisfy the same boundary conditions on $S_1, S_2, \cdots, S_N$ and S_o. To do that, we define a difference potential:

$$\varphi_d = \varphi_1 - \varphi_2 \tag{3.40}$$

From(3. 39a) and (3. 39b), it is obvious that φ_d must satisfy Laplace's equation in the volume V

$$\nabla^2 \varphi_d = 0 \tag{3.41}$$

Utilizing the vector identity(1. 148), in which let $\psi = \varphi_d$ and $\boldsymbol{A} = \nabla \varphi_d$, we have

$$\nabla \cdot (\varphi_d \nabla \varphi_d) = \varphi_d \nabla^2 \varphi_d + |\nabla \varphi_d|^2 \tag{3.42}$$

From(3. 41), the first term on the right side of(3. 42) vanishes. Integrating both sides of(3. 42) over the volume V and applying the divergence theorem to the left side, we have

$$\oint_S (\varphi_d \nabla \varphi_d) \cdot \boldsymbol{e}_n \mathrm{d}s = \int_V |\nabla \varphi_d|^2 \mathrm{d}v \tag{3.43}$$

where $\boldsymbol{e}_n$ denotes the unit normal outward from V, and the surface S consists of S_o as well as S_1, $S_2, \cdots, S_N$. Noticing that $\nabla \varphi_d \cdot \boldsymbol{e}_n = \partial \varphi_d / \partial n$, (3. 43) can be rewritten as

$$\oint_S \varphi_d \frac{\partial \varphi_d}{\partial n} \mathrm{d}s = \int_V |\nabla \varphi_d|^2 \mathrm{d}v \tag{3.44}$$

Now, we only need to show that(3. 44) implies φ_d must be zero if φ_1 and φ_2 satisfies the same boundary conditions. The boundary conditions can take different forms depending on the specific electrostatic problems. Typical forms of the boundary conditions include but not limited to the following.

(1) The potential φ is specified on some or all the boundaries. Then $\varphi_1 = \varphi_2$ on these boundaries, and therefore, φ_d on these boundaries is identically zero;

(2) $\partial \varphi / \partial n$ is specified on some or all the boundaries (which is equivalent to specified surface charge densities if these boundaries are conductor-dielectric interfaces). Then $\partial \varphi_1 / \partial n = \partial \varphi_2 / \partial n$ on these boundaries, and therefore, $\partial \varphi_d / \partial n$ on these boundaries is identically zero;

(3) If S_o (or partial S_o) is at infinity, it can be considered as the surface (or partial surface) of a sphere centered at origin with a radius r approaching infinity. As r increases, both φ_1 and φ_2 decrease as $1/r$ (if the charge distribution is within a bounded region, which is true for most practical problems). Hence φ_d decrease as $1/r$ and $\nabla\varphi_d$ decreases as $1/r^2$, making the integrand $\varphi_d(\partial\varphi_d/\partial n)$ decreases as $1/r^3$. As the surface area of S_o (or partial S_o) increases as r^2, the surface integral of $\varphi_d(\partial\varphi_d/\partial n)$ on S_o (or partial S_o) decreases as $1/r$ and approaches zero at infinity.

All the above cases lead to the conclusion that the surface integral on the left side of (3.44) is zero, and as a result, the volume integral on the right side of (3.44) must also be zero, i.e.,

$$\int_V |\nabla\varphi_d|^2 \mathrm{d}v = 0 \tag{3.45}$$

Since the integrand $|\nabla\varphi_d|^2$ is nonnegative everywhere, (3.45) can be satisfied only if $|\nabla\varphi_d|^2$ is zero everywhere inside the volume V. The gradient of φ_d is everywhere zero, meaning that φ_d is constant at all points in V. Therefore, φ_1 can be different from φ_2 by only a constant. However, as we know, a constant difference in potential distribution does not make any difference in electric fields. ① And the constant difference can be eliminated by selecting the same reference zero potential point in the solution of φ_1 and φ_2, in which case $\varphi_1 = \varphi_2$. This proves that there is only one possible solution. ②

3.4 Method of Images(镜像法)

There is a class of electrostatic problems that can be simplified by replacing bounding surfaces by appropriate image charges. This method is called the **method of images(镜像法)**. ③

3.4.1 Image with Respect to Planes(平面镜像)

To illustrate the method of images, we consider the problem of finding electrostatic field produced by a point charge in front of an infinitely large grounded conducting plane. As shown in Figure 3-3(a), a positive point charge Q is located at a distance d above conducting plane. Here, the objective is to solve for the potential everywhere above the conducting plane ($z>0$). It can be formulated as the boundary-value problem of solving Poisson's equation:

$$\nabla^2\varphi = \frac{\partial^2\varphi}{\partial x^2} + \frac{\partial^2\varphi}{\partial y^2} + \frac{\partial^2\varphi}{\partial z^2} = -\frac{Q\delta(\boldsymbol{r}-\boldsymbol{d})}{\varepsilon_0} \quad (z>0) \tag{3.46}$$

---

① 电位分布加上任意一个常数,其空间变化率不会发生变化,由电位分布确定的电场强度也不会变化。

② 唯一性定理的重要意义在于:①给出了静态场边值问题具有唯一解的条件;②为静态场边值问题的各种求解方法提供了理论依据;③为求解结果的正确性提供了判据。3.4 节和 3.5 节介绍基于唯一性定理的两种特殊且很重要的静电场边值问题求解方法。

③ 镜像法的基本思想是引入位于边界外虚设的较简单的镜像电荷分布来等效替代该边界上未知的较为复杂的电荷分布,从而将原本带有复杂电荷分布的边界值问题转换成无限大单一均匀媒质空间已知电荷分布求电场的问题,简化分析计算过程。镜像法的理论依据是唯一性定理。

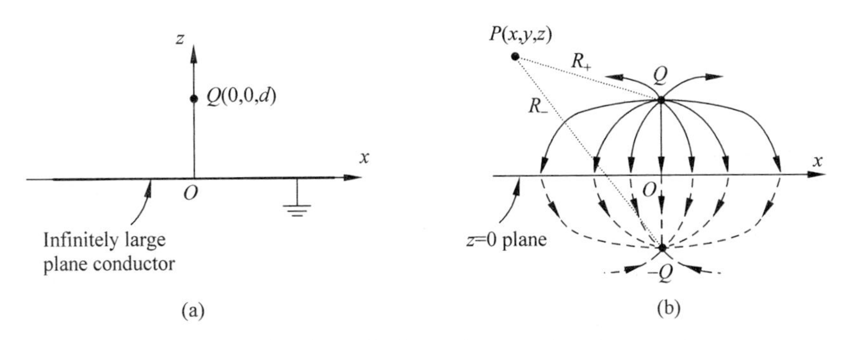

Figure 3-3 Point charge in front of a grounded plane conductor

subject to the boundary conditions

$$\varphi(x,y,0)=0 \tag{3.47}$$

and

$$\varphi(x,y,z)\to 0,\quad \text{as}\quad x\to\pm\infty, y\to\pm\infty \quad \text{or}\quad z\to+\infty \tag{3.48}$$

In(3.46), the volume charge density of the point charge Q is represented by $Q\delta(\boldsymbol{r}-\boldsymbol{d})$, where $\boldsymbol{d}=\boldsymbol{e}_z d$ is the position vector of the location of the point charge.

Obviously, φ in this problem is a field depending on all the three coordinates x, y and z. Therefore, we cannot construct its solution by direct integration of the equation(3.46).

From the physical point of view, the positive charge Q at $z=d$ induces negative charges on the surface of the conducting plane, resulting in a surface charge density ρ_s. Hence the potential to solve can be written as

$$\varphi(x,y,z)=\frac{Q}{4\pi\varepsilon_0\sqrt{x^2+y^2+(z-d)^2}}+\frac{1}{4\pi\varepsilon_0}\int_S\frac{\rho_s(x',y')}{\sqrt{(x-x')^2+(y-y')^2+z^2}}\mathrm{d}x'\mathrm{d}y'$$

where S is the surface of the plane conductor. Unfortunately, the induced surface charge distribution ρ_s is unknown. Moreover, it is quite difficult to evaluate the surface integral in the above expression even if ρ_s is found. However, with the method of images, this problem can be easily solved, which is demonstrated as follows.

As has been pointed out, it is the unknown ρ_s on the surface of the conducting plane that causes the trouble in solving this problem. In the method of images, we remove the conducting plane together with the induced charges and replace them with an image point charge $-Q$ at $z=-d$ as shown in Figure 3-3(b). Then, the potential at a point $P(x,y,z)$ in the $z>0$ region can be easily found as

$$\varphi(x,y,z)=\frac{Q}{4\pi\varepsilon}\left(\frac{1}{R_+}-\frac{1}{R_-}\right) \tag{3.49}$$

where R_+ and R_- are respectively the distances from $+Q$ and $-Q$ to the field point(x,y,z), i. e.,

$$R_+=\sqrt{x^2+y^2+(z-d)^2}\quad \text{and}\quad R_-=\sqrt{x^2+y^2+(z+d)^2}$$

Now we need to verify that the potential expression of(3.49) is exactly the solution of the electrostatic problem of Figure 3-3(a) in the $z>0$ region.

In the $z>0$ region, the medium and source distribution in the problem of Figure 3-3(a) are the same as those in Figure 3-3(b). Therefore, it is apparent that (3.49) satisfies the governing equation (3.46). It is also obvious that (3.49) satisfies the boundary conditions (3.47) and (3.48). Therefore, (3.49) gives a potential field that satisfies the same equation and the same boundary conditions in the $z>0$ region as specified in the problem of Figure 3-3(a). According to the uniqueness theorem, (3.49) must be the solution of the problem of Figure 3-3(a) in the $z>0$ region. ①

With the solution of potential φ, electric field intensity $\boldsymbol{E}$ in the $z>0$ region can be found by taking the negative gradient of φ. A few of the field lines are shown in Figure 3-3(b). The induced surface charge distribution ρ_s can be found by taking the negative directional derivative of φ along the normal direction on the conductor surface. Notice that, in the $z<0$ region, the potential field solution of Figure 3-3(b) is not the same as that of Figure 3-3(a). Apparently, the field is zero in the $z<0$ region in Figure 3-3(a). But in Figure 3-3(b), the field is nonzero as indicated by the dashed electric field lines.

Now we see that the method of images significantly simplifies the solution of this electrostatic problem of Figure 3-3(a). This is achieved by introducing a simple **image charge** (**镜像电荷**) that is equivalent to the unknown charge distribution on the boundary. It is important to realize that introduction of the image charge should not change anything within the region in which the field is to be determined ($z>0$ in this problem). In other words, the image charges must be located outside the region of interest ($z<0$ in this problem). Outside the region of interest, (3.49) is still the solution of the problem in Figure 3-3(b), but not the solution of the problem in Figure 3-3(a) anymore. As a matter of fact, both φ and $\boldsymbol{E}$ are zero in the $z<0$ region in Figure 3-3(a).

A similar problem is the electric field due to a line charge ρ_l above an infinite conducting plane, which can be found from ρ_l and its image $-\rho_l$ (with the conducting plane removed).

Example 3-5 As is shown in Figure 3-4(a), a positive point charge Q is located in the first quadrant ($x>0$, $y>0$) that is bounded by two orthogonal conducting planes that are grounded. The point charge is d_1 and d_2 from the two planes. Determine the potential distribution within the first quadrant.

Solution: To solve this problem by using the method of images, we need to find the image charges that can replace the effect of the two conducting half-planes. The image charges should be outside the first quadrant. After the conducting half-planes are replaced by the image charges, the potential at the locations of the half-planes should remain to be zero. To make the potential of the horizontal half-plane zero, we can first add an image charge $-Q$ in the fourth quadrant. Then to make the potential of the vertical half-plane zero, we add an image charge $-Q$ in the

① 图3-3(a)所示的问题中,点电荷 Q 在无限大接地导体平面感应出的电荷分布可以等效替换为距离导体平面相同距离的另一侧的镜像电荷 $-Q$,即图3-3(b)所示的问题。能够做上述等效替换是基于唯一性定理。具体而言,图3-3(a)与图3-3(b)所示的两个问题中的电位在 $y>0$ 的区域内满足同样的泊松方程和同样的边界条件,因此在该区域中电位的解必然是相同的。通过这个例子也可以看到应用镜像原理的两个原则:镜像电荷必须位于所求解的场区域以外;镜像电荷的个数、位置及电荷量的大小由满足所求解的场区域的边界条件确定。

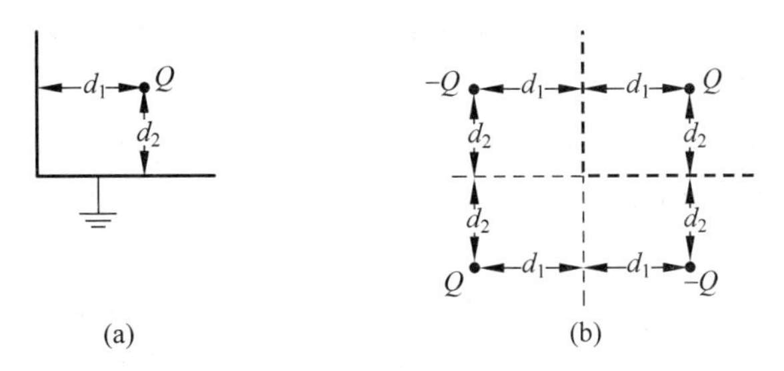

Figure 3-4 Point charge in front of two perpendicular conducting half planes

second quadrant. However, the image charge in the fourth quadrant produces a non-zero potential on the vertical half-plane, and the one in the second quadrant produces a non-zero potential on the horizontal half-plane. To balance out the non-zero potentials, we can introduce a third image charge $+Q$ in the third quadrant. With the three image charges as shown in Figure 3-4(b), it can be easily verified that the zero-potential boundary conditions on both half-planes are satisfied. According to the uniqueness theorem, the effect of the two conducting half-planes can be replaced by the image charges. The potential and electric field distribution in the first quadrant in Figure 3-4(b) is the same as that in Figure 3-4(a). Therefore, we have

$$\varphi(x,y,z)=\frac{Q}{4\pi\varepsilon_0\sqrt{(x-d_1)^2+(y-d_2)^2+z^2}}-\frac{Q}{4\pi\varepsilon_0\sqrt{(x+d_1)^2+(y-d_2)^2+z^2}}$$
$$-\frac{Q}{4\pi\varepsilon_0\sqrt{(x-d_1)^2+(y+d_2)^2+z^2}}+\frac{Q}{4\pi\varepsilon_0\sqrt{(x+d_1)^2+(y+d_2)^2+z^2}}$$

The electric field intensity in the first quadrant and the surface charge density induced on the two half-planes can also be found from the system of four charges.

As an extension of Example 3-5, if the angle α made by the two intersecting half planes are other than 90°, the method of image may still be used to find the solution of the fields due to a point charge. The number of image charges needed depends on the angle. Specifically, if the angle $\alpha=180°/n$ with n to be a positive integer, $(2n-1)$ image charges is needed to replace the conducting half planes. Otherwise, infinite number of image charges are required, in which case, an approximate solution can be found by ignoring those too far away from the region of interests.

3.4.2 Image with Respect to Spheres(球面镜像)

Here we consider the electrostatic problem of a point charge in front of a spherical conductor. As is shown in Figure 3-5(a), a positive point charge Q is located at a distance d from the center of a grounded conducting sphere of radius a($a<d$). The problem is to find the φ and $\boldsymbol{E}$ field distributions outside the sphere. Apparently, the difficulty of this problem lies in the unknown induced charge distribution on the surface of the conducting sphere. This difficulty can be circumvented if an image point charge Q_i can be found to replace the effect of the sphere. If this image charge Q_i exists, we must have the following:

(1) Q_i must be a negative charge inside the sphere and on the line OQ due to geometrical symmetry.

(2) After the conducting sphere is replaced by the image charge Q_i, the boundary condition on the spherical surface must remain unchanged. In other words, the potential at $r=a$ should be zero.

Now let's prove such image charge Q_i does exist as is illustrated in Figure 3-5(b). Q_i cannot be equal to $-Q$, because $-Q$ and the original Q do not make the spherical surface $r=a$ the zero-potential surface as required. Therefore, Q_i is an unknown. Another unknown is the distance between Q_i and the origin O, denoted by d_i. To find the solution of d_i and Q_i, we first write down the potential caused by Q and Q_i at a point M as

$$\varphi_M = \frac{1}{4\pi\varepsilon_0}\left(\frac{Q}{R} + \frac{Q_i}{R'}\right)$$

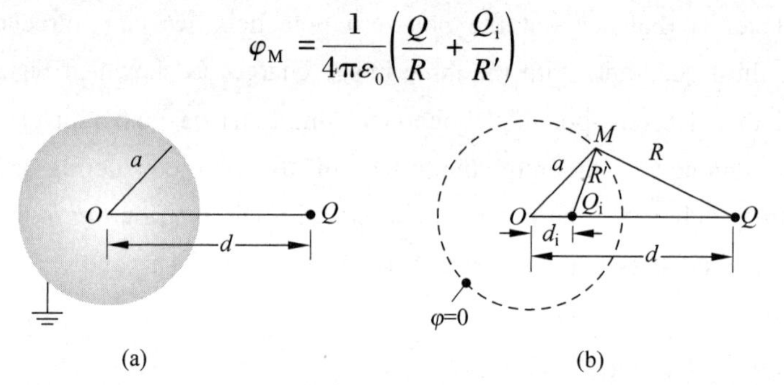

Figure 3-5 Point charge in front of a grounded sphere

where R and R' are respectively the distance from Q and Q_i to the point M. The boundary condition is $\varphi_M=0$ for any point M on the $r=a$ surface, which requires

$$\frac{R'}{R} = -\frac{Q_i}{Q} = \text{constant} \tag{3.50}$$

while the point M travels on the spherical surface. This condition can be satisfied by simply selecting d_i so that triangles $\angle OMQ_i$ and $\angle OQM$ are similar. Notice that the two triangles have one common angle $\angle MOQ_i = \angle QOM$, and the edges $\overline{OM}=a$, $\overline{OQ}=d$ are constant lengths. If we select $\overline{OQ_i}=d_i$ so that

$$\frac{\overline{OQ_i}}{\overline{OM}} = \frac{\overline{OM}}{\overline{OP}}$$

then the two triangles become similar, and we have

$$\frac{d_i}{a} = \frac{a}{d} = \frac{R'}{R} \tag{3.51}$$

from which we immediately find that

$$d_i = \frac{a^2}{d} \tag{3.52}$$

From(3.51), the constant ratio in(3.50) must be a/d, and hence

$$Q_i = -\frac{a}{d}Q \tag{3.53}$$

Now we see that, with d_i and Q_i given by (3.52) and (3.53), the potential field in Figure 3-5(b) satisfies the same boundary condition as in Figure 3-5(a). Therefore, Q_i must be the image charge of Q with respect to the spherical surface $r=a$. The φ and $\boldsymbol{E}$ of all points external to the grounded sphere can now be calculated as if they are produced by the point charges Q and Q_i. Specifically, as shown in Figure 3-6, the electric potential φ at an arbitrary point $P(r,\theta)$ is

$$\varphi(r,\theta)=\frac{1}{4\pi\varepsilon}\left(\frac{Q}{R}-\frac{a}{d}\frac{Q}{R'}\right) \tag{3.54}$$

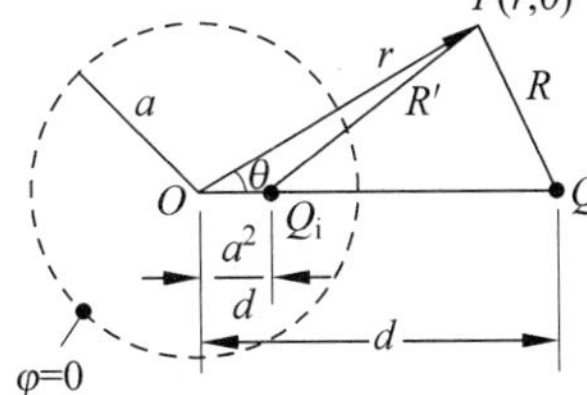

Figure 3-6 Image method solution of the problem in Figure 3-5

By the law of cosines,

$$R=\sqrt{r^2+d^2-2rd\cos\theta} \tag{3.55}$$

and

$$R'=\sqrt{r^2+(a^2/d)^2-2r(a^2/d)\cos\theta} \tag{3.56}$$

Substitute(3.55) and(3.56) into(3.54), then the r-component of the $\boldsymbol{E}$ field can be calculated as

$$E_r(r,\theta)=-\frac{\partial\varphi(r,\theta)}{\partial r}$$
$$=\frac{Q}{4\pi\varepsilon_0}\left\{\frac{r-d\cos\theta}{(r^2+d^2-2rd\cos\theta)^{3/2}}-\frac{a[r-(a^2/d)\cos\theta]}{d[r^2+(a^2/d)^2-2r(a^2/d)\cos\theta]^{3/2}}\right\} \tag{3.57}$$

With(3.57), we can find the induced surface charge on the sphere by letting $r=a$, and after some mathematical manipulation, we have

$$\rho_s=\varepsilon_0 E_r(a,\theta)=-\frac{Q(d^2-a^2)}{4\pi a(a^2+d^2-2ad\cos\theta)^{3/2}} \tag{3.58}$$

(3.58) tells us that the induced surface charge is negative and that its magnitude is maximum at $\theta=0$ and minimum at $\theta=\pi$, as expected.

The total charge induced on the sphere is an integration of ρ_s given by(3.58) over the surface of the sphere, i. e. ,

$$Q_{\text{induced}}=\oint\rho_s\mathrm{d}s=\int_0^{2\pi}\int_0^{\pi}\rho_s a^2\sin\theta\mathrm{d}\theta\mathrm{d}\phi=-\frac{a}{d}Q=Q_i \tag{3.59}$$

Note that the total induced charge is exactly equal to the image charge Q_i.

Example 3-6 A point charge Q is located outside an isolated conducting sphere with a distance d from the center of the sphere. As is illustrated in Figure 3-7(a), the conducting sphere has a radius a. Determine the image of the charge Q with respect to the surface of the conducting sphere.

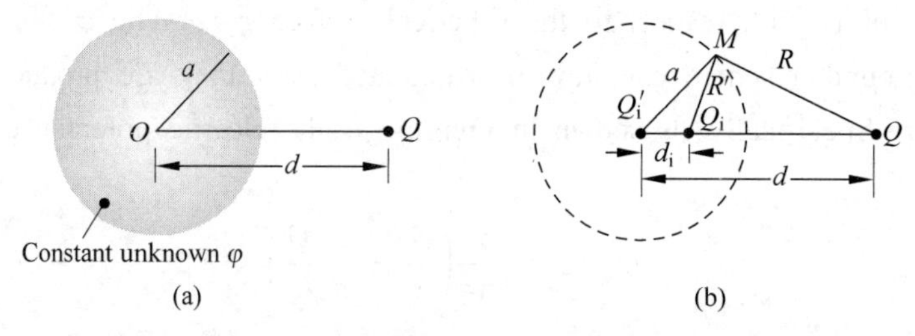

Figure 3-7 Point charge in front of an isolated conducting sphere

Solution: Different from the problem of Figure 3-5, the sphere is isolated, which means the potential on the sphere surface is not zero. Nevertheless, the sphere surface is still equipotential, which can be realized by the image charge Q_i and its location given by (3.53) and (3.52). However, Q_i and Q together make the sphere surface a zero-potential surface, whereas in this example, the potential on the sphere surface is a non-zero constant. This constant potential is unknown, but we know the isolated sphere is neutral, which means the total image charges must also be zero (why?). Therefore, as is shown in Figure 3-7(b), we can introduce an additional image charge

$$Q_i' = -Q_i = \frac{a}{d}Q \tag{3.60}$$

at the sphere center to make the net image charge zero. Q_i' must be located at the sphere center so that the potential on the $r=a$ surface remains constant. Then, the original problem can be solved as a problem with three point charges: Q_i' at $r=0$, Q_i at $r=a^2/d$, and the original Q at $r=d$. ①

3.4.3 Image in Cylinders(圆柱面镜像)

Consider a line charge ρ_l outside of a parallel, conducting, circular cylinder with radius a as shown in Figure 3-8(a). The distance between the line charge and the axis of the cylinder is d. The problem is to find the field distributions outside the cylinder. Again, the difficulty of this problem lies in the unknown induced charge distribution on the surface of the conducting

---

① 例3-6推断孤立导体球外点电荷 Q 关于球面的镜像电荷有两个 Q_i 和 Q_i'。其依据是 Q、Q_i 和 Q_i' 三个电荷共同产生的电位在球面上是常数，且球面包围的总电荷必须为零。然而，由于原问题中导体球表面的电位或者电位的法向导数均未知，上述依据实际上不足以说明这两个镜像电荷代替导体形成的电位分布与原问题中电位分布满足同样的边界条件。要严格地证明例题中镜像法得到的解就是真实的解，需要回到唯一性定理的证明：假设 φ_d 为镜像法得到的电位与真实电位之差，则 φ_d 在导体球外满足式(3.44)。由于镜像法得到的解与真实解在导体表面都为常数，φ_d 在导体表面必然也是常数，因此式(3.44)的左边可以写 $\varphi_d\oint_S(\partial\varphi_d/\partial n)\,ds = -\varphi_d\oint_S E_{nd}\,ds$，其中 E_{nd} 为镜像法得到的电场与真实电场在导体球表面的法向分量之差。根据高斯定律，镜像法得到的电场与真实电场的法向分量在导体球表面的通量都等于零(球面包含的总电荷量均为零)。因此，$\oint_S E_{nd}\,ds = 0$，故式(3.44)左边等于零，式(3.44)的右边 $\int_V |\nabla\varphi_d|^2 dv = 0$，从而证明了 φ_d 在导体球外的整个区域均为零。

cylinder, which can be solved by using the method of image. We first recognize the following:

(1) The image must be a parallel line charge (denoted by ρ_i) inside the cylinder, and it must lie somewhere along OP, due to the symmetry of the geometry.

(2) After the conducting cylinder is replaced by the image charge, the boundary condition on the cylindrical surface remains unchanged. Particularly, the potential at $\rho = a$ should be constant.

Let the distance between the image charge and the axis be d_i as shown in Figure 3-8(b). Then we need to determine the two unknowns, ρ_i and d_i.

Recall that, in Example 2-11, the equipotential surfaces of the field produced by two parallel line charges, ρ_l and $-\rho_l$, are circular cylindrical surfaces. If one of the equipotential surfaces coincides with the surface of the conducting cylinder in Figure 3-8(a), then according to the uniqueness theorem, the conducting cylinder can be replaced by the line charge $-\rho_l$ in Example 2-11. Therefore, we infer that the image of the line charge ρ_l in Figure 3-8(a) to be①

$$\rho_i = -\rho_l \tag{3.61}$$

To find d_i, we first write down the expression of the potential due to the line charges ρ_i and ρ_l. According to Example 2-11, at any point M on the cylindrical surface $\rho = a$, we have

$$\varphi_M = \frac{\rho_l}{2\pi\varepsilon_0}\ln\frac{\rho'}{\rho} \tag{3.62}$$

where ρ and ρ' are the distances from the point M to the line charges ρ_l and ρ_i respectively as is shown in Figure 3-8(b). Obviously, the boundary condition requires that ρ'/ρ maintain constant when the point M travels on the cylindrical surface. This condition can be satisfied by simply selecting a d_i value so that triangles $\angle OMP_i$ and $\angle OPM$ are similar. Note that these two triangles have one common angle $\angle MOP_i = \angle POM$. Hence the two triangles can be made similar by letting

$$\frac{\overline{OP_i}}{\overline{OM}} = \frac{\overline{OM}}{\overline{OP}}$$

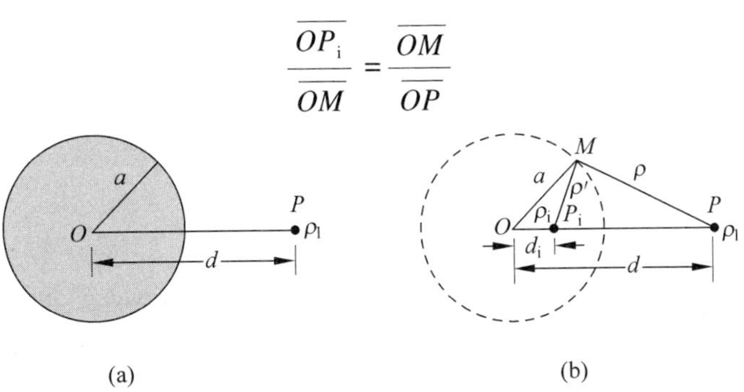

Figure 3-8 Line charge in front of a parallel conducting circular cylinder

① 这里假设无限长导体柱单位长度所带电荷为 $-\rho_l$。需要指出，由于导体柱无限长，导体外线电荷 ρ_l 也分布到无穷远处，该问题并没有设定导体柱是否接地，也没有给定导体柱的带电量。如果认为该导体柱接地，那么无穷远处就是参考零电位点，由例题 2-11 可以推断导体柱单位长度所带电荷必须为 $-\rho_l$。如果导体柱单位长度所带电荷不是 $-\rho_l$，就不能选取无穷远处为参考零电位点，而该问题的镜像电荷应该再加上一个位于导体柱轴线上的线电荷，其密度为导体柱单位长度实际所带电荷与 $-\rho_l$ 之差。而根据例题 2-10，该位于导体柱轴线上的镜像电荷的电位在无穷远处为无穷大。

Since $\overline{OM}=a$, $\overline{OP}=d$ and $\overline{OP_i}=d_i$, the above relation is satisfied by letting

$$d_i = \frac{a^2}{d} \tag{3.63}$$

And as a result,

$$\frac{\rho'}{\rho} = \frac{d_i}{a} = \frac{a}{d} = \text{constant} \tag{3.64}$$

for any point M on the cylindrical surface. By substituting (3.64) into (3.62), the constant potential on the cylindrical surface is

$$\varphi_M = \frac{\rho_l}{2\pi\varepsilon_0}\ln\frac{a}{d} \tag{3.65}$$

Now, it is verified that the line charge $\rho_i=-\rho_l$ is the image of the original line charge ρ_l with respect to the cylindrical conducting surface $\rho=a$, and the fields at any point outside the surface can be determined equivalently by ρ_l and ρ_i.

The above discussion demonstrates that a cylindrical conductor with surface charges induced by an external line charge can be replaced by an internal line charge. This conclusion is useful in determining the capacitance of two wire transmission lines as demonstrated in the following example.

Example 3-7 Two-wire transmission line: as shown in Figure 3-9(a), two infinitely long conducting wires of radius a are parallel to each other with a distance D between the axes. Determine the capacitance per unit length between the two wires.

Solution: As shown in Figure 3-9(b), the two conducting wires can be replaced by a pair of line charges $+\rho_l$ and $-\rho_l$, as long as the potential generated by the two line charges is constant on each of the cylindrical surfaces. Referring to the method of image used in the problem of Figure 3-8, the separation between the image charge within one cylinder and the axis of the other cylinder should be $d=D-d_i$. Using (3.63), we have

$$d = D - d_i = D - \frac{a^2}{d}$$

from which we obtain

$$d = \frac{1}{2}(D + \sqrt{D^2 - 4a^2}) \tag{3.66}$$

The potential difference between the two wires is that between any two points on the respective wires. Using (3.65), the potential on the cylinder surface surrounding positive line charge $+\rho_l$ is

$$\varphi_+ = -\frac{\rho_l}{2\pi\varepsilon_0}\ln\frac{a}{d}$$

The potential on the cylinder surface surrounding negative line charge $-\rho_l$ is

$$\varphi_- = \frac{\rho_l}{2\pi\varepsilon_0}\ln\frac{a}{d}$$

Then, the capacitance per unit length can be calculated as

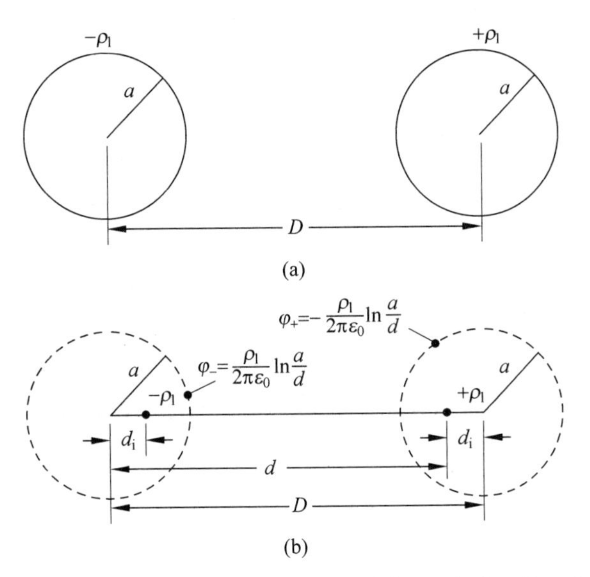

Figure 3-9 Two-wire transmission line and the equivalent line charges

$$C = \frac{\rho_l}{\varphi_+ - \varphi_-} = \frac{\pi\varepsilon_0}{\ln(d/a)} \tag{3.67}$$

Substituting(3.66) into(3.67) we have

$$C = \frac{\pi\varepsilon_0}{\ln[(D/2a) + \sqrt{(D/2a)^2 - 1}]} \tag{3.68}$$

Since

$$\ln[x + \sqrt{x^2 - 1}] = \cosh^{-1}x$$

for $x>1$,(3.68) can be written alternatively as

$$C = \frac{\pi\varepsilon_0}{\cosh^{-1}(D/2a)} \tag{3.69}$$

The potential distribution and electric field intensity around the two-wire line can be determined easily from the equivalent line charges.

The more general case of a two-wire transmission line of different radii can also be solved by using the method of image in a similar way. The key is to find the location of the equivalent line charges that make the wire surfaces equipotential.

3.5 Method of Separation of Variables(分离变量法)

The method of images is useful in solving certain types of electrostatic problems in which conducting boundaries can be replaced by equivalent charges. However, when the geometry of the boundaries is not simple, and/or the free charges are not known, the method of images cannot be used. In some problems, a system of conductors is maintained at specified potentials or specified normal derivatives of the potentials. If the boundaries of the conductors coincide with the coordinate surfaces of an orthogonal coordinate system, we may solve the problem by using the **method of separation of variables** (**分离变量法**).

In this section, the method of separation of variables is introduced as a method of solving Laplace's equations with given boundary conditions of the potential φ. Generally, problems formulated as partial differential equations with prescribed boundary conditions are called **boundary-value problems**(**边界值问题**). Boundary-value problems for electrostatic potential functions can be classified into three types: ① **Dirichlet problems**(**狄里赫利问题**,第一类边值问题), in which the value of the potential is specified everywhere on the boundaries; ② **Neumann problems**(**纽曼问题**,第二类边值问题), in which the normal derivative of the potential is specified everywhere on the boundaries; ③ **Mixed boundary-value problems**(**混合边值问题**), in which the potential is specified over some boundaries and the normal derivative of the potential is specified over the remaining ones. Different specified boundary conditions will require the choice of different potential functions, as will be demonstrated in this section. The solutions of Laplace's equation are often called harmonic functions(调和函数).[①]

Laplace's equation for scalar electric potential φ in Cartesian coordinates is

$$\frac{\partial^2 \varphi}{\partial x^2}+\frac{\partial^2 \varphi}{\partial y^2}+\frac{\partial^2 \varphi}{\partial z^2}=0 \tag{3.70}$$

To apply the method of separation of variables, we assume that the solution $\varphi(x,y,z)$ can be expressed as a product in the following form:

$$\varphi(x,y,z)=X(x)Y(y)Z(z) \tag{3.71}$$

where $X(x)$, $Y(y)$, and $Z(z)$ are functions of only x, y and z, respectively. Substituting (3.71) in (3.70), we have

$$Z(z)Y(y)\frac{\mathrm{d}^2X(x)}{\mathrm{d}x^2}+X(x)Z(z)\frac{\mathrm{d}^2Y(y)}{\mathrm{d}y^2}+X(x)Y(y)\frac{\mathrm{d}^2Z(z)}{\mathrm{d}z^2}=0$$

Divide both sides of the above equation by the product $X(x)Y(y)Z(z)$, we have

$$\frac{1}{X(x)}\frac{\mathrm{d}^2X(x)}{\mathrm{d}x^2}+\frac{1}{Y(y)}\frac{\mathrm{d}^2Y(y)}{\mathrm{d}y^2}+\frac{1}{Z(z)}\frac{\mathrm{d}^2Z(z)}{\mathrm{d}z^2}=0 \tag{3.72}$$

Notice that each of the three terms on the left side of (3.72) is a function of only one coordinate variable. In order for (3.72) to be satisfied for all values of x, y, z, each of the three terms must be a constant. For instance, if we differentiate (3.72) with respect to x, we have

$$\frac{\mathrm{d}}{\mathrm{d}x}\left[\frac{1}{X(x)}\frac{\mathrm{d}^2X(x)}{\mathrm{d}x^2}\right]=0 \tag{3.73}$$

This requires that

$$\frac{1}{X(x)}\frac{\mathrm{d}^2X(x)}{\mathrm{d}x^2}=-k_x^2 \tag{3.74}$$

---

① 分离变量法是求解边值问题的一种经典方法,其基本思想是将偏微分方程中含有 n 个自变量的待求函数表示成 n 个只含一个变量的函数的乘积,把偏微分方程分解成 n 个常微分方程,求出各常微分方程的通解后,将它们线性叠加得到级数形式解,并利用给定的边界条件确定待定常数。分离变量法的理论依据是唯一性定理。

where k_x^2 is a constant of integration that will be determined later from the boundary conditions of the problem. The negative sign on the right side of (3.74) as well as the square sign on k_x are employed only for mathematical convenience, which will be seen later. The separation constant k_x can be a real or an imaginary number. If k_x is imaginary, k_x^2 is a negative real number, making $-k_x^2$ a positive real number. Now we rewrite (3.74) as

$$\frac{d^2X(x)}{dx^2}+k_x^2X(x)=0 \tag{3.75}$$

Similarly, we can obtain equations for the functions $Y(y)$ and $Z(z)$ as

$$\frac{d^2Y(y)}{dy^2}+k_y^2Y(y)=0 \tag{3.76}$$

$$\frac{d^2Z(z)}{dz^2}+k_z^2Z(z)=0 \tag{3.77}$$

where the separation constants k_y and k_z are generally different from k_x, but due to (3.72), satisfy:

$$k_x^2+k_y^2+k_z^2=0 \tag{3.78}$$

Now the problem is reduced to finding the appropriate solutions $X(x)$, $Y(y)$ and $Z(z)$ from the second-order ordinary differential equations (3.75), (3.76) and (3.77) respectively. The possible solutions of (3.75) are well known and listed in Table 3-1.

Table 3-1 Solutions of equation $X''(x)+k_x^2X(x)=0$ ①

k_x	$X(x)$	Exponential form of $X(x)$
0	A_0x+B_0	
k	$A_1\sin kx+B_1\cos kx$	$C_1e^{jkx}+D_1e^{-jkx}$
jk	$A_2\sinh kx+B_2\cosh kx$	$C_2e^{kx}+D_2e^{-kx}$

The first possible solution, $X(x)=A_0x+B_0$, as listed in Table 3-1 is the result of $k_x=0$, in which case the potential function is a straight line with a slope A_0 and an intercept B_0 at $x=0$. Example 3-1 is an example of this solution.

When $k_x=k$ is a nonzero real number, the solution is a linear combination of sinkx and coskx, both of which have a period of $2\pi/k$. Generally, if the potential to be solved is periodic (usually with multiple zeros) along x-direction, a linear combination of sinkx and coskx should be chosen as the solution. In some special cases, if the boundary condition requires the potential to be zero at $x=0$, sinkx alone must be chosen; if the potential is expected to be symmetrical with respect to $x=0$, then coskx alone must be chosen. Sometimes it may be desirable to use

---

① 表3-1中列出的指数函数、三角函数以及双曲函数之间有如下转换关系：

$$e^{\pm jkx}=\cos kx\pm\sin kx,\cos kx=\frac{1}{2}(e^{jkx}+e^{-jkx}),\sin kx=\frac{1}{2j}(e^{jkx}-e^{-jkx})$$

$$e^{\pm kx}=\cosh kx\pm\sinh kx,\cosh kx=\frac{e^{kx}+e^{-kx}}{2},\sinh kx=\frac{1}{2}(e^{kx}-e^{-kx})$$

具体采用哪类函数作为方程的解取决于具体问题的边界条件。

$A_1\sin k(x-x_0)$ as the solution if a zero is found to be at $x=x_0$, whereas $B_1\cos k(x-x_0)$ should be used if the potential is symmetrical with respect to $x=x_0$.

If $k_x = jk$ is a purely imaginary number, the solution can take the form of a linear combination of hyperbolic functions $A_2\sinh kx+B_2\cosh kx$, or equivalently, exponential functions $C_2\mathrm{e}^{kx}+D_2\mathrm{e}^{-kx}$.

Hyperbolic and exponential functions are also plotted in Figure 3-10 for easy reference. These functions are non-periodic. The function $\sinh kx$ is an odd function of x and its value approaches $\pm\infty$ as x goes to $\pm\infty$. The function $\cosh kx$ is an even function of x. It equals unity at $x=0$, and approaches $+\infty$ as x goes to $+\infty$ or $-\infty$. The function e^{kx} approaches zero as x goes to $-\infty$ and approaches $+\infty$ as x goes to $+\infty$. The function e^{-kx} approaches $+\infty$ as x goes to $-\infty$ and approaches zero as x goes to $+\infty$.

The solutions of the equations (3.76) and (3.77) for $Y(y)$ and $Z(z)$ are similar. The choice of the proper form of the solution and the associated constants are determined by specified boundary conditions in specific problems, as shown in the following examples.

Example 3-8 As illustrated in Figure 3-11, two semi-infinite grounded conducting plates are parallel to the x-z plane and separated by a distance b. A third conducting plate in the y-z plane is insulated from the two grounded plates and maintained at a constant potential U_0. Determine the potential distribution in the region ($x>0, 0<y<b$) enclosed by the three plates.

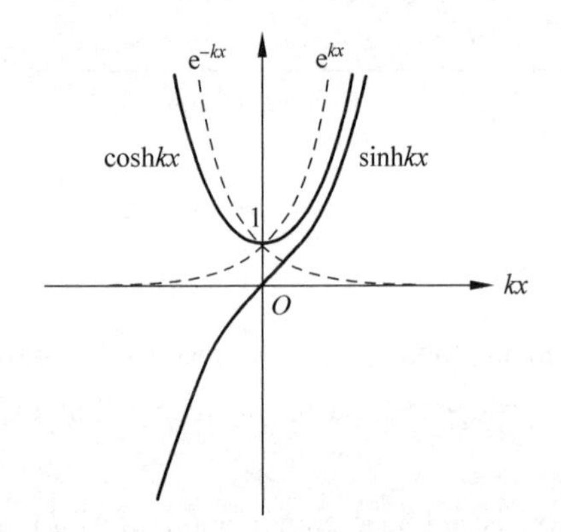

Figure 3-10 Illustration of different solutions of equation $X(x)+k_x^2 X(x)=0$

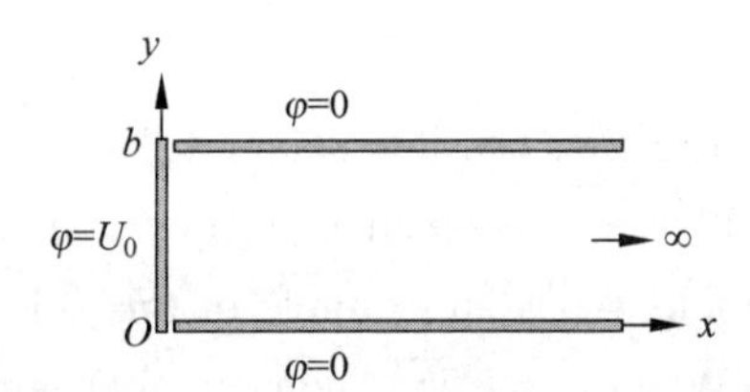

Figure 3-11 Illustration of electrostatic problem in Example 3-8

Solution: Referring to the coordinates in Figure 3-11, φ is independent of z, so we have

$$\varphi(x,y,z)=\varphi(x,y)=X(x)Y(y) \tag{3.79}$$

The boundary conditions for the potential are:

In the x-direction:

$$\varphi(0,y)=U_0 \tag{3.80a}$$

$$\varphi(\infty,y)=0 \tag{3.80b}$$

In the y-direction:

$$\varphi(x,0)=0 \tag{3.80c}$$

$$\varphi(x,b)=0 \tag{3.80d}$$

(3.79) implies that $k_z=0$ and from(3.78), we have

$$k_x^2+k_y^2=0 \tag{3.81}$$

We first notice that, according to the boundary condition(3.80b), $X(x)$ should approach zero as x approaches $+\infty$. Of all the possible solutions in Table 3-1, only the exponential function e^{-kx} meets this requirement, hence we can determine that $k_x=jk$ is imaginary and

$$X(x)=D_2e^{-kx} \tag{3.82}$$

where k is a real number. This choice of k_x implies that $k_y=k$ is real. So the function $Y(y)$ should be a combination of sine and cosine functions. Condition (3.80c) indicates that the proper choice for $Y(y)$ is

$$Y(y)=A_1\sin ky \tag{3.83}$$

Substitute (3.82) and (3.83) into (3.79), we obtain an appropriate solution of the following form:

$$\varphi(x,y)=(D_2A_1)e^{-kx}\sin ky=Ce^{-kx}\sin ky \tag{3.84}$$

where the product D_2A_1 has been combined into a single arbitrary constant C. Since (3.84) should satisfy(3.80d), we have,

$$\varphi(x,b)=Ce^{-kx}\sin kb=0 \tag{3.85}$$

which can be satisfied, for all values of x, only if

$$\sin kb=0$$

Therefore,

$$k=\frac{n\pi}{b},\quad n=1,2,3,\cdots \tag{3.86}$$

which means that k can only take discrete values. Substitute(3.86) into(3.84), we obtain

$$\varphi_n(x,y)=C_ne^{-n\pi x/b}\sin\frac{n\pi}{b}y \tag{3.87}$$

where the subscript n indicates the nth possible value of the constant k, and hence indicates the nth possible solution of φ. (Question: why n cannot be 0 or negative integral values?) Apparently, for any n value, the function $\varphi_n(x,y)$ in(3.87) satisfies the Laplace's equation and the boundary conditions (3.80b-d). But any $\varphi_n(x,y)$ alone cannot satisfy the remaining boundary condition(3.80a) at $x=0$ for all values of y from 0 to b. Nevertheless, since Laplace's equation is a linear partial differential equation, a linear combination of $\varphi_n(x,y)$ with all possible n values is also a solution, which could satisfy the boundary condition(3.80a). So, the desired solution can be written as

$$\varphi(x,y)=\sum_{n=1}^{\infty}\varphi_n(x,y)=\sum_{n=1}^{\infty}C_ne^{-n\pi x/b}\sin\frac{n\pi}{b}y \tag{3.88}$$

It is easy to verify that $\varphi(x,y)$ in(3.88) satisfies boundary conditions(3.80b-d). So now, we only need to let $\varphi(x,y)$ in(3.88) satisfy boundary condition(3.80a). This requires

$$\varphi(0,y)=\sum_{n=1}^{\infty}C_n\sin\frac{n\pi}{b}y=U_0,\quad \text{for}\quad 0<y<b \tag{3.89}$$

(3.89) is essentially a Fourier-series expansion of the periodic rectangular wave with a fundamental period of $2b$ shown in Figure 3-12, which has a constant value U_0 in the interval $0<y<b$. Notice that the $\sin(n\pi/b)y$ term in (3.89) is an odd function, and therefore, the rectangular wave has a constant value $-U_0$ in the interval $-b<y<0$.

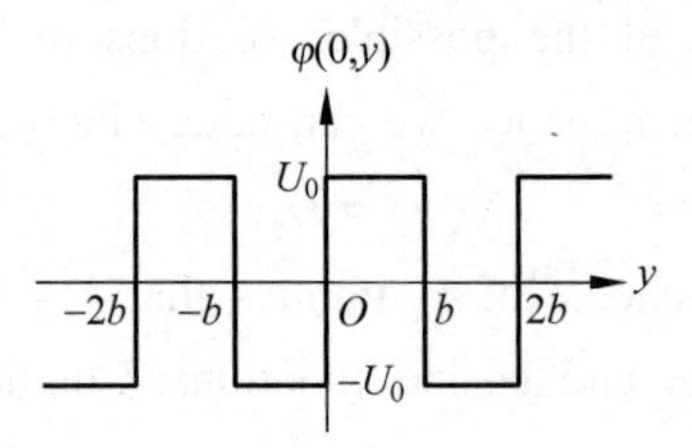

Figure 3-12 A periodic rectangular wave function

In order to evaluate the coefficients C_n, we multiply both sides of (3.89) by $\sin(m\pi/b)y$ and integrate the products from $y=0$ to $y=b$:

$$\sum_{n=1}^{\infty}\int_0^b C_n \sin\frac{n\pi}{b}y\sin\frac{m\pi}{b}y\mathrm{d}y = \int_0^b U_0 \sin\frac{m\pi}{b}y\mathrm{d}y \tag{3.90}$$

The integral on the right side of (3.90) is easily evaluated:

$$\int_0^b U_0\sin\frac{m\pi}{b}y\mathrm{d}y = \begin{cases}\dfrac{2bU_0}{m\pi} & \text{if } m \text{ is odd}\\ 0 & \text{if } m \text{ is even}\end{cases} \tag{3.91}$$

Each integral on the left side of (3.90) is

$$\begin{aligned}\int_0^b C_n\sin\frac{n\pi}{b}y\sin\frac{m\pi}{b}y\mathrm{d}y &= \frac{C_n}{2}\int_0^b\left[\cos\frac{(n-m)\pi}{b}y - \cos\frac{(n+m)\pi}{b}y\right]\mathrm{d}y\\ &= \begin{cases}\dfrac{C_n}{2}b & \text{if } m=n\\ 0 & \text{if } m\neq n\end{cases}\end{aligned} \tag{3.92}$$

Substituting (3.91) and (3.92) into (3.90), we obtain

$$\frac{C_m}{2}b = \begin{cases}\dfrac{2bU_0}{m\pi} & \text{if } m \text{ is odd}\\ 0 & \text{if } m \text{ is even}\end{cases}$$

which gives us the solution (with the index m replaced by n)

$$C_n = \begin{cases}\dfrac{4U_0}{n\pi} & \text{if } n \text{ is odd}\\ 0 & \text{if } n \text{ is even}\end{cases} \tag{3.93}$$

Substitute (3.93) into (3.88), we have the final solution of the potential distribution

$$\varphi(x,y) = \sum_{n=1,3,5,\cdots}^{\infty}\frac{4U_0}{n\pi}\mathrm{e}^{-n\pi x/b}\sin\frac{n\pi}{b}y \quad \text{for} \quad x>0, 0<y<b \tag{3.94}$$

The solution (3.94) is a rather complicated expression involving a summation of infinite series. However, since the terms in the series decreases as $1/n$ as n increases, only the first few terms are needed to obtain a good approximation.

Example 3-9 Consider a region enclosed by four conducting plates as illustrated in Figure 3-13. The top, right and bottom plates are grounded. The left plate is insulated from the others and maintaned at a constant potential U_0. All plates are infinite in extent in the z-direction. Determine the potential distribution within this region.

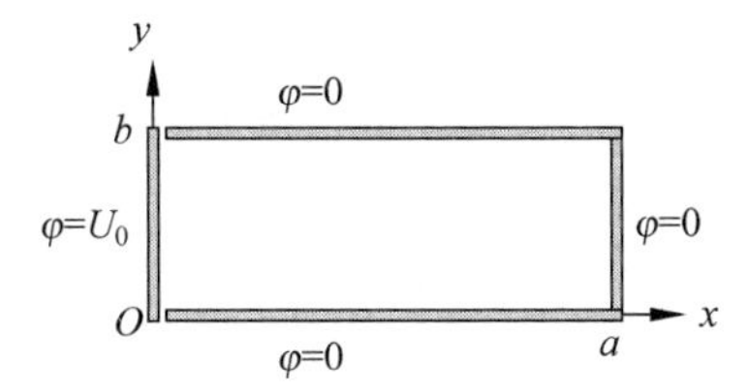

Figure 3-13 Illustration of electrostatic problem in Example 3-9

Solution: Like Example 3-8, the potential φ is independent of z, so we have

$$\varphi(x,y,z) = \varphi(x,y) = X(x)Y(y) \tag{3.95}$$

The boundary conditions are:

In the x-direction:

$$\varphi(0,y) = U_0 \tag{3.96a}$$

$$\varphi(a,y) = 0 \tag{3.96b}$$

In the y-direction:

$$\varphi(x,0) = 0 \tag{3.96c}$$

$$\varphi(x,b) = 0 \tag{3.96d}$$

(3.95) implies that $k_z=0$ and from (3.78),

$$k_x^2 + k_y^2 = 0 \tag{3.97}$$

which is the same as (3.81) in Example 3-8. The boundary conditions in the y-direction, (3.96c) and (3.96d), are also the same as those specified in Example 3-8. To make $\varphi(x,0)=0$ and $\varphi(x,b)=0$ for all values of x between 0 and a, $Y(0)$ and $Y(b)$ must be zero. Of the functions listed in Table 3-1, only sine and cosine functions are periodic with multiple zeros, so $Y(y)$ must be a linear combination of sine and cosine functions. With $Y(0)=Y(b)=0$, we have

$$Y(y) = A_1 \sin ky \tag{3.98}$$

which is the same as in (3.83), and k can take discrete values as

$$k = \frac{n\pi}{b}, \quad n = 1,2,3,\cdots \tag{3.99}$$

This means $k_y=k$ is real, and according to (3.97), $k_x=jk$. As a result, in the x-direction, $X(x)$ is a linear combination of sinh and cosh functions, i. e.,

$$X(x) = A_2 \sinh kx + B_2 \cosh kx \tag{3.100}$$

To determine A_2 and B_2, we apply the boundary condition (3.96b), which demands that $X(a)=0$; that is,

$$0 = A_2 \sinh ka + B_2 \cosh ka$$

or

$$B_2 = -A_2 \frac{\sinh ka}{\cosh ka}$$

Therefore, we have

$$X(x) = A_2\left(\sinh kx - \frac{\sinh ka}{\cosh ka}\cosh kx\right) = \frac{A_2}{\cosh ka}(\cosh ka \sinh kx - \sinh ka \cosh kx)$$
$$= A_3 \sinh k(x-a) \tag{3.101}$$

where $A_3 = A_2/\cosh ka$. Note that (3.101) is a shift in the argument of the sinh function. Now, we obtain the appropriate product solution

$$\varphi_n(x,y) = B_0 A_1 A_3 \sinh k(x-a)\sin ky = C'_n \sinh\frac{n\pi}{b}(x-a)\sin\frac{n\pi}{b}y \tag{3.102}$$

where $C'_n = B_0 A_1 A_3$. We have now used all of the boundary conditions except (3.96a), which may be satisfied by a Fourier-series expansion of $\varphi(0,y) = U_0$ over the interval from $y=0$ to $y=b$. We have

$$\sum_{n=1}^{\infty}\varphi_n(0,y) = -\sum_{n=1}^{\infty} C'_n \sinh\frac{n\pi}{b}a \sin\frac{n\pi}{b}y = U_0, \quad (0 < y < b) \tag{3.103}$$

We note that (3.103) is of the same form as (3.89), except that C_n is replaced by $-C'_n \sinh(n\pi a/b)$. The values for the coefficient C'_n can then be written down from (3.93):

$$C'_n = \begin{cases} -\dfrac{4U_0}{n\pi \sinh(n\pi a/b)} & \text{if } n \text{ is odd} \\ 0 & \text{if } n \text{ is even} \end{cases} \tag{3.104}$$

The potential solution is then the summation of $\varphi_n(x,y)$ in (3.102) with the coefficient C'_n given by (3.104), i.e.,

$$\varphi(x,y) = \frac{4U_0}{\pi}\sum_{n=1,3,5,\cdots}^{\infty}\frac{1}{n\sinh(n\pi a/b)}\sinh\frac{n\pi}{b}(a-x)\sin\frac{n\pi}{b}y$$
$$(0 < x < a \quad \text{and} \quad 0 < y < b) \tag{3.105}$$

The electric field distribution within the enclosure is obtained by the relation

$$\boldsymbol{E}(x,y) = -\nabla\varphi(x,y) = -\left(\boldsymbol{e}_x\frac{\partial}{\partial x} + \boldsymbol{e}_y\frac{\partial}{\partial y}\right)\varphi(x,y)$$

Summary

Concepts

Laplacian operator(拉普拉斯算子)

Poisson's equation(泊松方程)

Laplace's equation(拉普拉斯方程)

Laws & Theorems

Uniqueness theorem(唯一性定理)

Methods

Method of images(镜像法)　　Method of separation of variables(分离变量法)

Review Questions

3.1 写出简单媒质中静电场电位满足的泊松方程。对于一般的媒质,该泊松方程是什么形式?

3.2 什么是静电场的唯一性定理?

3.3 若已知给定区域内 $\nabla^2\varphi=0$,是否可以推断该区域内的电位为常数? 若否,还需要什么条件才能得出该结论?

3.4 无限长均匀线电荷对平行于该线电荷的导体圆柱面的镜像是什么?

3.5 点电荷对接地导体球面的镜像是什么?

3.6 简述应用分离变量法求解静电场问题的原理和基本步骤。分离变量法适合于求解哪些类型的静电场问题?

3.7 静电场问题的分离变量法中 3 个分离系数 k_x、k_y 和 k_z 能全为实数么? 能全为虚数么? 为什么?

Problems

3.1 A large parallel-plate capacitor with height d is filled with two layers of dielectric slabs. The dielectric constant of the layer between $z=0$ and $z=0.8d$ is ε_{r1} and the dielectric constant of the layer between $z=0.8d$ and $z=d$ is ε_{r2}. The bottom plate at $z=0$ is grounded and the top plate at $z=d$ has a constant potential U_0. Assuming negligible fringing effect, determine

(1) the potential φ and electric field $\boldsymbol{E}$ inside the capacitor,

(2) the surface charge densities on the top and bottom plates.

3.2 Prove that the potential φ due to a charge distribution given in (2.58) satisfies Poisson's equation.

3.3 Prove that, if a potential function φ satisfies Laplace's equation in a given region and φ is constant on the boundary of the region, then φ is constant throughout the region.

3.4 Prove that a potential function satisfying Laplace's equation in a given region possesses no maximum or minimum within the region.

3.5 A point charge Q exists at a distance d above a large grounded conducting plate. Determine

(1) the surface charge density ρ_s on the conducting plate,

(2) the total charge induced on the conducting plate.

3.6 A straight-line charge of ρ_l is parallel to and at a height h from the surface of an infinitely large grounded conducting plate. Referring to Figure 3-14, prove that the surface charge induced on the plane is

$$\rho_s = \frac{-\rho_l h}{\pi(x^2 + h^2)}$$

3.7 Two semi-infinitely large conducting plates are located in the y-z plane and x-z plane respectively, as illustrated in Figure 3-15. A point charge of 200 mC is placed at point $A(1,3,0)$. Determine the electric potential and the electric field intensity at point $B(3,2,0)$.

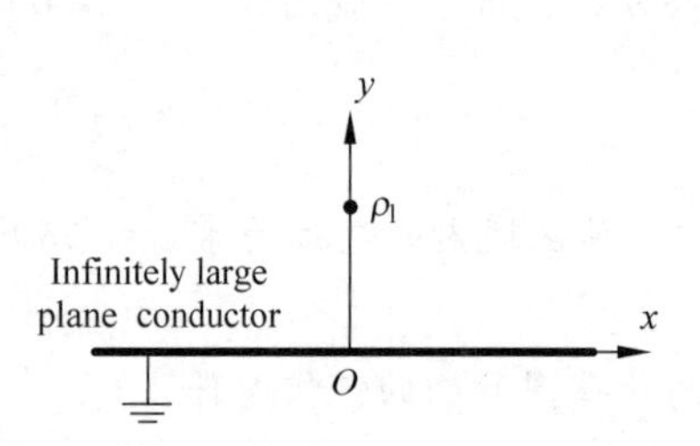

Figure 3-14 Illustration of Problem 3.6

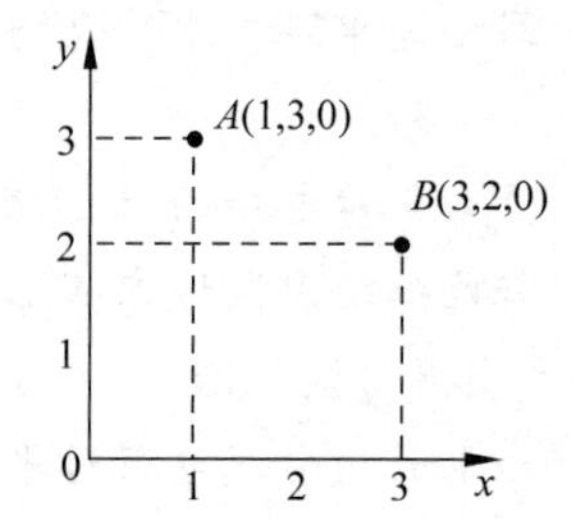

Figure 3-15 Illustration of Problem 3.7

3.8 Two semi-infinitely large grounded metal plates are located at $\phi = 0$ and $\phi = \pi/3$ respectively. A point charge q is situated at $\left(1, \frac{\pi}{6}, 0\right)$ in the cylindrical coordinate system. Find the potential at point $\left(3, \frac{\pi}{6}, 0\right)$.

3.9 A straight conducting wire of radius a is parallel to and at height h from the surface of the earth. Assuming that the earth is perfectly conducting, determine the capacitance between the wire and the earth.

3.10 A point charge Q resides inside a hollow spherical cavity with a grounded conducting shell. The radius of the cavity is a, and the point charge is at a distance d from the cavity center (where $a>d$). Use the method of images to determine ①the potential distribution inside the cavity, ②the charge density ρ_s induced on the inner surface of the shell.

3.11 A point charge Q is located at (x_0, y_0) outside a conducting hemisphere of radius a on top of an infinitely large conducting plate, as shown in Figure 3-16. Find the locations and values of the image charges that are needed for solving the fields outside the conductor.

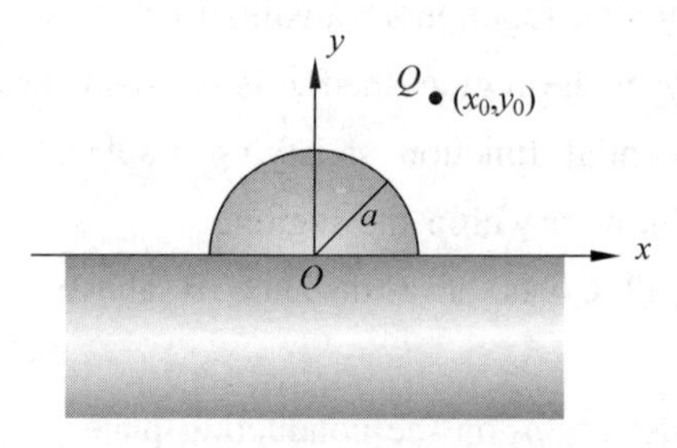

Figure 3-16 Illustration of Problem 3.11

3.12 Repeat solving the problem in Example 3-9 with the boundary conditions on the top, bottom, and right plates in Figure 3-13 changed to $\partial\varphi/\partial n = 0$.

3.13 For Example 3-9, if the top, bottom, and left plates in Figure 3-13 are grounded

($\varphi=0$) and the right plate is maintained at a constant potential U_0, prove that the potential distribution within the enclosed region is

$$\varphi(x,y)=\sum_{n=1}^{\infty}\frac{2U_0[1-(-1)^n]}{n\pi\sinh\left(\frac{n\pi a}{b}\right)}\sinh\frac{n\pi x}{b}\sin\frac{n\pi y}{b}$$

3.14　Consider the region enclosed by four conducting plates as shown in Figure 3-17 (the four plates are assumed to be infinitely long along the z-direction). The left and right plates are grounded, and the top and bottom plates have constant potentials U_1 and U_2 respectively. Find the potential distribution inside the enclosure.

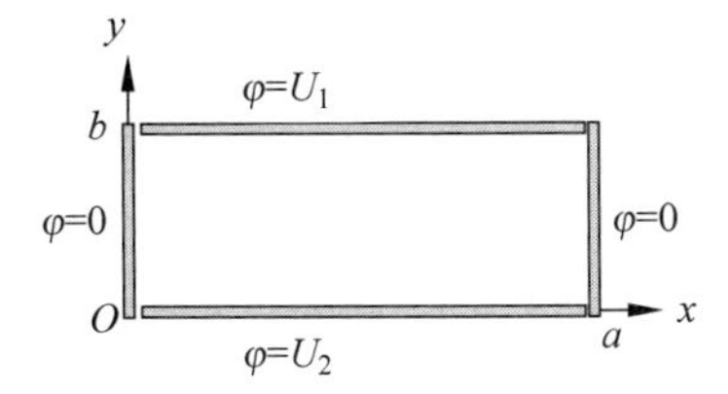

Figure 3-17　Illustration of Problem 3.14

3.15　Consider a metallic rectangular box with sides a and b and height c. The side walls and the bottom surface are grounded. The top surface is isolated and kept at a constant potential U_0. Determine the potential distribution inside the box.

Chapter 4 Steady Electric Currents

(恒定电流)

4.1 Introduction(引言)

In the electrostatic problems studied in Chapter 2 and 3, the electric charges are stationary. In this chapter, we study steady electric currents that comprise electric charges moving with constant speed. Steady currents are sources of static magnetic fields, which will be discussed in the next chapter. There are several types of electric currents depending on the mechanism that causes the motion of the electric charges. **Conduction currents**(传导电流) in conductors and semiconductors are caused by the drift motion of conduction electrons and/or holes. **Electrolytic currents**(电解电流) are the result of migration of **ions**(离子). **Convection currents**(运流电流) are results of the motion of electrons and/or ions in vacuum. In this chapter, we only discuss steady conduction currents that are governed by Ohm's law. The steady current satisfies the **equation of continuity**(连续性方程) that can be derived from the principle of conservation of charges. We introduce the concept of **electromotive force**(电动势) that is necessary to maintain a steady current within a conducting medium. We then derive the boundary conditions between two media of different conductivities. Finally, we show how to calculate the **resistance/conductance** (电阻/电导) of a conducting medium. ①

4.2 Current Density, Ohm's Law and Joule's Law (电流密度、欧姆定律与焦耳定律)

Consider the steady current due to the motion of **charge carriers**(电荷载流子), each carrying charge q (which is negative for electrons) and moving across a differential surface element d$\boldsymbol{s}$ with a velocity $\boldsymbol{u}$ as shown in Figure 4-1. If n is the number of charge carriers per

---

① 恒定电流是带电粒子(电荷载流子)定向移动形成的。导体或半导体中电子或空穴的定向移动形成的是传导电流,电解质中化学反应导致的离子定向移动形成的是电解电流,真空或者稀薄空气中(例如真空管中)电子或离子的定向移动形成的是运流电流。第6章将引入时变电磁场中极为重要的一个概念:位移电流。位移电流不是电荷定向移动生成的,且随时间是变化的,因此不属于恒定电流的范畴。本章仅讨论传导电流,即导电媒质中自由电荷在恒定电场力作用下定向移动形成的电流。

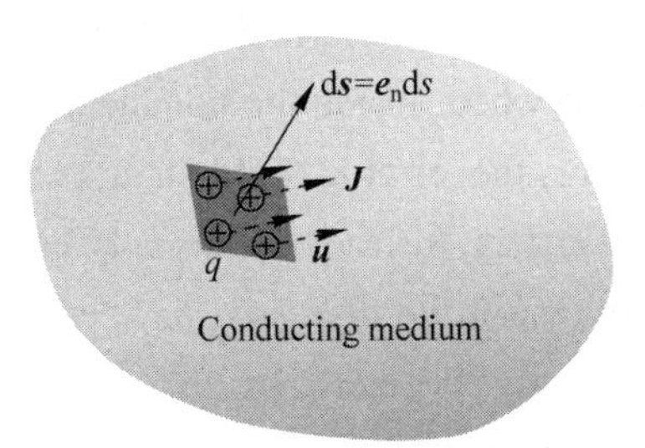

Figure 4-1 Electric current due to moving charge carriers

unit volume, then in time Δt, each charge carrier moves a distance $\boldsymbol{u}\Delta t$, and the amount of charge passing through the surface $\mathrm{d}\boldsymbol{s}$ is

$$\Delta Q = nq(\boldsymbol{u}\Delta t) \cdot \mathrm{d}\boldsymbol{s} \tag{4.1}$$

Then the current through the surface $\mathrm{d}\boldsymbol{s}$ is

$$\mathrm{d}I = \frac{\Delta Q}{\Delta t} = nq\boldsymbol{u} \cdot \mathrm{d}\boldsymbol{s} \tag{4.2}$$

We define a **volume current density**(**体电流密度**), or simply **current density**(**电流密度**), $\boldsymbol{J}$, to be

$$\boldsymbol{J} = nq\boldsymbol{u}(\mathrm{A/m^2}) \tag{4.3}$$

Then(4.2) can be written as

$$\mathrm{d}I = \boldsymbol{J} \cdot \mathrm{d}\boldsymbol{s} \tag{4.4}$$

and the total current I through an arbitrary surface S can be expressed as

$$I = \int_s \boldsymbol{J} \cdot \mathrm{d}\boldsymbol{s} \tag{4.5}$$

If the conducting medium has more than one kind of charge carriers(electrons, holes and ions) drifting with different velocities, (4.3) should be generalized to be

$$\boldsymbol{J} = \sum_i n_i q_i \boldsymbol{u}_i \tag{4.6}$$

where n_i is the number of the ith kind of charge carrier per unit volume.

Conduction currents are the result of the drift motion of charge carriers under the influence of an applied electric field. For most conducting materials, the average drift velocity is directly proportional to the electric field intensity. For metallic conductors we have

$$\boldsymbol{u} = -\mu_e \boldsymbol{E} \tag{4.7}$$

where μ_e is the electron mobility(迁移率) in($\mathrm{m^2/V \cdot s}$). The electron mobility is 3.2×10^{-3} ($\mathrm{m^2/V \cdot s}$) for copper(铜) and 5.2×10^{-3}($\mathrm{m^2/V \cdot s}$) for silver(银). From(4.3) and(4.7) we have

$$\boldsymbol{J} = -\rho_{ve}\mu_e \boldsymbol{E} \tag{4.8}$$

where $\rho_{ve} = -ne$ is the charge density of the drifting electrons with $e = -1.6\times10^{-19}$C to be the charge of an electron. Rewrite(4.8) as

$$\boldsymbol{J} = \sigma \boldsymbol{E} \tag{4.9}$$

where the proportionality constant, $\sigma = -\rho_{ve}\mu_e$, is called the **conductivity** (**电导率**) of the medium. The unit for σ is ampere per volt-meter($\mathrm{A/V \cdot m}$) or siemens per meter(S/m).

It is worth pointing out that, although there may exist drifting charges with a density of ρ_{ve} in a conductor, the net charge everywhere inside the conductor is still zero. This will be derived in detail in Section 4-2. Therefore, the electric field in a conducting medium is usually not produced by the moving charge carriers. Instead, it is usually established by the static charges at the terminals of a DC source, which will be discussed in Section 4-4. ①

For semiconductors, conductivity depends on the concentration and mobility of both electrons and holes, and can be expressed as

$$\sigma = -\rho_{ve}\mu_e + \rho_{vh}\mu_h \tag{4.10}$$

where μ_e and μ_h are respectively the mobility of the electrons and holes. For germanium(锗), typical values are $\mu_e = 0.38$, $\mu_h = 0.18$; for silicon(硅), $\mu_e = 0.12$, $\mu_h = 0.03$ ($m^2/V \cdot s$).

Equation (4.9) is the constitutive relation of a conducting medium. For isotropic materials, σ is a scalar, and these materials are also called **ohmic media**(**欧姆材料**). The conductivities of a few materials are listed in Table 4-1. Notice that most metal materials have a large conductivity in the order of 10^7. On the contrary, glass and rubber are usually considered as insulators.

Table 4-1 Conductivity of some materials

Medium	Conductivity/(S/m)	Medium	Conductivity/(S/m)
See water	4	Silver	6.2×10^7
Germanium	2.2	Copper	5.8×10^7
Silicon	1.6×10^{-3}	Gold	4.1×10^7
Dry soil	10^{-5}	Aluminum	3.53×10^7
Glass	10^{-12}	Brass	1.57×10^7
Rubber	10^{-15}	Iron	10^7

(4.9) can also be interpreted as the **Ohm's law**(**欧姆定律**) in the differential form. It can be derived from the normal form of Ohm's law in circuit theory. To show that, consider a piece of homogeneous material of conductivity σ, length L, and uniform cross section S, as shown in Figure 4-2. Ohm's law states that

$$U = RI = \left(\frac{L}{\sigma S}\right) I \tag{4.11}$$

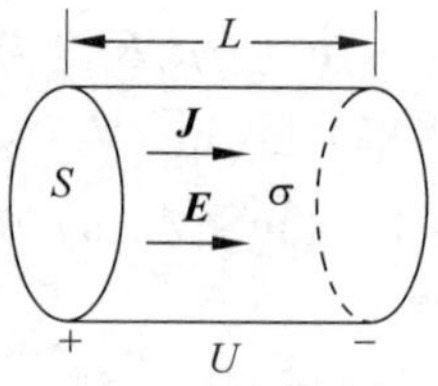

Figure 4-2 Homogeneous conductor with a constant cross section

① 这里需要注意导电媒质中电荷载流子的密度与净电荷的密度是两个不同的概念。电荷载流子通常不是导电媒质中电场的源,因为导电媒质中还同时存在与电荷载流子极性相反,但不发生移动的固定离子。在4.4节将证明,电流恒定状态下导电媒质内部的净电荷密度为零。

where

$$R = \frac{L}{\sigma S} \tag{4.12}$$

is the resistance of the conducting material. Consider the conductor piece to be small enough so that both $\boldsymbol{J}$ and $\boldsymbol{E}$ are uniform and along the same direction of the current flow. Then the potential difference or voltage between the two terminals is

$$U = \int_L \boldsymbol{E} \cdot \mathrm{d}\boldsymbol{l} = EL \tag{4.13}$$

The total current is

$$I = \int_S \boldsymbol{J} \cdot \mathrm{d}\boldsymbol{s} = JS \tag{4.14}$$

Substitute(4.13) and(4.14) into(4.11), we obtain

$$E = \frac{J}{\sigma}$$

which is the exactly the constitutive relation(4.9).

Example 4-1 Determine the DC resistance of a copper wire with ① a 1km length and a 1mm radius ② a 1cm length and a 0.1mm radius.

Solution: Assume the wire has a uniform cross section so that(4.12) can be used to find the resistance. The conductivity of copper is $\sigma_{cu} = 5.80 \times 10^7$ S/m.

(1) For $L = 10^3$ m, $S = 10^{-6}\pi$ m^2,

$$R = \frac{L}{\sigma_{cu} S} = \frac{10^3}{5.80 \times 10^7 \times 10^{-6}\pi} = 5.49\ (\Omega)$$

(2) For $L = 10^{-2}$ m, $S = (0.1 \times 10^{-3})^2\pi = 10^{-8}\pi(\mathrm{m}^2)$,

$$R = \frac{L}{\sigma_{cu} S} = \frac{10^{-2}}{5.80 \times 10^7 \times 10^{-8}\pi} = 0.00549(\Omega)$$

The reciprocal of resistance is **conductance** (**电导**), denoted by G. The unit for conductance is siemens(S). So, the conductance of a conductor with conductivity σ, length L, and cross section S is

$$G = \frac{1}{R} = \sigma \frac{S}{l} \tag{4.15}$$

From circuit theory, power dissipation(功率消耗) occurs whenever electric currents pass through a resistor. This can be explained as the result of the collision of the moving electrons with atoms, which transforms the electric energy to heat. The work Δw done by an electric field $\boldsymbol{E}$ in moving a charge q a distance $\mathrm{d}\boldsymbol{l}$ is $q\boldsymbol{E} \cdot \mathrm{d}\boldsymbol{l}$, which corresponds to a power

$$p = \lim_{\Delta t \to 0} \frac{\Delta w}{\Delta t} = \lim_{\Delta t \to 0} q\boldsymbol{E} \cdot \left(\frac{\mathrm{d}\boldsymbol{l}}{\Delta t}\right) = q\boldsymbol{E} \cdot \boldsymbol{u} \tag{4.16}$$

where $\boldsymbol{u}$ is the drift velocity of the charge. The total power delivered to all kinds of charge carriers in a volume $\mathrm{d}v$ is

$$\mathrm{d}p = \sum_i p_i = \boldsymbol{E} \cdot \left(\sum_i n_i q_i \boldsymbol{u}_i\right) \mathrm{d}v$$

which, by using (4.6), becomes

$$dp = \boldsymbol{E} \cdot \boldsymbol{J} dv$$

And therefore, the **power dissipation density**(损耗功率密度) is

$$\frac{dp}{dv} = \boldsymbol{E} \cdot \boldsymbol{J} \tag{4.17}$$

For a given volume V, the total electric power converted into heat is

$$P = \int_V \boldsymbol{E} \cdot \boldsymbol{J} dv \tag{4.18}$$

which is known as **Joule's law** (焦耳定律). Therefore, (4.17) is a differential form of Joule's law. Given the conductivity σ, (4.17) and (4.18) can also be written as

$$\frac{dp}{dv} = \sigma |\boldsymbol{E}|^2 = \frac{|\boldsymbol{J}|^2}{\sigma} \tag{4.19}$$

$$P = \int_V \sigma |\boldsymbol{E}|^2 dv = \int_V \frac{|\boldsymbol{J}|^2}{\sigma} dv \tag{4.20}$$

In a conductor of a constant cross section, take $dv = d\boldsymbol{s} \cdot d\boldsymbol{l}$, with $d\boldsymbol{s}$ and $d\boldsymbol{l}$ both along the same direction of $\boldsymbol{J}$ and $\boldsymbol{E}$. Then (4.18) can be written as

$$P = \int_L \boldsymbol{E} \cdot d\boldsymbol{l} \int_S \boldsymbol{J} \cdot d\boldsymbol{s} = UI \tag{4.21}$$

where I is the total current in the conductor. Since $U = RI$, we have

$$P = I^2 R \tag{4.22}$$

which is the expression of Joule's law in circuit theory.

4.3 Divergence of Current Density and Conservation of Charge(电流密度的散度与电荷守恒定律)

One of the fundamental postulates of physics is **the principle of conservation of charge** (电荷守恒原理), which states that electric charges may not be created or destroyed, and the net charge in an isolated system never changes. Consider an arbitrary volume V bounded by surface S, within which a net charge Q exists. If a net current I flows across the surface S out of this region, the charge in the volume must decrease at a time rate that equals the current. Conversely, if a net inward current I flows across the surface S, the charge in the volume V must increase at a time rate equal to the current. The current leaving the region is the total outward flux of the current density vector through the surface S. Therefore, we have

$$I = \oint_S \boldsymbol{J} \cdot d\boldsymbol{s} = -\frac{dQ}{dt} = -\frac{d}{dt}\int_V \rho_v dv \tag{4.23}$$

where ρ_v is the net charge density in the volume V. Using divergence theorem, the surface integral of $\boldsymbol{J}$ in (4.23) can be converted to a volume integral of $\nabla \cdot \boldsymbol{J}$. Therefore, we have

$$\int_V \nabla \cdot \boldsymbol{J} dv = -\int_V \frac{\partial \rho_v}{\partial t} dv \tag{4.24}$$

Since (4.24) must hold for any region V, the integrands must be equal. Thus, we have

$$\nabla \cdot \boldsymbol{J} = -\frac{\partial \rho_v}{\partial t} \tag{4.25}$$

(4.25) is called the **equation of continuity**(**连续性方程**), which is the result of charge conservation.[1]

In Section 2-5, it is concluded that there is no volume charge and $\boldsymbol{E}=0$ inside a conductor under equilibrium conditions. This can also be proved by using the equation of continuity(4.25). Substituting Ohm's law(4.9) into(4.25) and assuming a constant σ, we have

$$\sigma \nabla \cdot \boldsymbol{E} = -\frac{\partial \rho_v}{\partial t} \tag{4.26}$$

In a simple medium, $\nabla \cdot \boldsymbol{E}=\rho_v/\varepsilon$, and therefore, (4.26) becomes

$$\frac{\partial \rho_v}{\partial t} + \frac{\sigma}{\varepsilon}\rho_v = 0 \tag{4.27}$$

The solution of(4.27) is

$$\rho_v = \rho_{v0} e^{-(\sigma/\varepsilon)t} \tag{4.28}$$

where ρ_{v0} is the initial charge density at $t=0$. (4.28) states that the charge density at a given location will decrease with time exponentially. An initial charge density ρ_{v0} will decay to $1/e$ or 36.8% of its value in a time equal to

$$\tau = \frac{\varepsilon}{\sigma} \tag{4.29}$$

The time constant τ is called the **relaxation time**(**弛豫时间**). For a good conductor such as copper, $\sigma=5.80\times10^7$ S/m, $\varepsilon\approx\varepsilon_0=8.85\times10^{-12}$ F/m, and by(4.29), τ equals 1.52×10^{-12} s which is an extremely short time interval. Therefore, realistically, ρ_v can be considered zero inside a good conductor. Any charge carried by a good conductor can exist only on the surface. In contrast, the relaxation time for a good insulator(with very small σ) can be hours or days.

For a steady current, its divergence must be constant. From(4.25), this constant must be zero, because otherwise the charge density ρ_v will grow to infinity at a constant rate. Hence(4.25) becomes

$$\nabla \cdot \boldsymbol{J} = 0 \tag{4.30}$$

for steady currents. In other words, steady electric currents are divergenceless or solenoidal. Over any enclosed surface, (4.30) leads to the following integral form:

$$\oint_S \boldsymbol{J} \cdot \mathrm{d}\boldsymbol{s} = 0 \tag{4.31}$$

which means that the streamlines of steady currents must close upon themselves(if not end at infinity).

Apply(4.31) to a junction in an electric circuit, we have

$$\sum_j I_j = 0 \tag{4.32}$$

---

[1] 需要注意的是,式(4.23)~式(4.25)中的电荷密度 ρ_v 包含所有的电荷,包括静止的和移动的;而式(4.8)和式(4.10)中的 ρ_{ve} 和 ρ_{vh} 仅仅是载流子电荷的密度。

where I_j is the current flowing out of the junction in the jth branch. (4.32) is an expression of **Kirchhoff's current law** (基尔霍夫电流定律), which is the basis for node analysis in circuit theory. ①

4.4 Curl of Steady Electric Field and Electromotive Force(恒定电场的旋度与电动势)

According to(4.9), the steady current in a conducting medium is the result of a steady electric field that drifts the charge carriers. The steady electric field $\boldsymbol{E}$ is due to static charges, which is not different from an electrostatic field and must be conservative(Why?). Therefore, we have,

$$\nabla \times \boldsymbol{E} = 0 \tag{4.33}$$

Therefore, the scalar line integral of the steady electric intensity around any closed path is zero, i.e.,

$$\oint_C \boldsymbol{E} \cdot \mathrm{d}\boldsymbol{l} = 0 \tag{4.34}$$

For an ohmic material with conductivity σ, (4.33) together with(4.9) leads to

$$\nabla \times \left(\frac{\boldsymbol{J}}{\sigma}\right) = 0 \tag{4.35}$$

And(4.34) leads to

$$\oint_C \left(\frac{\boldsymbol{J}}{\sigma}\right) \cdot \mathrm{d}\boldsymbol{l} = 0 \tag{4.36}$$

For a homogeneous conducting medium, (4.35) simplifies to

$$\nabla \times \boldsymbol{J} = 0 \tag{4.37}$$

And(4.36) simplifies to

$$\oint_C \boldsymbol{J} \cdot \mathrm{d}\boldsymbol{l} = 0 \tag{4.38}$$

(4.36) or(4.38) tells us that a steady current cannot be maintained in the same direction in a closed circuit simply by a conservative electrostatic field. In other words, there must exist a non-conservative force that drives the steady flow of the electric current. ②

From the point of view of energy, a steady current in a circuit in a conductor with finite conductivity must dissipate power as indicated by(4.19). This power must come from a non-conservative field, because a charge carrier completing a closed circuit in a conservative field neither gains nor loses energy. The source of the non-conservative field may be electric batteries (converting chemical energy to electric energy), electric generators (converting mechanical

---

① 式(4.32)给出的基尔霍夫电流定律：进入电路中某节点的所有电流的代数和为零。

② 式(4.36)源自导电媒质中恒定电场的无旋性质，该式表明导电媒质中的恒定电流若仅由保守的恒定电场驱动，是无法形成闭合回路的。然而，式(4.31)表明恒定电流场的矢量线一般需要形成闭合回路，这是电荷守恒定律要求的。因此，恒定电流要形成闭合回路，必须存在一种非保守力。

energy to electric energy), thermocouples (converting thermal energy to electric energy), photovoltaic cells (converting light energy to electric energy), or other devices. These energy sources, when connected to an electric circuit, provide the necessary non-conservative force for the charge carriers. In the following analysis, this force is modeled as an **equivalent impressed electric field intensity** (**等效外加电场强度**) $\boldsymbol{E}_i$.[①]

Consider an electric battery with two electrodes, as shown in Figure 4-3. Chemical action creates an accumulation of positive and negative charges at the electrodes. These charges generate an electrostatic field intensity $\boldsymbol{E}$ both outside and inside the battery. Outside the battery (inside the conducting medium connecting the two electrodes), charge carriers move due to the $\boldsymbol{E}$ field, leading to a steady current $\boldsymbol{J}$. However, inside the battery, charge carriers move along the direction against $\boldsymbol{E}$, which is the result of a non-conservative $\boldsymbol{E}_i$ due to the chemical action. The line integral of the impressed field intensity $\boldsymbol{E}_i$ from the negative to the positive electrode inside the battery is referred to as the **electromotive force** (**emf**, **电动势**) of the battery. Apparently, the unit for emf is volt. The electromotive force, denoted by $\mathcal{E}$, is a measure of the strength of the non-conservative source. By definition, we have

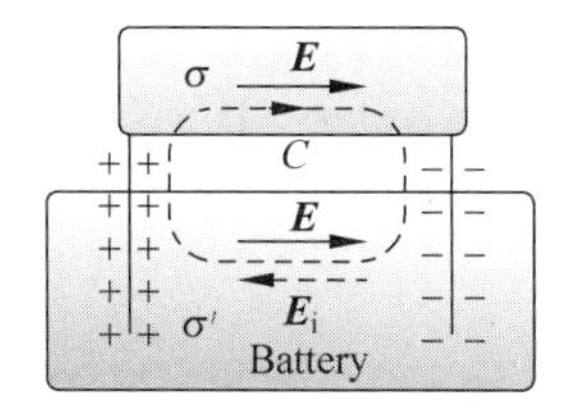

Figure 4-3 Electric fields in the circuit with a battery

$$\mathcal{E}=\int_{-}^{+} \boldsymbol{E}_i \cdot \mathrm{d}\boldsymbol{l} \tag{4.39}$$

Since $\boldsymbol{E}_i$ is zero outside the battery, (4.39) can be rewritten as

$$\mathcal{E}=\oint_C \boldsymbol{E}_i \cdot \mathrm{d}\boldsymbol{l} \tag{4.40}$$

where C is the closed path of the circuit in Figure 4-3. As the conservative electrostatic field intensity $\boldsymbol{E}$ satisfies (4.34), we can also rewrite (4.40) as

$$\mathcal{E}=\oint_C (\boldsymbol{E}+\boldsymbol{E}_i) \cdot \mathrm{d}\boldsymbol{l} \tag{4.41}$$

(4.40) and (4.41) provide the general definition of the emf in a close path. Generally, the emf of a closed path is not necessarily produced by a battery, and $\boldsymbol{E}_i$ is not necessarily due to chemical action. For example, the emf can be produced by the time-varying magnetic flux that links the closed path C, which will be discussed in Chapter 6. In that case, $\boldsymbol{E}_i$ is a non-conservative time-varying electric field due to the time-varying magnetic field.[②]

In a DC circuit, (4.41) shows that the line integral of the total electric field (including conservative and non-conservative components) over a closed path is nonzero if the path goes through a battery, and the integral value equals the emf of the battery. If the closed path does not

---

① 从能量守恒的角度出发，导电媒质中的恒定电流必然会消耗功率，而该消耗的功率不可能由恒定电场提供。这是因为恒定电场是保守场，沿着恒定电流形成的闭合回路，恒定电场力对电荷载流子做功必然为零。因此，要维持恒定电流，必须存在某种非保守力，该非保守力将其他物理形式的能量转换为恒定电流消耗的能量。

② 式(4.40)或式(4.41)给出了电动势的一般性定义，沿某一闭合曲线 C 的总电场(包括保守的静态场 $\boldsymbol{E}$ 和非保守的等效电场 $\boldsymbol{E}_i$)的环量积分等于该闭合曲线上的电动势，其参考正方向为线积分的方向。

go through a battery, the integral must be zero.

In the conducting medium outside the battery, we know that $\boldsymbol{J}=\sigma\boldsymbol{E}$. Inside the battery, a similar relation between the current and electric fields exists, that is, the charge carriers are drifted by combined forces of $\boldsymbol{E}$ and $\boldsymbol{E}_{\mathrm{i}}$. Let σ' be an equivalent conductivity inside the battery, so that we can write

$$\boldsymbol{J}=\sigma'(\boldsymbol{E}+\boldsymbol{E}_{\mathrm{i}}) \tag{4.42}$$

inside the battery. Then (4.41) becomes

$$\begin{aligned}\mathcal{E}&=\underbrace{\int_{-}^{+}(\boldsymbol{E}+\boldsymbol{E}_{\mathrm{i}})\cdot \mathrm{d}\boldsymbol{l}}_{\text{Inside the battery}}+\underbrace{\int_{+}^{-}\boldsymbol{E}\cdot \mathrm{d}\boldsymbol{l}}_{\text{Outside the battery}}\\&=\underbrace{\int_{-}^{+}\left(\frac{\boldsymbol{J}}{\sigma'}\right)\cdot \mathrm{d}\boldsymbol{l}}_{\text{Inside the battery}}+\underbrace{\int_{+}^{-}\left(\frac{\boldsymbol{J}}{\sigma}\right)\cdot \mathrm{d}\boldsymbol{l}}_{\text{Outside the battery}}\end{aligned} \tag{4.43}$$

Assume both the battery and the conductor outside have a uniform cross section, then (4.43) becomes

$$\mathcal{E}=R'I+RI \tag{4.44}$$

where R is the resistance of the conductor outside the battery and R' is the internal resistance of the battery. R and R' are respectively associated with σ and σ' through the relation of (4.12).

If a closed path in an electric circuit contains multiple sources of emf and multiple resistors, (4.44) is generalized to be

$$\sum_{i=1}^{M}\mathcal{E}_i=\sum_{j=1}^{M}R_j'I_j+\sum_{k=1}^{N}R_kI_k \tag{4.45}$$

where M is the number of batteries, R_j' is the internal resistance of the jth battery, and N is the number of resistors outside the batteries. Notice that, in (4.45), we assume the closed-path circuit is composed of only M batteries and N resistors, and the term R_kI_k represents the voltage drop across the kth branch in the circuit. Nevertheless, not all the branches in the closed-path circuit must be a resistor; not all the branches contain electric currents. In that case, a more general version of (4.45) which can also be derived from (4.43) directly is

$$\sum_{i=1}^{M}\mathcal{E}_i=\sum_{j=1}^{M}R_j'I_j+\sum_{k=1}^{N}U_k \tag{4.46}$$

where U_k is the voltage drop across the kth branch. (4.46) is an expression of **Kirchhoff's voltage law (基尔霍夫电压定律)**. It states that, around a closed path in an electric circuit, the algebraic sum of the emfs is equal to the algebraic sum of the internal resistance times current inside the batteries and the voltage drops outside the batteries. ①

In the circuit of Figure 4-3, if the conductor outside the battery is removed, leaving the battery open-circuited, the current in (4.46) becomes zero. The only branch outside the battery has a voltage drop U_{open} from the positive to the negative electrode. Therefore, from (4.46), we have

$$\mathcal{E}=U_{\mathrm{open}}$$

---

① 基尔霍夫电压定律的一种形式：沿任意闭合电路回路，所有电动势的代数和等于所有电源内部电流与内阻的乘积与电源外部电压降的代数和。

This result can also be derived directly from(4.43), where the conservative field $\boldsymbol{E}$ inside the battery must be equal in magnitude and opposite in direction to $\boldsymbol{E}_i$. This is because no current flows in the open-circuited battery, and thus the net force acting on the charge carriers must be zero. Hence we have

$$\mathcal{E}=\int_{-}^{+} -\boldsymbol{E}_{\text{open}} \cdot \mathrm{d}\boldsymbol{l} = \int_{+}^{-} \boldsymbol{E}_{\text{open}} \cdot \mathrm{d}\boldsymbol{l} = U_{\text{open}} \tag{4.47}$$

(4.47) means that the emf of a battery equals the open output voltage U_{open} of the battery. Notice that $\boldsymbol{E}_{\text{open}}$ in (4.47) is generally larger than $\boldsymbol{E}$ in (4.41), unless the internal resistance of the battery is zero(in other words, unless the battery is an **ideal voltage source**,**理想电压源**).①

4.5 Boundary Conditions for Current Density (电流密度的边界条件)

When a current flows across an interface between two different media, the current density vector changes in both direction and magnitude. A set of boundary conditions can be derived for $\boldsymbol{J}$ and $\boldsymbol{E}$ in a way like that used in Section 2.8 for $\boldsymbol{D}$ and $\boldsymbol{E}$. Here we limit the analysis in the region outside the non-conservative energy sources, then the governing equations for $\boldsymbol{J}$ in the integral form have been given as(4.31) and(4.36). By applying(4.31) and(4.36) at the interface between two ohmic media with conductivities σ_1 and σ_2 respectively, we obtain the boundary conditions for the normal and tangential components of $\boldsymbol{J}$. The derivation process is very similar to that in Section 2.8. Specifically, from(4.31), we have

$$J_{1n} = J_{2n} \tag{4.48}$$

which states that the normal component of $\boldsymbol{J}$ is continuous across an interface. From(4.36), we conclude that J_t/σ is continuous, i.e.,

$$\frac{J_{1t}}{J_{2t}} = \frac{\sigma_1}{\sigma_2} \tag{4.49}$$

(4.49) states that the ratio of the tangential components of $\boldsymbol{J}$ at two sides of an interface is equal to the ratio of the conductivities.

Example 4-2 Figure 4-4 illustrates the steady current density at an interface between two conducting media with different conductivities σ_1 and σ_2. The current density at a point on the interface in the medium 1 side has a magnitude J_1 and makes an angle α_1 with the normal direction $\boldsymbol{e}_{n2}$. Determine the magnitude and direction of the current density at the same point in the medium 2 side.

Solution: Using(4.48) and(4.49), we have

$$J_1\cos\alpha_1 = J_2\cos\alpha_2 \tag{4.50}$$

and

---

① 本节引入的非保守等效电场、电动势以及电压源等概念,是形成恒定电流所必需的。但本章主要讨论的还是导电媒质中(即电源外部)的恒定电场和恒定电流,而不涉及非保守等效电场和电动势。

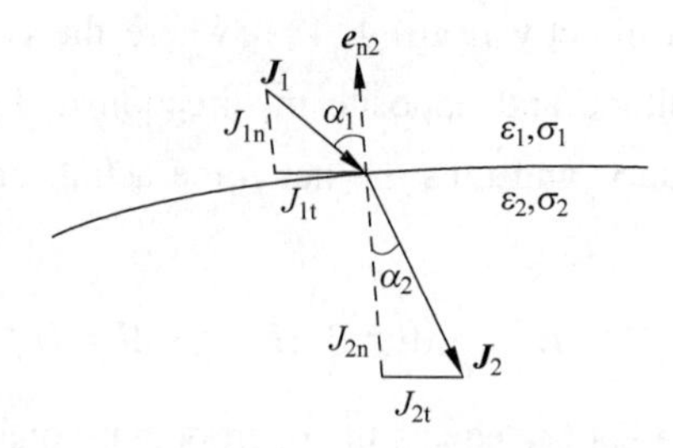

Figure 4-4 Boundary conditions at the interface between two conducting media (Example 4-2)

$$\frac{J_1 \sin\alpha_1}{\sigma_1} = \frac{J_2 \sin\alpha_2}{\sigma_2} \tag{4.51}$$

Division of (4.51) by (4.50) yields

$$\frac{\tan\alpha_2}{\tan\alpha_1} = \frac{\sigma_2}{\sigma_1} \tag{4.52}$$

If $\sigma_1 \gg \sigma_2$, α_2 approaches zero, and $\boldsymbol{J}_2$ is approximately perpendicularly to the interface. That means if the current flows outward from a good conductor to a poor conductor, the current as well as the electric field in the poor conductor is normal to the interface.

Generally, the magnitude of $\boldsymbol{J}_2$ can be found as

$$J_2 = \sqrt{J_{2t}^2 + J_{2n}^2} = \sqrt{(J_2 \sin\alpha_2)^2 + (J_2 \cos\alpha_2)^2} = \left[\left(\frac{\sigma_2}{\sigma_1} J_1 \sin\alpha_1\right)^2 + (J_1 \cos\alpha_1)^2\right]^{1/2}$$

or

$$J_2 = J_1 \left[\left(\frac{\sigma_2}{\sigma_1} \sin\alpha_1\right)^2 + \cos^2\alpha_1\right]^{1/2} \tag{4.53}$$

When a steady current flows across the boundary between two different lossy (有损的) dielectrics with permittivities ε_1, ε_2 and finite conductivities σ_1, σ_2 respectively. ① The tangential component of the electric field is continuous across the interface as usual. That is, $E_{1t} = E_{2t}$. The normal component of the electric field, however, must simultaneously satisfy (4.48) and (2.106), i.e.,

$$J_{1n} = J_{2n} \rightarrow \sigma_1 E_{1n} = \sigma_2 E_{2n} \tag{4.54}$$

$$D_{1n} - D_{2n} = \rho_s \rightarrow \varepsilon_1 E_{1n} - \varepsilon_2 E_{2n} = \rho_s \tag{4.55}$$

where the reference unit normal is outward from medium 2. Hence, unless $\sigma_2/\sigma_1 = \varepsilon_2/\varepsilon_1$, a surface charge must exist at the interface. ②

From (4.54) and (4.55) we find

$$\rho_s = \left(\varepsilon_1 \frac{\sigma_2}{\sigma_1} - \varepsilon_2\right) E_{2n} = \left(\varepsilon_1 - \varepsilon_2 \frac{\sigma_1}{\sigma_2}\right) E_{1n} \tag{4.56}$$

---

① 导电媒质通常也称为有损媒质，因为导电媒质中的传导电流必然带来损耗。

② 与静电场不同，存在恒定电流时，两种不同的导电媒质分界面上有可能积累自由电荷，最终平衡状态下，分界面上的面电荷密度使得电场强度的法向分量同时满足式(4.54)和式(4.55)两个边界条件。在达到平衡状态之前，分界面上的面电荷密度处于不断积累变化的动态过程，该过程中电场强度也会不断变化，其法向分量仍然满足式(4.55)，但不满足式(4.54)。

Notice that the subscript n in above expressions represent the normal direction pointing from medium 2 to medium 1. If medium 2 is a much better conductor than medium 1(i. e. ,$\sigma_2 \gg \sigma_1$ or $\sigma_1/\sigma_2 \to 0$),(4.56) becomes approximately

$$\rho_s = \varepsilon_1 E_{1n} = D_{1n} \tag{4.57}$$

which is the same as(2.107).

Example 4-3 A DC voltage U is applied across a parallel-plate capacitor of area S as is shown in Figure 4-5. Between the two conducting plates, there are two different lossy dielectrics with thicknesses d_1 and d_2, permittivities ε_1 and ε_2, and conductivities σ_1 and σ_2, respectively. Determine ① the current density between the plates, ② the electric field intensities in both dielectrics, and ③ the surface charge densities on the plates and at the interface.

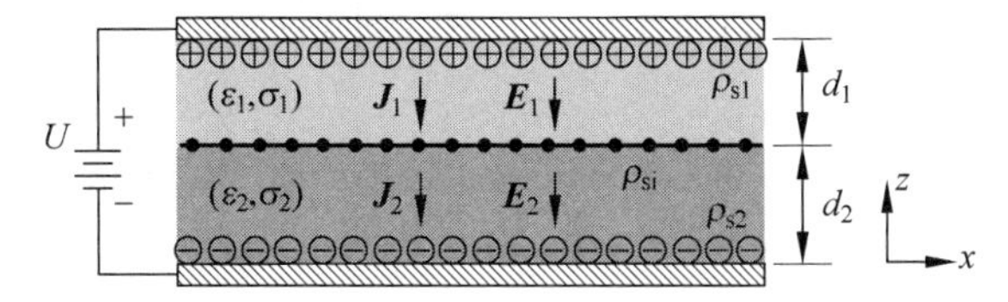

Figure 4-5 Parallel-plate capacitor with two lossy dielectrics(Example 4-3)

Solution: Neglecting fringing effects, we can assume that the current and electric field are independent of x and y coordinates and only have z component within the capacitor.

(1) Let the current density $\boldsymbol{J} = -\boldsymbol{e}_z J$. The continuity of the normal component of the current assures that the $\boldsymbol{J}$ in both media are the same. Then we have

$$U = -\int_{-}^{+} \boldsymbol{E} \cdot d\boldsymbol{l} = -\int_0^{d_2} \frac{\boldsymbol{J}}{\sigma_2} \cdot \boldsymbol{e}_z dz - \int_{d_2}^{d_1+d_2} \frac{\boldsymbol{J}}{\sigma_1} \cdot \boldsymbol{e}_z dz = \frac{J}{\sigma_2} d_2 + \frac{J}{\sigma_1} d_1$$

Hence,

$$J = \frac{U}{(d_1/\sigma_1) + (d_2/\sigma_2)} = \frac{\sigma_1 \sigma_2 U}{\sigma_2 d_1 + \sigma_1 d_2} \tag{4.58}$$

and

$$\boldsymbol{J} = -\boldsymbol{e}_z \frac{\sigma_1 \sigma_2 U}{\sigma_2 d_1 + \sigma_1 d_2} \tag{4.59}$$

(2) Using Ohm's law, the electric field intensities in both media are

$$\boldsymbol{E}_1 = \frac{\boldsymbol{J}}{\sigma_1} = -\boldsymbol{e}_z \frac{\sigma_2 U}{\sigma_2 d_1 + \sigma_1 d_2} \tag{4.60}$$

and

$$\boldsymbol{E}_2 = \frac{\boldsymbol{J}}{\sigma_2} = -\boldsymbol{e}_z \frac{\sigma_1 U}{\sigma_2 d_1 + \sigma_1 d_2} \tag{4.61}$$

(3) The surface charge densities on the upper and lower plates can be determined by using the boundary conditions of conductors, i. e. ,

$$\rho_{s1} = \varepsilon_1 \boldsymbol{E}_1 \cdot \boldsymbol{e}_n = \varepsilon_1 \boldsymbol{E}_1 \cdot \boldsymbol{e}_z = \frac{\varepsilon_1 \sigma_2 U}{\sigma_2 d_1 + \sigma_1 d_2} \tag{4.62}$$

$$\rho_{s2} = \varepsilon_2 \boldsymbol{E}_2 \cdot \boldsymbol{e}_n = \varepsilon_2 \boldsymbol{E}_2 \cdot \boldsymbol{e}_z = -\frac{\varepsilon_2 \sigma_1 U}{\sigma_2 d_1 + \sigma_1 d_2} \tag{4.63}$$

(4.56) can be used to find the surface charge density at the interface of the dielectrics. We have

$$\begin{aligned}\rho_{si} &= \left(\varepsilon_1 - \varepsilon_2 \frac{\sigma_1}{\sigma_2}\right) E_{1n} = \left(\varepsilon_1 - \varepsilon_2 \frac{\sigma_1}{\sigma_2}\right)(\boldsymbol{e}_z \cdot \boldsymbol{E}_1) \\ &= \left(\varepsilon_1 - \varepsilon_2 \frac{\sigma_1}{\sigma_2}\right)\left(-\frac{\sigma_2 U}{\sigma_2 d_1 + \sigma_1 d_2}\right) = \frac{(\varepsilon_2 \sigma_1 - \varepsilon_1 \sigma_2)}{\sigma_2 d_1 + \sigma_1 d_2} U\end{aligned} \tag{4.64}$$

From these results we see that $\rho_{s2} \neq -\rho_{s1}$, but $\rho_{s1} + \rho_{s2} + \rho_{si} = 0$

4.6 Resistance Calculations(电阻计算)

In Section 2.9, we calculated the capacitance between two arbitrary shaped conductors in terms of electric field quantities with the basic formula

$$C = \frac{Q}{U} = \frac{\oint_S \boldsymbol{D} \cdot d\boldsymbol{s}}{-\int_-^+ \boldsymbol{E} \cdot d\boldsymbol{l}} = \frac{\oint_S \varepsilon \boldsymbol{E} \cdot d\boldsymbol{s}}{-\int_-^+ \boldsymbol{E} \cdot d\boldsymbol{l}} \tag{4.65}$$

where the surface integral in the numerator is carried out over a surface enclosing the positive conductor and the line integral in the denominator is from the negative (lower-potential) conductor to the positive (higher-potential) conductor.

When the dielectric medium between the two conductors is lossy (having a small but nonzero conductivity), a leakage current $\boldsymbol{J}$ will flow from the positive to the negative conductor as is shown in Figure 4-6. According to Ohm's law, $\boldsymbol{J} = \sigma \boldsymbol{E}$, the streamlines for $\boldsymbol{J}$ and $\boldsymbol{E}$ will be the same in an isotropic medium. Then the resistance between the conductors is

$$R = \frac{U}{I} = \frac{-\int_-^+ \boldsymbol{E} \cdot d\boldsymbol{l}}{\oint_S \boldsymbol{J} \cdot d\boldsymbol{s}} = \frac{-\int_-^+ \boldsymbol{E} \cdot d\boldsymbol{l}}{\oint_S \sigma \boldsymbol{E} \cdot d\boldsymbol{s}} \tag{4.66}$$

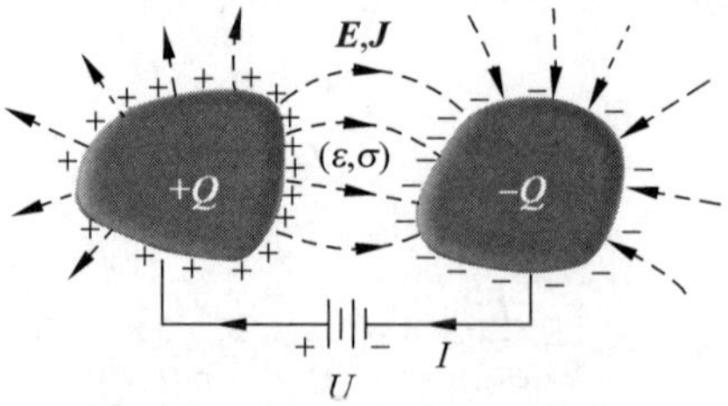

Figure 4-6 Two conductors in a lossy dielectric medium

where the line and surface integrals are taken over the same path and S as those in (4.65). Comparison of (4.65) and (4.66) shows the following relationship between R and C of the same conductor system:

$$RC = \frac{C}{G} = \frac{\varepsilon}{\sigma} \tag{4.67}$$

(4.67) holds if ε and σ of the medium are constants or have the same space dependence.

Example 4-4 Find the **leakage resistance**(**漏电阻**) per unit length ① between the inner and outer conductors of a **coaxial cable**(**同轴电缆**) that has an inner conductor of radius a, an outer conductor of inner radius b, and a medium with conductivity σ, and ② of a parallel-wire transmission line consisting of wires of radius a separated by a distance D in a medium with conductivity σ.

Solution:

(1) The capacitance per unit length of a coaxial cable has been derived as (2.117) in Example 2-17:

$$C = \frac{2\pi\varepsilon}{\ln(b/a)}$$

Hence the leakage resistance per unit length is, from (4.67)

$$R = \frac{\varepsilon}{\sigma}\left(\frac{1}{C}\right) = \frac{1}{2\pi\sigma}\ln\left(\frac{b}{a}\right) \tag{4.68}$$

The conductance per unit length is $G = 1/R$.

(2) For the parallel-wire transmission line, Example 3-7 gives the capacitance per unit length to be

$$C = \frac{\pi\varepsilon}{\cosh^{-1}\left(\frac{D}{2a}\right)}$$

Therefore, the leakage resistance per unit length is

$$R = \frac{\varepsilon}{\sigma}\left(\frac{1}{C}\right) = \frac{1}{\pi\sigma}\cosh^{-1}\left(\frac{D}{2a}\right) = \frac{1}{\pi\sigma}\ln\left[\frac{D}{2a} + \sqrt{\left(\frac{D}{2a}\right)^2 - 1}\right] \tag{4.69}$$

The conductance per unit length is $G = 1/R$.

Notice that the analogy between the electrostatic and steady-current problems become not accurate when there exist fringing effects. This is because current flow can be confined strictly within a conductor surrounded by a medium with $\sigma = 0$, whereas electric flux usually cannot be contained within a dielectric since the surrounding medium cannot have a zero permittivity. As an example, while (4.12) is usually accurate for calculating the DC resistance of a piece of conducting material, the counterpart expression (2.114) is less accurate for calculating the capacitance unless the fringing effect is negligibly small.

Figure 4-7 illustrates the procedures for calculating the resistance between specified equipotential surfaces (or terminals) of a resistor.[①] As shown in Figure 4-7(a), we can start by assuming a potential difference U between the conductor terminals, which sets up a boundary

---

① 电阻计算问题中,通常假设电阻的两个端点为导电率足够大的等位体。两个端点之间的导电媒质的导电率远小于端点导体。

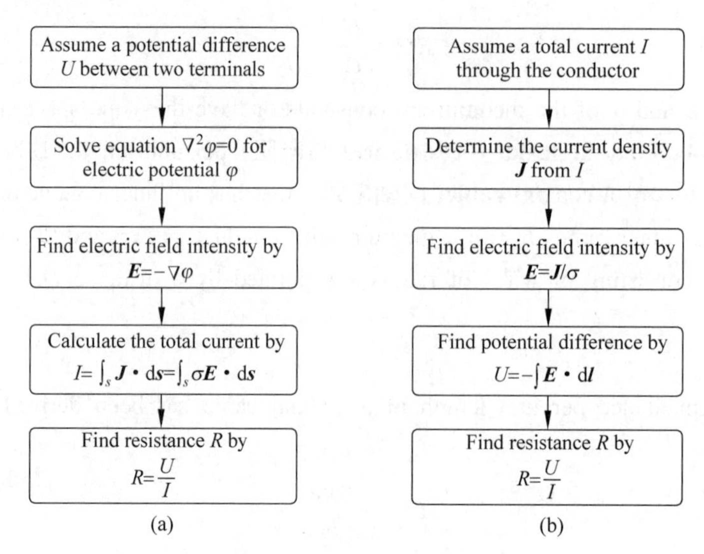

Figure 4-7 Procedures for calculating the resistance between two equipotential surfaces

value problem of the potential φ. For homogeneous material, the solution of φ is obtained by solving the Laplace's equation. [①]Then the electric field $\boldsymbol{E}$ can be found by taking the negative gradient of φ. The total current I is calculated as the surface integration of the current density $\boldsymbol{J}=\sigma\boldsymbol{E}$. The resistance R is found by taking the ratio U/I.

For problems in which $\boldsymbol{J}$ can be determined easily from a total current I, we can start the solution by assuming a total current I as is shown in Figure 4-7(b). After finding $\boldsymbol{J}$ from I, the electric field can be easily found as $\boldsymbol{E}=\boldsymbol{J}/\sigma$. Then the potential difference U is determined by a line integration along a path between the two resistor terminals. The resistance is found as $R=U/I$.

Example 4-5 A conducting material of uniform thickness h and conductivity σ has the shape of a quarter of a flat ring, with inner radius a and outer radius b, as shown in Figure 4-8. Determine the resistance between the end faces.

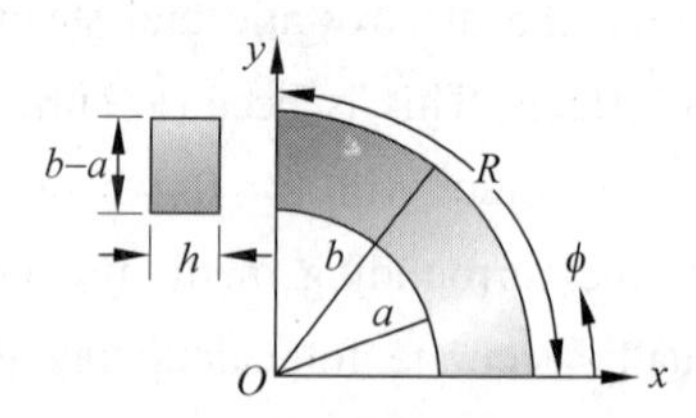

Figure 4-8 A quarter of a flat conducting circular ring (Example 4-5)

Solution: Using the cylindrical coordinate system, we first assume a potential difference U between the end faces. Then the potential φ satisfies the Laplace's equation subject to the following boundary conditions:

---

① 更一般地,若导电媒质中是非均匀的,即 σ 是空间位置的函数,则电位 φ 满足的方程不再是拉普拉斯方程,而应该是 $\nabla\cdot(\sigma\nabla\varphi)=0$。参见习题 4.10 和习题 4.11。

$$\varphi|_{\phi=0}=0 \tag{4.70a}$$

$$\varphi|_{\phi=\frac{\pi}{2}}=U \tag{4.70b}$$

Since potential φ is a function of ϕ only, Laplace's equation of φ simplifies to

$$\frac{\mathrm{d}^2\varphi}{\mathrm{d}\phi^2}=0 \tag{4.71}$$

The general solution of (4.71) is

$$\varphi=c_1\phi+c_2$$

which, upon using the boundary conditions in (4.70a) and (4.70b), becomes

$$\varphi=\frac{2U}{\pi}\phi \tag{4.72}$$

The current density is

$$\boldsymbol{J}=\sigma\boldsymbol{E}=-\sigma\nabla\varphi=-\boldsymbol{e}_\phi\sigma\frac{\partial\varphi}{\rho\partial\phi}=-\boldsymbol{e}_\phi\frac{2\sigma U}{\pi\rho} \tag{4.73}$$

The total current I can be found by integrating $\boldsymbol{J}$ over the $\phi=\pi/2$ surface at which $\mathrm{d}\boldsymbol{s}=-\boldsymbol{e}_\phi h\mathrm{d}\rho$.

$$I=\int_S\boldsymbol{J}\cdot\mathrm{d}\boldsymbol{s}=\frac{2\sigma U}{\pi}h\int_a^b\frac{\mathrm{d}\rho}{\rho}=\frac{2\sigma hU}{\pi}\ln\frac{b}{a} \tag{4.74}$$

Therefore,

$$R=\frac{U}{I}=\frac{\pi}{2\sigma h\ln(b/a)} \tag{4.75}$$

Note that, for this problem, it is not convenient to begin by assuming a total current I because it is not obvious how $\boldsymbol{J}$ varies with ρ for a given I. Without $\boldsymbol{J}$, $\boldsymbol{E}$ and U cannot be determined.

Summary

Concepts

Current density(电流密度)

Power dissipation(功率消耗)

Electromotive force(电动势)

Resistance(电阻)

Laws & Theorems

Ohm's law(欧姆定律)

Kirchhoff's voltage/current law(基尔霍夫电压定律)

Equation of continuity(连续性方程)

Boundary conditions of steady currents(恒定电流场的边界条件)

Methods

Find resistance of typical conducting bodies

Review Questions

4.1 什么是迁移率？什么是电导率？两者之间有什么关系？

4.2 什么是微分形式的欧姆定律？

4.3 简述焦耳定律，写出微分和积分形式的焦耳定律表达式。

4.4 连续性方程的物理意义是什么？

4.5 对于非均匀导体，式(4.26)应如何做修改？

4.6 什么是弛豫时间？金属铜的弛豫时间是什么数量级？

4.7 简述基尔霍夫电流定律。该定律的电磁场理论基础是什么？

4.8 非均匀导体中，$\nabla\times\boldsymbol{J}=0$ 是否仍然成立？为什么？

4.9 电动势是如何定义的？

4.10 什么是等效外加电场？等效外加电场与静电场的区别是什么？

4.11 简述包含电动势概念的基尔霍夫电压定律。该定律的电磁场理论基础是什么？

4.12 理想电压源的特征是什么？

4.13 电导率不同的两种媒质分界面上，恒定电流密度的法向和切向边界条件分别是什么？

4.14 恒定电流场中的电流密度矢量和电导率分别与静电场中的什么物理量可比拟？

4.15 简述静电场中的电容问题和恒定电流场中的电阻问题的比拟关系。该比拟关系成立的条件是什么？

Problems

4.1 The two plates of a parallel-plate capacitor with an area of 25cm^2 are separated by a distance of 0.4cm. The dielectric slab between the plates has a dielectric constant $\varepsilon_r = 2$ and conductivity $= 4\times10^{-5}$S/m. A potential difference of 150V is applied between the two plates. With the fringing effect neglected, determine the electric field intensity, the volume current density, the power density, the power dissipation, the current, and the resistance in the dielectric slab.

4.2 A total charge of 1mC is uniformly deposited within a lossy dielectric ($\varepsilon = 1.2\varepsilon_0$, $\sigma = 10$S/m) sphere of radius 0.1m due to a lightning strike at time $t=0$.

(1) Determine the electric field intensity and the current density inside the sphere for all t.

(2) Find the time it takes for the charge density in the sphere to drop to 1% of its initial value.

(3) Find the change of the electrostatic energy stored in the sphere as the charge density drops to 1% of its initial value. What happens to this energy?

(4) Determine the electrostatic energy stored in the space outside the sphere. Does this energy change with time?

4.3 Consider a planar interface between two media as shown in Figure 4-9. Medium 1 has a relative permittivity of 3 and a conductivity of 50μS/m, and medium 2 has a relative permittivity of 5 and a conductivity of 30μS/m. If the current density $\boldsymbol{J}_2$ has a magnitude of 3A/m^2, and makes an angle of $\theta_2=\dfrac{\pi}{3}$ with the normal direction on the interface. Determine the magnitude of the current density $\boldsymbol{J}_1$ and the angle θ_1. What is the surface charge density at the interface?

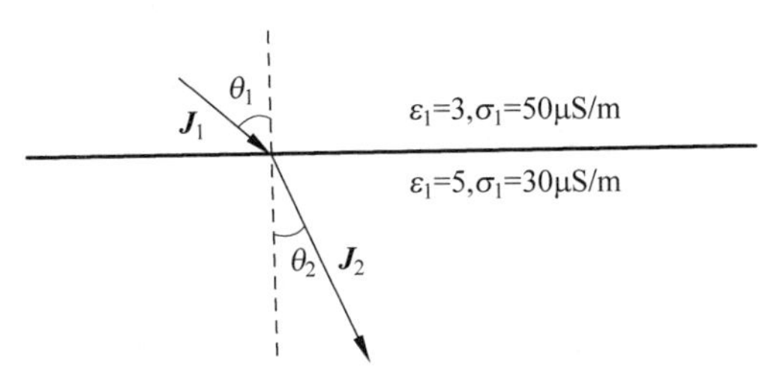

Figure 4-9 Interface between two lossy dielectric media(Problem 4.3)

4.4 Two lossy dielectric media with permittivities and conductivities(ε_1, σ_1) and(ε_2, σ_2) are in contact. The electric field in medium 1 has a magnitude E_1 and makes an angle α_1 with the normal direction of the interface as shown in Figure 4-10.

(1) Find the magnitude and direction of $\boldsymbol{E}_2$ in medium 2.

(2) Find the surface charge density on the interface.

(3) Compare the results in parts(1) and(2) with the case in which both media are perfect dielectrics(i. e. ,$\sigma_1=\sigma_2=0$).

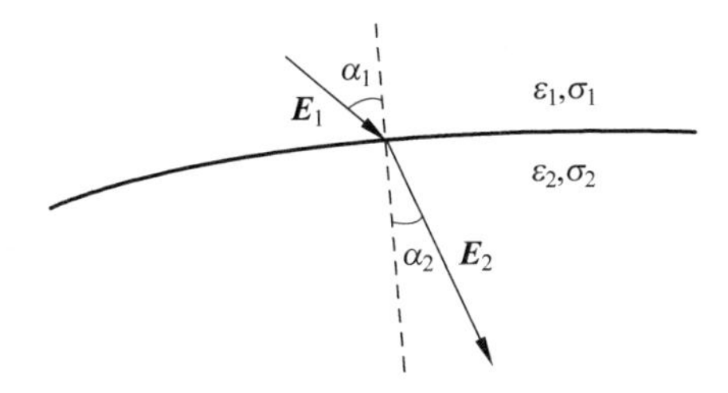

Figure 4-10 Interface between two lossy dielectric media(Problem 4.4)

4.5 Refer to Example 4-3. If a voltage U_0 is applied across the parallel-plate capacitor at $t=0$,

(1) express the surface charge density ρ_{si} on the dielectric interface as a function of t,

(2) express the electric field intensities $\boldsymbol{E}_1$ and $\boldsymbol{E}_2$ as functions of t.

4.6 The space between two parallel conducting plates each having a surface area S is filled with an inhomogeneous ohmic medium whose conductivity varies linearly from σ_1 at one plate($y=0$) to σ_2 at the other plate($y=d$). A DC voltage U_0 is applied across the plates as in Figure 4-11. Determine

(1) the total resistance between the plates,

(2) the surface charge densities on the plates,

(3) the volume charge density and the total amount of charge between the plates.

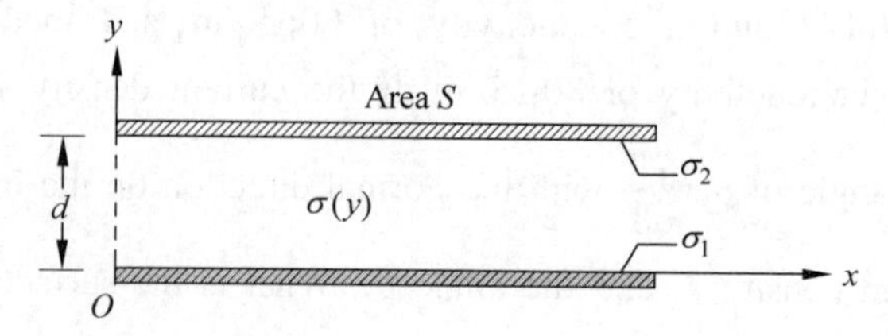

Figure 4-11 Inhomogeneous ohmic medium with conductivity $\sigma(y)$ (Problem 4.6)

4.7 A DC voltage U_0 is applied across a cylindrical capacitor of length L. The radii of the inner and outer conductors are a and b, respectively. The permittivity and conductivity in the region $a<\rho<c$ is $(\varepsilon_1, \sigma_1)$. The permittivity and conductivity in the region $c<\rho<b$ is $(\varepsilon_2, \sigma_2)$. Determine

(1) the current density in each region,

(2) the surface charge densities on the inner and outer conductors and at the interface between the two dielectrics.

4.8 Refer to the flat conducting quarter ring in Example 4-5 and Figure 4-8. Find the resistance between the curved sides, i. e., the resistance between the surfaces of $\rho=a$ and $\rho=b$.

4.9 Find the resistance between two concentric spherical surfaces of radii r_1 and r_2 ($r_1 < r_2$) given that the space between the surfaces is filled with

(1) a simple medium with a constant conductivity σ,

(2) a material with conductivity profile $\sigma=\sigma_0(1+k/r)$, with k as a constant (Note: Laplace's equation for φ does not apply here).

4.10 Given the electric field distribution $\boldsymbol{E}(\boldsymbol{r})$ in a conducting medium with a non-uniform conductivity $\sigma(\boldsymbol{r})$ and a constant permittivity ε_0, find an expression for the free charge density ρ_v within the medium.

4.11 Repeat solving Problem 4.10 for a conducting medium with a constant conductivity σ and a non-uniform permittivity $\varepsilon(\boldsymbol{r})$.

Chapter 5 Static Magnetic Fields

(恒定磁场)

5.1 Introduction(引言)

In previous Chapters, we dealt with static electric fields produced by stationary electric charges. The definition of the electric field intensity is based on the experimental law that specifies the force experienced by a stationary test charge. When the test charge q is in motion with a velocity $\boldsymbol{u}$, experiments show that it may experience another force even if no electric fields exist. This force, denoted by $\boldsymbol{F}_m$, has a magnitude proportional to q and $\boldsymbol{u}$. The direction of $\boldsymbol{F}_m$ at any point is perpendicular to $\boldsymbol{u}$ and a fixed direction at that point. Therefore, we can introduce a vector field, **magnetic flux density**(磁通密度), denoted by $\boldsymbol{B}$, so that the force $\boldsymbol{F}_m$ can be expressed as ①

$$\boldsymbol{F}_m = q\boldsymbol{u} \times \boldsymbol{B} \tag{5.1}$$

The unit of $\boldsymbol{B}$ is weber per square meter(Wb/m^2) or tesla(T). When both electric field and magnetic field exist, the total electromagnetic force on a charge q is

$$\boldsymbol{F} = q(\boldsymbol{E} + \boldsymbol{u} \times B) \tag{5.2}$$

which is called **Lorentz's force**(洛伦兹力) equation.

In this chapter, we derive the source-field relation of the static magnetic fields based on Ampere's law of force law(安培力定律) and Biot-Savart law(比奥萨法定律). We derive the divergence and curl of the magnetic fields and introduce the fundamental Ampere's circuital law of magnetostatics(静磁场的安培环路定律). Magnetization(磁化) of magnetic materials is studied, based on which the magnetic field intensity $\boldsymbol{H}$ is introduced. The boundary conditions of

---

① 与静电场类似,恒定磁场的概念也可以由实验引入。磁感应强度是描述磁场的基本矢量。从实验的角度出发,磁感应强度可以理解为单位测试电荷在磁场运动时受到的磁场力。该磁场力的大小与测试电荷的大小成正比,也与测试电荷的运动速率成正比,但同时,磁场力的大小还受电荷运动方向的影响。在磁场中的特定位置,当电荷沿特定方向运动时,受到的磁场力为零,而垂直于该特定方向运动时,磁场力达到最大,且磁场力的方向既与电荷运动方向垂直也和该特定方向垂直。总结这些实验现象,并引入磁感应强度矢量 $\boldsymbol{B}$,即得到磁场力的表达式(5.1)和洛伦兹力的表达式(5.2)。

$\boldsymbol{B}$ and $\boldsymbol{H}$ are derived. Finally, **inductance**(电感) and magnetic energy are discussed. ①

5.2 Magnetic Flux Density and Biot-Savart Law (磁通密度与比奥萨法定律)

Another important law related to magnetic force is **Ampere's law of force** (安培力定律). It states that the magnetic force between two circuits with current I_1 and I_2 can be expressed as

$$\boldsymbol{F}_{21} = \frac{\mu_0}{4\pi} I_1 I_2 \oint_{C_1} \oint_{C_2} \frac{\mathrm{d}\boldsymbol{l}_1 \times (\mathrm{d}\boldsymbol{l}_2 \times \boldsymbol{e}_{R_{21}})}{R_{21}^2} \tag{5.3}$$

where $\boldsymbol{F}_{21}$ is the magnetic force exerted by the circuit I_2 on the circuit I_1, $\mu_0 = 4\pi\times10^{-7}$ (H/m) is the **permeability**(磁导率) of free space, and $\boldsymbol{R}_{21} = \boldsymbol{e}_{R_{21}} R_{21}$ is the distance vector pointing from the differential current element $I_2\mathrm{d}\boldsymbol{l}_2$ to $I_1\mathrm{d}\boldsymbol{l}_1$ as shown in Figure 5-1. ②

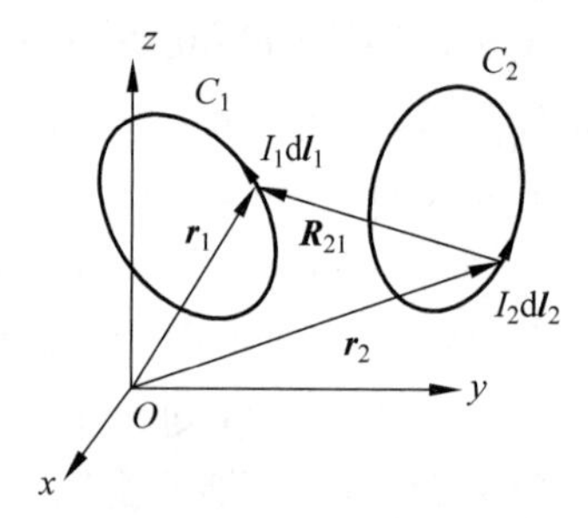

Figure 5-1 Two circuits carrying current I_1 and I_2 respectively

(5.3) can be rewritten as

$$\boldsymbol{F}_{21} = \oint_{C_1} I_1 \mathrm{d}\boldsymbol{l}_1 \times \left[\frac{\mu_0}{4\pi} \oint_{C_2} \frac{(I_2 \mathrm{d}\boldsymbol{l}_2 \times \boldsymbol{e}_{R_{21}})}{R_{21}^2} \right] = \oint_{C_1} I_1 \mathrm{d}\boldsymbol{l}_1 \times \boldsymbol{B}_{21} \tag{5.4}$$

The second equal sign in (5.4) is due to **Biot-Savart law**(比奥萨法定律), according to which the term within the square bracket is equal to $\boldsymbol{B}_{21}$, the magnetic flux density produced by the entire current I_2 at the location of the current element $I_1\mathrm{d}\boldsymbol{l}_1$. ③Generally, Biot-Savart law states that $\boldsymbol{B}$ produced by a current I in a closed path C' can be expressed as

$$\boldsymbol{B}(\boldsymbol{r}) = \frac{\mu_0 I}{4\pi} \oint_{C'} \frac{\mathrm{d}\boldsymbol{l}' \times \boldsymbol{e}_R}{R^2} = \frac{\mu_0 I}{4\pi} \oint_{C'} \frac{\mathrm{d}\boldsymbol{l}' \times \boldsymbol{R}}{R^3} = \frac{\mu_0 I}{4\pi} \oint_{C'} \frac{\mathrm{d}\boldsymbol{l}' \times (\boldsymbol{r} - \boldsymbol{r}')}{|\boldsymbol{r} - \boldsymbol{r}'|^3} \tag{5.5}$$

---

① 本章从安培力定律和比奥萨法定律出发，得到恒定磁场的场源关系，在此基础上推导磁感应强度的散度和旋度，得到静磁场的安培环路定律。然后讨论磁介质在磁场中的磁化现象，引入磁场强度的概念以及磁场中的边界条件。介绍典型电流生成的磁场的求解方法，并应用于求解电感问题，最后讨论了磁场储能的定义和应用。

② 式(5.3)给出的 $\boldsymbol{F}_{21}$ 是载流回路2对载流回路1的安培力。将式(5.3)中的下标1和2互换，则可以得到载流回路1对载流回路2的作用力 $\boldsymbol{F}_{12}$。根据牛顿第三运动定律，应该有 $\boldsymbol{F}_{21}=-\boldsymbol{F}_{12}$。然而由于矢量叉积不满足交换律，从式(5.3)推导出 $\boldsymbol{F}_{21}=-\boldsymbol{F}_{12}$ 不是显而易见的，其证明留作习题。

③ 式(5.4)的第二个等号右边的表达式也可以从磁感应强度的定义得出：对于载流回路 C_1 中的一小段电流元，$I_1\mathrm{d}\boldsymbol{l}_1=(\mathrm{d}q_1/\mathrm{d}t)\mathrm{d}\boldsymbol{l}_1=(\mathrm{d}\boldsymbol{l}_1/\mathrm{d}t)\mathrm{d}q_1=\boldsymbol{u}_1\mathrm{d}q_1$，相当于以速度 $\boldsymbol{u}_1$ 运动的电荷 $\mathrm{d}q_1$。根据式(5.1)，其受到的磁场力为 $\mathrm{d}q_1\boldsymbol{u}_1\times\boldsymbol{B}_{21}=I_1\mathrm{d}\boldsymbol{l}_1\times\boldsymbol{B}_{21}$。因此，比奥萨法定律与安培力定律是等价的。

where $\boldsymbol{r}$ is the position vector of a field point. $\boldsymbol{r}'$ is the position vector of a source point on the closed path C', and $\boldsymbol{R}$ is the distance vector pointing from $\boldsymbol{r}'$ to $\boldsymbol{r}$. ①

Example 5-1 A DC current I flows in a straight line of length L between $z=-L/2$ and $z=L/2$. Find the magnetic flux density $\boldsymbol{B}$ at an arbitrary point $P(\rho,\phi,z)$. (Notice that currents exist in closed circuits. This example is just to show how to calculate $\boldsymbol{B}$ by using Biot-Savart law. In practice, other part of the current-carrying circuit must be included in finding the actual $\boldsymbol{B}$ field.)

Solution: Due to symmetry of the problem, we know that the $\boldsymbol{B}$ field is not a function of ϕ. Therefore, we only need to calculate the field at a point $P(\rho,0,z)$. Referring to Figure 5-2, a differential length on the wire is

$$\mathrm{d}\boldsymbol{l}' = \boldsymbol{e}_z \mathrm{d}z'$$

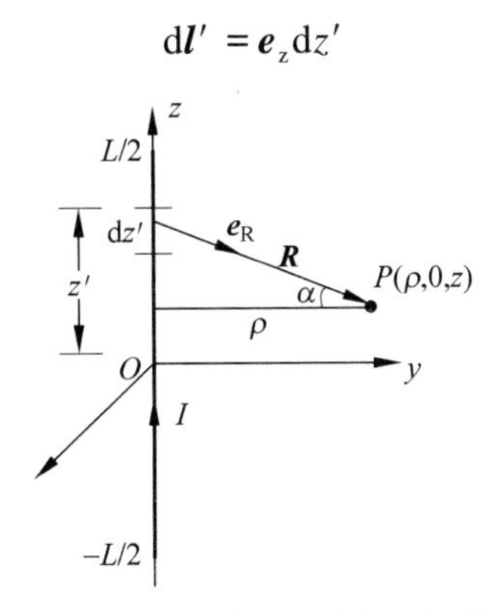

Figure 5-2 A current-carrying straight wire (Example 5-1)

The distance vector from $\mathrm{d}\boldsymbol{l}'$ to the field point P is

$$\boldsymbol{R} = \boldsymbol{e}_\rho \rho + \boldsymbol{e}_z (z - z')$$

Therefore,

$$\mathrm{d}\boldsymbol{l}' \times \boldsymbol{R} = (\boldsymbol{e}_z \mathrm{d}z') \times (\boldsymbol{e}_\rho \rho) = \boldsymbol{e}_\phi \rho \mathrm{d}z'$$

Let α be the elevation angle of the differential length $\mathrm{d}\boldsymbol{l}'$ observed at the point P. Then

$$z' = z + \rho\tan\alpha, \quad \mathrm{d}z' = \rho\sec^2\alpha \mathrm{d}\alpha, \quad R = \rho\sec\alpha$$

Substitute the above expression into (5.5),

$$\boldsymbol{B} = \boldsymbol{e}_\phi \frac{\mu_0 I}{4\pi}\int_{-L/2}^{L/2} \frac{\rho \mathrm{d}z'}{R^3} = \boldsymbol{e}_\phi \frac{\mu_0 I}{4\pi\rho}\int_{\alpha_-}^{\alpha_+} \cos\alpha \mathrm{d}\alpha = \boldsymbol{e}_\phi \frac{\mu_0 I}{4\pi\rho}(\sin\alpha_+ - \sin\alpha_-) \tag{5.6}$$

where α_+ and α_- are respectively the elevation angles of the wire ends at $(0,0,L/2)$ and $(0,0,-L/2)$ observed from the point P. Specifically,

$$\sin\alpha_+ = \left(\frac{L}{2} - z\right)\Big/\sqrt{\rho^2 + \left(\frac{L}{2} - z\right)^2}, \quad \sin\alpha_- = \left(-\frac{L}{2} - z\right)\Big/\sqrt{\rho^2 + \left(\frac{L}{2} + z\right)^2}$$

---

① 比奥萨法定律给出的磁场表达式与静电荷产生的静电场表达式类似，均为包含 $1/R^2$ 项的卷积积分。需要注意的是，由于电流连续性，除了理论上存在的起始和终止于无穷远处的无限长电流，产生恒定磁场的电流通常形成闭合回路，这两种情况下生成的恒定磁场在远区(距离源电流足够远的区域)随距离的变化通常并不是平方反比关系。接下来的三个例子很好地说明了这一点。

When $z=0$, (5.6) reduces to

$$B = e_\phi \frac{\mu_0 IL}{2\pi\rho\sqrt{L^2 + 4\rho^2}} \tag{5.7}$$

Further, if $L>>\rho$, we have

$$B = e_\phi \frac{\mu_0 I}{2\pi\rho} \tag{5.8}$$

Example 5-2 Find the magnetic flux density at axis of a square loop, with side length w and a direct current I.

Solution: Let the loop lie in the xy-plane, as shown in Figure 5-3. The magnetic flux density at the axis of the square loop is the sum of four $\boldsymbol{B}$ vectors, each due to the straight current in one side of the loop. According to Example 5-1, the current in each side of the loop produces a $\boldsymbol{B}$ field that can be found by letting

$$L = w \quad \text{and} \quad \rho = \sqrt{z^2 + \frac{w^2}{4}}$$

in (5.7). At the axis of the square loop, the horizontal components of the four $\boldsymbol{B}$ vectors cancel each other due to geometric symmetry, and the four z components add up to be

$$\begin{aligned} \boldsymbol{B}(0,0,z) &= \boldsymbol{e}_z \left(\frac{\mu_0 I w}{2\pi\rho\sqrt{w^2 + 4\rho^2}}\right)\left(\frac{w/2}{\rho}\right) \times 4 \\ &= \boldsymbol{e}_z \frac{\mu_0 I w^2}{2\pi\left(z^2 + \frac{w^2}{4}\right)\sqrt{z^2 + \frac{w^2}{2}}} \end{aligned} \tag{5.9}$$

Example 5-3 Find the magnetic flux density at a point on the axis of a circular loop of radius a that carries a direct current I.

Solution: For a field point $P(0,0,z)$ on the z axis as shown in Figure 5-4, $\mathrm{d}\boldsymbol{l}' = \boldsymbol{e}'_\phi b\mathrm{d}\phi'$, $\boldsymbol{R} = \boldsymbol{e}_z z - \boldsymbol{e}'_\rho a$, and $R = (z^2 + a^2)^{1/2}$. Therefore,

$$\mathrm{d}\boldsymbol{l}' \times \boldsymbol{R} = e'_\phi a\mathrm{d}\phi' \times (\boldsymbol{e}_z z - \boldsymbol{e}'_\rho a) = \boldsymbol{e}'_\rho az\mathrm{d}\phi' + \boldsymbol{e}_z a^2 \mathrm{d}\phi'$$

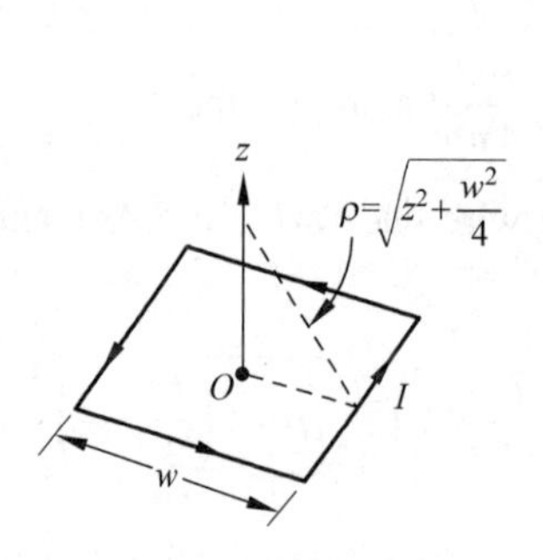

Figure 5-3 A square loop carrying current I (Example 5-2)

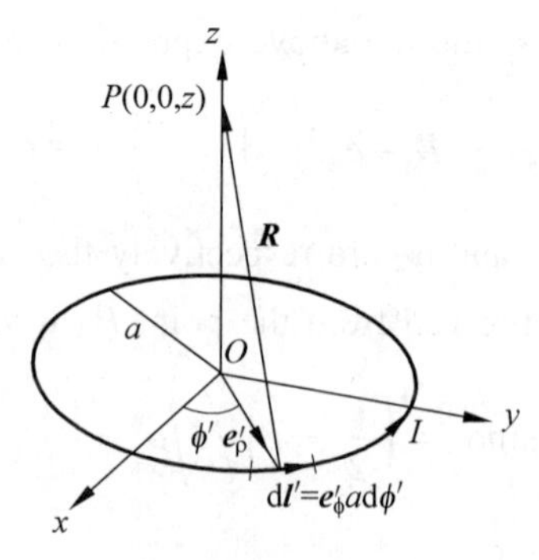

Figure 5-4 A circular loop carrying current I (Example 5-3)

Because of the cylindrical symmetry, it is easy to see that the ρ component is canceled by the contribution of the element located diametrically opposite to $\mathrm{d}\boldsymbol{l}'$. So, we only need to consider the z component of this cross product, which gives

$$\boldsymbol{B}(0,0,z) = \frac{\mu_0 I}{4\pi}\int_0^{2\pi} \boldsymbol{e}_z \frac{a^2 \mathrm{d}\phi'}{(z^2 + a^2)^{3/2}} = \boldsymbol{e}_z \frac{\mu_0 I a^2}{2(z^2 + a^2)^{3/2}} \tag{5.10}$$

Notice that, in the above examples, if we let $|z| \to \infty$, both expressions (5.9) and (5.10) become

$$\boldsymbol{B}(0,0,z) = \boldsymbol{e}_z \frac{\mu_0 I S}{2\pi |z|^3} \tag{5.11}$$

where S is the area of the current carrying loop. $S = w^2$ in Example 5-2 and $S = \pi a^2$ in Example 5-3. ①

The Biot-Savart law (5.5) describes the relation between static magnetic fields and line currents. If the current is continuously distributed within a region V' with volume current density $\boldsymbol{J}$, then (5.5) should be modified accordingly to be

$$\boldsymbol{B}(\boldsymbol{r}) = \frac{\mu_0}{4\pi}\int_{V'} \frac{\boldsymbol{J}(\boldsymbol{r}') \times \boldsymbol{e}_R}{R^2}\mathrm{d}v' = \frac{\mu_0}{4\pi}\int_{V'} \frac{\boldsymbol{J}(\boldsymbol{r}') \times \boldsymbol{R}}{R^3}\mathrm{d}v' = \frac{\mu_0}{4\pi}\int_{V'} \frac{\boldsymbol{J}(\boldsymbol{r}') \times (\boldsymbol{r} - \boldsymbol{r}')}{|\boldsymbol{r} - \boldsymbol{r}'|^3}\mathrm{d}v' \tag{5.12}$$

The current may also exist as a distribution of **surface current**(面电流) with a **surface current density**(面电流密度) $\boldsymbol{J}_S$. The surface current is a current confined in a surface layer. The surface layer is thin enough so that the current density variation along the thickness direction can be ignored. Therefore, the surface current density $\boldsymbol{J}_S$ has a unit of A/m. For a distribution of surface current $\boldsymbol{J}_S$ on the surface S', the magnetic flux density can be calculated as

$$\boldsymbol{B}(\boldsymbol{r}) = \frac{\mu_0}{4\pi}\int_{S'} \frac{\boldsymbol{J}_S(\boldsymbol{r}') \times \boldsymbol{e}_R}{R^2}\mathrm{d}s' = \frac{\mu_0}{4\pi}\int_{S'} \frac{\boldsymbol{J}_S(\boldsymbol{r}') \times \boldsymbol{R}}{R^3}\mathrm{d}s' = \frac{\mu_0}{4\pi}\int_{S'} \frac{\boldsymbol{J}_S(\boldsymbol{r}') \times (\boldsymbol{r} - \boldsymbol{r}')}{|\boldsymbol{r} - \boldsymbol{r}'|^3}\mathrm{d}s' \tag{5.13}$$

5.3 Divergence of Magnetic Flux Density and Vector Magnetic Potential(磁通密度的散度与矢量磁位)

In this section, we derive the divergence of magnetic flux density based on Biot-Savart Law. We consider the general expression of $\boldsymbol{B}$ in (5.12). By noticing (1.78), (5.12) can be rewritten as

$$\boldsymbol{B}(\boldsymbol{r}) = \frac{\mu_0}{4\pi}\int_{V'} \nabla\left(\frac{1}{R}\right) \times \boldsymbol{J}(\boldsymbol{r}')\mathrm{d}v' \tag{5.14}$$

By using the vector identity (1.149), (5.14) becomes

$$\boldsymbol{B}(\boldsymbol{r}) = \frac{\mu_0}{4\pi}\int_{V'} \left[\nabla \times \left(\frac{\boldsymbol{J}(\boldsymbol{r}')}{R}\right) - \frac{\nabla \times \boldsymbol{J}(\boldsymbol{r}')}{R}\right]\mathrm{d}v' \tag{5.15}$$

① 例 5-2 和例 5-3 均由比奥萨法定律表达式(5.5)计算得到,式(5.12)表明,闭合载流回路生成的远区磁场与载流回路所围面积成正比,与距离的 3 次方成反比。这可以理解为,在距离足够远时,流向相反电流元对观察点的磁场贡献趋近于相互抵消,从而使得磁场衰减比平方反比更快。而例 5-1 中长直电流模型中没有反方向电流,当电流长度 L 有限时,式(5.8)表明其远区磁场与距离 ρ^2 成反比,而当长度 L 趋于无穷大时,式(5.8)表明远区磁场与距离 ρ 成反比。

Since $\nabla\times$ is a differential operator with respect to the field coordinates, and $\boldsymbol{J}(\boldsymbol{r}')$ is independent of the field coordinates (not a function of $\boldsymbol{r}$), $\nabla\times\boldsymbol{J}(\boldsymbol{r}')$ must be zero. By taking the operator $\nabla\times$ outside the integral, we have

$$\boldsymbol{B}(\boldsymbol{r}) = \nabla\times\frac{\mu_0}{4\pi}\int_{V'}\frac{\boldsymbol{J}(\boldsymbol{r}')}{R}\mathrm{d}v' \tag{5.16}$$

which means that $\boldsymbol{B}$ can be expressed as the curl of another vector. From (1.142), we know that $\boldsymbol{B}$ must be divergenceless, i. e.

$$\nabla\cdot\boldsymbol{B} = 0 \tag{5.17}$$

Integrating both sides of (5.17) over any region, and using divergence theorem, we have

$$\oint_S \boldsymbol{B}\cdot\mathrm{d}\boldsymbol{s} = 0 \tag{5.18}$$

where S is the bounding surface of any volume. (5.18) is called the **Law of conservation of magnetic flux** (磁通守恒定律). It means that there are no magnetic flow sources and the magnetic flux lines always close upon themselves. ①

Since $\boldsymbol{B}$ is a divergenceless field that can be expressed in terms of the curl of another vector field, we define a **vector magnetic potential** (矢量磁位) $\boldsymbol{A}$ that satisfies

$$\boldsymbol{B} = \nabla\times\boldsymbol{A} \tag{5.19}$$

Compare (5.19) with (5.16), we have

$$\boldsymbol{A} = \frac{\mu_0}{4\pi}\int_{V'}\frac{\boldsymbol{J}(\boldsymbol{r}')}{R}\mathrm{d}v' + \nabla\psi \tag{5.20}$$

where ψ is an arbitrary scalar field. By noticing the null identity (1.138), we know that the curl of $\boldsymbol{A}$ in (5.20) must be equal to $\boldsymbol{B}$ for any scalar field ψ. Therefore, the vector potential $\boldsymbol{A}$ is not uniquely determined by (5.19). This is because (5.19) only specifies the curl of $\boldsymbol{A}$. According to Helmholtz's theorem introduced in Section 1-14, to determine $\boldsymbol{A}$, we also need to specify its divergence. For static magnetic fields produced by currents within finite regions, we usually adopt the **Coulomb gauge** (库伦规范), which requires

$$\nabla\cdot\boldsymbol{A} = 0 \tag{5.21}$$

Substitute (5.20) into (5.21), we have

$$\nabla\cdot\boldsymbol{A} = \frac{\mu_0}{4\pi}\int_{V'}\nabla\cdot\left(\frac{\boldsymbol{J}(\boldsymbol{r}')}{R}\right)\mathrm{d}v' + \nabla^2\psi = 0 \tag{5.22}$$

Now we need to determine ψ by using (5.22). To do that, we first focus on the volume integral involving $\boldsymbol{J}(\boldsymbol{r}')$ in (5.22). By using the vector identity (1.148), the integral can be rewritten as

$$\int_{V'}\nabla\cdot\left(\frac{\boldsymbol{J}(\boldsymbol{r}')}{R}\right)\mathrm{d}v' = \int_{V'}\left[\frac{\nabla\cdot\boldsymbol{J}(\boldsymbol{r}')}{R} + \nabla\left(\frac{1}{R}\right)\cdot\boldsymbol{J}(\boldsymbol{r}')\right]\mathrm{d}v' \tag{5.23}$$

Again, by noticing that $\boldsymbol{J}(\boldsymbol{r}')$ is independent of the field coordinates, $\nabla\cdot\boldsymbol{J}(\boldsymbol{r}')$ must be zero. And by noticing that

① 这里从比奥萨法定律出发证明磁场是无散场，从而得到结论：磁场不存在通量源（即磁荷），这也意味着磁力线必须形成闭合回路，没有起点与终点，也被称为磁通连续性原理。

$$\nabla\left(\frac{1}{R}\right) = -\nabla'\left(\frac{1}{R}\right)$$

(5.23) becomes

$$\int_{V'} \nabla\cdot\left(\frac{\boldsymbol{J}(\boldsymbol{r}')}{R}\right)\mathrm{d}v' = -\int_{V'} \nabla'\left(\frac{1}{R}\right)\cdot \boldsymbol{J}(\boldsymbol{r}')\mathrm{d}v'$$
$$= \int_{V'}\left[\frac{\nabla'\cdot \boldsymbol{J}(\boldsymbol{r}')}{R} - \nabla'\cdot\left(\frac{\boldsymbol{J}(\boldsymbol{r}')}{R}\right)\right]\mathrm{d}v' \qquad (5.24)$$

The second equal sign in (5.24) is the result of applying the vector identity (1.148). Since $\boldsymbol{J}(\boldsymbol{r}')$ is a steady current, we know from the previous chapter that $\nabla'\cdot\boldsymbol{J}(\boldsymbol{r}')=0$. Then, by using the divergence theorem, (5.24) becomes

$$\int_{V'} \nabla\cdot\left(\frac{\boldsymbol{J}(\boldsymbol{r}')}{R}\right)\mathrm{d}v' = -\int_{S'} \frac{\boldsymbol{J}(\boldsymbol{r}')}{R}\cdot \mathrm{d}\boldsymbol{s}' \qquad (5.25)$$

where S' is the boundary surface of the volume V'. Since V' contains all the currents represented by $\boldsymbol{J}(\boldsymbol{r}')$, the normal component of $\boldsymbol{J}(\boldsymbol{r}')$ on the surface S' must be zero (unless S' extends to infinity). Otherwise, there will be nonzero current outside the surface S' due to continuity of the normal component of current densities. Therefore, the surface integration in (5.25) must be zero, and eventually we have

$$\int_{V'} \nabla\cdot\left(\frac{\boldsymbol{J}(\boldsymbol{r}')}{R}\right)\mathrm{d}v' = 0 \qquad (5.26)$$

Substitute (5.26) back into (5.22), we have

$$\nabla\cdot\boldsymbol{A} = \nabla^2\psi = 0 \qquad (5.27)$$

(5.27) shows that ψ satisfies the Laplace equation everywhere under Coulomb gauge. Particularly, (5.27) can be satisfied if $\nabla\psi$ is an arbitrary constant vector $\boldsymbol{C}$. Then, by (5.20), the vector potential is determined to within an additive constant vector $\boldsymbol{C}$, which is consistent to Helmholtz' theorem. We then reasonably select $\boldsymbol{C}=0$, which leads to a concise expression of $\boldsymbol{A}$ as ①

$$\boldsymbol{A} = \frac{\mu_0}{4\pi}\int_{V'} \frac{\boldsymbol{J}(\boldsymbol{r}')}{R}\mathrm{d}v' \qquad (5.28)$$

For a surface current with density $\boldsymbol{J}_S$, (5.28) becomes

$$\boldsymbol{A} = \frac{\mu_0}{4\pi}\int_{S'} \frac{\boldsymbol{J}_S(\boldsymbol{r}')}{R}\mathrm{d}s' \qquad (5.29)$$

For a line current I, (5.28) becomes

$$\boldsymbol{A} = \frac{\mu_0 I}{4\pi}\oint_{C'} \frac{\mathrm{d}\boldsymbol{l}'}{R} \qquad (5.30)$$

---

① 比较式(5.19)和式(5.16)似乎可以直接得到矢量磁位 $\boldsymbol{A}$ 的表达式(5.28),但仍然经过较为复杂的推导才明确得到了式(5.28)。这是因为根据亥姆霍兹定理,式(5.19)只规定了 $\boldsymbol{A}$ 的旋度,仍然需要规定 $\boldsymbol{A}$ 的散度才能唯一确定矢量场 $\boldsymbol{A}$。库仑规范就明确规定了 $\boldsymbol{A}$ 的散度为零。这里的看似较为复杂的推导实际上证明了库仑规范的合理性,即该规范使得由电流计算磁位 $\boldsymbol{A}$ 的表达式简化为式(5.28)。需要注意的是,库仑规范导致式(5.28)成立有一个前提,即电流分布在有限区域内,否则将无法从式(5.25)得到式(5.26),也就不能由库仑规范得出 $\boldsymbol{A}$ 的表达式(5.28)。

Notice that the line integration is over a closed path C', as steady currents usually exist in closed circuits.

Given a distribution of steady currents in free space, it may be much easier to first find the vector potential $\boldsymbol{A}$ using (5.28)-(5.30), and then determine the magnetic flux density $\boldsymbol{B}$ by taking the curl of $\boldsymbol{A}$.

Vector potential $\boldsymbol{A}$ can also be used to find the **magnetic flux**(**磁通**) Φ through a surface S, which is defined as

$$\Phi = \int_S \boldsymbol{B} \cdot \mathrm{d}\boldsymbol{s} \tag{5.31}$$

The unit for magnetic flux is weber(Wb), or tesla-square meter($\mathrm{T \cdot m^2}$). Using (5.19) and Stokes's theorem, we have another way of calculating the magnetic flux:

$$\Phi = \oint_C \boldsymbol{A} \cdot \mathrm{d}\boldsymbol{l} \tag{5.32}$$

where C is the closed path that bounds the surface S. ①

A **magnetic dipole** (**磁偶极子**) is a small circular loop with a diameter that is much smaller than its distance from the observation point. In Example 5-3, we have found the exact expression for the $\boldsymbol{B}$ field on the axis of a circular current loop. The exact expression of $\boldsymbol{B}$ off the axis will be much more complicated. However, if the observation point is far away from the current loop, the current loop can be seen as a magnetic dipole, and a simplified expression for the $\boldsymbol{B}$ field can be derived as follows.

Suppose a small circular loop with radius a carries a current I. As shown in Figure 5-5, the distance r between the observation point and the center of the loop satisfies $r \gg b$. We first find the vector magnetic potential $\boldsymbol{A}$ by using (5.30). Referring to Figure 5-5, we have

$$\mathrm{d}\boldsymbol{l}' = a\boldsymbol{e}'_\phi \mathrm{d}\phi' \tag{5.33}$$

and

$$\begin{aligned} R^2 &= r^2 + a^2 - 2ar\cos\psi \\ &= r^2 + a^2 - 2ar\sin\theta\cos(\phi - \phi') \end{aligned} \tag{5.34}$$

where ψ is the angle between $\boldsymbol{r}$ and $\boldsymbol{e}'_\rho$(the direction pointing from the origin toward the source point at $\mathrm{d}\boldsymbol{l}'$). In (5.34), we use the fact that $r\cos\psi = r\sin\theta\cos(\phi - \phi')$ is the projection of the vector $\boldsymbol{r}$ onto the direction $\boldsymbol{e}'_\rho$. Notice that the unit vectors $\boldsymbol{e}'_\rho$ and $\boldsymbol{e}'_\phi$ in (5.33) are primed because they are associated with the source point $(a, \phi', 0)$. $\boldsymbol{e}'_\rho$ and $\boldsymbol{e}'_\phi$ are different from $\boldsymbol{e}_\rho$ and $\boldsymbol{e}_\phi$ associated with the field point $P(\rho, \phi, z)$.

Referring to Figure 5-5, $\boldsymbol{e}'_\phi$ can be decomposed into two components along the directions of $\boldsymbol{e}_\rho$ and $\boldsymbol{e}_\phi$ respectively, i.e.,

$$\boldsymbol{e}'_\phi = \boldsymbol{e}'_\phi\cos(\phi - \phi') + \boldsymbol{e}'_\rho\sin(\phi - \phi') \tag{5.35}$$

① 磁通量 Φ 的符号可正可负，取决于式(5.31)中微分面元 d$\boldsymbol{s}$ 方向的选取。式(5.32)中线积分的积分路径 C 的方向与微分面元 d$\boldsymbol{s}$ 的方向遵循右手定则。

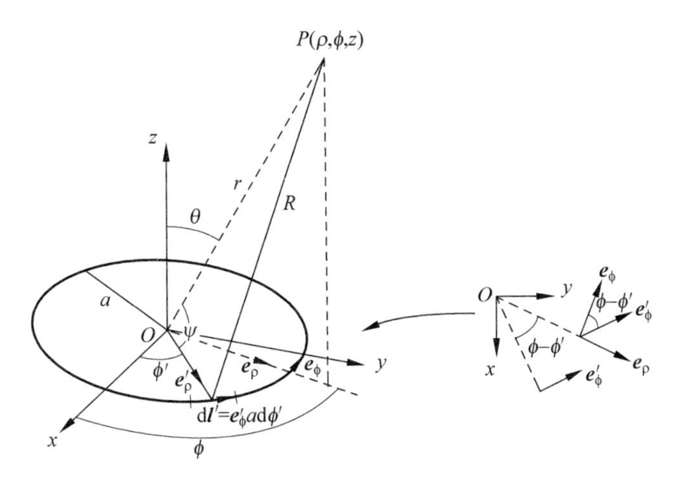

Figure 5-5 A small circular loop carrying current I

From(5.34), we have

$$\frac{1}{R}=\frac{1}{r}\left[1+\frac{a^2}{r^2}-\frac{2a}{r}\sin\theta\cos(\phi-\phi')\right]^{-\frac{1}{2}} \tag{5.36}$$

For simplification, we neglect the a^2/r^2 term since $a \ll r$, which gives

$$\frac{1}{R}\approx\frac{1}{r}\left[1-\frac{2a}{r}\sin\theta\cos(\phi-\phi')\right]^{-\frac{1}{2}}\approx\frac{1}{r}\left[1+\frac{a}{r}\sin\theta\cos(\phi-\phi')\right] \tag{5.37}$$

Substitute(5.33) and(5.37) into(5.30), we have

$$\boldsymbol{A}\approx\frac{\mu_0 I}{4\pi}\oint_{C'}\frac{a\boldsymbol{e}'_\phi\mathrm{d}\phi'}{r}\left[1+\frac{a}{r}\sin\theta\cos(\phi-\phi')\right]=\frac{\mu_0 Ia}{4\pi r}\oint_{C'}\left[\boldsymbol{e}'_\phi+\frac{a}{r}\sin\theta\cos(\phi-\phi')\boldsymbol{e}'_\phi\right]\mathrm{d}\phi' \tag{5.38}$$

From(5.35), it is easy to see that

$$\oint_{C'}\boldsymbol{e}'_\phi\mathrm{d}\phi'=0$$

Therefore, (5.38) becomes

$$\boldsymbol{A}\approx\frac{\mu_0 Ia^2\sin\theta}{4\pi r^2}\oint_{C'}\boldsymbol{e}'_\phi\cos(\phi-\phi')\mathrm{d}\phi' \tag{5.39}$$

Substitute(5.35) into(5.39), we have

$$\begin{aligned}\boldsymbol{A}&\approx\frac{\mu_0 Ia^2\sin\theta}{4\pi r^2}\oint_{C'}[\boldsymbol{e}_\phi\cos(\phi-\phi')+\boldsymbol{e}_\rho\sin(\phi-\phi')]\cos(\phi-\phi')\mathrm{d}\phi'\\&=\frac{\mu_0 Ia^2\sin\theta}{4\pi r^2}\oint_{C'}\left[\boldsymbol{e}_\phi\frac{1+\cos2(\phi-\phi')}{2}+\boldsymbol{e}_\rho\frac{\sin2(\phi-\phi')}{2}\right]\mathrm{d}\phi'\\&=\boldsymbol{e}_\phi\frac{\mu_0 Ia^2}{4r^2}\sin\theta\end{aligned} \tag{5.40}$$

The magnetic flux density can be derived as

$$B = \nabla \times A = \frac{\mu_0 I a^2}{4r^3}(e_r 2\cos\theta + e_\theta \sin\theta) \tag{5.41}$$

(5.39) can be rearranged as

$$A = e_\phi \frac{\mu_0 (I\pi a^2)}{4\pi r^2}\sin\theta = e_\phi \frac{\mu_0 m}{4\pi r^2}\sin\theta$$

$$= \frac{\mu_0 m \times e_r}{4\pi r^2} \tag{5.42}$$

where

$$m = e_z m = e_z I\pi a^2 = e_z IS(\mathrm{A}\cdot\mathrm{m}^2) \tag{5.43}$$

is defined as the **magnetic dipole moment**(磁偶极距,磁矩). The magnitude of the magnetic dipole moment is the product of the area of the loop and the current in it. The direction of the magnetic dipole moment is the direction of the thumb as the fingers of the right hand follow the direction of the current. Similarly, we can also rewrite(5.41) as

$$B = \frac{\mu_0 m}{4\pi r^3}(e_r 2\cos\theta + e_\theta \sin\theta) \tag{5.44}$$

Although the above derivation is based on a circular loop, it can be shown that the same expressions(5.42) and(5.44) can be obtained when the loop has a rectangular shape. Also notice that, by letting $\theta=0$ in(5.44), we have an expression of $\boldsymbol{B}$ far away in the z axis, which is equivalent to(5.11).

Compare(5.42) and(5.44) with the expressions for $\boldsymbol{E}$ and φ of an electric dipole in (2.57) and(2.16), which is repeated as below:

$$\varphi = \frac{p \cdot e_r}{4\pi\varepsilon_0 r^2} \tag{5.45}$$

$$E = \frac{p}{4\pi\varepsilon_0 r^3}(e_r 2\cos\theta + e_\theta \sin\theta) \tag{5.46}$$

We see that, except for the change of p to m and ε_0 to $1/\mu_0$, (5.42) has the same form as (5.45), and(5.44) has the same form as(5.46). Hence the magnetic fields of a magnetic dipole lying in the xy-plane have the same far-field form as that of the electric fields of an electric dipole positioned along the z-axis. That is why we call a small current-carrying loop a magnetic dipole.

The $\boldsymbol{B}$ lines of a magnetic dipole are compared with the $\boldsymbol{E}$ lines of an electric dipole in Figure 5-6. At distant points, flux lines of both types of dipoles have the same distribution patterns. However, near the dipoles, the flux lines of a magnetic dipole are continuous, whereas the field lines of an electric dipole originate from the positive charges and terminate on the negative charge.

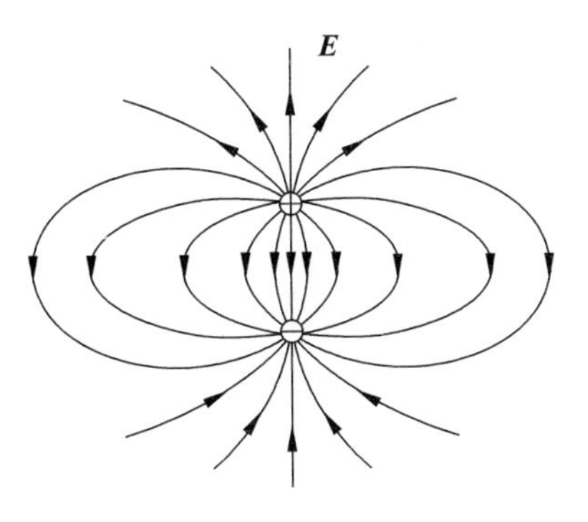

(a) the electric field of an electric dipole

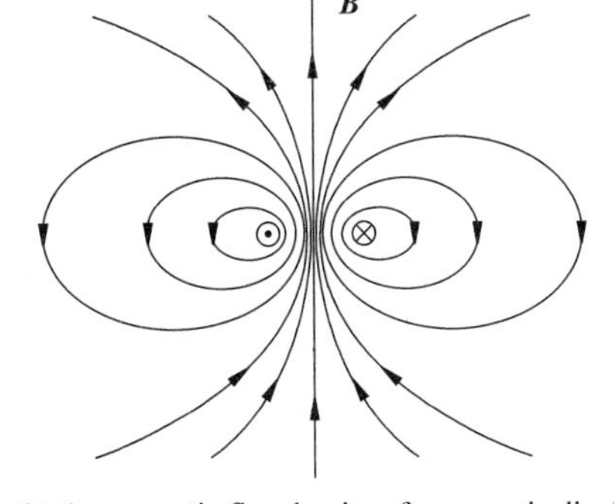

(b) the magnetic flux density of a magnetic dipole

Figure 5-6 Streamlines

5.4 Curl of Magnetic Flux Density and Ampere's Circuital Law(磁通密度的旋度与安培环路定律)

To derive the curl of magnetic flux density, we start by taking curl on both sides of(5.16), which gives

$$\nabla \times \boldsymbol{B}(\boldsymbol{r}) = \frac{\mu_0}{4\pi}\int_{V'} \nabla\times\nabla\times\left[\frac{\boldsymbol{J}(\boldsymbol{r}')}{R}\right]\mathrm{d}v' \tag{5.47}$$

With vector identity(1.134), (5.47) becomes

$$\nabla \times \boldsymbol{B}(\boldsymbol{r}) = \frac{\mu_0}{4\pi}\int_{V'} \nabla\left(\nabla\cdot\left[\frac{\boldsymbol{J}(\boldsymbol{r}')}{R}\right]\right)\mathrm{d}v' - \frac{\mu_0}{4\pi}\int_{V'} \nabla^2\left[\frac{\boldsymbol{J}(\boldsymbol{r}')}{R}\right]\mathrm{d}v' \tag{5.48}$$

Notice that ∇ is with respect to the field coordinates. Therefore, the divergence and gradient operator in the first term in the right-hand side of (5.48) can be taken out of the volume integral. Moreover, considering that $\boldsymbol{J}(\boldsymbol{r}')$ is independent of the field coordinates, we can apply (1.137) to the second term in the right-hand side of(5.48). Then(5.48) becomes

$$\nabla \times \boldsymbol{B}(\boldsymbol{r}) = \nabla\left[\nabla\cdot\left(\frac{\mu_0}{4\pi}\int_{V'}\frac{\boldsymbol{J}(\boldsymbol{r}')}{R}\mathrm{d}v'\right)\right] - \frac{\mu_0}{4\pi}\int_{V'}\boldsymbol{J}(\boldsymbol{r}')\,\nabla^2\left(\frac{1}{R}\right)\mathrm{d}v' \tag{5.49}$$

From(5.21) and(5.28), we know that

$$\nabla\cdot\left(\frac{\mu_0}{4\pi}\int_{V'}\frac{\boldsymbol{J}(\boldsymbol{r}')}{R}\mathrm{d}v'\right) = \nabla\cdot\boldsymbol{A} = 0 \tag{5.50}$$

Substituting(5.50) into(5.49), we have

$$\nabla \times \boldsymbol{B}(\boldsymbol{r}) = -\frac{\mu_0}{4\pi}\int_{V'}\boldsymbol{J}(\boldsymbol{r}')\,\nabla^2\left(\frac{1}{R}\right)\mathrm{d}v' \tag{5.51}$$

By substituting(1.133) into(5.51) and applying the sampling property of δ function(1.104), we have

$$\nabla \times \boldsymbol{B}(\boldsymbol{r}) = \mu_0\boldsymbol{J}(\boldsymbol{r}) \tag{5.52}$$

Since the divergence of the curl of any vector field is zero, from(5.52) we immediately obtain $\nabla\cdot\boldsymbol{J}=0$, which is consistent with(4.30) for steady currents.

The integral form of the curl relation in(5.52) is obtained by integrating both sides over an open surface and applying Stokes's theorem, which gives

$$\oint_C \boldsymbol{B} \cdot \mathrm{d}\boldsymbol{l} = \mu_0 \int_S \boldsymbol{J} \cdot \mathrm{d}\boldsymbol{s} = \mu_0 I \tag{5.53}$$

where the path C for the line integral is the contour bounding the surface S, and I is the total current through S. The integration direction of $\mathrm{d}\boldsymbol{l}$ and the reference direction of current I in(5.53) follow the right-hand rule. (5.53) is the **Ampere's circuital law**(**安培环路定律**) for steady magnetic fields in free space. It states that the circulation of the magnetic flux density in free space over an arbitrary closed path is equal to μ_0 times the total current flowing through the surface bounded by the path. ①

Ampere's circuital law is useful in determining static magnetic fields when certain symmetry conditions are satisfied. Specifically, when the magnetic flux density $\boldsymbol{B}$ has a constant tangential component over the closed path C, the line integral in(5.53) can be converted to the product of the length of the path C and the tangential component of $\boldsymbol{B}$. ② A few examples are as follows.

Example 5-4 An infinitely long wire carries a steady current I uniformly distributed over the cross section with radius a. Determine the magnetic flux density inside and outside the wire.

Solution: Due to the cylindrical symmetry, if the axis of the conductor is aligned along the z-axis, the magnetic flux density $\boldsymbol{B}$ must be along the ϕ-direction and the magnitude of $\boldsymbol{B}$ must be constant along any concentric circular path around the z-axis. Therefore, the total circulation of $\boldsymbol{B}$ over any circular path C with radius ρ is

$$\oint_C \boldsymbol{B} \cdot \mathrm{d}\boldsymbol{l} = \int_0^{2\pi} B_\phi \rho \mathrm{d}\phi = 2\pi\rho B_\phi \tag{5.54}$$

Now we apply Ampere's circuital law to concentric circular paths inside and outside the conductor separately as follows. (Notice that the direction of C and the reference direction of I follow the right-hand rule)

(1) Inside the conductor, as shown in Figure 5-7(a), the current through the area enclosed by concentric circle C_1 is$(\rho^2/a^2)I$. By Ampere's circuital law,

$$2\pi\rho B_\phi = \mu_0 \left(\frac{\rho}{a}\right)^2 I \tag{5.55}$$

The $\boldsymbol{B}$ field inside the conductor is

$$\boldsymbol{B} = \boldsymbol{e}_\phi B_\phi = \boldsymbol{e}_\phi \frac{\mu_0 \rho I}{2\pi a^2} \quad (\rho \leqslant a) \tag{5.56a}$$

(2) Outside the conductor, as shown in Figure 5-7(b), the current through the area

① 自由空间中的安培环路定律：磁通密度在任意闭合路径上的环量等于μ_0乘以穿过该路径所围曲面的总的电流。需要注意，闭合路径所围曲面可以是任意形状的，只要闭合路径给定，在该曲面上的总电流，也即电流密度的通量积分不变，与如何选取该曲面无关。

② 在磁场分布具有一定对称性的情况下，利用安培环路定律计算磁通密度比较方便，其前提是能够根据对称性判断在特定的闭合回路上磁通密度的切向分量是否均匀，从而将式(5.53)中的磁通密度环量积分转换为磁通密度切向分量的大小与闭合回路长度的乘积，继而直接得到磁通密度的大小。

enclosed by concentric circle C_2 is I. So, we have the $\boldsymbol{B}$ field outside the conductor to be

$$\boldsymbol{B} = \boldsymbol{e}_\phi B_\phi = \boldsymbol{e}_\phi \frac{\mu_0 I}{2\pi\rho} \quad (\rho \geqslant a) \tag{5.56b}$$

(5.56a) and (5.56b) shows that the magnitude of $\boldsymbol{B}$ increases linearly when ρ grows from 0 to a. After that, the magnitude of $\boldsymbol{B}$ decreases inversely with ρ. The variation of B_ϕ versus ρ is sketched in Figure 5-7(c).

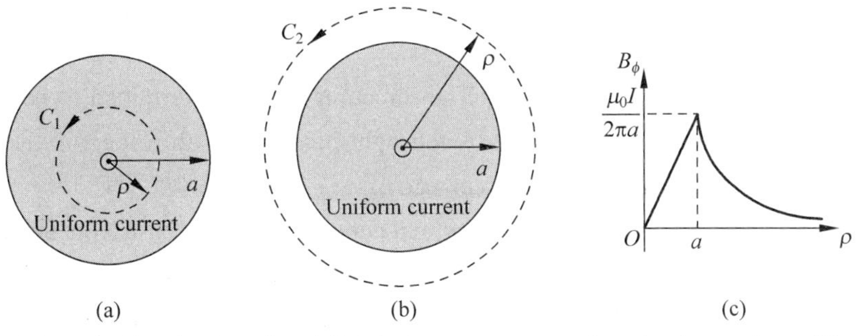

Figure 5-7 Magnetic flux density of an infinitely long, current-carrying circular conductor (Example 5-4)

If the current-carrying conductor is a very thin cylindrical tube and the steady current I is distributed on the surface of the tube, we can still use Ampere's circuital law to easily find the solution that $\boldsymbol{B}=0$ inside the tube and $\boldsymbol{B}$ is given by (5.56b) outside the tube. Thus, for an infinitely long, hollow cylinder carrying a surface current density $\boldsymbol{J}_s = \boldsymbol{e}_z J_s$, $I = 2\pi a J_s$ we have

$$\boldsymbol{B} = \begin{cases} 0, & \rho < a \\ \boldsymbol{e}_\phi \dfrac{\mu_0 a}{\rho} J_s, & \rho > a \end{cases} \tag{5.57}$$

Example 5-5 Figure 5-8 illustrates a toroid with an inner radius a, an outer radius b. The core of the toroid is air, around which an N-turn coil carrying a current I is closely wounded. Determine the magnetic flux density inside the toroid.

Solution: From geometric symmetry, $\boldsymbol{B}$ has only a ϕ-component which is constant along any circular path around the axis of the toroid. Consider a circular contour C with radius ρ as shown in the figure. For $a < \rho < b$, the Ampere's circuital law (5.53) leads directly to

$$\oint \boldsymbol{B} \cdot d\boldsymbol{l} = 2\pi\rho B_\phi = \mu_0 N I$$

Therefore,

$$\boldsymbol{B} = \boldsymbol{e}_\phi B_\phi = \boldsymbol{e}_\phi \frac{\mu_0 N I}{2\pi\rho}, \quad (a < \rho < b) \tag{5.58}$$

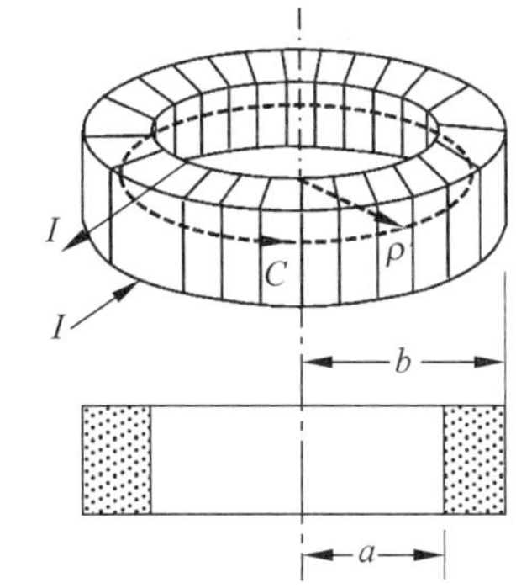

Figure 5-8 A current-carrying toroid coil (Example 5-5)

Apparently $\boldsymbol{B} = 0$ for $\rho < a$ and $\rho > b$, since the net current enclosed by a circular contour in these two regions is zero.

Example 5-6 As is shown in Figure 5-9, an infinitely long solenoid with air core has n closely wounded turns per unit length. The coil carries a current I. Determine the magnetic flux density inside the solenoid.

Solution: Due to the geometric symmetry, the magnetic field outside of the solenoid must be zero, and the **B**-field inside must be parallel to the axis. Applying Ampere's circuital law to the rectangular contour C of length l as is shown in Figure 5-9, we have

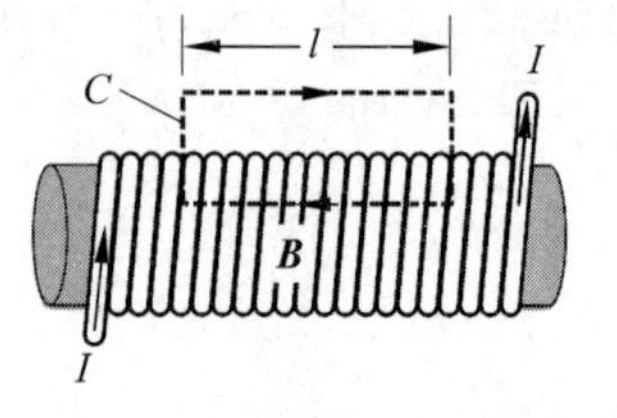

Figure 5-9 A current-carrying long solenoid (Example 5-6)

$$Bl = \mu_0 nlI$$

which gives

$$B = \mu_0 nI \tag{5.59}$$

The direction of **B** is determined to be from right to left by the right-hand rule with respect to the direction of the current I.

In Section 5.3, we introduced the vector magnetic potential **A** based on the fact that the magnetic flux density **B** is a solenoidal (divergenceless) field. Subsection 5.3.1 demonstrates that, in a magnetostatic problem, it may be much easier to first find the potential **A** and then determine the **B** field by taking the curl of **A**. This is very similar to the introduction of the scalar electric potential φ for the solving for the **E** field in electrostatics. We learned in Chapter 3 that φ in electrostatics satisfies the Poisson's/Laplace's equation, and therefore, electrostatic problems can be formulated as solving a Poisson's/Laplace's equation with boundary conditions. Similarly, magnetostatic problems can also be formulated as a Poisson's/Laplace's equation, but in the vector form. Here we derive the Poisson's equation satisfied by **A**, which is completely based on the two fundamental properties of the **B** field, (5.17) and (5.52), and the definition of **A** vector, (5.19).

Substituting (5.19) into (5.52), we have

$$\nabla \times \nabla \times \boldsymbol{A} = \mu_0 \boldsymbol{J} \tag{5.60}$$

By using the vector identity (1.134), (5.60) becomes

$$\nabla(\nabla \cdot \boldsymbol{A}) - \nabla^2 \boldsymbol{A} = \mu_0 \boldsymbol{J} \tag{5.61}$$

With Coulomb gauge (5.21), the term with $\nabla \cdot \boldsymbol{A}$ vanishes, and we have

$$\nabla^2 \boldsymbol{A} = -\mu_0 \boldsymbol{J} \tag{5.62}$$

which is a **vector Poisson's equation (矢量泊松方程)**①. In Cartesian coordinates, according to (1.136), the vector Poisson's equation (5.62) is equivalent to three scalar Poisson's equations:

$$\nabla^2 A_x = -\mu_0 J_x \tag{5.63a}$$

$$\nabla^2 A_y = -\mu_0 J_y \tag{5.63b}$$

$$\nabla^2 A_z = -\mu_0 J_z \tag{5.63c}$$

Each of these three equations is mathematically the same as the scalar Poisson's equation (3.6) in electrostatics. Since the equation (3.6) has a particular solution of (2.58) in free space, the solution for (5.63a) in free space is

---

① 这里也可以看到恒定磁场的库仑规范令 $\nabla \cdot \boldsymbol{A} = 0$ 的另一个理由，即库仑规范使得磁位 **A** 满足的方程简化为矢量泊松方程(5.62)。

$$A_x = \frac{\mu_0}{4\pi}\int_{V'} \frac{J_x}{R} \mathrm{d}v'$$

We can write similar solutions for A_y and A_z. Combining the three components, we have the solution for (5.62) as[①]

$$\boldsymbol{A} = \frac{\mu_0}{4\pi}\int_{V'} \frac{\boldsymbol{J}}{R} \mathrm{d}v' \tag{5.64}$$

which is the same as (5.28) that is previously derived.

5.5 Magnetization and Equivalent Current Densities (磁化与等效电流密度)

Till now, we have only discussed static magnetic fields due to steady currents in free space. In reality, presence of material media will affect the magnetic field due to currents, just as they will affect the electric field due to charges. In return, the magnetic field will also affect the atoms/molecules of the media. Just as dielectrics will be polarized due to the electric fields, the magnetic media will be magnetized due to the magnetic fields, which is called **magnetization** (磁化).

To explain magnetization, we consider the elementary atomic model of matter, i. e., all materials are composed of atoms, each with a positively charged nucleus and a number of orbiting negatively charged electrons. The orbiting electrons can be considered as circulating currents, which forms microscopic magnetic dipoles. In the absence of an external magnetic field, the atomic magnetic dipoles of most media (except permanent magnets) have random orientations, resulting in no net magnetic dipole moment. However, when an external magnetic field is applied, the atomic magnetic dipoles are aligned, which leads to a nonzero induced magnetic dipole moment.

To analyze the macroscopic effect of induced magnetic dipole moment, let $\boldsymbol{m}_k$ be the magnetic dipole moment of a kth atom. We define a **magnetization vector** (磁化强度矢量), $\boldsymbol{M}$, as

$$\boldsymbol{M} = \lim_{\Delta v \to 0} \frac{\sum_{k=1}^{n\Delta v} \boldsymbol{m}_k}{\Delta v} \ (\mathrm{A/m}) \tag{5.65}$$

where n is the number of atoms per unit volume. Apparently, $\boldsymbol{M}$ represents the volume density of the magnetic dipole moment. Then, the magnetic dipole moment $\mathrm{d}\boldsymbol{m}$ of an elemental volume $\mathrm{d}v'$ is $\mathrm{d}\boldsymbol{m} = \boldsymbol{M}\mathrm{d}v'$. According to (5.42), $\mathrm{d}\boldsymbol{m}$ at a source point $\boldsymbol{r}'$ will produce a vector magnetic potential $\mathrm{d}\boldsymbol{A}$ at a field point $\boldsymbol{r}$:

$$\mathrm{d}\boldsymbol{A} = \frac{\mu_0 \boldsymbol{M} \times \boldsymbol{e}_R}{4\pi R^2} \mathrm{d}v' \tag{5.66}$$

---

① 磁位的表达式(5.64)与式(5.28)也可以由1.14节的亥姆霍兹定理直接得到(见习题5.11)。

where $\boldsymbol{e}_R$ is the unit vector of $\boldsymbol{R}=\boldsymbol{r}-\boldsymbol{r}'$. Using(1.82), we can write(5.66) as

$$\mathrm{d}\boldsymbol{A} = \frac{\mu_0}{4\pi}\boldsymbol{M} \times \nabla'\left(\frac{1}{R}\right)\mathrm{d}v'$$

Thus

$$\boldsymbol{A} = \int_{V'} \mathrm{d}\boldsymbol{A} = \frac{\mu_0}{4\pi}\int_{V'} \boldsymbol{M} \times \nabla'\left(\frac{1}{R}\right)\mathrm{d}v' \quad (5.67)$$

where V' is the volume of the magnetized material. By the vector identity(1.149), we have

$$\boldsymbol{M} \times \nabla'\left(\frac{1}{R}\right) = \frac{1}{R}\nabla' \times \boldsymbol{M} - \nabla' \times \left(\frac{\boldsymbol{M}}{R}\right) \quad (5.68)$$

and hence(5.67) becomes

$$\boldsymbol{A} = \frac{\mu_0}{4\pi}\int_{V'} \frac{\nabla' \times \boldsymbol{M}}{R}\mathrm{d}v' - \frac{\mu_0}{4\pi}\int_{V'} \nabla' \times \left(\frac{\boldsymbol{M}}{R}\right)\mathrm{d}v' \quad (5.69)$$

By using the vector identity(see Problem1.18),

$$\int_{V'} \nabla' \times \boldsymbol{F}\mathrm{d}v' = -\oint_{S'} \boldsymbol{F} \times \mathrm{d}\boldsymbol{s}' \quad (5.70)$$

where V' is an arbitrary volume bounded by surface S', the second volume integral in(5.69) can be converted into a surface integral. So we have

$$\boldsymbol{A} = \frac{\mu_0}{4\pi}\int_{V'} \frac{\nabla' \times \boldsymbol{M}}{R}\mathrm{d}v' + \frac{\mu_0}{4\pi}\oint_{S'} \frac{\boldsymbol{M} \times \boldsymbol{e}_n'}{R}\mathrm{d}s' \quad (5.71)$$

where $\boldsymbol{e}_n'$ is the unit outward normal vector on ds'.

Comparing the two integral terms on the right side of(5.71) with the expressions of $\boldsymbol{A}$ in (5.28) and(5.29), we can conclude that the effect of the magnetization represented by the $\boldsymbol{M}$ vector is equivalent to a volume current density

$$\boldsymbol{J}_m = \nabla \times \boldsymbol{M} \quad (\mathrm{A/m^2}) \quad (5.72)$$

and a surface current density

$$\boldsymbol{J}_{ms} = \boldsymbol{M} \times \boldsymbol{e}_n \quad (\mathrm{A/m}) \quad (5.73)$$

$\boldsymbol{J}_m$ and $\boldsymbol{J}_{ms}$ are respectively called the **equivalent magnetization volume current density**(**等效磁化体电流密度**) and **equivalent magnetization surface current density**(**等效磁化面电流密度**)①. Notice that in(5.72) and(5.73), we have omitted the primes(′) on ∇ and $\boldsymbol{e}_n$ for simplicity, because the two equations involve only source coordinates and there is no need to discriminate the source coordinates from field coordinates.

The expressions of equivalent magnetization current densities(5.72) and(5.73) are established by mathematical derivation. They can also be interpreted physically by referring to Figure 5-10. Assuming that an externally applied magnetic field magnetizes the material. Each atom is modeled as a magnetic dipole with the direction of the dipole moment vector aligned with the magnetization vector $\boldsymbol{M}$. To interpret the volume magnetization current density, we first exam the total atomic current through an arbitrary surface S bounded by the contour C as shown

---

① 注意 $\boldsymbol{J}_{ms}$ 是面电流密度,磁化强度 $\boldsymbol{M}$ 的量纲和面电流密度一致,单位均为 A/m.

in Figure 5-10(a). Apparently, the circulating current of an atom does not contribute to the total current unless it links with the contour C. Take a differential length $\mathrm{d}\boldsymbol{l}$ along the contour C. The current linking with $\mathrm{d}\boldsymbol{l}$ is contributed by the atoms centered within an oblique cylinder with volume $\Delta V = (\Delta S)(\mathrm{d}\boldsymbol{l} \cdot \boldsymbol{e}_{\mathrm{M}})$, where ΔS is the area of the atomic magnetic dipole, and $\boldsymbol{e}_{\mathrm{M}}$ is the direction of the magnetization vector $\boldsymbol{M}$, which also is the direction of the magnetic dipole moment. Therefore, the current contribution linked with $\mathrm{d}\boldsymbol{l}$ is $\mathrm{d}I = n\Delta S \mathrm{d}\boldsymbol{l} \cdot \boldsymbol{e}_{\mathrm{M}} \Delta I = \mathrm{d}\boldsymbol{l} \cdot \boldsymbol{M}$, where ΔI is the circulating current of the orbiting electrons of a single atom and $\boldsymbol{M} = \boldsymbol{e}_{\mathrm{M}} n \Delta S \Delta I$. Then, the total magnetization current through the surface S can be calculated as

$$I_{\mathrm{m}} = \oint_C \boldsymbol{M} \cdot \mathrm{d}\boldsymbol{l} = \int_S (\nabla \times \boldsymbol{M}) \cdot \mathrm{d}\boldsymbol{s} \tag{5.74}$$

where Stokes's theorem has been used. Since (5.74) holds for any surface S bounded by the contour C, $\nabla\times\boldsymbol{M}$ must be equivalent to volume magnetization current density $\boldsymbol{J}_{\mathrm{m}}$. This leads to (5.72).

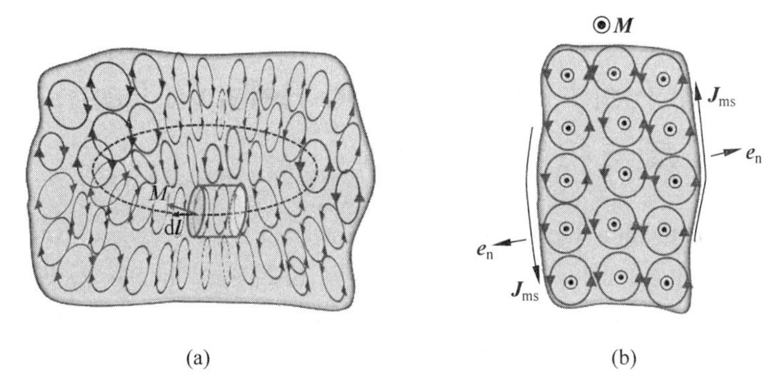

Figure 5-10 A cross section of a magnetized material

On the surface of the material, there will be a surface current density $\boldsymbol{J}_{\mathrm{ms}}$ contributed by the atoms on the surface. As shown in Figure 5-10(b), the direction of $\boldsymbol{J}_{\mathrm{ms}}$ is perpendicular to both $\boldsymbol{M}$ and the normal vector $\boldsymbol{e}_{\mathrm{n}}$. It is easy to find out that $\boldsymbol{J}_{\mathrm{ms}}$ can be expressed as the cross-product $\boldsymbol{M}\times\boldsymbol{e}_{\mathrm{n}}$, which leads to (5.73).

Example 5-7 A cylinder magnet with radius a and length L is uniformly magnetized with $\boldsymbol{M} = \boldsymbol{e}_z M_0$. Determine the magnetic flux density on the axis of the cylinder.

Solution: For convenience, we can align the axis of the magnetized cylinder along the z-axis as shown in Figure 5-11. Since the magnetization $\boldsymbol{M}$ is a constant within the magnet, $\boldsymbol{J}_{\mathrm{m}} = \nabla'\times\boldsymbol{M} = 0$, which means no equivalent magnetization volume current exists. The equivalent magnetization surface current density on the side wall is

$$\boldsymbol{J}_{\mathrm{ms}} = \boldsymbol{M} \times \boldsymbol{e}'_{\mathrm{n}} = (\boldsymbol{e}_z M_0) \times \boldsymbol{e}_\rho = \boldsymbol{e}_\phi M_0 \tag{5.75}$$

There is no surface current on the top and bottom faces. To find $\boldsymbol{B}$ at $P(0,0,z)$, we consider a differential length $\mathrm{d}z'$ with a current $\boldsymbol{e}_\phi M_0\, \mathrm{d}z'$ and use (5.10) to obtain

$$\mathrm{d}\boldsymbol{B} = \boldsymbol{e}_z \frac{\mu_0 M_0 a^2 \mathrm{d}z'}{2[(z-z')^2 + a^2]^{3/2}}$$

and

$$\boldsymbol{B} = \int \mathrm{d}\boldsymbol{B} = \boldsymbol{e}_z \int_0^L \frac{\mu_0 M_0 a^2}{2[(z-z')^2 + a^2]^{3/2}} \mathrm{d}z'$$

$$= \boldsymbol{e}_z \frac{\mu_0 M_0}{2} \left[\frac{z}{\sqrt{z^2 + a^2}} - \frac{z - L}{\sqrt{(z-L)^2 + a^2}} \right] \tag{5.76}$$

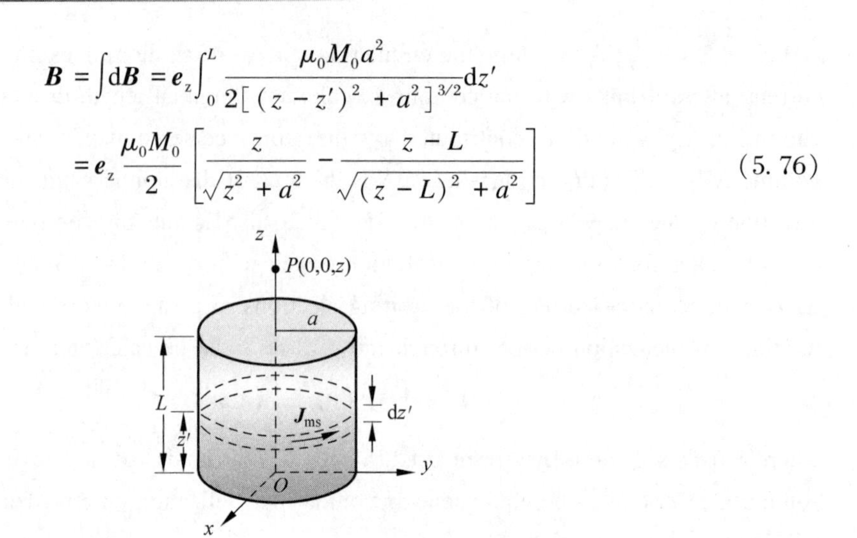

Figure 5-11 A uniformly magnetized circular cylinder(Example 5-7)

5.6 Magnetic Field Intensity and Relative Permeability (磁场强度与相对磁导率)

Because of magnetization, the magnetic flux density in the presence of a magnetic material will be different from its value in free space. As discussed in the previous section, the macroscopic effect of magnetization is equivalent to an induced volume current density $\boldsymbol{J}_{\mathrm{m}}$ given by(5.72). Therefore, we can rewrite the curl equation(5.52) as

$$\frac{1}{\mu_0} \nabla \times \boldsymbol{B} = \boldsymbol{J} + \boldsymbol{J}_{\mathrm{m}} = \boldsymbol{J} + \nabla \times \boldsymbol{M}$$

or

$$\nabla \times \left(\frac{\boldsymbol{B}}{\mu_0} - \boldsymbol{M} \right) = \boldsymbol{J} \tag{5.77}$$

We now define a new fundamental field quantity, the **magnetic field intensity**(**磁场强度**) $\boldsymbol{H}$ as

$$\boldsymbol{H} = \frac{\boldsymbol{B}}{\mu_0} - \boldsymbol{M} \tag{5.78}$$

Substitute(5.78) into(5.77), we obtain the new equation

$$\nabla \times \boldsymbol{H} = \boldsymbol{J} \tag{5.79}$$

where $\boldsymbol{J}$ is the volume density of **free current**(**自由电流**).① This new equation relates the magnetic field and the distribution of free currents without involving any magnetization vector $\boldsymbol{M}$ or equivalent magnetization currents $\boldsymbol{J}_{\mathrm{m}}$.

---

① 这里所谓“自由电流”指的是与媒质磁化无关的电流,也可以理解为产生外加磁场的源。自由电流可能存在于磁性媒质外部,也可能存在于磁性媒质内,但一定不是媒质磁化的结果,而是媒质磁化的原因。相反,磁化电流则是媒质磁化的结果。

(5.17) and (5.79) are the two fundamental governing differential equations for magnetostatics. The permeability of the medium does not appear explicitly in these two equations.

The corresponding integral form of(5.79) is obtained by taking the surface integral of both sides and using Stokes's theorem, which gives

$$\oint_C \boldsymbol{H} \cdot d\boldsymbol{l} = I \tag{5.80}$$

where C is the contour (closed path) bounding the arbitrary surface S and I is the total free current passing through S. The relative directions of C and current flow I follow the right-hand rule. (5.80) is another form of **Ampere's circuital law**(**安培环路定律**). It states that the circulation of the magnetic field intensity around any closed path is equal to the free current flowing through the surface bounded by the path. ①

As we indicated in Section 5.4, Ampere's circuital law is most useful in determining the magnetic field caused by a current when cylindrical symmetry exists, that is, when there is a closed path around the current over which the tangential component of the magnetic field is constant.

However, there still exists the problem of how to find the magnetic flux density $\boldsymbol{B}$ from the magnetic field intensity $\boldsymbol{H}$, which is apparently dependent the magnetization characteristic of the material. Fortunately, when the magnetic properties of the medium are linear and isotropic, the magnetization vector is directly proportional to the magnetic field intensity:

$$\boldsymbol{M} = \chi_m \boldsymbol{H} \tag{5.81}$$

where χ_m is a dimensionless quantity called **magnetic susceptibility**(**磁化率**). Substitution of (5.81) in(5.78) yields the following constitutive relation:

$$\boldsymbol{B} = \mu_0(1 + \chi_m)\boldsymbol{H} = \mu_0\mu_r\boldsymbol{H} = \mu\boldsymbol{H} \tag{5.82a}$$

or

$$\boldsymbol{H} = \frac{1}{\mu}\boldsymbol{B} \tag{5.82b}$$

where

$$\mu_r = 1 + \chi_m = \frac{\mu}{\mu_0} \tag{5.83}$$

is another dimensionless quantity known as the **relative permeability**(**相对磁导率**) of the medium. The parameter $\mu = \mu_0\mu_r$(H/m) is the **absolute permeability**(**绝对磁导率**)(or just **permeability**) of the medium. χ_m and therefore μ_r can be a function of space coordinates. For a simple medium(linear, isotropic, and homogeneous), χ_m and μ_r are constant scalars.

The permeability of most materials is very close to that of free space(μ_0). These materials are called **nonmagnetic materials**(**非磁性材料**). The materials with $\mu < \mu_0$ is called **diamagnetic materials**(**抗磁性材料**), and those with $\mu > \mu_0$ are called **paramagnetic materials**

① 恒定磁场的安培环路定律：磁场强度在任意闭合回路上的环量等于穿过该闭合回路所围的任意曲面的自由电流。

(顺磁性材料). For **ferromagnetic materials**(铁磁性材料) such as iron, nickel, cobalt and some special alloys, μ_r could be very large(50-5000 and up to 10^6 or even more). Moreover, the permeability of ferromagnetic materials depends on the magnitude of $\boldsymbol{H}$ as well as the previous history of the material, which will not be discussed here.

Example 5-8 Find the magnetic fields($\boldsymbol{B}$ and $\boldsymbol{H}$) and the magnetization vector($\boldsymbol{M}$) distributions due to a straight, infinitely long line current I along the axis of a coaxial magnetic cylinder with radius a and permeability μ.

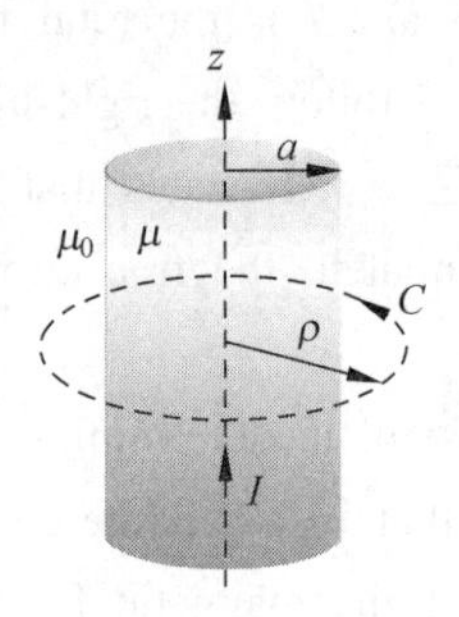

Figure 5-12 A straight infinitely long line current (Example 5-8)

Solution: Due to the symmetry, on any coaxial cylindrical surface around the current, the $\boldsymbol{H}$ field has only a constant ϕ component. As shown in Figure 5-12, applying the Ampere's circuital law to a contour C with radius ρ around the line current I, we have

$$\oint_C \boldsymbol{H}\cdot \mathrm{d}\boldsymbol{l} = 2\pi\rho H_\phi = I$$

Therefore,

$$H_\phi = \frac{I}{2\pi\rho}$$

both inside or outside the magnetic cylinder. The magnetic flux density is then found to be

$$\boldsymbol{B} = \begin{cases} \boldsymbol{e}_\phi \dfrac{\mu I}{2\pi\rho} & 0 < \rho < a \\ \boldsymbol{e}_\phi \dfrac{\mu_0 I}{2\pi\rho} & a < \rho < \infty \end{cases}$$

The magnetization vector distribution is

$$\boldsymbol{M} = \frac{\boldsymbol{B}}{\mu_0} - \boldsymbol{H} = \begin{cases} \boldsymbol{e}_\phi \dfrac{\mu - \mu_0}{\mu_0}\dfrac{I}{2\pi\rho} & \rho < a \\ 0 & a < \rho < \infty \end{cases}$$

Now we notice that there exist some analogous relations between the quantities in electrostatics and those in magnetostatics as listed in Table 5-1. With the analogous relations, most of the equations relating the basic quantities in electrostatics can be converted into corresponding analogous ones in magnetostatics, and vice versa.

Table 5-1 Analogous relations between the quantities in electrostatics and magnetostatics

Electrostatics	Magnetostatics	Electrostatics	Magnetostatics
$\boldsymbol{E}$	$\boldsymbol{B}$	$\boldsymbol{D}$	$\boldsymbol{H}$
ε	$1/\mu$	$\boldsymbol{P}$	$-\boldsymbol{M}$
ρ	$\boldsymbol{J}$	φ	$\boldsymbol{A}$
$\cdot$	$\times$	$\times$	$\cdot$

5.7 Boundary Conditions for Magnetostatic Fields (恒定磁场的边界条件)

When magnetic problems involve multiple media with different permeabilities, the **B** and **H** vectors may exhibit discontinuity across the interfaces, and as a result, the differential equations (5.17) and (5.79) may not apply. Instead, we need to use boundary conditions of **B** and **H**, which can be derived from the integral equations they satisfy. The derivation of magnetostatic boundary conditions is similar to that of electrostatic fields as is done in Section 2-8.

As shown in Figure 5-13(a), by applying the law of conservation of magnetic flux (5.18) to the small box, it is easy to conclude that the normal component of **B** is continuous across an interface ①, i. e.,

$$B_{1n} = B_{2n} \tag{5.84}$$

For linear media, $\boldsymbol{B}_1 = \mu_1 \boldsymbol{H}_1$ and $\boldsymbol{B}_2 = \mu_2 \boldsymbol{H}_2$, (5.84) becomes

$$\mu_1 H_{1n} = \mu_2 H_{2n} \tag{5.85}$$

The boundary condition for the tangential components of magnetostatic field is obtained by applying the integral equation (5.80) to the closed path *abcda* in Figure 5-13(b). By letting $bc = da = \Delta h$ approach zero, we have

$$\oint_{abcda} \boldsymbol{H} \cdot \mathrm{d}\boldsymbol{l} = \boldsymbol{H}_1 \cdot \Delta \boldsymbol{w} + \boldsymbol{H}_2 \cdot (-\Delta \boldsymbol{w}) = J_{sn} \Delta w$$

or

$$H_{1t} - H_{2t} = J_{sn} \tag{5.86}$$

where $\Delta \boldsymbol{w}$ is the vector-pointing from node a to b. J_{sn} is the surface current density on the interface normal to the integral path *abcda*. The direction of J_{sn} is that of the thumb when the fingers of the right hand follow the direction of the path. The vector expression of this boundary condition is

$$\boldsymbol{e}_{n2} \times (\boldsymbol{H}_1 - \boldsymbol{H}_2) = \boldsymbol{J}_s \tag{5.87}$$

where $\boldsymbol{e}_{n2}$ is the outward unit normal from medium 2 at the interface. Thus, the tangential component of the **H** field is discontinuous across an interface where a free surface current exists, and the amount of discontinuity equals the density of the surface current. ②

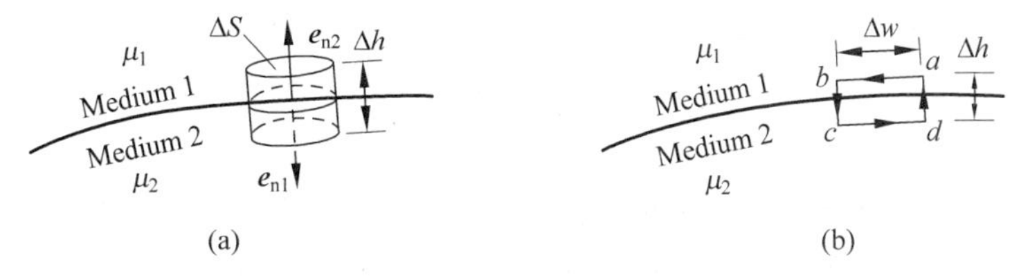

Figure 5-13 Geometry for determining the boundary condition of magnetostatic fields

① 恒定磁场法向边界条件：不同媒质分界面上磁感应强度 **B** 的法向分量连续。

② 恒定磁场切向边界条件：不同媒质分界面上磁场强度 **H** 的切向分量突变量等于该分界面上的面电流密度。

The surface current can exist only on the surface of an ideal perfect conductor or a superconductor($\sigma=\infty$). For dielectric media, there is no surface current. Therefore, the tangential component of $\boldsymbol{H}$ field is continuous across the boundary. For conducting media with finite conductivity, currents are defined by volume current densities, and free surface currents do not exist on the interface, hence the tangential component of $\boldsymbol{H}$ field is also continuous across the boundary.

Example 5-9 Two magnetic media with permeabilities μ_1 and μ_2 have a common boundary, as shown in Figure 5-14. The magnetic field intensity in medium 1 at an interface point has a magnitude H_1 and makes an angle α_1 with the normal. Determine the magnitude and the direction of the magnetic field intensity at the same interface point in medium 2.

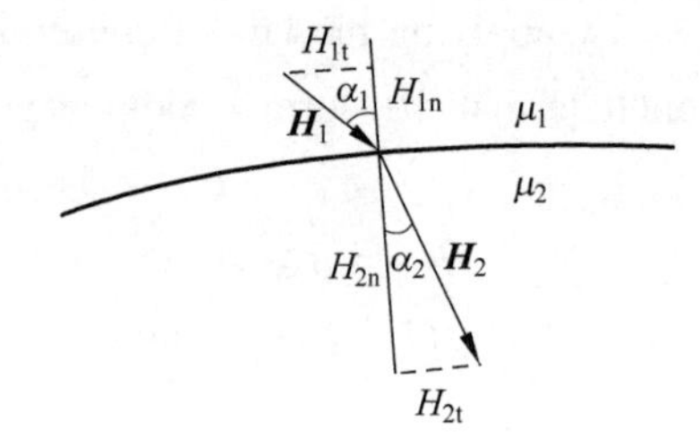

Figure 5-14 $\boldsymbol{H}$ fields at an interface(Example 5-9)

Solution: Continuity of the normal component of $\boldsymbol{B}$ field requires, from(5.85),

$$\mu_2 H_2 \cos\alpha_2 = \mu_1 H_1 \cos\alpha_1 \tag{5.88}$$

Since neither of the media is a perfect conductor, the tangential component of $\boldsymbol{H}$ field is continuous. We have

$$H_2 \sin\alpha_2 = H_1 \sin\alpha_1 \tag{5.89}$$

Division of(5.89) by(5.88) gives

$$\frac{\tan\alpha_2}{\tan\alpha_1} = \frac{\mu_2}{\mu_1} \tag{5.90}$$

or

$$\alpha_2 = \arctan\left(\frac{\mu_2}{\mu_1}\tan\alpha_1\right) \tag{5.91}$$

which describes the refraction property of the magnetic field. The magnitude of $\boldsymbol{H}_2$ is

$$H_2 = \sqrt{H_{2t}^2 + H_{2n}^2} = \sqrt{(H_2\sin\alpha_2)^2 + (H_2\cos\alpha_2)^2}$$

From(5.88) and(5.89) we obtain

$$H_2 = H_1\left[\sin^2\alpha_1 + \left(\frac{\mu_1}{\mu_2}\cos\alpha_1\right)^2\right]^{1/2} \tag{5.92}$$

In Example 5-9, if medium 1 is nonmagnetic(like air) and medium 2 is ferromagnetic(like iron), then $\mu_2 \gg \mu_1$ and from(5.90), α_2 will be nearly 90°. This means that for any arbitrary angle α_1 that is not close to zero, the magnetic field in a ferromagnetic medium runs almost parallel to the interface. On the other hand, if medium 1 is ferromagnetic and medium 2 is air ($\mu_1 \gg \mu_2$), then α_2 will be nearly zero. That is, if a magnetic field originates in a ferromagnetic medium, the flux lines will emerge into air in a direction almost normal to the interface.

The above discussion indicates that materials with high permeability have the ability to confine magnetic flux just as conductors in an electric circuit are able to confine currents. This property leads to the concept of **magnetic circuits**(磁路), which is discussed in the following section.

5.8 Magnetic Circuits(磁路)

The operation of most electric power machinery, such as transformers, generators, and motors can by analyzed by using the concept of magnetic circuits. A magnetic circuit consists of a structure with high-permeability magnetic material that confines the magnetic flux to a closed path. The closed path of magnetic flux is called a magnetic circuit. There is a lot of similarity between the magnetic and electric circuits. Electric circuit problems are those to find the voltages and the currents in an electric network that is excited by voltage or current sources. Analogously, magnetic circuit problems are those to find the magnetic fluxes and magnetic field intensities in a ferromagnetic structure due to current-carrying windings.

Analysis of magnetic circuits is based on Ampere's circuital law. Consider the simple example of magnetic circuits as shown in Figure 5-15(a). An N-turn winding around the magnetic core with high permeability carries a current I that excites a magnetic field $\boldsymbol{H}$ within the core. Applying Ampere's circuital law to the closed path C along the axis of the magnetic core, we have

$$\oint_C \boldsymbol{H} \cdot \mathrm{d}\boldsymbol{l} = NI \tag{5.93}$$

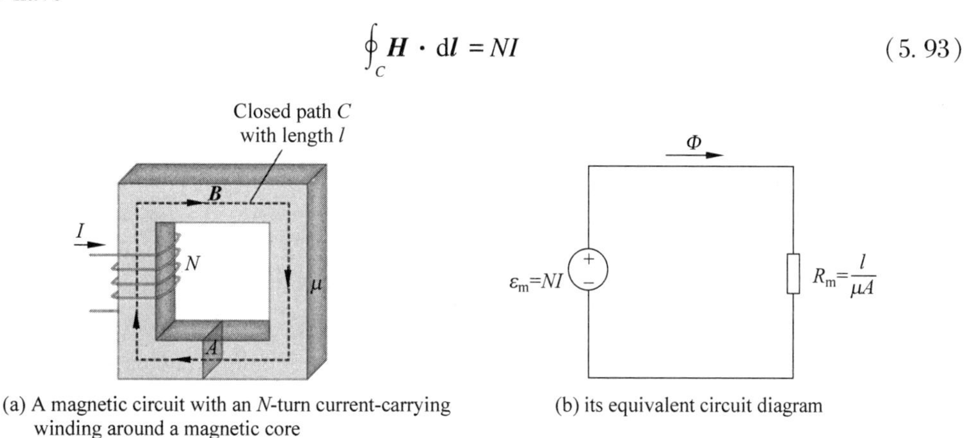

(a) A magnetic circuit with an N-turn current-carrying winding around a magnetic core

(b) its equivalent circuit diagram

Figure 5-15 Simple example of magnetic circuits

Assume the magnetic flux leakage can be neglected, and assume the magnetic flux density is uniform over any cross-sectional plane of the magnetic core.[①] Then a constant magnetic flux Φ will flow throughout the magnetic core, and the magnitude of the magnetic flux density can be evaluated as $B=\Phi/A$, with A to be the cross-section area. If A is constant along the closed path

---

① 磁路分析通常基于两个假设：磁芯中的磁通泄露可以忽略；磁力线在磁芯横截面上的分布是均匀的。基于这两个假设，围绕磁芯的多匝载流线圈在磁芯内产生的磁通就类似于电动势在电路中产生的电流。然而，由于电路周围的空气电导率几乎为零，电路中的电流基本上不会有泄露。而磁芯周围的空气磁导率不为零，磁路的磁通泄露相对于电路的电流泄露大很多。

C,(5.93) can be rewritten as

$$\Phi\left(\frac{l}{\mu A}\right) = NI \tag{5.94}$$

where l is the length of the closed path C. Notice that the term($l/\mu A$) in(5.94) resembles the expression of a resistor with length l and cross-sectional area A except that the conductivity σ is replaced by the permeability μ. Hence we define a **reluctance**(磁阻), R_{m}, of a piece of homogeneous material with length l, uniform cross-sectional area A and permeability μ to be

$$R_{\mathrm{m}} = \frac{l}{\mu A} \tag{5.95}$$

The term(NI) on the right side of(5.94) is the cause of the flux Φ in the magnetic material with reluctance R_{m}, and hence, is analogous to electromotive force(emf) that causes current flow in an electric circuit. In magnetic circuits, NI is called a **magnetomotive force**(mmf,**磁动势**), denoted by $\mathcal{E}_{\mathrm{m}}$, i.e.,

$$\mathcal{E}_{\mathrm{m}} = NI \tag{5.96}$$

Apparently, the unit of mmf is ampere(A) or ampere-turns to emphasize the multiple turns of the winding. With the concept of reluctance and mmf, (5.94) can be rewritten as

$$\Phi R_{\mathrm{m}} = \mathcal{E}_{\mathrm{m}} \tag{5.97}$$

which is analogous to the expression of the Kirchhoff's voltage law in an electric circuit, with Φ analogous to the current, R_{m} analogues to the resistance of a resistor, and $\mathcal{E}_{\mathrm{m}}$ analogous to the emf of an ideal voltage source connected to the resistor. Therefore, the magnetic circuit in Figure 5-15(a) can be represented by a circuit diagram shown in Figure 5-15(b).

(5.97) can be extended to a basic equation for magnetic circuits that corresponds to the Kirchhoff's voltage law for electric circuits. That is, for any closed path containing M current-carrying windings and N homogeneous magnetic materials in a magnetic circuit,

$$\sum_{i}^{M} \mathcal{E}_{\mathrm{m}i} = \sum_{j}^{N} R_{\mathrm{m}j} \Phi_j \tag{5.98}$$

where $\mathcal{E}_{\mathrm{m}i} = N_i I_i$ is the mmf of the ith winding with N_i turns and current I_i, Φ_j is the magnetic flux within the jth magnetic material with a reluctance $R_{\mathrm{m}j}$.

Analogous to the Kirchhoff's current law for electric circuits, another basic equation for magnetic circuits can be derived from(5.18), the law of conservation of magnetic flux. That is, the algebraic sum of all the magnetic fluxes flowing out of a junction in a magnetic circuit is zero, which can be written as

$$\sum_{i}^{M} \Phi_i = 0 \tag{5.99}$$

where Φ_i is the ith of the M magnetic fluxes flowing out of the junction. With(5.98) and(5.99), we can perform the loop and node analysis of more complicated magnetic circuits as demonstrated in the next examples. ①

---

① 实际应用中,准确的磁路分析还需要考虑一系列非理想因素的影响:载流线圈和磁芯周边存在少量分散的磁通,这部分磁通称为漏磁通;磁性材料的磁导率与其中存在的磁场强度之间存在非线性关系;铁磁质的磁化过程存在磁滞现象。

Example 5-10 N turns of wire carrying current I are wound around a toroidal ferromagnetic core with permeability $\mu(\mu \gg \mu_0)$. As shown in Figure 5-16, the core has a mean radius ρ_0, a circular cross section with radius a $(a \ll \rho_0)$, and a very narrow air gap of length $g(g \ll a)$. Determine the $\boldsymbol{B}$ and $\boldsymbol{H}$ fields in the ferromagnetic core as well as in the air gap.

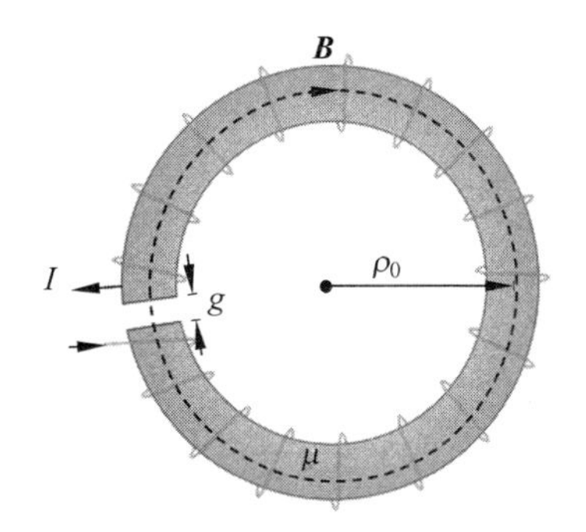

Figure 5-16 Magnetic circuit of Example 5-10

Solution: Since the air gap is very narrow, we can assume the magnetic flux leakage from the air gap is negligable. That is, the magnetic flux coming out from one end face of the ferromagnetic core flows though the air gap and enters the other end face entirely. Then ferromagnetic core with the air gap can be treated as two reluctances in series. From(5.98), we have

$$NI = \Phi R_{\mathrm{mc}} + \Phi R_{\mathrm{mg}}$$

where

$$R_{\mathrm{mc}} = \frac{2\pi\rho_0 - g}{\mu\pi a^2}$$

is the reluctance of the ferromagnetic core and

$$R_{\mathrm{mg}} = \frac{g}{\mu_0\pi a^2}$$

is the reluctance of the air gap. Therefore, the magnetic flux is

$$\Phi = \frac{NI}{R_{\mathrm{mc}} + R_{\mathrm{mg}}} = \frac{NI}{\dfrac{2\pi\rho_0 - g}{\mu\pi a^2} + \dfrac{g}{\mu_0\pi a^2}} = \frac{\pi a^2\mu\mu_0 NI}{(2\pi\rho_0 - g)\mu_0 + g\mu}$$

The magnetic flux density in both the ferromagnetic core and the air gap is

$$\boldsymbol{B} = \boldsymbol{e}_\phi \frac{\Phi}{\pi a^2} = \boldsymbol{e}_\phi \frac{\mu\mu_0 NI}{(2\pi\rho_0 - g)\mu_0 + g\mu}$$

The magnetic field intensity is

$$\boldsymbol{H} = \begin{cases} \dfrac{\boldsymbol{B}}{\mu} = \boldsymbol{e}_\phi \dfrac{\mu_0 NI}{(2\pi\rho_0 - g)\mu_0 + g\mu} & \text{(in the ferromagnetic core)} \\[2ex] \dfrac{\boldsymbol{B}}{\mu_0} = \boldsymbol{e}_\phi \dfrac{\mu NI}{(2\pi\rho_0 - g)\mu_0 + g\mu} & \text{(in the air gap)} \end{cases}$$

Since $\mu \gg \mu_0$, the magnetic field intensity in the air gap is much stronger than that in the ferromagnetic core.

Example 5-11 In the magnetic circuit shown in Figure 5-17, steady currents I_1 and I_2 flow in two wires with N_1 and N_2 turns respectively. The ferromagnetic core has a cross-sectional area A and a permeability μ. Determine the magnitude of magnetic flux density in the middle branch of the magnetic circuit.

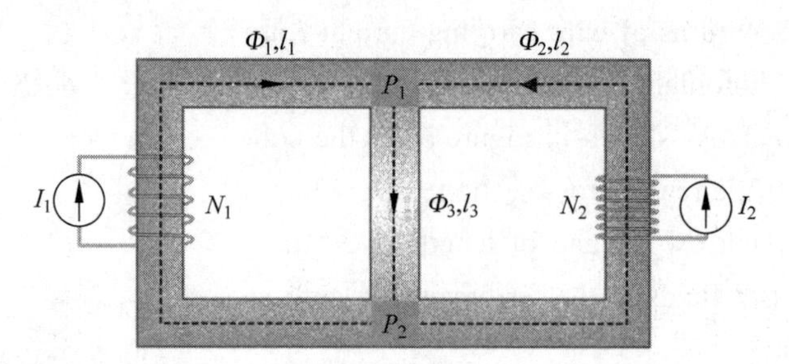

Figure 5-17 Magnetic circuit of Example 5-11

Solution: As is shown in Figure 5-17, the circuit consists of three branches between the two junctions P_1 and P_2. The left branch with length l_1 has a reluctance

$$R_{m1} = \frac{l_1}{\mu A} \tag{5.100a}$$

The right branch with l_2 has a reluctance

$$R_{m2} = \frac{l_2}{\mu A} \tag{5.100b}$$

The middle branch with length l_3 has a reluctance

$$R_{m3} = \frac{l_3}{\mu A} \tag{5.100c}$$

The two sources of mmf, N_1I_1 and N_2I_2, are in series with reluctances R_{m1} and R_{m2} respectively. Assume the magnetic fluxes in the left, right, and middle branches are Φ_1, Φ_2 and Φ_3 respectively. Applying (5.98) to the loop formed by the left and middle branches gives

$$N_1I_1 = R_{m1}\Phi_1 + R_{m3}\Phi_3 \tag{5.101a}$$

Applying (5.98) to the loop formed by the right and middle branches gives

$$N_2I_2 = R_{m2}\Phi_2 + R_{m3}\Phi_3 \tag{5.101b}$$

Applying (5.99) to junction P_1 gives

$$-\Phi_1 - \Phi_2 + \Phi_3 = 0 \tag{5.101c}$$

By solving the three equations (5.101a)-(5.101c), we obtain

$$\Phi_3 = \frac{R_{m2}N_1I_1 + R_{m1}N_2I_2}{R_{m1}R_{m2} + R_{m2}R_{m3} + R_{m1}R_{m3}} \tag{5.102}$$

By substituting (5.100a)-(5.100c) into (5.102), we obtain the magnitude of the magnetic flux density in the middle branch as

$$B = \frac{\Phi_3}{A} = \mu\frac{l_2N_1I_1 + l_1N_2I_2}{l_1l_2 + l_2l_3 + l_1l_3}$$

5.9 Inductances and Inductors (电感)

5.9.1 Self-inductances of thin wires (细线自感)

Consider a closed path, C bounding an arbitrary surface S as is shown in Figure 5-18(a). If a current I flows in C, a magnetic flux Φ through the surface S will be created. The magnetic flux Φ links with the circuit C, and therefore, is equal to the **magnetic flux linkage**(磁链) for a single-turn of C. If C has N turns, Φ is considered to link with C by N times, and the magnetic flux linkage(denoted by Λ) due to Φ will be

$$\Lambda = N\Phi \tag{5.103}$$

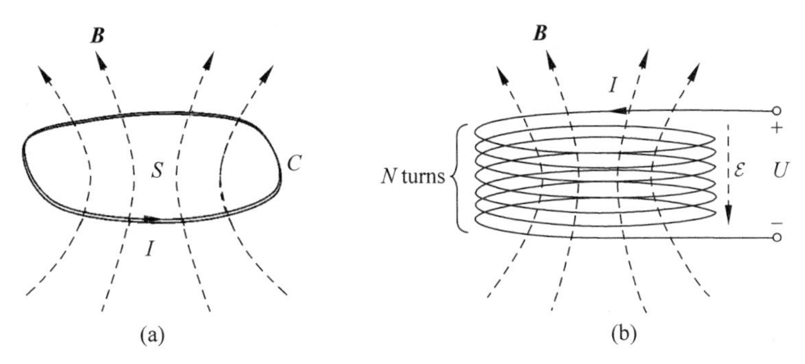

Figure 5-18 Inductance and inductors

From Biot-Savart law(5.5), $\boldsymbol{B}$ is directly proportional to I, and hence Φ and Λ must also be proportional to I. Therefore, we can define the **self-inductance**(自感), or simply **inductance**(电感), of a closed-loop circuit as

$$L = \frac{\Lambda}{I} \tag{5.104}$$

The unit of inductance is henry(H, 亨). A more general definition for the inductance L that suits time-varying problems is

$$L = \frac{\mathrm{d}\Lambda}{\mathrm{d}I} \tag{5.105}$$

The inductance calculated from(5.104) is usually equal to that from(5.105) up to very high frequencies. Notice that the inductance of a given conducting loop depends on the geometrical size and shape of the loop, as well as on the permeability of the medium. With a linear medium, the inductance does not depend on the current in the loop. Whether or not the current in the loop exists does not affect the inductance.

Why should we introduce the concept of inductance and why is it defined as(5.105)? The answer is that, the definition of inductance bridges the **Faraday's law of electromagnetic induction**(法拉第电磁感应定律) and the **current-voltage characteristic**(伏安特性) of an **inductor**(电感器,简称电感). An inductor is a circuit element usually made of a conducting wire in certain shape to provide an inductance. According to Faraday's law of electromagnetic induction, if the current flowing through an inductor is time varying, an electromotive force (emf) will be induced in the inductor. For an N-turn inductor as is shown in Figure 5-18(b), an emf of $-\mathrm{d}\Phi/\mathrm{d}t$ is induced on each turn, and the total emf is

$$\mathcal{E} = -N\frac{\mathrm{d}\Phi}{\mathrm{d}t} = -\frac{\mathrm{d}\Lambda}{\mathrm{d}t} \tag{5.106}$$

Notice that the negative sign in(5.106) means $\mathcal{E}$ is preventing the magnetic flux (and the current as well) from increasing. The reference direction of $\mathcal{E}$ is the same as that of the current I, which is consistent with the definition of $\mathcal{E}$ in (4.40). According to the Kirchhoff's voltage law, if the internal resistance of the inductor can be ignored, $\mathcal{E}$ is equal to the voltage drop from the negative to the positive terminal of the inductor (outside the inductor). With U to be the voltage drop across the inductor from the positive to the negative terminal, we must have $\mathcal{E}=-U$. Then, by substituting (5.104) or (5.105) into (5.106), we have

$$U=-\mathcal{E}=\frac{\mathrm{d}\Lambda}{\mathrm{d}t}=L\frac{\mathrm{d}I}{\mathrm{d}t} \tag{5.107}$$

which is identical to the current-voltage characteristic across an inductor in the circuit theory.

Given a conducting wire or circuital loop, the method of determining the inductance is quite straightforward by using its definition. We usually start by assuming a current I in the conducting wire. Find the magnetic flux density $\boldsymbol{B}$ from I by using Ampere's circuital law or Biot-Savart law. Then calculate the magnetic flux Φ by integrating the $\boldsymbol{B}$ field over the surface enclosed by the loop that links with $\boldsymbol{B}$. Obtain the flux linkage Λ by multiplying Φ with the number of turns of the loop. And finally, find L by taking the ratio $L=\Lambda/I$.

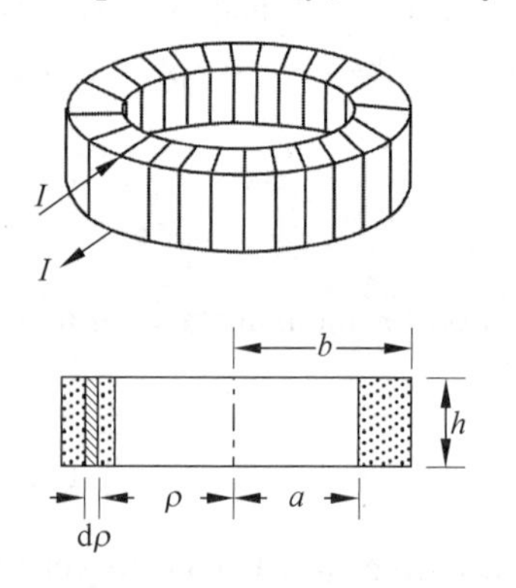

Figure 5-19 A closely wound toroidal coil (Example 5-12)

Example 5-12 N turns of wire wound on a toroid with a rectangular cross section is as shown in Figure 5-19. Assuming the permeability of the core of the toroid is μ, find the self-inductance of the toroidal coil.

Solution: The toroid structure is the same as that in Example 5-5. Using cylindrical coordinate system with the axis of the toroid along z-direction, we have $\boldsymbol{B}=\boldsymbol{e}_\phi B_\phi$ due to symmetry. By applying Ampere's circuital law, we can obtain the solution of B_ϕ as

$$B_\phi=\frac{\mu NI}{2\pi\rho}$$

Then the magnetic flux through a cross-sectional area S of the toroid is

$$\Phi=\int_S \boldsymbol{B}\cdot \mathrm{d}\boldsymbol{s}=\int_S\left(\boldsymbol{e}_\phi\frac{\mu NI}{2\pi\rho}\right)\cdot(\boldsymbol{e}_\phi h\mathrm{d}\rho)=\frac{\mu NIh}{2\pi}\int_a^b\frac{\mathrm{d}\rho}{\rho}=\frac{\mu NIh}{2\pi}\ln\frac{b}{a}$$

The flux linkage Λ is $N\Phi$, so

$$\Lambda=\frac{\mu N^2 Ih}{2\pi}\ln\frac{b}{a}$$

Finally, we obtain

$$L=\frac{\Lambda}{I}=\frac{\mu N^2 h}{2\pi}\ln\frac{b}{a} \tag{5.108}$$

Example 5-13 A long solenoid with n turns per unit length has a magnetic core with permeability μ. Find the inductance per unit length of the solenoid.

Solution The magnetic flux density inside a long air-core solenoid has been found in Example 5-6. With a permeability different from that in Example 5-6, the magnetic flux density in the solenoid can be obtained by modifying(5.59), which gives

$$B = \mu n I$$

which is constant inside the solenoid. Hence,

$$\Phi = BS = \mu n S I \tag{5.109}$$

where S is the cross-sectional area of the solenoid. The flux linkage per unit length is

$$\Lambda' = n\Phi = \mu n^2 S I \tag{5.110}$$

Therefore the inductance per unit length is

$$L' = \mu n^2 S(\text{H/m}) \tag{5.111}$$

A significant observation about the results of the previous two examples is that, the self-inductance of wire-wound inductors is proportional to the square of the number of turns. [①] This phenomenon is manifested by(5.5) and(5.106). For an N-turn wire, the length of the integral contour C' in(5.5) is N times as long as that of a one-turn wire of the same shape. Therefore, the magnetic flux density $\boldsymbol{B}$, as well as Φ produced by the current is proportional to N. [②] If the current in the wire varies with time, the induced emf on each turn of the wire is also proportional to N. The emfs on all the N turns add up, leading to a total induced emf proportional to N^2 as indicated by(5.106).

5.9.2 Internal and external inductances(内电感与外电感)

The above discussions and examples are based on the assumption that the wire of the inductor carrying a current is infinitely thin and each turn of the wires links with the same amount of the magnetic flux. In reality, the current-carrying conductor cannot be infinitely thin and there might be magnetic flux density field penetrating the conductor. In other words, the current is continuously distributed within a conductor as a volume current. To calculate the inductance in this case, we can divide the volume current into infinite number of infinitesimal line currents. As is shown in Figure 5-20(a), a differential current $\mathrm{d}I$ flowing through the differential area $\mathrm{d}s'$ links with a magnetic flux Φ' that can be obtained by integrating the $\boldsymbol{B}$ field over the surface enclosed by the path of $\mathrm{d}I$. Since the magnetic flux Φ' only links with a small fraction($\mathrm{d}I/I$) of the total current, its contribution to the total magnetic flux linkage is counted as $\mathrm{d}\Lambda = (\mathrm{d}I/I)\Phi'$. Hence, the total magnetic flux linkage can be calculated as

$$\Lambda = \int \mathrm{d}\Lambda = \int \frac{\Phi'}{I}\mathrm{d}I = \frac{1}{I}\int_{S'} \Phi' J \mathrm{d}s' \tag{5.112}$$

where S' is the cross section of the conductor that carries the current I, and J is the magnitude of the current density on the differential element $\mathrm{d}s'$ of the surface S'. Then the inductance is

---

① 多匝线圈的电感与线圈匝数的平方成正比是磁场相关的一个重要的物理现象，提供了低频电路应用中实现大电感的一个重要方法。

② 这里假设 N 匝线圈中的电流均匀分布。该假设成立的前提是线圈中的电流恒定或者随时间变化足够缓慢，即时变电流变化频率对应的电磁波波长远大于线圈尺寸。否则线圈中的电流将不是均匀分布，无法定义单个集总参数电感。

$$L = \frac{\Lambda}{I} = \frac{1}{I^2}\int_{S'} \Phi' J \mathrm{d}s' \tag{5.113}$$

If the current is uniformly distributed within the conductor that has a uniform cross section, $J = I/A_{S'}$, where $A_{S'}$ is the area of the cross section of the conductor. Then (5.113) becomes

$$L = \frac{1}{A_{S'} I}\int_{S'} \Phi' \mathrm{d}s' \tag{5.114}$$

The inductor can also be calculated by first dividing the entire magnetic flux into infinite number of differential fluxes. As is shown in Figure 5-20(b), the differential flux $\mathrm{d}\Phi$ links with only a partial current I'. The total magnetic flux linkage Λ is the summation (integration) of all the differential fluxes $\mathrm{d}\Phi$ weighted by the ratio (I'/I). Therefore, the inductance can be calculated as

$$L = \frac{1}{I}\int\left(\frac{I'}{I}\right)\mathrm{d}\Phi = \frac{1}{I^2}\int_S I' B \mathrm{d}s \tag{5.115}$$

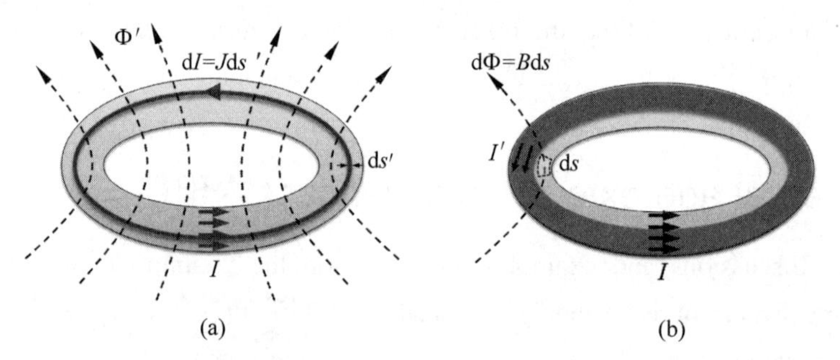

Figure 5-20 Calculation of the inductor of a conductor carrying continuous volume current

Notice that the surface S of the integration in (5.115) is different from the surface S' in (5.112)-(5.114) (How?). However, the calculated result using (5.115) should be the same as that using (5.113).

Example 5-14 An air-filled coaxial transmission line has an inner conductor of radius a and a very thin outer conductor of radius b. Determine the inductance per unit length of the line.

Solution: Refer to Figure 5-21. Assume that a current I flows in the inner conductor and returns via the outer conductor in the opposite direction. Let the axis of the transmission line be along z-direction. Because of the cylindrical symmetry, $\boldsymbol{B}$ has only a ϕ-component. Also assume that the current I is uniformly distributed over the cross section of the inner conductor. Then, we can find the distribution of $\boldsymbol{B}$ inside and outside the inner conductor by using Ampere's circuital law as is done in Example 5-4.

(1) Inside the inner conductor ($0 \leqslant \rho \leqslant a$), from (5.56a),

$$\boldsymbol{B}_1 = \boldsymbol{e}_\phi B_{\phi 1} = \boldsymbol{e}_\phi \frac{\mu_0 \rho I}{2\pi a^2} \tag{5.116}$$

(2) Between the inner and outer conductors ($a \leqslant \rho \leqslant b$), from (5.56b),

$$\boldsymbol{B}_2 = \boldsymbol{e}_\phi B_{\phi 2} = \boldsymbol{e}_\phi \frac{\mu_0 I}{2\pi \rho} \tag{5.117}$$

We first use(5. 114) to calculate the inductance. The surface of integration S' is selected to be a cross section of the inner conductor($0 \leqslant \rho \leqslant a$ and $0 \leqslant \phi \leqslant 2\pi$, as shown on the right side of Figure 5-21). Consider a differential element $ds' = \rho d\rho d\phi$ on S', which carries current dI. In a unit length along z, dI is linked with the flux Φ' that can be obtained by integrating the magnetic flux density in(5. 116) and(5. 117), i. e.,

$$\Phi' = \int_{\rho}^{a} B_{\phi 1} d\rho + \int_{a}^{b} B_{\phi 2} d\rho = \frac{\mu_0 I}{2\pi a^2}\int_{\rho}^{a} \rho d\rho + \frac{\mu_0 I}{2\pi}\int_{a}^{b} \frac{d\rho}{\rho}$$

$$= \frac{\mu_0 I}{4\pi a^2}(a^2 - \rho^2) + \frac{\mu_0 I}{2\pi}\ln\frac{b}{a} \tag{5.118}$$

Substitute(5. 118) into(5. 114), the inductance per unit length is

$$L' = \frac{1}{(\pi a^2) I}\int_0^{2\pi}\int_0^a \left[\frac{\mu_0 I}{4\pi a^2}(a^2 - \rho^2) + \frac{\mu_0 I}{2\pi}\ln\frac{b}{a}\right]\rho d\rho d\phi$$

$$= \frac{\mu_0}{2\pi}\left(\frac{1}{4} + \ln\frac{b}{a}\right) \tag{5.119}$$

We can also use(5. 115) to solve this problem. To do that, we first determine the surface of integration S in(5. 115) to be a rectangular with width $b(0 \leqslant \rho \leqslant b)$ and unit length along the z-direction, as is shown in the left side of Figure 5-21. Any differential magnetic flux $d\Phi = B_2\, ds$ in region $a \leqslant \rho \leqslant b$ is linked with the entire current I, whereas a magnetic flux $d\Phi = B_1 ds$ in region $0 \leqslant \rho \leqslant a$ is linked with I' which is only partial of the current I. Specifically,

$$I' = \frac{I}{\pi a^2}(\pi \rho^2) = \frac{\rho^2}{a^2} I$$

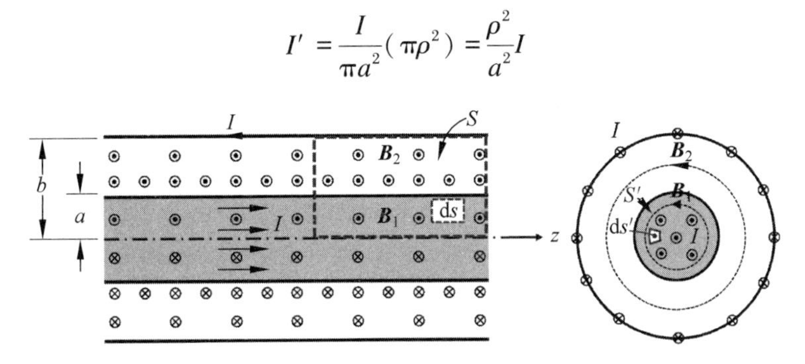

Figure 5-21 Two views of a coaxial transmission line(Example 5-14)

Therefore, according to(5. 115), the inductance per unit length is

$$L' = \frac{1}{I^2}\left[\int_0^a I' B_1 d\rho + \int_a^b I B_2 d\rho\right] = \frac{1}{I^2}\left[\int_0^a \left(\frac{\rho^2}{a^2} I\right)\left(\frac{\mu_0 \rho I}{2\pi a^2}\right) d\rho + \int_a^b I\left(\frac{\mu_0 I}{2\pi \rho}\right) d\rho\right]$$

$$= \frac{\mu_0}{8\pi} + \frac{\mu_0}{2\pi}\ln\frac{b}{a}\,(\mathrm{H/m}) \tag{5.120}$$

which is the same as(5. 119). The first term $\mu_0/8\pi$ is contributed by the flux linkage internal to the solid inner conductor, and therefore, is considered as the **internal inductance**(**内电感**) per unit length of the inner conductor. The second term is contributed by the linkage of the flux that exists between the inner and the outer conductors, and therefore, is called the **external inductance**(**外电感**) per unit length of the coaxial line.

It should be pointed out that the results in the above example are accurate in static conditions. In the time-varying case that will be discussed in Chapter 7, the current distribution in a conductor is not even, and thereby leads to a change in the internal inductance. In high frequencies, the current concentrates in the "skin" of a good conductor as a surface current, and the internal self-inductance tends to be zero. ①

Example 5-15 Find the inductance per unit length of a transmission line consisting of two long parallel conducting wires with radius a and separated by a distance D (assume that $D \gg a$).

Solution: The internal self-inductance per unit length of each wire is, from Example 5-14, $\mu_0/8\pi$. So for two wires we have

$$L_i' = 2 \times \frac{\mu_0}{8\pi} = \frac{\mu_0}{4\pi} \text{(H/m)} \tag{5.121}$$

To find the external self-inductance per unit length, we first calculate the magnetic flux per unit length linking with an assumed current I in the wires. In the xz-plane as shown in Figure 5-22, the $\boldsymbol{B}$ fields due to the equal and opposite currents in the two wires have only a y-component, and can be found easily by using Ampere's circuital law as

$$B_{y1} = \frac{\mu_0 I}{2\pi x} \tag{5.122}$$

$$B_{y2} = \frac{\mu_0 I}{2\pi(D - x)} \tag{5.123}$$

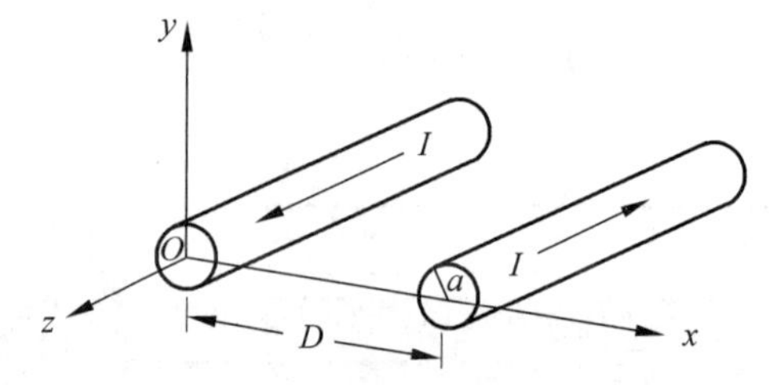

Figure 5-22 A two-wire transmission line (Example 5-15)

The flux linkage per unit length is then

$$\Phi' = \int_a^{D-a} (B_{y1} + B_{y2})\,\mathrm{d}x = \int_a^{D-a} \frac{\mu_0 I}{2\pi}\left[\frac{1}{x} + \frac{1}{D-x}\right]\mathrm{d}x$$
$$= \frac{\mu_0 I}{\pi}\ln\left(\frac{D-a}{a}\right) \approx \frac{\mu_0 I}{\pi}\ln\frac{D}{a} \text{ (Wb/m)}$$

Therefore

$$L_e' = \frac{\Phi'}{I} = \frac{\mu_0}{\pi}\ln\frac{D}{a} \text{(H/m)} \tag{5.124}$$

and the total self-inductance per unit length of the two-wire line is

① 通常情况下，内电感比外电感小很多。尤其在时变条件下，电流在导体内不再是均匀分布，长直导体的内电感值 $\mu_0/8\pi$ 不再精确。第 7 章将讨论高频条件下电磁场的趋肤效应，即电流仅集中分布于导体的表层，从而导致内电感趋近于零。

$$L' = L_i' + L_e' = \frac{\mu_0}{\pi}\left(\frac{1}{4} + \ln\frac{D}{a}\right) \text{ (H/m)} \tag{5.125}$$

5.9.3 Mutual inductances and the Neumann Formula (互感与纽曼公式)

Now consider two neighboring closed loops, C_1 and C_2 bounding surfaces S_1 and S_2, respectively, as shown in Figure 5-23. If a current I_1 flows in C_1, a magnetic field $\boldsymbol{B}_1$ will be produced, which creates a magnetic flux Φ_{12} linking with C_2. Φ_{12} is a mutual magnetic flux through the surface S_2 due to $\boldsymbol{B}_1$, and by definition,

$$\Phi_{12} = \int_{S_2} \boldsymbol{B}_1 \cdot \mathrm{d}\boldsymbol{s}_2 \tag{5.126}$$

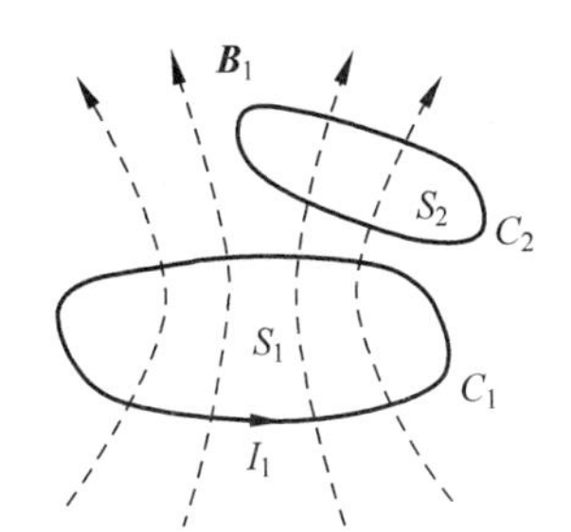

Figure 5-23 Two magnetically coupled loops

If the loop C_2 has N_2 turns, the magnetic flux linkage Λ_{12} due to Φ_{12} is

$$\Lambda_{12} = N_2 \Phi_{12} \tag{5.127}$$

Since $\boldsymbol{B}_1$ is directly proportional to I_1, Λ_{12} is also proportional to I_1. So, we can write

$$\Lambda_{12} = L_{12} I_1 \tag{5.128}$$

or

$$L_{12} = \frac{\Lambda_{12}}{I_1} \tag{5.129}$$

where the proportionality constant L_{12} is called the **mutual inductance**(互感) between loops C_1 and C_2. A more general definition for L_{12} is

$$L_{12} = \frac{\mathrm{d}\Lambda_{12}}{\mathrm{d}I_1} \tag{5.130}$$

If the current I_1 in C_1 varies in time, the magnetic flux density $\boldsymbol{B}_1$ will also vary with time, leading to a time-varying magnetic flux Φ_{12} linking with C_2. According to Faraday's law of induction, there will be an induced emf in the close loop C_2. The induced emf can be written as

$$\begin{aligned} \mathcal{E}_{12} &= -N_2 \frac{\mathrm{d}\Phi_{12}}{\mathrm{d}t} = -\frac{\mathrm{d}\Lambda_{12}}{\mathrm{d}t} = -\frac{\mathrm{d}\Lambda_{12}}{\mathrm{d}I_1}\frac{\mathrm{d}I_1}{\mathrm{d}t} \\ &= -L_{12}\frac{\mathrm{d}I_1}{\mathrm{d}t} \end{aligned} \tag{5.131}$$

in which the mutual inductance expression(5.130) is used.

The magnetic flux produced by I_1 also links with C_1 itself, and the total flux linkage is

$$\Lambda_{11} = N_1 \Phi_{11} = N_1 \int_{S_1} \boldsymbol{B}_1 \cdot \mathrm{d}\boldsymbol{s}_1 \tag{5.132}$$

The self-inductance of loop C_1(in a multi-coil circumstance) is

$$L_{11} = \frac{\Lambda_{11}}{I_1} \tag{5.133}$$

or more generally,

$$L_{11} = \frac{\mathrm{d}\Lambda_{11}}{\mathrm{d}I_1} \tag{5.134}$$

The above discussion only considers a current I_1 flows in C_1 as shown in Figure 5-23. If a current I_2 flows in C_2, a magnetic field $\boldsymbol{B}_2$ will be produced, which creates a magnetic flux Φ_{22} linking with C_2 itself and a mutual magnetic flux Φ_{21} linking with C_1. In a similar fashion, we can define a self-inductance L_{22} and a mutual inductance L_{21}. If the medium around the loops are reciprocal, it can be proved that

$$L_{12} = L_{21} \tag{5.135}$$

as a result of the **principle of reciprocity**（互易原理）. To prove (5.135), we combine (5.126), (5.127) and (5.129) to obtain

$$L_{12} = \frac{N_2}{I_1}\int_{S_2} \boldsymbol{B}_1 \cdot \mathrm{d}\boldsymbol{s}_2$$

By using (5.32), we have

$$L_{12} = \frac{N_2}{I_1}\oint_{C_2} \boldsymbol{A}_1 \cdot \mathrm{d}\boldsymbol{l}_2 \tag{5.136}$$

Now from (5.30),

$$\boldsymbol{A}_1 = \frac{\mu_0 N_1 I_1}{4\pi}\oint_{C_1} \frac{\mathrm{d}\boldsymbol{l}_1}{R} \tag{5.137}$$

Substitution of (5.137) in (5.136) yields

$$L_{12} = \frac{\mu_0 N_1 N_2}{4\pi}\oint_{C_1}\oint_{C_2} \frac{\mathrm{d}\boldsymbol{l}_1 \cdot \mathrm{d}\boldsymbol{l}_2}{R} \tag{5.138a}$$

where R is the distance between the differential lengths $\mathrm{d}\boldsymbol{l}_1$ and $\mathrm{d}\boldsymbol{l}_2$. A more general version of the mutual inductance expression of (5.138a) is

$$L_{12} = \frac{\mu_0}{4\pi}\oint_{C_1}\oint_{C_2} \frac{\mathrm{d}\boldsymbol{l}_1 \cdot \mathrm{d}\boldsymbol{l}_2}{R} \tag{5.138b}$$

where C_1 and C_2 are the entire paths of the current flow in the two circuits. In other words, the contour integrals in (5.138b) are evaluated N_1 times over the loop of C_1 and N_2 times over the loop of C_2, whereas in (5.137) and (5.136), the contour integrals are evaluated only once over the loops of C_1 and C_2. By interchanging the subscripts 1 and 2 in (5.138a) or (5.138b), we immediately arrive at the conclusion that $L_{21} = L_{12}$, which proves (5.135).

(5.138b) is usually called the **Neumann formula**（纽曼公式） for mutual inductances. It indicates that the mutual inductance is a property of the geometrical shape and the physical arrangement of coupled circuits. For a linear medium, mutual inductance is proportional to the medium's permeability and is independent of the currents in the circuits.

To find the mutual inductance between two closed loops, we can use Neumann formula directly, which requires the evaluation of a double line integral. However, the double line integral is usually cumbersome to evaluate. For problems with symmetry conditions, a better approach of finding the mutual inductance is to use Ampere's circuital law to determine the magnetic flux density due to an assumed current. Then calculate the mutual flux linkage and find the mutual inductance by using its definition(5.129). By following the latter approach, we can either calculate the flux linkage to C_2 with an assumed current in C_1 (which leads to an evaluation of L_{12}) or calculate the flux linkage to C_1 with an assumed current in C_2 (which leads to an evaluation of L_{21}). Due to the relation(5.135) as ensured by the Neumann formula, we have the freedom to choose a simpler way (finding L_{12} or L_{21}) to determine the mutual inductance.

Example 5-16 Two coils of N_1 and N_2 turns are wound concentrically on a straight cylindrical core of radius a and permeability μ. The windings have lengths l_1 and l_2, respectively. Find the mutual inductance between the coils.

Solution: As shown in Figure 5-24, assume current I_1 flows in the inner coil. From (5.109), we find that the flux Φ_{12} in the solenoid core that links with the outer coil is

$$\Phi_{12} = \mu\left(\frac{N_1}{l_1}\right)(\pi a^2) I_1$$

Since the outer coil has N_2 turns, we have

$$\Lambda_{12} = N_2\Phi_{12} = \frac{\mu}{l_1}N_1 N_2 \pi a^2 I_1$$

Hence the mutual inductance is

$$L_{12} = \frac{\Lambda_{12}}{I_1} = \frac{\mu}{l_1}N_1 N_2 \pi a^2 (\text{H}) \tag{5.139}$$

Example 5-17 Determine the mutual inductance between a conducting rectangular loop and a very long straight wire as shown in Figure 5-25.

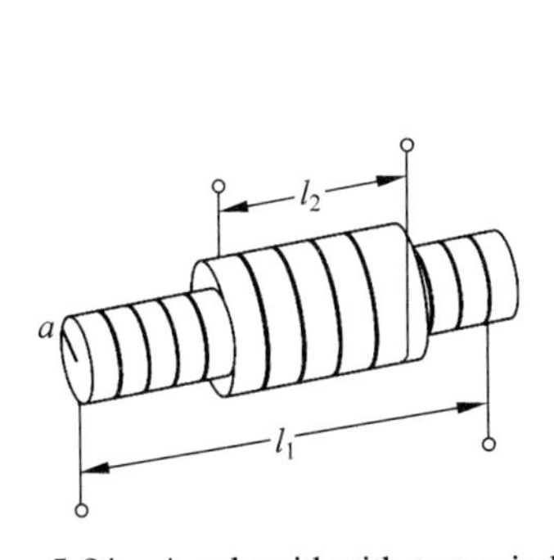

Figure 5-24 A solenoid with two windings (Example 5-16)

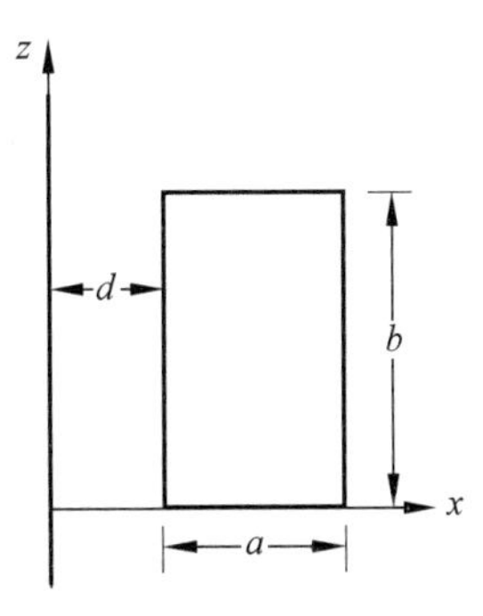

Figure 5-25 A conducting rectangular loop and a long straight wire(Example 5-17)

Solution: We can start with either assuming a current I_1 in the rectangular loop or assuming a current I_2 in the long straight wire. Apparently, the magnetic flux density $\boldsymbol{B}_2$ due to the current I_2 in the straight wire is much easier to find, which can be written as

$$\boldsymbol{B}_2 = \boldsymbol{e}_\phi \frac{\mu_0 I_2}{2\pi\rho} \tag{5.140}$$

The flux linkage to the rectangular loop is

$$\Lambda_{21} = \Phi_{21} = \int_{S_1} \boldsymbol{B}_2 \cdot \mathrm{d}\boldsymbol{s}_1$$
$$= \int_0^b \int_d^{d+a} \left(\boldsymbol{e}_\phi \frac{\mu_0 I_2}{2\pi\rho} \right) \cdot \boldsymbol{e}_\phi \mathrm{d}\rho \mathrm{d}z$$
$$= \frac{\mu_0 I_2 b}{2\pi} \ln\left(1 + \frac{a}{d}\right) \tag{5.141}$$

Therefore, the mutual inductance is

$$L_{21} = \frac{\Lambda_{21}}{I_2} = \frac{\mu_0 b}{2\pi} \ln\left(1 + \frac{a}{d}\right) \tag{5.142}$$

5.10 Magnetic Energy(磁能)

5.10.1 Magnetic Energy in Terms of Currents and Magnetic Fluxes(电流和磁通表示的磁能)

In Section 2.10, we derived the expression of electrostatic energy as the work required to assemble a group of charges. Similarly, work is also needed to establish the current in the conducting loops and the work will be stored as **magnetic energy**(磁能). Consider a single closed loop with a self-inductance L_1 in which the current is initially zero. A current source is connected to the loop, which increases the current i_1 from 0 to I_1. From Faraday's law of induction, the change of current in a closed loop induces an emf that creates a voltage across the inductor. As discussed in section 5.9.1, the voltage across the loop as an inductor is $v_1 = L_1 \mathrm{d}i/\mathrm{d}t$. From circuit theory, the work required for i_1 to increase from 0 to I_1 is

$$W_1 = \int v_1 i_1 \mathrm{d}t = L_1 \int_0^{I_1} i_1 \mathrm{d}i_1 = \frac{1}{2} L_1 I_1^2 \tag{5.143}$$

Since $L_1 = \Phi_1 / I_1$ for linear media, (5.143) can be written alternatively in terms of flux linkage as

$$W_1 = \frac{1}{2} I_1 \Phi_1 \tag{5.144}$$

which is stored as magnetic energy.

Now consider an additional closed loop C_2. We let the current i_2 carried in C_2 increase from zero to I_2 while keeping i_1 in C_1 at constant I_1. During this process, a work W_2 must be done in loop C_2 to counteract the induced emf. W_2 can be derived in a similar way that is used to derive W_1, which leads to an expression similar to(5.143):

$$W_2 = \frac{1}{2} L_2 I_2^2 \tag{5.145}$$

At the same time, due to the **mutual coupling**(互耦), some of the magnetic flux produced by i_2 links with loop C_1 and causes an induced emf. Similar to the derivation of(5.131), the induced emf in C_1 can be expressed as $\mathcal{E}_{21} = -L_{21} \mathrm{d}i_2/\mathrm{d}t$. To keep i_1 constant at I_1, $\mathcal{E}_{21}$ must be overcome by a voltage $v_{21}=L_{21}\,\mathrm{d}i_2/\mathrm{d}t$, which requires a work

$$W_{21}=\int v_{21}I_1\mathrm{d}t=L_{21}I_1\int_0^{I_2}\mathrm{d}i_2=L_{21}I_1I_2 \tag{5.146}$$

The total amount of work done in raising the currents to I_1 and I_2, respectively, is then the sum of W_1, W_2 and W_{21}:

$$\begin{aligned}W_{\mathrm{m}}&=\frac{1}{2}L_1I_1^2+L_{21}I_1I_2+\frac{1}{2}L_2I_2^2\\&=\frac{1}{2}\sum_{j=1}^{2}\sum_{k=1}^{2}L_{jk}I_jI_k\end{aligned} \tag{5.147}$$

where L_1 and L_2 is rewritten as L_{11} and L_{22} in the two-loop system, and the fact $L_{12}=L_{21}$ has been used. (5.147) can be generalized to N loops, which gives the expression of the magnetic energy of a system of N loops carrying currents $I_1, I_2, \cdots, I_N$:

$$W_{\mathrm{m}}=\frac{1}{2}\sum_{j=1}^{N}\sum_{k=1}^{N}L_{jk}I_jI_k \tag{5.148}$$

For a current I flowing in a single inductor with inductance L, the stored magnetic energy is

$$W_{\mathrm{m}}=\frac{1}{2}LI^2 \tag{5.149}$$

An alternative way of deriving (5.148) is to apply the Faraday's law of induction directly to all the closed loops at the same time. Let i_k be the current in the kth loop, and ϕ_k be the magnetic flux linking with the kth loop. While i_k increases, an output voltage $v_k=\mathrm{d}\phi_k/\mathrm{d}t$ provided by the current source is needed to overcome the induced emf. As a result, the work done by the current source to the kth loop in time $\mathrm{d}t$ is

$$\mathrm{d}W_k=v_ki_k\mathrm{d}t=i_k\mathrm{d}\phi_k \tag{5.150}$$

Notice that the change, $\mathrm{d}\phi_k$, in the flux ϕ_k linking with the kth loop is the result of the changes of the currents in all the coupled loops. The differential work done to all the N loops is

$$\mathrm{d}W_{\mathrm{m}}=\sum_{k=1}^{N}\mathrm{d}W_k=\sum_{k=1}^{N}i_k\mathrm{d}\phi_k \tag{5.151}$$

The total stored energy is the integration of $\mathrm{d}W_{\mathrm{m}}$ in which i_k increases from zero to its final value I_k and ϕ_k increases from zero to its final value Φ_k. Assume that all the currents and fluxes increase to their final values synchronously, which can be modeled by letting $i_k=\alpha I_k$, and $\phi_k=\alpha\Phi_k$ with the factor α increasing from 0 to 1. Then we obtain the total stored magnetic energy to be

$$\begin{aligned}W_{\mathrm{m}}&=\int\mathrm{d}W_{\mathrm{m}}=\sum_{k=1}^{N}I_k\Phi_k\int_0^1\alpha\mathrm{d}\alpha\\&=\frac{1}{2}\sum_{k=1}^{N}I_k\Phi_k\end{aligned} \tag{5.152}$$

where the flux Φ_k linking with the kth loop is due to the currents in all the coupled loops, i.e.,

$$\Phi_k=\sum_{j=1}^{N}L_{jk}I_j \tag{5.153}$$

By substituting (5.153) into (5.152), we immediately obtain (5.148). Notice that, for $N=1$,

(5.152) is simplified to (5.144) as expected.

5.10.2 Magnetic Energy in Terms of Field Quantities (场量表示的磁能)

For a continuous distribution of the current within a volume, the expression of stored energy can be derived based on (5.152). To do that, we divide the volume current into many thin current-carrying loops. Then the magnetic flux Φ_k linking with the kth loop C_k is

$$\Phi_k = \int_{S_k} \boldsymbol{B} \cdot \boldsymbol{e}_{\mathrm{n}} \mathrm{d}s'_k = \oint_{C_k} \boldsymbol{A} \cdot \mathrm{d}\boldsymbol{l}'_k \tag{5.154}$$

where S_k is the surface bounded by C_k. Substituting (5.154) in (5.152), and assuming the loop C_k carries a current ΔI_k, we have

$$W_{\mathrm{m}} = \frac{1}{2} \sum_{k=1}^{N} \Delta I_k \oint_{C_k} \boldsymbol{A} \cdot \mathrm{d}\boldsymbol{l}'_k \tag{5.155}$$

where N is the total number of current-carrying loops. Now, denote the cross-sectional area of C_k as Δa_k,

$$\Delta I_k \mathrm{d}\boldsymbol{l}'_k = J(\Delta a_k)\mathrm{d}\boldsymbol{l}'_k = \boldsymbol{J}\Delta v'_k \tag{5.156}$$

where $\Delta v'_k = \Delta a_k \mathrm{d}l'$ is the volume of the differential current element $\Delta I_k \mathrm{d}\boldsymbol{l}'_k$. Substitute (5.156) into (5.155) and let $N \to \infty$. $\Delta v'_k$ becomes $\mathrm{d}v'$ and the summation in (5.155) becomes an integral, which leads to

$$W_{\mathrm{m}} = \frac{1}{2}\int_{V'} \boldsymbol{A} \cdot \boldsymbol{J} \mathrm{d}v' \tag{5.157}$$

where V' is the volume of the conducting medium in which $\boldsymbol{J}$ is nonzero.

Now, by using the differential form of Ampere's circuital law (5.79) and the vector identity (1.150), we can rewrite the integrand in (5.157) as

$$\begin{aligned} \boldsymbol{A} \cdot \boldsymbol{J} &= \boldsymbol{A} \cdot (\nabla \times \boldsymbol{H}) = \boldsymbol{H} \cdot (\nabla \times \boldsymbol{A}) - \nabla \cdot (\boldsymbol{A} \times \boldsymbol{H}) \\ &= \boldsymbol{H} \cdot \boldsymbol{B} - \nabla \cdot (\boldsymbol{A} \times \boldsymbol{H}) \end{aligned} \tag{5.158}$$

in which (5.19) is used. Substituting (5.158) into (5.157), we obtain

$$\begin{aligned} W_{\mathrm{m}} &= \frac{1}{2}\int_{V'} \boldsymbol{H} \cdot \boldsymbol{B} \mathrm{d}v' - \frac{1}{2}\int_{V'} \nabla \cdot (\boldsymbol{A} \times \boldsymbol{H}) \mathrm{d}v' \\ &= \frac{1}{2}\int_{V'} \boldsymbol{H} \cdot \boldsymbol{B} \mathrm{d}v' - \frac{1}{2}\oint_{S'} (\boldsymbol{A} \times \boldsymbol{H}) \cdot \boldsymbol{e}_{\mathrm{n}} \mathrm{d}s' \end{aligned} \tag{5.159}$$

in which the divergence theorem has been applied with S' being the closed surface bounding V'. The volume V' has been stipulated in (5.157). Nevertheless, V' can be extended to include all space without changing W_{m} because in the extended region $\boldsymbol{J} = 0$. When V' is extended to the entire space, the surface S' can be taken as a sphere with a large radius $r \to \infty$. In that case, $|\boldsymbol{A}|$ on the surface S' due to the finite distribution of the current $\boldsymbol{J}$ decreases as $1/r$, and $|\boldsymbol{H}|$ decreases as $1/r^2$, as can be seen from (5.28) and (5.12), respectively. As a result, the magnitude of $(\boldsymbol{A}\times\boldsymbol{H})$ decreases as $1/r^3$ in the surface integral in (5.159). Since the surface S' increases only as r^2, the surface integral in (5.159) vanishes as V' is extended to the entire space. Then(5.159) becomes

$$W_{\mathrm{m}} = \frac{1}{2}\int_{\infty} \boldsymbol{H} \cdot \boldsymbol{B} \mathrm{d}v' \tag{5.160a}$$

where the subscript ∞ indicates that the limit of integration is the entire space(or at least covers all the region in which the $\boldsymbol{B}$ and $\boldsymbol{H}$ field is nonzero). With $\boldsymbol{H}=\boldsymbol{B}/\mu$, we can rewrite(5.160a) in two alternative forms:

$$W_{\mathrm{m}} = \frac{1}{2}\int_{\infty} \frac{B^2}{\mu} \mathrm{d}v' \tag{5.160b}$$

and

$$W_{\mathrm{m}} = \frac{1}{2}\int_{\infty} \mu H^2 \mathrm{d}v' \tag{5.160c}$$

Define a **magnetic energy density**(磁能密度), w_{m}, such that

$$W_{\mathrm{m}} = \int_{\infty} w_{\mathrm{m}} \mathrm{d}v' \tag{5.161}$$

Then, by comparing(5.161) with(5.160), we can express w_{m} as

$$w_{\mathrm{m}} = \frac{1}{2}\boldsymbol{H} \cdot \boldsymbol{B} \tag{5.162a}$$

And in an linear isotropic medium, we have

$$w_{\mathrm{m}} = \frac{1}{2}\frac{B^2}{\mu} \tag{5.162b}$$

$$w_{\mathrm{m}} = \frac{1}{2}\mu H^2 \tag{5.162c}$$

By noticing(5.149), we have another way of finding the inductance of an inductor. That is, we can first calculate the stored magnetic energy W_{m} by using(5.160), and then determine the inductance as

$$L = \frac{2W_{\mathrm{m}}}{I^2} \tag{5.163}$$

which directly results from(5.149). Finding self-inductance in this way could be much easier than doing so via calculating the flux linkage as is demonstrated in the following example.

Example 5-18 By using stored magnetic energy, determine the inductance per unit length of an air coaxial transmission line that has a solid inner conductor of radius a and a very thin outer conductor of inner radius b.

Solution: This is the same problem as that in Example 5-14, in which the self-inductance was determined via quite complex calculation of the flux linkages. Refer to Figure 5-21, in which a uniform current I flows in the inner conductor and returns in the outer conductor. From (5.116) and(5.160b), the magnetic energy per unit length stored in the inner conductor is,

$$W'_{\mathrm{m1}} = \frac{1}{2\mu_0}\int_0^a B_{\phi 1}^2 2\pi\rho \mathrm{d}\rho = \frac{\mu_0 I^2}{4\pi a^4}\int_0^a \rho^3 \mathrm{d}\rho = \frac{\mu_0 I^2}{16\pi} \tag{5.164a}$$

From(5.117) and(5.160b), the magnetic energy per unit length stored in the region between the inner and outer conductors is,

$$W'_{m2} = \frac{1}{2\mu_0}\int_a^b B_{\phi 2}^2 2\pi\rho \mathrm{d}\rho = \frac{\mu_0 I^2}{4\pi}\int_a^b \frac{1}{\rho}\mathrm{d}\rho = \frac{\mu_0 I^2}{4\pi}\ln\frac{b}{a} \tag{5.164b}$$

Now from(5.163), the inductance per unit length is

$$L' = \frac{2}{I^2}(W'_{m1} + W'_{m2}) = \frac{\mu_0}{8\pi} + \frac{\mu_0}{2\pi}\ln\frac{b}{a}\ (\mathrm{H/m}) \tag{5.165}$$

which is the same as(5.120).

Summary

Concepts

Magnetic flux density(磁通密度)

Vector magnetic potential(矢量磁位)

Magnetic dipole/magnetic dipole moment(磁偶极子/磁偶极距)

Magnetic field density(磁场强度)

Magnetization(磁化)

Magnetization vector(磁化强度)

Magnetization current density(磁化电流密度)

Inductance(电感)

Magnetic energy/magnetic energy density(磁能/磁能密度)

Laws & Theorems

Biot-Savart law(比奥萨法定律)

Ampere's circuital law(安培环路定律)

Law of conservation of magnetic flux(磁通守恒定律)

Field and potential distribution of magnetic dipole(磁偶极子产生的磁场和矢量磁位分布)

Boundary conditions of magnetostatic fields(恒定磁场的边界条件)

Methods

Find ***B***/***H*** field distribution given current distributions using direct integration/superposition;

Find ***B***/***H*** field distribution given current distributions using Ampere's circuital law;

Find ***B***/***H*** field distribution given charge distributions using boundary conditions;

Find ***B***/***H*** field distribution given vector potential distributions;

Find inductance of typical inductors.

Review Questions

5.1 写出洛伦兹力方程。

5.2 写出比奥萨法定律公式。

5.3 为什么说不存在磁荷?

5.4 什么是磁通守恒定律?

5.5 产生恒定磁场的源是什么? 恒定磁场的源是通量源还是漩涡源?

5.6 什么是安培环路定律？安培环路定律适用于求解什么条件下的恒定磁场问题？

5.7 无限长均匀直线电流产生的磁感应强度是怎样的？

5.8 导体中能存在恒定磁场么？

5.9 定义矢量磁位 $\boldsymbol{A}$。如何唯一确定矢量 $\boldsymbol{A}$？

5.10 已知矢量磁位 $\boldsymbol{A}$ 的分布，如何计算穿过某给定曲面的磁通？

5.11 什么是磁偶极子？什么是磁偶极矩？画出磁偶极子产生的磁力线。

5.12 定义磁化强度矢量，其单位是什么？

5.13 解释等效磁化面电流和等效磁化体电流的物理意义。

5.14 定义磁场强度矢量。其单位什么？

5.15 恒定磁场在两种不同媒质分界面上的边界条件是什么？

5.16 为什么铁磁体外侧的磁力线垂直于铁磁体的表面？

5.17 定义自感和互感。

5.18 绕线电感的自感与其匝数是什么关系？

5.19 写出以 $\boldsymbol{B}$ 和 $\boldsymbol{H}$ 表示的磁场储能表达式。

Problems

5.1 Prove that the magnetic force governed by Ampere's law of force satisfies Newton's Third Law. In other words, the magnetic force between two circuits with current I_1 and I_2 as given by(5.3) satisfies $\boldsymbol{F}_{21}=-\boldsymbol{F}_{12}$.

5.2 A very thin wire extends from $z=0$ to $z=\infty$ and carries a current I. Derive an expression for the magnetic flux density at any point in the $z=0$ plane.

5.3 A current I flows along the inner conductor of an infinitely long coaxial line and returns via the outer conductor. The radius of the inner conductor is a, and the inner and outer radii of the outer conductor are b and c respectively. Find the magnetic flux density $\boldsymbol{B}$ and plot B versus ρ.

5.4 Figure 5-26 shows an infinitely long solenoid with air core having a radius b and n closely wound turns per unit length. The windings are slanted at an angle α and carry a current I. Determine the magnetic flux density both inside and outside the solenoid. (Hint: decompose the current into two components −one along z-direction, the other along ϕ-direction.)

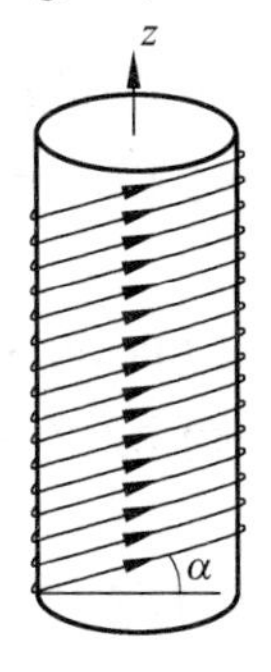

Figure 5-26 A long solenoid with closely wound windings carrying a current I(Problem 5.4)

5.5 An air core with radius a and length L is closely wounded with an N-turn coil that carries a current I. Determine the magnetic flux density at a point on the axis of the core. Show that the result reduces to that given in(5.59) when L approaches infinity.

5.6 A very long, thin conducting strip of width w resides in the xz-plane between $x=\pm w/2$. A surface current $\boldsymbol{J}_s=\boldsymbol{e}_z J_{s0}$ flows over the strip. Find the magnetic flux density outside the strip.

5.7 Figure 5-27 illustrates the **Helmholtz coils**, which are used to generate a nearly uniform magnetic field in a three-dimensional region between the coils. The two identical N-turn coils with radius a are coaxially aligned and separated by a distance d. A current I flows in each coil in the same direction.

(1) Find the magnetic flux density $\boldsymbol{B}=\boldsymbol{e}_z B_z$ at a point midway between the coils.

(2) Show that $\mathrm{d}B_z/\mathrm{d}z$ vanishes at the midpoint.

(3) Find the relation between a and d such that $d^2 B_z/\mathrm{d}z^2$ also vanishes at the midpoint.

5.8 Find the magnetic flux density at point P surrounded by a wire loop carrying a 75A current as shown in Figure 5-28.

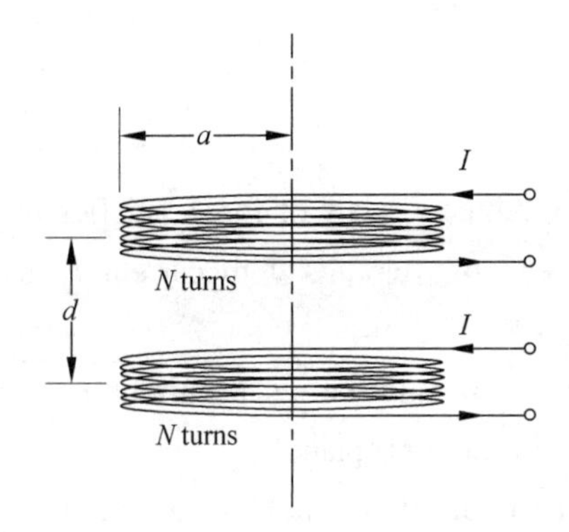

Figure 5-27 Helmholtz coils(Problems 5.7)

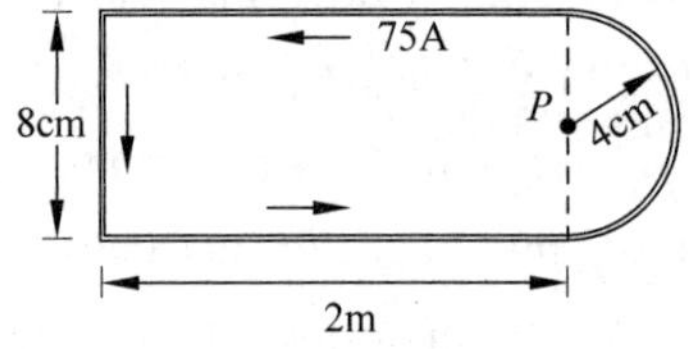

Figure 5-28 Illustration of Problems 5.8

5.9 As shown in Figure 5-29, an incomplete circular loop with very long leads carries a current of 24 A. Determine the magnetic field intensity and the magnetic flux density at the center of the circular loop.

5.10 Derive the expression of magnetic potential due to a continuous current distribution (5.28) by using Helmholtz's theorem.

5.11 As shown in Figure 5-30, a cylindrical hollow cavity resides in a long cylindrical conductor carrying a uniform current density $\boldsymbol{J}=\boldsymbol{e}_z J$. Find the magnitude and direction of the magnetic flux density $\boldsymbol{B}$ in the cavity. (Hint: Decompose the zero current in the cavity region into two current components, $\boldsymbol{J}$ and $-\boldsymbol{J}$.)

5.12 A very large slab of material of thickness d resides parallel to the $x-y$ plane, with the presence of a uniform magnetic field intensity $\boldsymbol{H}_0=\boldsymbol{e}_z H_0$. Ignoring edge effect, determine the magnetic field intensity inside the slab:

(1) if the slab material has a permeability μ,

(2) if the slab is a permanent magnet having a magnetization vector $\boldsymbol{M}_i=\boldsymbol{e}_z M_i$.

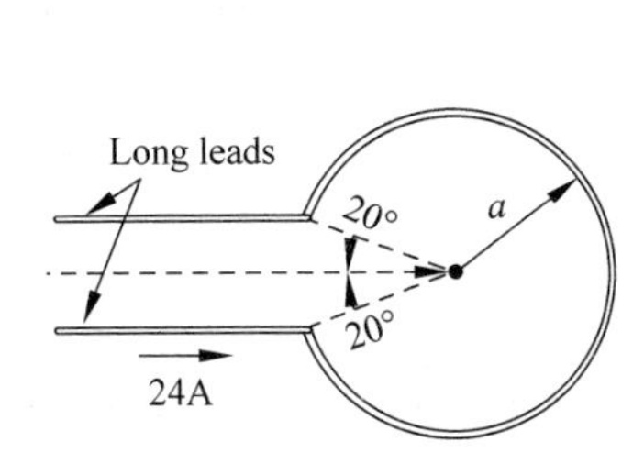

Figure 5-29 Illustration of Problems 5. 9

Figure 5-30 Cross section of a long cylindrical conductor with a hollow cavity(Problem 5. 11)

5. 13 A circular magnetic core with radius a and permeability μ is inserted coaxially in the solenoid of Figure 5-9. If the inner radius of the solenoid is b and $a<b$.

(1) Determine $\boldsymbol{B}$, $\boldsymbol{H}$, and $\boldsymbol{M}$ inside the solenoid in the regions $\rho<a$ and $a<\rho<b$.

(2) Find the equivalent magnetization current densities $\boldsymbol{J}_{\mathrm{m}}$ and $\boldsymbol{J}_{\mathrm{ms}}$ in the magnetic core.

5. 14 Consider a planar interface at $y=0$ between air(region 1, $\mu_{\mathrm{r1}}=1$) and iron(region 2, $\mu_{\mathrm{r2}}=5000$).

(1) Assuming $\boldsymbol{B}_1=\boldsymbol{e}_x 0.5-\boldsymbol{e}_y 10$ (mT) in region 1, find $\boldsymbol{B}_2$ in region 2 and the angle it makes with the normal direction of the interface.

(2) Assuming $\boldsymbol{B}_2=\boldsymbol{e}_x 10+\boldsymbol{e}_y 0.5$ (mT) in region 2, find $\boldsymbol{B}_1$ in region 1 and the angle it makes with the normal direction of the interface.

5. 15 A long straight current I resides on the infinitely-large planar interface between free space and a certain medium. Find ① the $\boldsymbol{B}$ and $\boldsymbol{H}$ everywhere, and ② the equivalent magnetization surface current at the interface.

5. 16 Determine the self-inductance of a toroidal coil of N turns of wire wound on an air frame with mean radius r_0 and a circular cross section of radius a. Obtain an approximate expression assuming $a \ll r_0$.

5. 17 Determine the inductance per unit length of the coaxial transmission line described in Problem 5. 3.

5. 18 Determine the mutual inductance between a very long, straight wire and a conducting circular loop, as shown in Figure 5-31. (Hint: use the following identity)

$$\int_0^{2\pi} \frac{\mathrm{d}\theta}{a+b\cos\theta}=\frac{2\pi}{\sqrt{a^2-b^2}}$$

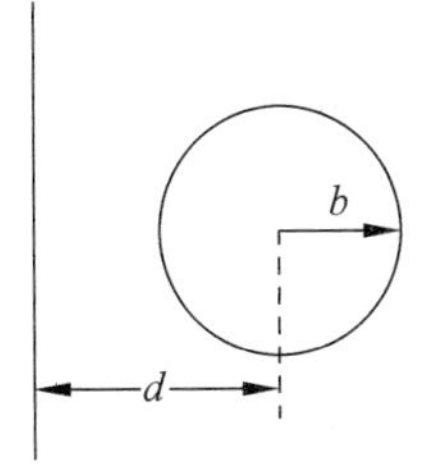

Figure 5-31 A long, straight wire and a conducting circular loop(Problem 5. 18)

5.19 Consider two coupled current-carrying loops. They have self-inductances L_1 and L_2 respectively, and they carry currents I_1 and I_2 respectively. The mutual inductance between them is L_{12}.

(1) Given the current I_1, use (5.148) to find the current I_2 that minimizes the stored magnetic energy W_m.

(2) Show that $M \leqslant \sqrt{L_1 L_2}$.

5.20 Refer to Example 5-12. Calculate the magnetization vector $\boldsymbol{M}$, the magnetization volume current density $\boldsymbol{J}$, and the magnetization surface current density $\boldsymbol{J}_s$ in the core material. Calculate the stored magnetic energy and find the inductance of the toroidal winding.

Chapter 6 Time-Varying Electromagnetic Fields(时变电磁场)

6.1 Introduction(引言)

In previous chapters, static electric fields and static magnetic fields are studied separately. In electrostatics, the electric field intensity vector $\boldsymbol{E}$ and the electric flux density (electric displacement) vector $\boldsymbol{D}$ satisfy the following governing equations

$$\nabla\times\boldsymbol{E}=0 \quad \text{and} \quad \nabla\cdot\boldsymbol{D}=\rho_{\mathrm{v}}$$

which states that static electric field intensity is an irrotational field, and the static charge is the flow source of static electric fields. For simple media, $\boldsymbol{E}$ and $\boldsymbol{D}$ are related by the constitutive relation

$$\boldsymbol{D}=\varepsilon\boldsymbol{E}$$

In magnetostatics, the magnetic flux density vector $\boldsymbol{B}$ and the magnetic field intensity vector $\boldsymbol{H}$ satisfy the following governing equations

$$\nabla\cdot\boldsymbol{B}=0 \quad \text{and} \quad \nabla\times\boldsymbol{H}=\boldsymbol{J}$$

which state that steady magnetic flux density is a solenoidal field, and the steady current is the vortex source of static magnetic fields. For simple media, $\boldsymbol{B}$ and $\boldsymbol{H}$ are related by a constitutive relation

$$\boldsymbol{H}=\frac{1}{\mu}\boldsymbol{B}$$

Apparently, the above governing equations form two independent groups, one involving $\boldsymbol{E}$ and $\boldsymbol{D}$, and the other involving $\boldsymbol{B}$ and $\boldsymbol{H}$. The two groups of equations appear to describe two independent physical phenomena. The static electric field is completely determined by the static electric charge density distribution ρ_{v}, whereas the static magnetic field is completely determined by the steady current density distribution $\boldsymbol{J}$. $\boldsymbol{J}$ and ρ_{v} are not related to each other in the static (or DC) scenario. However, in the time-varying (or AC) scenario, the electric and magnetic phenomena are no longer independent of each other, which essentially leads to a new physical phenomenon, the **electromagnetic fields**(电磁场). ①

Starting from this chapter, we focus on the time-varying electromagnetic fields(时变电磁场) due to time-varying charges and currents. Additional experimental laws and theoretical derivations are used to formulate the famous Maxwell's equations, which describe all the macroscopic electromagnetic phenomena precisely and completely. It is shown that that time-varying electric fields and magnetic fields must co-exist and have definite relation to each other. The interaction between time-varying electric fields and time-varying magnetic fields results in **electromagnetic waves**(电磁波).

6.2 Curl of Time-Varying Electric Field and Faraday's Law of Induction(电场的旋度与法拉第电磁感应定律)

In Section 4.4, we introduced the concept of **electromotive force**(**emf**, denoted by $\mathcal{E}$) that is necessary for maintaining a steady current in circuits. The emf cannot result from static charges because static charges can only produce conservative electric fields. It must come from other physical mechanism that causes an impressed electric field and makes the total electric field non-conservative. For example, electric batteries(essentially chemical actions) provide the necessary emf to form steady currents. A time-varying magnetic field can also provide an emf, leading to a time-varying electric field that is non-conservative. This phenomenon was first discovered experimentally by Michael Faraday in 1831. Faraday found that the electromotive force induced in a closed circuit is equal to the negative rate of increase of the magnetic flux linking the circuit. This is known as **Faraday's law of electromagnetic induction**(**法拉第电磁感应定律**), which can be formulated as

$$\mathcal{E}=-\frac{\partial \Phi}{\partial t} \tag{6.1}$$

The negative sign in(6.1) means that the induced emf will cause the current in the circuit to flow in a direction along which the current will counteract the change of the linking magnetic flux, which is also known as **Lenz's law**(**楞次定律**).

We derived an expression relating the emf $\mathcal{E}$ to the conservative and non-conservative electric fields as in(4.41). Now in the time-varying case, there is no need to discriminate between conservative and non-conservative fields. For the emf induced in a loop C, we can simply rewrite(4.41) as

$$\mathcal{E}=\oint_C \boldsymbol{E} \cdot \mathrm{d}\boldsymbol{l} \tag{6.2}$$

---

① 时变条件下的电磁场与静态条件下电场和磁场之所以不同,是因为在时变条件下出现了与静态场所不具备的物理规律。具体而言,6.2节指出,由于法拉第电磁感应定律的存在,静电场的基本公式 $\nabla\times\boldsymbol{E}=0$ 不再适用于时变场。6.3节指出,由于静态条件下电流连续性定律 $\nabla\cdot\boldsymbol{J}=0$ 不适用于时变条件,静磁场的基本公式 $\nabla\times\boldsymbol{H}=\boldsymbol{J}$ 在时变条件下也需要被修正,并且在修正过程中引入了位移电流的概念。修正后的基本方程中,电场和磁场不再相互独立,必须作为同一个物理现象进行研究。

where $\boldsymbol{E}$ includes both the conservative and non-conservative components if they exist. By equating(6.1) with(6.2), we have

$$\oint_C \boldsymbol{E} \cdot \mathrm{d}\boldsymbol{l} = -\frac{\partial \Phi}{\partial t} = -\frac{\partial}{\partial t}\int_S \boldsymbol{B} \cdot \mathrm{d}\boldsymbol{s} \tag{6.3}$$

where S is any surface enclosed by the closed contour(the loop) C. (6.3) shows that a time-varying magnetic flux through any surface S can produce electric fields (must also be time varying).

Although Faraday's law was discovered experimentally by observing an induced current in a conducting loop due to the changing magnetic flux linking the loop, (6.3) turns out to be a general relation that is valid even if there is no conducting loop along the contour C. According to(6.3), it should be noticed that an induced emf can be the result of either a time-varying magnetic field or a moving conductor(or circuit) that essentially changes the integration surface S and its boundary C. Based on the latter case, alternating current generators can be developed. In this chapter, we concentrate on the case of stationary system. Then the magnetic flux variation is caused only by time-varying magnetic fields, and we can interchange the time-derivative with the surface integral in(6.3), which leads to①

$$\oint_C \boldsymbol{E} \cdot \mathrm{d}\boldsymbol{l} = -\int_S \frac{\partial}{\partial t}\boldsymbol{B} \cdot \mathrm{d}\boldsymbol{s} \tag{6.4}$$

(6.4) is valid for any surface S bounded by the contour C. By applying the Stokes's theorem to the left side of(6.4), we then have

$$\nabla \times \boldsymbol{E} = -\frac{\partial}{\partial t}\boldsymbol{B} \tag{6.5}$$

which can be seen as the differential form of Faraday's law of electromagnetic induction. (6.5) is a modification to(2.43), the governing equation of static electric field. It clearly states that time-varying magnetic field is the vortex source of the time-varying electric field, and the time-varying electric field must be non-conservative.

6.3 Curl of Time-Varying Magnetic Field and Displacement Current(时变磁场的旋度与位移电流)

In the time-varying case, another equation that obviously needs to be modified is(4.30), the governing equation of $\boldsymbol{J}$ in the static case. The modified version has already been given in Chapter 4 as(4.25), which is repeated below:

① 法拉第电磁感应定律可以表述为闭合回路上的感应电动势等于链接到该闭合回路的磁通增加速率的相反数。作为一个实验定律,其中的感应电动势可以通过导体回路中感应形成的电流得到验证,但作为更一般的物理规律,式(6.3)表示的法拉第电磁感应定律并不要求闭合曲线 C 的位置上存在一个导体回路。另一方面,如果存在导体,式(6.3)表明,法拉第电磁感应定律中感应电动势的来源可能是变化的磁场,也可能是运动的导体或者电路。当包含运动导体的闭合回路 C 所围的曲面 S 随时间变化时,闭合回路 C 上也会出现感应电动势,这也是发电机的基本原理。本章仅讨论静态系统条件下的电磁问题,即假设不存在运动的导体。此时,式(6.3)便可以改写为式(6.4)。

$$\nabla \cdot \boldsymbol{J} = -\frac{\partial \rho_v}{\partial t} \tag{6.6}$$

(6.6) is known as the **equation of continuity**(**连续性方程**) or the **principle of conservation of charges** (**电荷守恒原理**), which is a basic physical law that must be satisfied. Since(4.30) is modified to(6.6), we realize that(5.79), the differential form of Ampere's circuital law in the static case, also needs to be modified. By taking the divergence of both sides of(5.79) and applying the null identity(1.142), we have

$$\nabla \cdot \boldsymbol{J} = \nabla \cdot (\nabla \times \boldsymbol{H}) = 0$$

which obviously contradicts(6.6). To make the modified(5.79) consistent with(6.6), an additional term $\partial \boldsymbol{D}/\partial t$ is added to the right side of(5.79), which leads to

$$\nabla \times \boldsymbol{H} = \boldsymbol{J} + \frac{\partial \boldsymbol{D}}{\partial t} \tag{6.7}$$

Now we see that(6.7) is consistent with(6.6), which can be verified by taking the divergence of both sides of(6.7) and then using(2.91). $\partial \boldsymbol{D}/\partial t$ is called **displacement current density**(**位移电流密度**) as it has the dimension of a current density(A/m^2). The introduction of the term $\partial \boldsymbol{D}/\partial t$ was one of the major contributions of James Clerk Maxwell(1831-1879).

(6.7) indicates that a time-varying electric field will produce time-varying magnetic fields, even without current flow. It can be converted into integral-form by taking the surface integral on both sides over an arbitrary surface S with a contour C, and then, applying Stokes's theorem, which leads to

$$\oint_C \boldsymbol{H} \cdot d\boldsymbol{l} = \int_S \left(\boldsymbol{J} + \frac{\partial \boldsymbol{D}}{\partial t}\right) \cdot d\boldsymbol{s} = \int_S \boldsymbol{J} \cdot d\boldsymbol{s} + \int_S \boldsymbol{J}_d \cdot d\boldsymbol{s} \tag{6.8}$$

$$= I + I_d$$

which is called the **generalized Ampere's circuital law** (**广义安培电路定律**) in the time-varying case. Note that $\boldsymbol{J}$ in(6.8) represents the current density due to the motion of charge carriers, whereas $\boldsymbol{J}_d = \partial \boldsymbol{D}/\partial t$ represents the current density due to the time-varying electric displacement vector. The surface integral of $\boldsymbol{J}$ is the free current I(including convection and conduction currents) flowing through the surface S, whereas the surface integral of $\boldsymbol{J}_d$ is the displacement current I_d flowing through the surface S.①

Example 6-1 An AC voltage source with output $U(t) = U_0\cos\omega t$ is applied across a parallel-plate capacitor C_0, as illustrated in Figure 6-1. Find the displacement current in the capacitor compared with the conduction current in the wire.

Solution: Here we assume fringing effects of the capacitor can be neglected, and the frequency ω is low enough that the wire in the circuit can be considered as a short circuit. The conduction current in the connecting wire is

$$I(t) = C_0 \frac{dU(t)}{dt} = -C_0 U_0 \omega \sin\omega t$$

---

① 位移电流$\partial \boldsymbol{D}/\partial t$的引入解决了时变条件下电荷守恒原理(也即电流连续性原理)与恒定磁场安培环路定律不相容的问题,同时补全了时变条件下电流的可能形式,因此时变条件下的广义安培电路定律也被称为**全电流定律**。

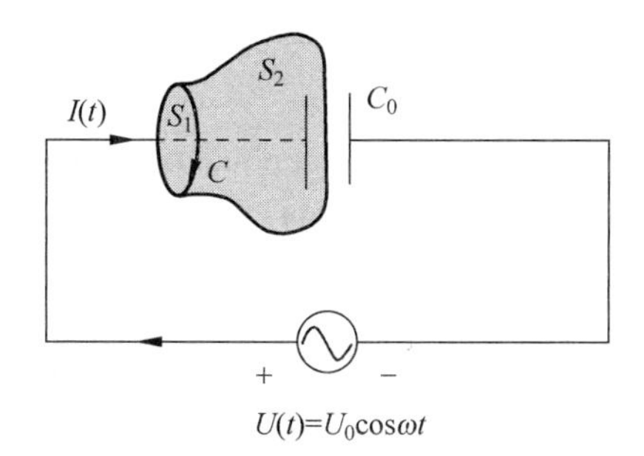

Figure 6-1 A parallel-plate capacitor connected to an AC voltage source(Example 6-1)

For a parallel-plate capacitor with an area A, plate separation d and a dielectric medium of permittivity ε, the capacitance is

$$C_0 = \varepsilon \frac{A}{d}$$

With a voltage $U(t)$ appearing between the plates, the uniform electric field intensity E in the dielectric is equal to (neglecting fringing effects) $E = U(t)/d$, hence

$$D = \varepsilon E = \varepsilon \frac{U_0}{d}\cos\omega t$$

The displacement current is then

$$I_D(t) = \int_A \frac{\partial \boldsymbol{D}}{\partial t}\cdot \mathrm{d}\boldsymbol{s} = -\left(\varepsilon \frac{A}{d}\right) U_0 \omega \sin\omega t = -C_0 U_0 \omega \sin\omega t$$

which is the same as the conduction current $I(t)$. This is also a necessary result of the generalized Ampere's circuital law(6.8). To show that, consider the contour C in Figure 6-1. Two typical open surfaces bounded by the same contour C are also illustrated in the figure, including ①a planar disk surface S_1 intersecting with the circuit wire outside the capacitor, and ②a curved surface S_2 passing through the dielectric medium within the capacitor. According to the generalized Ampere's circuital law, we should have

$$\int_{S_1}\left(\boldsymbol{J} + \frac{\partial \boldsymbol{D}}{\partial t}\right)\cdot \mathrm{d}\boldsymbol{s} = \int_{S_2}\left(\boldsymbol{J} + \frac{\partial \boldsymbol{D}}{\partial t}\right)\cdot \mathrm{d}\boldsymbol{s} = \oint_C \boldsymbol{H}\cdot \mathrm{d}\boldsymbol{l}$$

For the surface S_1, only the conduction current $\boldsymbol{J}$ is nonzero because the wire is a short circuit and there is no $\boldsymbol{E}$ field along the wire, and consequently, $\boldsymbol{D}=0$. The first surface integral of the above equation is evaluated as

$$\int_{S_1}\left(\boldsymbol{J} + \frac{\partial \boldsymbol{D}}{\partial t}\right)\cdot \mathrm{d}\boldsymbol{s} = \int_{S_1} \boldsymbol{J}\cdot \mathrm{d}\boldsymbol{s} = I(t)$$

For the surface S_2 that passes through the dielectric medium, no conduction current flows through any part of S_2. If the displacement current was not introduced (without the $\partial \boldsymbol{D}/\partial t$ term), the right side of (6.8) would be zero. This would result in a contradiction. The inclusion of the displacement-current term by Maxwell eliminates this contradiction. We have

$$\int_{S_2}\left(\boldsymbol{J} + \frac{\partial \boldsymbol{D}}{\partial t}\right)\cdot \mathrm{d}\boldsymbol{s} = \int_{S_2} \frac{\partial \boldsymbol{D}}{\partial t}\cdot \mathrm{d}\boldsymbol{s} = I_D(t)$$

As is previously shown, $I_D(t)=I(t)$. Hence, we obtain the same result whether surface S_1 or surface S_2 is chosen. ①

Example 6-2 Find the magnitude ratio of conduction current density and the displacement current density in the sea water at 1MHz and 1GHz. Assume the sea water has a conductivity $\sigma=4\text{S/m}$ and a permittivity $\varepsilon_r=81$.

Solution: Suppose the electric field varies at a single frequency ω at a point in the sea water:

$$\boldsymbol{E}=\boldsymbol{e}_E E_m \cos\omega t$$

Then the conduction current density at this point is

$$\boldsymbol{J}_c=\sigma\boldsymbol{E}=\boldsymbol{e}_E\sigma E_m\cos(\omega t)$$

The displacement current density at this point is

$$\boldsymbol{J}_d=\frac{\partial\boldsymbol{D}}{\partial t}=\varepsilon_0\varepsilon_r\frac{\partial\boldsymbol{E}}{\partial t}=-\boldsymbol{e}_E\omega\varepsilon_0\varepsilon_r E_m\sin(\omega t)$$

The magnitude ratio is

$$\frac{|\boldsymbol{J}_c|}{|\boldsymbol{J}_d|}=\frac{\sigma}{\omega\varepsilon_0\varepsilon_r} \tag{6.9}$$

At 1MHz, $\omega=2\pi\times10^6$ rad/min, and (6.9) is evaluated as

$$\frac{|\boldsymbol{J}_c|}{|\boldsymbol{J}_d|}\bigg|_{@1\text{MHz}}=\frac{4}{2\pi\times10^6\times\dfrac{1}{36\pi}\times10^{-9}\times81}=889$$

Obviously, the conduction current is much larger than the displace current, which means that the sea water behaves more like a good conductor instead of a dielectric. ②At 1GHz, $\omega=2\pi\times10^9$ rad/min, and (6.9) is evaluated as

$$\frac{|\boldsymbol{J}_c|}{|\boldsymbol{J}_d|}\bigg|_{@1\text{GHz}}=\frac{4}{2\pi\times10^9\times\dfrac{1}{36\pi}\times10^{-9}\times81}=0.89$$

Obviously, increasing the frequency increases the displacement current and hence makes it comparable with the conduction current. At 1GHz, the sea water is not considered as a good conductor anymore. Also notice that the displacement current is along the same direction as the conduction current but has a phase 90° ahead of the conduction current for the time-harmonic field. ③

---

① 例 6-1 从另一个角度表明了引入位移电流的必要性。如果不考虑位移电流,将恒定磁场的安培环路定律直接应用到 Figure 6-1 所示的电路中,那么通过曲面 S_1 的传导电流应与通过 S_2 的传导电流相等,都等于磁场强度沿闭合曲线 C 的环量。然而穿过曲面 S_1 的传导电流 $I(t)$ 不为 0,穿过曲面 S_2 的传导电流却恒等于零,两者显然矛盾。但如果引入位移电流,则可以发现穿过曲面 S_2 的位移电流不为零,且与传导电流 $I(t)$ 相等,这样就解决了上述矛盾。

② 例 6-2 表明,海水对于 1MHz 的电磁波表现得更像导体,其中的位移电流可忽略。但对于更高频率的电磁波(例如 100MHz 以上),位移电流将与传导电流可比,不能忽略。

③ 对于一般的简单媒质而言,该结论都是成立的,即位移电流和传导电流的方向一致,但位移电流的相位领先传导电流 90°。从本章稍后引入的时谐场的概念出发,将很容易验证该结论。

Example 6-3 Find the magnitude ratio of the conduction current density and displacement current density in copper($\sigma=5.8\times10^7$ S/m and $\varepsilon_r=1$) as a function of the frequency.

Solution: Similar to Example 6-2, the magnitude ratio can be calculated by using(6.9):

$$\frac{|J_c|}{|J_d|}=\frac{\sigma}{\omega\varepsilon_0\varepsilon_r}=\frac{5.8\times10^7}{2\pi f\times\frac{1}{36\pi}\times10^{-9}\times1}=\frac{10^{18}}{f}$$

which means that the displacement current in copper is negligible for frequencies as high as terahertz (THz). However, for frequencies near or above 10^{18}Hz, copper and most metal materials cannot be considered as conductors.

6.4 Maxwell's Equations(麦克斯韦方程组)

In the previous section, the two curl equations of the electric and magnetic fields are modified respectively for the time-varying situation. Nevertheless, the two governing divergence equations, (2.91) and (5.17) turn out to be still valid in the time-varying case. (2.91) and (5.17) together with the two modified curl equations (6.5) and (6.7) form the famous **Maxwell's Equations**(麦克斯韦方程组), which are summarized as below:

$$\nabla\times\boldsymbol{E}=-\frac{\partial\boldsymbol{B}}{\partial t} \tag{6.10a}$$

$$\nabla\times\boldsymbol{H}=\boldsymbol{J}+\frac{\partial\boldsymbol{D}}{\partial t} \tag{6.10b}$$

$$\nabla\cdot\boldsymbol{D}=\rho_v \tag{6.10c}$$

$$\nabla\cdot\boldsymbol{B}=0 \tag{6.10d}$$

Notice that ρ_v in(6.10c) is the volume density of free charges, and $\boldsymbol{J}$ in(6.10b) is the density of free currents, which may comprise both convection current and conduction current. These four equations, together with the equation of continuity in(6.6) and Lorentz's force equation in(5.2), lay the foundation of electromagnetic theory, and can be used to explain and predict all macroscopic electromagnetic phenomena.

The four Maxwell's equations in(6.10a)-(6.10c) are not all independent. In fact, the two divergence equations, (6.10c), (6.10d), can be derived from the two curl equations, (6.10a), (6.10b), by making use of the equation of continuity(6.6). The derivation is left as an exercise to the reader(see Problem 6.2). The four fundamental field vectors $\boldsymbol{E},\boldsymbol{D},\boldsymbol{B},\boldsymbol{H}$ altogether have twelve components. Twelve scalar equations are required for the determination of these twelve unknowns. The required equations are supplied by the two vector curl equations and the two constitutive relations $\boldsymbol{D}=\varepsilon\boldsymbol{E}$ and $\boldsymbol{H}=\boldsymbol{B}/\mu$, each vector equation being equivalent to three scalar equations.①

① 麦克斯韦方程组包含两个旋度方程(6.10a),(6.10b)和两个散度方程(6.10c),(6.10d),其中两个旋度方程来自对静电场和恒定磁场的旋度方程的修正。从修正后的两个旋度方程出发,结合连续性方程,可以推导出两个散度方程,因此麦克斯韦方程组的四个方程并不是相互独立的。通常,两个旋度方程(6.10a),(6.10b)被认为是更基本的方程。两个散度方程(6.10c),(6.10d)则与静态电场和静态磁场的两个散度方程一致。

The four Maxwell's equations in(6.10) are differential equations that are valid at every point in space. The equivalent integral-form Maxwell's equations are already given in(6.4), (6.8),(2.92) and(5.18),which are summarized as below:

$$\oint_C \boldsymbol{E} \cdot \mathrm{d}\boldsymbol{l} = -\int_S \frac{\partial \boldsymbol{B}}{\partial t} \cdot \mathrm{d}\boldsymbol{s} \tag{6.11a}$$

$$\oint_C \boldsymbol{H} \cdot \mathrm{d}\boldsymbol{l} = \int_S \left(\boldsymbol{J} + \frac{\partial \boldsymbol{D}}{\partial t}\right) \cdot \mathrm{d}\boldsymbol{s} \tag{6.11b}$$

$$\oint_S \boldsymbol{D} \cdot \mathrm{d}\boldsymbol{s} = \int_V \rho_v \mathrm{d}v \tag{6.11c}$$

$$\oint_S \boldsymbol{B} \cdot \mathrm{d}\boldsymbol{s} = 0 \tag{6.11d}$$

(6.11a) is the same as(6.4), which is an expression of Faraday's law of electromagnetic induction. (6.11b) is the generalized Ampere's circuital law. (6.11c) is the Gauss's law which remains the same in the time-varying case. (6.11d) is still the law of conservation of magnetic flux, from which we conclude that there are no isolated magnetic charges and that the total outward magnetic flux through any closed surface is zero. Both the differential and the integral forms of Maxwell's equations are summarized in Table 6-1 for easy reference.

Table 6-1 Maxwell's equations

Differential Form	Integral Form	Physical Significance
$\nabla \times \boldsymbol{E} = -\frac{\partial \boldsymbol{B}}{\partial t}$	$\oint_C \boldsymbol{E} \cdot \mathrm{d}\boldsymbol{l} = -\int_S \frac{\partial \boldsymbol{B}}{\partial t} \cdot \mathrm{d}\boldsymbol{s}$	Faraday's law
$\nabla \times \boldsymbol{H} = \boldsymbol{J} + \frac{\partial \boldsymbol{D}}{\partial t}$	$\oint_C \boldsymbol{H} \cdot \mathrm{d}\boldsymbol{l} = \int_S \left(\boldsymbol{J} + \frac{\partial \boldsymbol{D}}{\partial t}\right) \cdot \mathrm{d}\boldsymbol{s}$	Generalized Ampere's circuital law
$\nabla \cdot \boldsymbol{D} = \rho_v$	$\oint_S \boldsymbol{D} \cdot \mathrm{d}\boldsymbol{s} = \int_V \rho_v \mathrm{d}v$	Gauss's law
$\nabla \cdot \boldsymbol{B} = 0$	$\oint_S \boldsymbol{B} \cdot \mathrm{d}\boldsymbol{s} = 0$	Conservation of magnetic flux (No isolated magnetic charge)

6.5 Electromagnetic Boundary Conditions (电磁场边界条件)

Boundary conditions of electromagnetic fields in the time-varying case can be derived in the same way as was done for electrostatic fields(Section 2.8) and magnetostatic fields(Section 4.5). Applying the integral form of the two curl equations(6.11a),(6.11b) to a flat rectangular contour across the boundary yields the boundary conditions for the tangential components of $\boldsymbol{E}$ and $\boldsymbol{H}$. Applying the integral form of the two divergence equations(6.11c), (6.11d) to a flat cylinder across the interface gives the boundary conditions for the normal components of $\boldsymbol{D}$ and $\boldsymbol{B}$. (6.11c),(6.11d) are the same integral equations used to derive normal boundary conditions in static cases. The two additional surface integral terms involving

$\partial \boldsymbol{B}/\partial t$ and $\partial \boldsymbol{D}/\partial t$ in(6.11a),(6.11b) are the only difference between the time-varying and static cases. However, in deriving the boundary conditions, these two terms vanish when the height of the rectangular contour (*abcda* in Figure 2-22 and Figure 5-13) approaches zero. Therefore, we have exactly the same formulation of the boundary conditions for $\boldsymbol{E}$, $\boldsymbol{H}$, $\boldsymbol{D}$ and $\boldsymbol{B}$ fields as

$$E_{1t} = E_{2t} \tag{6.12a}$$

$$\boldsymbol{e}_{n2} \times (\boldsymbol{H}_1 - \boldsymbol{H}_2) = \boldsymbol{J}_s \tag{6.12b}$$

$$\boldsymbol{e}_{n2} \cdot (\boldsymbol{D}_1 - \boldsymbol{D}_2) = \rho_s \tag{6.12c}$$

$$B_{1n} = B_{2n} \tag{6.12d}$$

Notice that the $\boldsymbol{e}_{n2}$ in (6.12b) and (6.12c) denotes the normal unit vector pointing from medium 2 to medium 1.

From the above boundary conditions, we have the following conclusions.

(1) The tangential component of an $\boldsymbol{E}$ field is always continuous across any interface.

(2) A surface current on an interface causes discontinuity in the tangential component of the $\boldsymbol{H}$ field across the interface. The amount of discontinuity is equal to the magnitude of the surface current density.

(3) A surface charge on an interface causes discontinuity in the normal component of the $\boldsymbol{D}$ field across the interface. The amount of discontinuity is equal to the magnitude of the surface charge density.

(4) The normal component of a $\boldsymbol{B}$ field is always continuous across any interface.

Since the two divergence equations can be derived from the two curl equations together with the equation of continuity, the four boundary conditions in (6.12) are not completely independent①. Specifically, for the time-varying case, the two boundary conditions (6.12d), (6.12c) that are obtained from(6.11d),(6.11c) can also be derived directly from(6.11a), (6.11b) together with the boundary conditions(6.12a),(6.12b), which is left as an exercise (see problem 6.6). Therefore, in most cases, we only need the tangential boundary conditions of $\boldsymbol{E}$ and $\boldsymbol{H}$ to solve time-varying electromagnetic problems.

We now examine the important special cases of ① a boundary between two perfect dielectrics, and ② a boundary between a perfect dielectric and a perfect conductor.

6.5.1 Interface between Two Perfect Dielectrics (两种理想介质分界面)

A perfect dielectric is characterized by a permittivity ε, a permeability μ, and $\sigma = 0$. There are usually no free charges and no surface currents at the interface between two perfect dielectrics. By letting $\rho_s = 0$ and $\boldsymbol{J}_s = 0$ in(6.12), we obtain the boundary conditions as listed in

① 时变场边界条件的推导方法与表达式与静态场相同，但由于时变场的电场和磁场不是相互独立的，旋度方程和散度方程不是相互独立的，边界条件也不是相互独立的。两个法向边界条件可以由两个切向边界条件推导得出。因此，时变场问题求解中，常常只需要用到两个切向边界条件。

Table 6-2. ①

Table 6-2 Boundary conditions between two perfect dielectrics

Scalar Form	Vector Form	
$E_{1t}=E_{2t}$ or $\frac{D_{1t}}{\varepsilon_1}=\frac{D_{2t}}{\varepsilon_2}$	$\boldsymbol{e}_{n2}\times\boldsymbol{E}_1=\boldsymbol{e}_{n2}\times\boldsymbol{E}_2$	(6.13a)
$H_{1t}=H_{2t}$ or $\frac{B_{1t}}{\mu_1}=\frac{B_{2t}}{\mu_2}$	$\boldsymbol{e}_{n2}\times\boldsymbol{H}_1=\boldsymbol{e}_{n2}\times\boldsymbol{H}_2$	(6.13b)
$D_{1n}=D_{2n}$ or $\varepsilon_1 E_{1n}=\varepsilon_2 E_{2n}$	$\boldsymbol{e}_{n2}\cdot\boldsymbol{D}_1=\boldsymbol{e}_{n2}\cdot\boldsymbol{D}_2$	(6.13c)
$B_{1n}=B_{2n}$ or $\mu_1 H_{1n}=\mu_2 H_{2n}$	$\boldsymbol{e}_{n2}\cdot\boldsymbol{B}_1=\boldsymbol{e}_{n2}\cdot\boldsymbol{B}_2$	(6.13d)

6.5.2 Interface between a Dielectric and a Perfect Conductor (理想介质与理想导体分界面)

A **perfect electric conductor**(**PEC,理想导体**), or simply called **perfect conductor** is an ideal medium with an infinite conductivity. Superconductors are known to have nearly infinite (larger than 10^{20} S/m) conductivities, and therefore, can be considered as perfect conductor. Usually, metals such as silver, copper, gold, and aluminum, with conductivities of the order of 10^7(S/m) can also be treated as perfect conductor in practical situations in the radio frequency range.

In the interior of a PEC, the electric field is zero(otherwise, it would produce an infinitely large current due to its infinitely large conductivity), which means that the volume charge density is zero and any net charges carried by the PEC can only reside on its surface. According to Maxwell's equations and constitutive relations, the time-varying $\boldsymbol{D}$, $\boldsymbol{B}$ and $\boldsymbol{H}$ fields must also be zero in the interior of a PEC. ②

Consider an interface between a perfect dielectric (medium 1) and a perfect conductor (medium 2), as shown in Figure 6-2. Since $\boldsymbol{E}_2=0$, $\boldsymbol{H}_2=0$, $\boldsymbol{D}_2=0$, and $\boldsymbol{B}_2=0$ in medium 2, the general boundary conditions in (6.12) reduce to (6.14) as listed in Table 6-3, in which the subscript "1" is omitted. The fields involved in the boundary conditions (6.14) are all on the dielectric side of the boundary, and the unit vector $\boldsymbol{e}_n$ is an outward normal from the PEC. It is worth pointing out that the surface current density $\boldsymbol{J}_s$ is defined for currents flowing through an infinitesimal thickness, which is made possible by the infinitely large conductivity of the PEC.

---

① 理想介质是指电导率 $\sigma=0$ 的媒质。显然,在理想介质中不存在移动的电荷,即不存在体电流,因此在分界面上也不会出现自由电荷的积累,即 $\rho_s=0$。对于一般的媒质($\sigma>0$),由于可能存在体电流,分界面上就有可能出现自由电荷的积累,因此式(6.13c)不再适用,电位移矢量 $\boldsymbol{D}$ 的法向边界条件必须由式(6.12c)给定。然而,边界条件式(6.13b)仍然适用,这是因为,除非 σ 趋向于无穷大,否则一般的媒质分界面上不可能存在面电流 $\boldsymbol{J}_s$。

② 理想导体是指电导率 $\sigma\to+\infty$ 的媒质。理想导体内部不存在时变场。注意这与一般导体在静电平衡条件下不存在静电场的结论的区别。

For a conductor with finite conductivity, $\boldsymbol{J}_s$ is zero, the tangential component of $\boldsymbol{H}$ is continuous across an interface and the fields within the conductor are not necessarily zero.

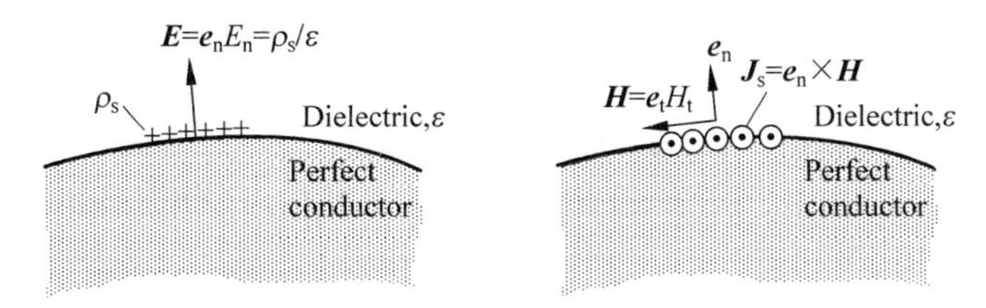

Figure 6-2 Boundary conditions at the interface between a perfect dielectric and a perfect conductor

Table 6-3 Boundary conditions between a perfect dielectric and a perfect conductor(dielectric side)

Scalar Form	Vector Form	
$E_t=0$	$\boldsymbol{e}_n\times\boldsymbol{E}=0$	(6.14a)
$H_t=J_s$	$\boldsymbol{e}_n\times\boldsymbol{H}=\boldsymbol{J}_s$	(6.14b)
$D_n=\rho_s(\varepsilon E_n=\rho_s$ for scalar $\varepsilon)$	$\boldsymbol{e}_n\cdot\boldsymbol{D}=\rho_s$	(6.14c)
$B_n=0(H_n=0$ for scalar $\mu)$	$\boldsymbol{e}_n\cdot\boldsymbol{B}=0$	(6.14d)

From the boundary conditions(6.14a) and(6.14c), it is concluded that the electric field intensity $\boldsymbol{E}$ is normal to the PEC surface, and is related to ρ_s by

$$\boldsymbol{E}=\boldsymbol{e}_nE_n=\boldsymbol{e}_n\frac{\rho_s}{\varepsilon} \tag{6.15}$$

(6.14b), (6.14d) show that the magnetic field intensity $\boldsymbol{H}$ is tangential to the interface with a magnitude equal to that of the surface current density, i.e.,

$$\boldsymbol{H}=\boldsymbol{e}_tH_t=\boldsymbol{e}_tJ_s=-\boldsymbol{e}_n\times\boldsymbol{J}_s \tag{6.16}$$

The three vectors, $\boldsymbol{H}$, $\boldsymbol{J}_s$ and $\boldsymbol{e}_n$, are mutually perpendicular to one another.

6.6 Potential Functions and Wave Equations (位函数与波动方程)

6.6.1 Potential functions for time-varying fields (时变场的位函数)

In Section 5.3, the concept of the vector magnetic potential $\boldsymbol{A}$ was introduced because of the solenoidal nature of $\boldsymbol{B}$(i.e., $\nabla\cdot\boldsymbol{B}=0$). Since $\boldsymbol{B}$ is still divergenceless in the time-varying case, we can also introduce the vector magnetic potential $\boldsymbol{A}$ for time-varying fields, and the time-varying $\boldsymbol{A}$ field also satisfies

$$\boldsymbol{B}=\nabla\times\boldsymbol{A} \tag{6.17}$$

Substitute(6.17) into the differential form of Faraday's law(6.10a), we get

$$\nabla\times\boldsymbol{E}=-\frac{\partial}{\partial t}(\nabla\times\boldsymbol{A})$$

which can be rewritten as

$$\nabla\times\left(\boldsymbol{E}+\frac{\partial \boldsymbol{A}}{\partial t}\right)=0 \tag{6.18}$$

Since the sum of the two terms in the parentheses of (6.18) is curl-free, it can be expressed as the gradient of a scalar. To be consistent with the definition of the scalar electric potential φ in electrostatics, we define φ in the time-varying case as

$$\boldsymbol{E}+\frac{\partial \boldsymbol{A}}{\partial t}=-\nabla\varphi \tag{6.19}$$

from which we obtain

$$\boldsymbol{E}=-\nabla\varphi-\frac{\partial \boldsymbol{A}}{\partial t} \tag{6.20}$$

Notice that, in the static case, $\partial\boldsymbol{A}/\partial t=0$ and (6.20) reduces to $\boldsymbol{E}=-\nabla\varphi$, and the $\boldsymbol{E}$ field is decoupled from the $\boldsymbol{B}$ field given by (6.17). For time-varying fields, however, both $\boldsymbol{E}$ and $\boldsymbol{B}$ are dependent on the vector potential $\boldsymbol{A}$ and they are coupled through the term $-\partial\boldsymbol{A}/\partial t$ in (6.20). [①]

With the introduction of potential functions, we can solve time-varying electromagnetic problems by first finding the solution of $\boldsymbol{A}$ and φ, and then obtaining the $\boldsymbol{E}$ and $\boldsymbol{B}$ fields with (6.17) and (6.20). To do that, we first find the governing equations satisfied by the $\boldsymbol{A}$ and φ fields.

Substitute (6.17) and (6.20) into (6.10b) and make use of the constitutive relations $\boldsymbol{H}=\boldsymbol{B}/\mu$ and $\boldsymbol{D}=\varepsilon\boldsymbol{E}$. We have

$$\nabla\times\nabla\times\boldsymbol{A}=\mu\boldsymbol{J}+\mu\varepsilon\frac{\partial}{\partial t}\left(-\nabla\varphi-\frac{\partial \boldsymbol{A}}{\partial t}\right) \tag{6.21}$$

Here, we assume that the fields are within a simple medium with a constant ε and μ. By using the vector identity (1.134), we can rewrite (6.21) as

$$\nabla(\nabla\cdot\boldsymbol{A})-\nabla^2\boldsymbol{A}=\mu\boldsymbol{J}-\nabla\left(\mu\varepsilon\frac{\partial\varphi}{\partial t}\right)-\mu\varepsilon\frac{\partial^2\boldsymbol{A}}{\partial t^2}$$

or

$$\nabla^2\boldsymbol{A}-\mu\varepsilon\frac{\partial^2\boldsymbol{A}}{\partial t^2}=-\mu\boldsymbol{J}+\nabla\left[(\nabla\cdot\boldsymbol{A})+\mu\varepsilon\frac{\partial\varphi}{\partial t}\right] \tag{6.22}$$

In Chapter 5, we determine $\boldsymbol{A}$ by using Coulomb gauge (5.21) that specifies the divergence of $\boldsymbol{A}$ to be zero. In time-varying electromagnetics, it is not necessary to use the same gauge. In fact, a better way to specify the divergence of $\boldsymbol{A}$ is to let

$$(\nabla\cdot\boldsymbol{A})+\mu\varepsilon\frac{\partial\varphi}{\partial t}=0 \tag{6.23}$$

which makes the second term on the right side of (6.22) vanish. Then we have

---

① 恒定磁场中引入矢量磁位 $\boldsymbol{A}$ 的数学基础是磁感应强度 $\boldsymbol{B}$ 的无散特性。在时变场中，$\boldsymbol{B}$ 仍然具有无散特性，因此时变场中 $\boldsymbol{A}$ 的定义式与恒定磁场中相同。静电场中引入标量电位的数学基础是电场强度 $\boldsymbol{E}$ 的无旋特性。然而，在时变场中，$\boldsymbol{E}$ 不再具备无旋特性，因此时变场中的标量位函数定义式，也即 $\boldsymbol{E}$ 与位函数的关系与静电场中的关系式不同。严格意义上，时变场中的标量位函数也不具备电位的物理意义。

$$\nabla^2 \boldsymbol{A} - \mu\varepsilon \frac{\partial^2 \boldsymbol{A}}{\partial t^2} = -\mu \boldsymbol{J} \tag{6.24}$$

(6.24) is the **nonhomogeneous wave equation**(**非齐次波动方程**) for vector potential $\boldsymbol{A}$. It is called a wave equation because its solutions represent waves traveling with a velocity equal to the speed of light, which will be demonstrated later. The nonhomogeneous wave equation is also called **d'Alembert equation**(**达朗贝尔方程**). The relation between $\boldsymbol{A}$ and φ in (6.23) is called the **Lorentz condition**(**洛伦兹条件**) or **Lorentz gauge**(**洛伦兹规范**). It reduces to the condition $\nabla \cdot \boldsymbol{A} = 0$ for static fields.

Now, by substituting(6.20) into(6.10c). We have

$$-\nabla \cdot \left[\varepsilon\left(\nabla\varphi + \frac{\partial \boldsymbol{A}}{\partial t}\right)\right] = \rho_v$$

which, for a constant ε, can be rewritten as

$$\nabla^2 \varphi + \frac{\partial}{\partial t}(\nabla \cdot \boldsymbol{A}) = -\frac{\rho_v}{\varepsilon}$$

Using Lorentz condition(6.23), we have

$$\nabla^2 \varphi - \mu\varepsilon \frac{\partial^2 \varphi}{\partial t^2} = -\frac{\rho_v}{\varepsilon} \tag{6.25}$$

which is the nonhomogeneous wave equation for scalar potential φ. Notice that, in a **source-free region**(**无源区域**) where no charge or current exists(i. e., $\rho_v = 0$ and $\boldsymbol{J} = 0$), (6.24) and(6.25) become the following homogeneous wave equations

$$\nabla^2 \boldsymbol{A} - \mu\varepsilon \frac{\partial^2 \boldsymbol{A}}{\partial t^2} = 0 \tag{6.26}$$

$$\nabla^2 \varphi - \mu\varepsilon \frac{\partial^2 \varphi}{\partial t^2} = 0 \tag{6.27}$$

It appears that the two wave equations(6.24) and(6.25), for $\boldsymbol{A}$ and φ respectively, are uncoupled from each other. However, it does not mean that the two potential functions $\boldsymbol{A}$ and φ are independent of each other as they must satisfy the Lorentz condition(6.23). Physically, $\boldsymbol{J}$ (the source of $\boldsymbol{A}$) and ρ_v(the source of φ) must satisfy the equation of continuity.[①]

6.6.2 Solution of Wave Equations for Potentials (位函数波动方程的解)

In an electromagnetic problem, if the charge and current distributions, ρ_v and $\boldsymbol{J}$ are known everywhere, we can first solve the nonhomogeneous wave equations, (6.24) and(6.25), for potentials $\boldsymbol{A}$ and φ. With $\boldsymbol{A}$ and φ determined, $\boldsymbol{E}$ and $\boldsymbol{B}$ can be found from(6.20) and(6.17),

① 位函数 $\boldsymbol{A}$ 和 φ 的引入，最大的作用之一是辅助求解已知电流和电荷分布产生的电场和磁场问题。在洛伦兹规范下，位函数 $\boldsymbol{A}$ 和 φ 满足的非齐次波动方程形式简单，$\boldsymbol{A}$ 和 φ 的解的形式也相对简洁。可以证明，简单媒质中时变电磁场 $\boldsymbol{E}$ 和 $\boldsymbol{H}$ 也满足相同形式的非齐次波动方程，但其中的非齐次项包含电流和电荷的较为复杂的表达式。由已知电流和电荷分布直接计算电、磁场的表达式更为复杂且不易使用。相较而言，先求解 $\boldsymbol{A}$ 和 φ，再通过式(6.20)和式(6.17)计算 $\boldsymbol{E}$ 和 $\boldsymbol{H}$ 更为简便。

respectively, by differentiation.

Consider solving the nonhomogeneous wave equation(6.25) for scalar electric potential φ. We can do this by first finding the solution for an elemental point charge, $Q(t)$, located at the origin of the coordinate system, and then obtaining the field due to all the charge elements with the principle of superposition. Since the point charge is at the origin, we have a problem with spherical symmetry, which means φ depends only on r and t, and the wave equation (6.25) becomes

$$\frac{1}{r^2}\frac{\partial}{\partial r}\left(r^2\frac{\partial\varphi}{\partial r}\right)-\mu\varepsilon\frac{\partial^2\varphi}{\partial t^2}=-\frac{Q(t)}{\varepsilon}\delta(\boldsymbol{r}) \tag{6.28}$$

Notice that the right side of the equation(6.28) is zero everywhere except at the origin. A general form of the solution to(6.28) everywhere except at the origin is

$$\varphi(r,t)=\frac{1}{r}f(t\pm r\sqrt{\mu\varepsilon}) \tag{6.29}$$

where f is an arbitrary twice-differentiable function. This can be easily proved by substituting (6.29) into(6.28), which verifies that(6.29) satisfies(6.28) except at the origin. Now, let

$$\Delta r=\frac{\Delta t}{\sqrt{\mu\varepsilon}} \tag{6.30}$$

We obviously have

$$f[(t+\Delta t)-(r+\Delta r)\sqrt{\mu\varepsilon}]=f(t-r\sqrt{\mu\varepsilon}) \tag{6.31}$$

which means $f(t-r\sqrt{\mu\varepsilon})$ travels in the $+r$-direction with a velocity $1/\sqrt{\mu\varepsilon}$.① Similarly, we find that $f(t+r\sqrt{\mu\varepsilon})$ represents a wave traveling along $-r$-direction. Since the source $Q(t)$ is located at the origin, it is nonphysical for the source to produce a wave traveling along $-r$-direction. Therefore, we only consider the solution

$$\varphi(r,t)=\frac{1}{r}f(t-r\sqrt{\mu\varepsilon})=\frac{1}{r}f(t-r/u) \tag{6.32}$$

where

$$u=\frac{1}{\sqrt{\mu\varepsilon}} \tag{6.33}$$

is essentially the **speed of light**(**光速**) in the medium(ε,μ).

To determine the specific form of the function $f(t-r/u)$, we compare(6.32) with the potential of a static point charge Q at the origin, which is, according to(2.50),

$$\varphi(r)=\frac{Q}{4\pi\varepsilon r} \tag{6.34}$$

In the time-varying case, the potential φ must be a function of both the location and the time. Intuitively, the static charge Q in(6.34) should be replaced by the time-varying charge $Q(t)$. However, the resulted expression is not in the form of $f(t-r/u)/r$. To obtain the solution of φ in

---

① 式(6.31)的物理意义为：在 r 代表的任何位置，任何时间 t 的场值，在经过 Δt 之后，必然会出现在 $r+\Delta r$ 的位置。这个现象就代表了沿 $+r$ 方向传播的波。其传播的速度为 $\Delta r/\Delta t$。由式(6.30)得出该速度即为光速。

the form of $f(t-r/u)/r$, it is reasonable to further replace $Q(t)$ with $Q(t-r/u)$, which gives the solution of the scalar potential due to a time-varying point charge at the origin: ①

$$\varphi(r,t)=\frac{Q(t-r/u)}{4\pi\varepsilon r} \tag{6.35}$$

Now, due to the property of spatial translational invariance, we can write the scalar potential at a field point $\boldsymbol{r}$ due to a point charge located at an arbitrary position $\boldsymbol{r}'$ as

$$\varphi(\boldsymbol{r},t)=\frac{Q(t-|\boldsymbol{r}-\boldsymbol{r}'|/u)}{4\pi\varepsilon|\boldsymbol{r}-\boldsymbol{r}'|} \tag{6.36}$$

For a continuous charge distribution $\rho_v(\boldsymbol{r}',t)$ over a volume V', we can treat each differential volume charge $\rho_v(\boldsymbol{r}',t)\mathrm{d}v'$ as a point charge located at $\boldsymbol{r}'$, which produces a differential potential at the field point $\boldsymbol{r}$ as

$$\mathrm{d}\varphi(\boldsymbol{r},t)=\frac{\rho_v(\boldsymbol{r}',t-|\boldsymbol{r}-\boldsymbol{r}'|/u)\mathrm{d}v'}{4\pi\varepsilon|\boldsymbol{r}-\boldsymbol{r}'|} \tag{6.37}$$

Using the principle of superposition, we can express the potential due to the charge distribution $\rho_v(\boldsymbol{r}',t)$ by integrating both sides of (6.37), which leads to

$$\varphi(\boldsymbol{r},t)=\frac{1}{4\pi\varepsilon}\int_{V'}\frac{\rho_v(\boldsymbol{r}',t-|\boldsymbol{r}-\boldsymbol{r}'|/u)}{|\boldsymbol{r}-\boldsymbol{r}'|}\mathrm{d}v'=\frac{1}{4\pi\varepsilon}\int_{V'}\frac{\rho_v(\boldsymbol{r}',t-R/u)}{R}\mathrm{d}v' \tag{6.38}$$

Notice that φ in (6.38) is not only a function of r but a function of $\boldsymbol{r}$ since the charge distribution may not have spherical symmetry.

(6.38) indicates that the scalar potential at a distance R from the source at time t depends on the value of the charge density at an earlier time $(t-R/u)$. It takes time R/u for the change in ρ_v to take effect at distance R. For this reason, $\varphi(\boldsymbol{r},t)$ in (6.38) is called the **retarded scalar potential**(**标量推迟位**).

The solution of the nonhomogeneous wave equation (6.24) for vector potential $\boldsymbol{A}$ can be derived similarly, which gives the following expression for the **retarded vector potential**(**矢量推迟位**)

$$\boldsymbol{A}(\boldsymbol{r},t)=\frac{\mu}{4\pi}\int_{V'}\frac{\boldsymbol{J}(\boldsymbol{r}',t-|\boldsymbol{r}-\boldsymbol{r}'|/u)}{|\boldsymbol{r}-\boldsymbol{r}'|}\mathrm{d}v'=\frac{\mu}{4\pi}\int_{V'}\frac{\boldsymbol{J}(\boldsymbol{r}',t-R/u)}{R}\mathrm{d}v' \tag{6.39}$$

Now we see that, if the charge and current distributions are known, the potential fields $\boldsymbol{A}$ and φ can be obtained directly from (6.38) and (6.39). Then the electric and magnetic fields can be derived from $\boldsymbol{A}$ and φ with (6.17) and (6.20), which will also be retarded in time. Physically, it means that the variation of the sources ρ_v and $\boldsymbol{J}$ will cause the change in electromagnetic fields at distant points at a later time. Therefore, time-varying electromagnetic fields must also be electromagnetic waves, and from previous discussions, the electromagnetic

① 时变点电荷的标量位表达式(6.35)与静态点电荷电位的表达式相比,区别是以 $Q(t-r/u)$ 代替了常数 Q,而不是直觉上简单地以 $Q(t)$ 代替常数 Q。这一方面是因为时变场的标量位作为波动方程的解,必须符合 $f(t-r/u)/r$ 的形式。另一方面,从物理的角度,坐标原点的电荷变化必须在经过一段时间 r/u 后才会影响到与原点距离为 r 的位置的标量位。

waves travel at the speed of light. [①]

In the **quasi-static**(准静态) approximation, we ignore this time-retardation effect and assume instant response. This assumption is implicit in dealing with circuit problems. [②]

6.7 Homogeneous Wave Equations in Source-Free Region(无源区域的齐次波动方程)

In the previous section, we derived the expressions of potential functions $\boldsymbol{A}$ and φ due to given sources ρ_v and $\boldsymbol{J}$ by solving the nonhomogeneous wave equations. These expressions describe how the sources produce electromagnetic waves. In problems of **wave propagation**(**波的传播问题**), we are concerned with the behavior of an electromagnetic wave in a source-free **region**(无源区域) where ρ_v and $\boldsymbol{J}$ are both zero. To study how electromagnetic waves propagate in the space, we use the **homogeneous wave equation**(齐次波动方程) satisfied by $\boldsymbol{E}$ and $\boldsymbol{H}$ fields, which is derived as below.

Consider a simple (linear, isotropic, and homogeneous) nonconducting medium characterized by ε, μ, and $\sigma=0$. Maxwell's equations(6.10) reduce to

$$\nabla\times\boldsymbol{E}=-\mu\frac{\partial\boldsymbol{H}}{\partial t} \tag{6.40a}$$

$$\nabla\times\boldsymbol{H}=\varepsilon\frac{\partial\boldsymbol{E}}{\partial t} \tag{6.40b}$$

$$\nabla\cdot\boldsymbol{E}=0 \tag{6.40c}$$

$$\nabla\cdot\boldsymbol{H}=0 \tag{6.40d}$$

Taking the curl of (6.40a) and use(6.40b), we have

$$\nabla\times\nabla\times\boldsymbol{E}=-\mu\frac{\partial}{\partial t}(\nabla\times\boldsymbol{H})=-\mu\varepsilon\frac{\partial^2\boldsymbol{E}}{\partial t^2}$$

Now $\nabla\times\nabla\times\boldsymbol{E}=\nabla(\nabla\cdot\boldsymbol{E})-\nabla^2\cdot\boldsymbol{E}=-\nabla^2\boldsymbol{E}$ because of (6.40c). Hence, we have

$$\nabla^2\boldsymbol{E}-\mu\varepsilon\frac{\partial^2\boldsymbol{E}}{\partial t^2}=0 \tag{6.41}$$

Similarly, we obtain the same equation for $\boldsymbol{H}$:

$$\nabla^2\boldsymbol{H}-\mu\varepsilon\frac{\partial^2\boldsymbol{H}}{\partial t^2}=0 \tag{6.42}$$

---

① 时变位函数的表达式(6.38)和式(6.39)被称为“**推迟位**”,其物理含义体现在表达式中时变电荷源 ρ_v 和电流源 $\boldsymbol{J}$ 的时间参量$(t-R/u)$,即源点 $\boldsymbol{r}'$ 处 t 时刻的源的变化必须在经过一段时间 R/u 之后才能对场点 $\boldsymbol{r}$ 处的场产生影响,其中 $R=|\boldsymbol{r}-\boldsymbol{r}'|$。这也说明,时变条件下,空间电磁场分布并不取决于同一时刻的源的分布。即便某一时刻,所有的源都消失,该时刻之前源产生的电磁场仍然存在。也就是说,时变电磁场能够脱离源,以电磁波的形式存在于空间中,这种现象称为**电磁辐射**。

② 所谓准静态,就是指源和场的变化“足够缓慢”的状态。而所谓“足够缓慢”,在电路问题中,就是指信号的波长远大于电路尺寸(波长大意味着频率低,也就意味着变化缓慢)。对应到场的理论,可以理解为式(6.38)和式(6.39)中的推迟项$(t-R/u)$可以近似为 t。

with(6.33),(6.41) and(6.42) can also be written as

$$\nabla^2 \boldsymbol{E} - \frac{1}{u^2}\frac{\partial^2 \boldsymbol{E}}{\partial t^2} = 0 \tag{6.43}$$

$$\nabla^2 \boldsymbol{H} - \frac{1}{u^2}\frac{\partial^2 \boldsymbol{H}}{\partial t^2} = 0 \tag{6.44}$$

(6.43) and (6.44) are **homogeneous vector wave equations** (齐次矢量波动方程). In Cartesian coordinates, (6.43) and (6.44) can each be decomposed into three one-dimensional, homogeneous, scalar wave equations.

Example 6-4 As shown in Figure 6-3, the electric field in the air between two infinitely large parallel PEC plates is

$$\boldsymbol{E} = \boldsymbol{e}_y E_0 \sin\left(\frac{\pi}{d}z\right)\cos(\omega t - k_x x)$$

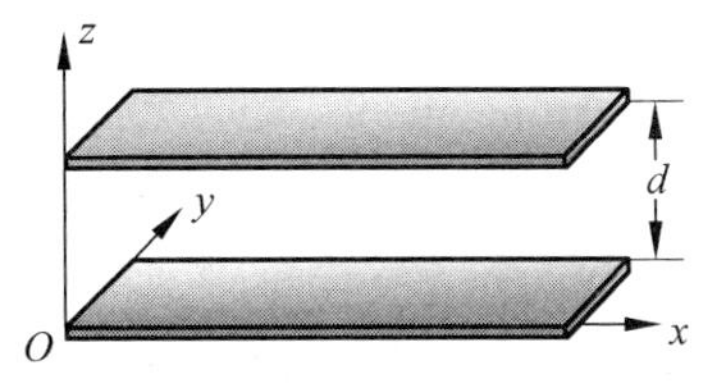

Figure 6-3 Two infinitely large parallel PEC plates (Example 6-4)

Find ①the relation between k_x and ω, ②the $\boldsymbol{H}$ field, and ③the surface current density on the PEC surfaces.

Solution:

(1) There is no source in between the two PEC plates, therefore, $\boldsymbol{E}$ satisfies the homogeneous wave equation(6.41). With the expression of $\boldsymbol{E}$, we have

$$\nabla^2 \boldsymbol{E} = \boldsymbol{e}_y\left[-\left(\frac{\pi}{d}\right)^2 - k_x^2\right]E_0 \sin\left(\frac{\pi}{d}z\right)\cos(\omega t - k_x x)$$

$$\frac{\partial^2 \boldsymbol{E}}{\partial t^2} = \boldsymbol{e}_y(-\omega^2)E_0 \sin\left(\frac{\pi}{d}z\right)\cos(\omega t - k_x x)$$

Substitute the above expressions into(6.41), we have

$$\left[-\left(\frac{\pi}{d}\right)^2 - k_x^2 + \mu_0\varepsilon_0\omega^2\right]E_0 \sin\left(\frac{\pi}{d}z\right)\cos(\omega t - k_x x) = 0$$

Therefore, we have the relation between k_x and ω as

$$k_x^2 + \left(\frac{\pi}{d}\right)^2 = \omega^2\mu_0\varepsilon_0$$

(2) The $\boldsymbol{E}$ field satisfies Maxwell's equation(6.40a). Therefore, we have

$$\frac{\partial \boldsymbol{H}}{\partial t} = -\frac{1}{\mu_0}\nabla\times\boldsymbol{E} = -\frac{1}{\mu_0}\left(-\boldsymbol{e}_x\frac{\partial E_y}{\partial z} + \boldsymbol{e}_z\frac{\partial E_y}{\partial x}\right)$$

$$= \frac{E_0}{\mu_0}\left[\boldsymbol{e}_x\frac{\pi}{d}\cos\left(\frac{\pi}{d}z\right)\cos(\omega t - k_x x) - \boldsymbol{e}_z k_x \sin\left(\frac{\pi}{d}z\right)\sin(\omega t - k_x x)\right]$$

The magnetic field can be obtained by integration of the above expression

$$H = \int \frac{\partial \boldsymbol{H}}{\partial t} dt = \boldsymbol{e}_x \frac{\pi E_0}{\omega \mu_0 d} \cos\left(\frac{\pi}{d} z\right) \sin(\omega t - k_x x) + \boldsymbol{e}_z \frac{k_x E_0}{\omega \mu_0} \sin\left(\frac{\pi}{d} z\right) \cos(\omega t - k_x x)$$

(3) From the $\boldsymbol{H}$ field obtained above, we can find the surface currents on the PEC plates by using the boundary conditions(6.14b). On the top surface of the lower plate($z=0$), we have

$$\boldsymbol{J}_s = \boldsymbol{e}_z \times \boldsymbol{H}\big|_{z=0} = \boldsymbol{e}_y \frac{\pi E_0}{\omega \mu_0 d} \sin(\omega t - k_x x)$$

On the bottom surface of the upper plate($z=d$), we have

$$\boldsymbol{J}_s = (-\boldsymbol{e}_z) \times \boldsymbol{H}\big|_{z=d} = \boldsymbol{e}_y \frac{\pi E_0}{\omega \mu_0 d} \sin(\omega t - k_x x)$$

6.8 Poynting Vector and Flow of Electromagnetic Power (坡印廷矢量与电磁功率流)

In this section, we show that electromagnetic waves carry electromagnetic power. We derive the relation between the electromagnetic fields and power flow, and then introduce the power density vector associated with electromagnetic waves. ①

We begin with the vector identity(1.150), which is rewritten here in terms of $\boldsymbol{E}$ and $\boldsymbol{H}$ fields:

$$\nabla \cdot (\boldsymbol{E} \times \boldsymbol{H}) = \boldsymbol{H} \cdot (\nabla \times \boldsymbol{E}) - \boldsymbol{E} \cdot (\nabla \times \boldsymbol{H}) \tag{6.45}$$

Substitute(6.10a) and(6.10b) in(6.45), then

$$\nabla \cdot (\boldsymbol{E} \times \boldsymbol{H}) = -\boldsymbol{H} \cdot \frac{\partial \boldsymbol{B}}{\partial t} - \boldsymbol{E} \cdot \frac{\partial \boldsymbol{D}}{\partial t} - \boldsymbol{E} \cdot \boldsymbol{J} \tag{6.46}$$

In a simple medium with constant ε, μ and σ, we have

$$\boldsymbol{H} \cdot \frac{\partial \boldsymbol{B}}{\partial t} = \boldsymbol{H} \cdot \frac{\partial (\mu \boldsymbol{H})}{\partial t} = \frac{1}{2} \frac{\partial (\mu \boldsymbol{H} \cdot \boldsymbol{H})}{\partial t} = \frac{\partial}{\partial t}\left(\frac{1}{2} \mu H^2\right)$$

$$\boldsymbol{E} \cdot \frac{\partial \boldsymbol{D}}{\partial t} = \boldsymbol{E} \cdot \frac{\partial (\varepsilon \boldsymbol{E})}{\partial t} = \frac{1}{2} \frac{\partial (\varepsilon \boldsymbol{E} \cdot \boldsymbol{E})}{\partial t} = \frac{\partial}{\partial t}\left(\frac{1}{2} \varepsilon E^2\right)$$

$$\boldsymbol{E} \cdot \boldsymbol{J} = \boldsymbol{E} \cdot (\sigma \boldsymbol{E}) = \sigma E^2$$

Then(6.46) can then be written as

$$\nabla \cdot (\boldsymbol{E} \times \boldsymbol{H}) = -\frac{\partial}{\partial t}\left(\frac{1}{2} \varepsilon E^2 + \frac{1}{2} \mu H^2\right) - \sigma E^2 \tag{6.47}$$

Integrate both sides of(6.47) over an arbitrary volume V,

$$-\oint_S (\boldsymbol{E} \times \boldsymbol{H}) \cdot d\boldsymbol{s} = \frac{\partial}{\partial t} \int_V \left(\frac{1}{2} \varepsilon E^2 + \frac{1}{2} \mu H^2\right) dv + \int_V \sigma E^2 dv \tag{6.48}$$

where the divergence theorem has been applied to convert the volume integral of $\nabla \cdot (\boldsymbol{E} \times \boldsymbol{H})$ over V to the closed surface integral of $(\boldsymbol{E} \times \boldsymbol{H})$ over the S. Notice that, the negative surface

---

① 与静电场、恒定磁场一样，时变电磁场也可以储能。时变场特有的现象是电场和磁场的储能会随时间变化，并且伴随能量的流动。本节将引入坡印廷矢量的概念来描述时变场中能量的流动。恒定电场中由于存在恒定电流，电场和磁场同时存在，也存在以坡印廷矢量描述的能量的流动，但电场和磁场的储能不随时间变化。

integral on the left side of(6.48) represents the total flux of the field($\boldsymbol{E}\times\boldsymbol{H}$) entering the closed surface S.

According to the expressions of electric and magnetic energies(2.136) and(5.161), the first integral on the right side of(6.48) involving E^2 and H^2 represents the time-rate of increase of the energy stored in the electric and magnetic fields. The second term is the ohmic power dissipated in the volume with conductivity σ. According to the law of conservation of energy, there must be an equal amount of power entering the volume through the surface S to supply the stored energy increase and the ohmic power dissipation. Thus, the left side of(6.48) must represent the power flow into the volume V through S, and($\boldsymbol{E}\times\boldsymbol{H}$) represents the power flow per unit area. Define

$$\boldsymbol{S}=\boldsymbol{E}\times\boldsymbol{H} \tag{6.49}$$

$\boldsymbol{S}$ is known as the **Poynting vector**(**坡印廷矢量**), which is a power density vector associated with an electromagnetic field. Note that the Poynting vector $\boldsymbol{S}$ is normal to both $\boldsymbol{E}$ and $\boldsymbol{H}$.

(6.48) states that the negative surface integral of $\boldsymbol{S}$ over a closed surface equals the power entering the enclosed volume(闭合曲面坡印廷矢量的通量积分的负值为进入该曲面所围区域的功率), which is referred to as **Poynting's theorem**(**坡印廷定理**). If the region of concern is lossless($\sigma=0$), then the last term in(6.48) vanishes, and the total power flowing into a closed surface is equal to the rate of increase of the stored electric and magnetic energies in the enclosed volume. In a static situation, the first term on the right side of(6.48) vanishes, and the total power flowing into a closed surface is equal to the ohmic power dissipated in the enclosed volume.

Example 6-5 Verify Poynting's theorem for a resistor carrying a DC current I. Assume that the resistor is a cylindrical conductor with radius a, length l and conductivity σ.

Solution: In the DC case, the current in the wire is uniformly distributed within the resistor. The current flows in the length direction(z-direction) as shown in Figure 6-4. Then we have

$$\boldsymbol{J}=\boldsymbol{e}_z\frac{I}{\pi a^2} \quad \text{and} \quad \boldsymbol{E}=\frac{\boldsymbol{J}}{\sigma}=\boldsymbol{e}_z\frac{I}{\sigma\pi a^2}$$

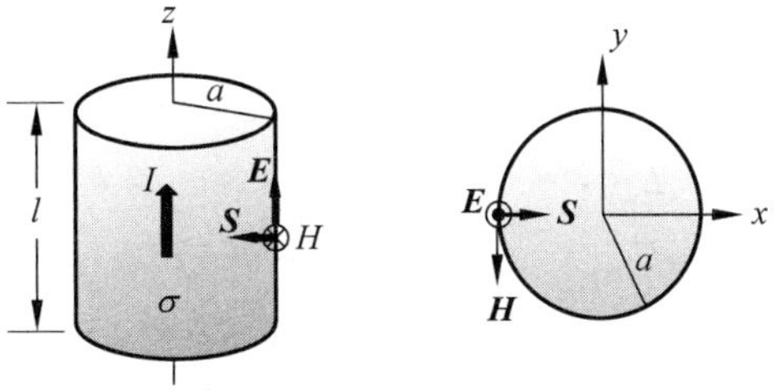

Figure 6-4 Illustrating Poynting's theorem(Example 6-5)

On the surface of the wire,

$$\boldsymbol{H}=\boldsymbol{e}_\phi\frac{I}{2\pi a}$$

Thus, the Poynting vector at the surface of the wire is

$$S = E \times H = (e_z \times e_\phi)\frac{I^2}{2\sigma\pi^2 a^3} = -e_\rho \frac{I^2}{2\sigma\pi^2 a^3}$$

which is directed everywhere into the wire surface. To verify Poynting's theorem, we integrate S over the wall of the wire segment in Figure 6-4

$$-\oint_S S\cdot ds = -\oint_S S\cdot e_\rho ds = \left(\frac{I^2}{2\sigma\pi^2 a^3}\right)2\pi bl = I^2\left(\frac{l}{\sigma\pi a^2}\right) = I^2 R$$

where $R = l/(\sigma\pi a^2)$ is the resistance of the cylindrical conductor. This verifies Poynting's theorem which states that the negative surface integral of the Poynting vector is exactly equal to I^2R, the ohmic power loss in the resistor.

6.9 Time-Harmonic Fields and Waves(时谐场与波)

Time-varying electromagnetic fields are vector functions of four variables, one of which is time t. The dependency of the fields on time is generally arbitrary, which makes the time-varying electromagnetic problems more complex to solve than the static problems. To simplify the problems, a common practice in engineering is to assume a sinusoidal variation in time for all the fields and obtain a solution to the problem at particular frequencies. Then for an arbitrary time dependence, time-varying fields can be determined by using Fourier integrals of the solutions at all the frequencies based on the principle of superposition. In this section, we introduce the **time-harmonic**(时谐) fields and waves, which are the solution to electromagnetic problems in the steady-state at a single frequency. Time-harmonic form of all the equations governing the time-varying electromagnetic fields will be derived.

6.9.1 Phasor Expressions of Sinusoidal Field Quantities (正弦场量的相量表示)

The instantaneous (time-dependent) expression of a sinusoidal quantity can be written as either a cosine or a sine function. Here, we follow the convention of using the cosine function. Take the E field as an example. The sinusoidal vector E can be written as

$$\begin{aligned} E(x,y,z,t) &= e_x E_x(x,y,z,t) + e_y E_y(x,y,z,t) + e_z E_z(x,y,z,t) \\ &= e_x E_{xm}(x,y,z)\cos[\omega t + \phi_x(x,y,z)] + \\ &\quad e_y E_{ym}(x,y,z)\cos[\omega t + \phi_y(x,y,z)] + \\ &\quad e_z E_{zm}(x,y,z)\cos[\omega t + \phi_z(x,y,z)] \end{aligned} \tag{6.50}$$

Obviously, the sinusoidal $E(x,y,z,t)$ is composed of three sinusoidal scalars, $E_x(x,y,z,t)$, $E_y(x,y,z,t)$, and $E_z(x,y,z,t)$, each determined by their amplitude (E_{xm}, E_{ym} and E_{zm}), frequency(ω), and initial phase(ϕ_x, ϕ_y and ϕ_z). The frequency is usually prescribed and fixed for all the quantities. But the three amplitudes and three phases are functions of locations. ①

By using **Euler's formula**(欧拉公式), we know that a cosine function can be written as the real part of a complex exponential function, i. e.

$$A\cos(\omega t+\phi)=\mathrm{Re}[A\mathrm{e}^{\mathrm{j}\phi}\mathrm{e}^{\mathrm{j}\omega t}] \tag{6.51}$$

where $A\mathrm{e}^{\mathrm{j}\phi}$ is called the scalar **phasor**(相量) of the sinusoidal scalar function $A\cos(\omega t+\phi)$. ② The phasor contains the amplitude and the initial phase of the time-harmonic quantity. Generally, a time-harmonic scalar field $\varphi(x,y,z,t)=\varphi_{\mathrm{m}}(x,y,z)\cos[\omega t+\phi(x,y,z)]$ can be represented by its phasor $\varphi(x,y,z)=\varphi_{\mathrm{m}}(x,y,z)\mathrm{e}^{\mathrm{j}\phi(x,y,z)}$, and the two forms of the scalar field are related by

$$\varphi(x,y,z,t)=\mathrm{Re}[\varphi(x,y,z)\mathrm{e}^{\mathrm{j}\omega t}] \tag{6.52}$$

Notice that, here we use the same function name φ for its instantaneous expression $\varphi(x,y,z,t)$ and phasor expression $\varphi(x,y,z)$. We will mostly deal with time-harmonic fields (and therefore with phasors) in later chapters. Whether a notation represents phasor or instantaneous quantity can be judged from the context. Specifically, if the notation represents a real quantity and is a function of time, then it must be in the instantaneous form. If the notation is not a function of time and if it has a complex value, it must be in the phasor form. ③

Now we use (6.52) to rewrite the $\boldsymbol{E}$ field expression in (6.50) as

$$\begin{aligned}\boldsymbol{E}(x,y,z,t)&=\mathrm{Re}[\boldsymbol{e}_{\mathrm{x}}E_{\mathrm{xm}}(x,y,z)\mathrm{e}^{\mathrm{j}\phi_{\mathrm{x}}(x,y,z)}\mathrm{e}^{\mathrm{j}\omega t}]+\\&\quad\mathrm{Re}[\boldsymbol{e}_{\mathrm{y}}E_{\mathrm{ym}}(x,y,z)\mathrm{e}^{\mathrm{j}\phi_{\mathrm{y}}(x,y,z)}\mathrm{e}^{\mathrm{j}\omega t}]+\\&\quad\mathrm{Re}[\boldsymbol{e}_{\mathrm{z}}E_{\mathrm{zm}}(x,y,z)\mathrm{e}^{\mathrm{j}\phi_{\mathrm{z}}(x,y,z)}\mathrm{e}^{\mathrm{j}\omega t}]\\&=\mathrm{Re}[\boldsymbol{e}_{\mathrm{x}}E_{\mathrm{x}}(x,y,z)\mathrm{e}^{\mathrm{j}\omega t}+\boldsymbol{e}_{\mathrm{y}}E_{\mathrm{y}}(x,y,z)\mathrm{e}^{\mathrm{j}\omega t}+\boldsymbol{e}_{\mathrm{z}}E_{\mathrm{z}}(x,y,z)\mathrm{e}^{\mathrm{j}\omega t}]\end{aligned} \tag{6.53}$$

where

$$\begin{aligned}E_{\mathrm{x}}(x,y,z)&=E_{\mathrm{xm}}(x,y,z)\mathrm{e}^{\mathrm{j}\phi_{\mathrm{x}}(x,y,z)}\\E_{\mathrm{y}}(x,y,z)&=E_{\mathrm{ym}}(x,y,z)\mathrm{e}^{\mathrm{j}\phi_{\mathrm{y}}(x,y,z)}\\E_{\mathrm{z}}(x,y,z)&=E_{\mathrm{zm}}(x,y,z)\mathrm{e}^{\mathrm{j}\phi_{\mathrm{z}}(x,y,z)}\end{aligned} \tag{6.54}$$

are the scalar phasors of the three components of $\boldsymbol{E}$ in the x-, y-, and z-directions. The three scalar phasors can be combined into one vector phasor $\boldsymbol{E}(x,y,z)$ given by

$$\boldsymbol{E}(x,y,z)=\boldsymbol{e}_{\mathrm{x}}E_{\mathrm{x}}(x,y,z)+\boldsymbol{e}_{\mathrm{y}}E_{\mathrm{y}}(x,y,z)+\boldsymbol{e}_{\mathrm{z}}E_{\mathrm{z}}(x,y,z) \tag{6.55}$$

---

① 时变场不仅是三维空间位置的函数,也是时间的函数,因而数学上是一个四元函数。所谓时谐场,就是假设场量随时间正弦或者余弦变化,也即按照单一频率时谐变化的时变场。这样,四元变量的电磁场就简化为一个三元函数,是仅与位置有关的场。但是时谐场只是假定了随时间变化的频率,要确定一个场量,还需要确定其振幅和初始相位。场量在不同位置可以有不同的振幅和相位,因此一个时谐变化的标量场对应于两个三元函数:振幅函数和相位函数。将这两个函数合并写成复数形式,即得到本节引入的相量函数。时谐变化的矢量场的三个分量分别对应于一个标量相量,三个标量相量合成该时谐矢量场的矢量相量。

② 注意:这里的 ϕ 代表时谐函数的初始相位,而不是柱坐标和球坐标下的坐标变量。这从上下文可以判断。

③ 式(6.52)以及随后的式(6.55)~式(6.57)中,同样的符号(φ 和 $\boldsymbol{E}$)既代表瞬变实函数,又代表相应的相量复函数。这是一种简化的标记方法,以后出现的场量符号都通过上下文判断其代表的是瞬变实函数还是相量复函数。

Then, according to (6.53), we have

$$\boldsymbol{E}(x,y,z,t)=\text{Re}[\boldsymbol{E}(x,y,z)\text{e}^{\text{j}\omega t}] \tag{6.56}$$

or generally (not necessarily in Cartesian coordinates),

$$\boldsymbol{E}(\boldsymbol{r},t)=\text{Re}[\boldsymbol{E}(\boldsymbol{r})\text{e}^{\text{j}\omega t}] \tag{6.57}$$

The phasor $\boldsymbol{E}(\boldsymbol{r})$ is a complex vector that contains information on the direction, magnitude, and phase of the real instantaneous vector $\boldsymbol{E}(\boldsymbol{r},t)$. Note that phasors are generally complex quantities, whereas instantaneous (time-dependent) quantities must be real.

By taking the derivative and integral of (6.51) respectively, we can easily obtain the following relations:

$$\frac{\text{d}}{\text{d}t}[A\cos(\omega t+\varphi)]=-\omega A\sin(\omega t+\varphi)=\omega A\cos\left(\omega t+\varphi+\frac{\pi}{2}\right)=\text{Re}[\text{j}\omega(A\text{e}^{\text{j}\varphi})\text{e}^{\text{j}\omega t}] \tag{6.58a}$$

$$\int A\cos(\omega t+\varphi)\text{d}t=\frac{1}{\omega}[A\sin(\omega t+\varphi)]=-\frac{1}{\omega}A\cos\left(\omega t+\varphi+\frac{\pi}{2}\right)=\text{Re}\left[\frac{1}{\text{j}\omega}(A\text{e}^{\text{j}\varphi})\text{e}^{\text{j}\omega t}\right] \tag{6.58b}$$

From (6.58), we conclude that, if $\varphi(\boldsymbol{r})$ is the phasor of $\varphi(\boldsymbol{r},t)$, then $\text{j}\omega\varphi(\boldsymbol{r})$ is the phasor of $\partial\varphi(\boldsymbol{r},t)/\partial t$, and $\varphi(\boldsymbol{r})/\text{j}\omega$ is the phasor of $\int\varphi(\boldsymbol{r},t)\text{d}t$. Similarly, if $\boldsymbol{E}(\boldsymbol{r})$ is the phasor of $\boldsymbol{E}(\boldsymbol{r},t)$, then $\text{j}\omega\boldsymbol{E}(\boldsymbol{r})$ is the phasor of $\partial\boldsymbol{E}(\boldsymbol{r},t)/\partial t$, and $\boldsymbol{E}(\boldsymbol{r})/\text{j}\omega$ is the phasor of $\int\boldsymbol{E}(\boldsymbol{r},t)\text{d}t$. In other words, differentiations and integrations of sinusoidal fields with respect to t can be represented by multiplication and division of the corresponding phasors with $\text{j}\omega$. Higher-order differentiations and integrations would be represented respectively by multiplications and divisions of the phasor with $\text{j}\omega$ for multiple times.

Example 6-6 Find the phasors of the following instantaneous expressions.

(1) $\boldsymbol{E}(z,t)=\boldsymbol{e}_{\text{x}}E_{\text{xm}}\cos(\omega t-kz+\phi_{\text{x}})+\boldsymbol{e}_{\text{y}}E_{\text{ym}}\sin(\omega t-kz+\phi_{\text{y}})$

(2) $\boldsymbol{H}(x,z,t)=\boldsymbol{e}_{\text{x}}H_{\text{m}}k\left(\frac{a}{\pi}\right)\sin\left(\frac{\pi x}{a}\right)\sin(kz-\omega t)+\boldsymbol{e}_{\text{z}}H_{\text{m}}\cos\left(\frac{\pi x}{a}\right)\cos(kz-\omega t)$

Solution:

(1) To find the phasor of a sinusoidal quantity, we first express each term using cosine functions. So we convert the sine term in the expression into cosine, and we have

$$\boldsymbol{E}(z,t)=\boldsymbol{e}_{\text{x}}\boldsymbol{E}_{\text{xm}}\cos(\omega t-kz+\phi_{\text{x}})+\boldsymbol{e}_{\text{y}}E_{\text{ym}}\cos(\omega t-kz+\phi_{\text{y}}-\pi/2)$$

Then, using (6.51), we have

$$\boldsymbol{E}(z,t)=\boldsymbol{e}_{\text{x}}E_{\text{xm}}\text{Re}[\text{e}^{\text{j}(\omega t-kz+\phi_{\text{x}})}]+\boldsymbol{e}_{\text{y}}E_{\text{ym}}\text{Re}[\text{e}^{\text{j}(\omega t-kz+\phi_{\text{x}}-\pi/2)}]$$

The phasor expression can be obtained by dropping the $\text{e}^{\text{j}\omega t}$ terms and the taking real operators:

$$\boldsymbol{E}(z)=\boldsymbol{e}_{\text{x}}E_{\text{xm}}\text{e}^{\text{j}(-kz+\phi_{\text{x}})}+\boldsymbol{e}_{\text{y}}E_{\text{ym}}\text{e}^{\text{j}(-kz+\phi_{\text{x}}-\pi/2)}=(\boldsymbol{e}_{\text{x}}E_{\text{xm}}\text{e}^{\text{j}\phi_{\text{x}}}-\boldsymbol{e}_{\text{y}}\text{j}E_{\text{ym}}\text{e}^{\text{j}\phi_{\text{y}}})\text{e}^{-\text{j}kz}$$

(2) By following the similar procedure above, we have

$$\boldsymbol{H}(x,z,t)=\boldsymbol{e}_{\mathrm{x}}H_{\mathrm{m}}k\left(\frac{a}{\pi}\right)\sin\left(\frac{\pi x}{a}\right)\cos\left(\omega t-kz+\frac{\pi}{2}\right)+\boldsymbol{e}_{\mathrm{z}}H_{\mathrm{m}}\cos\left(\frac{\pi x}{a}\right)\cos(\omega t-kz)$$

$$=\boldsymbol{e}_{\mathrm{x}}H_{\mathrm{m}}k\left(\frac{a}{\pi}\right)\sin\left(\frac{\pi x}{a}\right)\mathrm{Re}\left[\mathrm{e}^{\mathrm{j}(\omega t-kz+\pi/2)}\right]+\boldsymbol{e}_{\mathrm{z}}H_{\mathrm{m}}\cos\left(\frac{\pi x}{a}\right)\mathrm{Re}\left[\mathrm{e}^{\mathrm{j}(\omega t-kz)}\right]$$

and

$$\boldsymbol{H}(x,z)=\boldsymbol{e}_{\mathrm{x}}\mathrm{j}H_{\mathrm{m}}k\left(\frac{a}{\pi}\right)\sin\left(\frac{\pi x}{a}\right)\mathrm{e}^{-\mathrm{j}kz}+\boldsymbol{e}_{\mathrm{z}}H_{\mathrm{m}}\cos\left(\frac{\pi x}{a}\right)\mathrm{e}^{-\mathrm{j}kz}$$

Example 6-7 Write the instantaneous expression of the following phasor.

$$\boldsymbol{E}(z)=\boldsymbol{e}_{\mathrm{x}}\mathrm{j}E_{\mathrm{xm}}\cos(k_{\mathrm{z}}z)$$

where k_{z} and E_{xm} are real constants

Solution: The instantaneous expression of a phasor can always be found by using(6.47). Notice that there is no complex quantities in an instantaneous expression. Therefore, all the complex quantities in the phasor expression should be converted into exponential form. In this problem, we notice that

$$j=\mathrm{e}^{\mathrm{j}\pi/2}$$

Therefore, we have

$$\boldsymbol{E}(z,t)=\mathrm{Re}\left[\boldsymbol{e}_{\mathrm{x}}\mathrm{j}E_{\mathrm{xm}}\cos(k_{\mathrm{z}}z)\mathrm{e}^{\mathrm{j}\omega t}\right]=\mathrm{Re}\left[\boldsymbol{e}_{\mathrm{x}}E_{\mathrm{xm}}\cos(k_{\mathrm{z}}z)\mathrm{e}^{\mathrm{j}\omega t+\mathrm{j}\pi/2}\right]$$

$$=\boldsymbol{e}_{\mathrm{x}}E_{\mathrm{xm}}\cos(k_{\mathrm{z}}z)\cos\left(\omega t+\frac{\pi}{2}\right)=-\boldsymbol{e}_{\mathrm{x}}E_{\mathrm{xm}}\cos(k_{\mathrm{z}}z)\sin\omega t$$

6.9.2 Phasor Form of Maxwell's Equations (相量形式的麦克斯韦方程)

We now convert the Maxwell's Equations in (6.10) into the time-harmonic form as follows.

$$\nabla\times\boldsymbol{E}=-\mathrm{j}\omega\boldsymbol{B} \tag{6.59a}$$

$$\nabla\times\boldsymbol{H}=\boldsymbol{J}+\mathrm{j}\omega\boldsymbol{D} \tag{6.59b}$$

$$\nabla\cdot\boldsymbol{D}=\rho_{\mathrm{v}} \tag{6.59c}$$

$$\nabla\cdot\boldsymbol{B}=0 \tag{6.59d}$$

Here, the fields $\boldsymbol{E},\boldsymbol{H},\boldsymbol{D},\boldsymbol{B}$ and sources ρ_{v} and $\boldsymbol{J}$ are all phasors.

The integral form of phasor Maxwell's equation can be obtained by rewriting(6.11) as

$$\oint_C\boldsymbol{E}\cdot\mathrm{d}\boldsymbol{l}=-\mathrm{j}\omega\int_S\boldsymbol{B}\cdot\mathrm{d}\boldsymbol{s} \tag{6.60a}$$

$$\oint_C\boldsymbol{H}\cdot\mathrm{d}\boldsymbol{l}=\int_S(\boldsymbol{J}+\mathrm{j}\omega\boldsymbol{D})\cdot\mathrm{d}\boldsymbol{s} \tag{6.60b}$$

$$\oint_S\boldsymbol{D}\cdot\mathrm{d}\boldsymbol{s}=\int_V\rho_{\mathrm{v}}\mathrm{d}v \tag{6.60c}$$

$$\oint_S\boldsymbol{B}\cdot\mathrm{d}\boldsymbol{s}=0 \tag{6.60d}$$

6.9.3 Nonhomogeneous Helmholtz's Equations and Phasor Form of Retarded Potential(非齐次亥姆霍兹方程,推迟位的相量形式)

By replacing the second derivatives with respective to t with$(\mathrm{j}\omega)^2=-\omega^2$ in(6.24) and(6.25), we obtain the time-harmonic wave equations for scalar potential φ and vector potential $\boldsymbol{A}$ as

$$\nabla^2\varphi + k^2\varphi = -\frac{\rho_v}{\varepsilon} \tag{6.61}$$

$$\nabla^2\boldsymbol{A} + k^2\boldsymbol{A} = -\mu\boldsymbol{J} \tag{6.62}$$

where

$$k = \omega\sqrt{\mu\varepsilon} = \frac{\omega}{u} \tag{6.63}$$

is called the **wavenumber**(波数). (6.61) and(6.62) are referred to as **nonhomogeneous Helmholtz's equations**(非齐次亥姆霍兹方程). The Lorentz condition for potentials, (6.23), is now

$$\nabla\cdot\boldsymbol{A} + \mathrm{j}\omega\mu\varepsilon\varphi = 0 \tag{6.64}$$

The solutions of(6.61) and(6.62) are the phasor forms of(6.38) and(6.39), respectively:

$$\varphi(\boldsymbol{r}) = \frac{1}{4\pi\varepsilon}\int_{V'}\frac{\rho(\boldsymbol{r}')\mathrm{e}^{-\mathrm{j}kR}}{R}\mathrm{d}v' \tag{6.65}$$

$$\boldsymbol{A}(\boldsymbol{r}) = \frac{\mu}{4\pi}\int_{V'}\frac{\boldsymbol{J}(\boldsymbol{r}')\mathrm{e}^{-\mathrm{j}kR}}{R}\mathrm{d}v' \tag{6.66}$$

These are the expressions for the retarded scalar and vector potentials due to time-harmonic sources. Now the Taylor-series expansion for the exponential factor $\mathrm{e}^{-\mathrm{j}kR}$ is

$$\mathrm{e}^{-\mathrm{j}kR} = 1 - \mathrm{j}kR + \frac{k^2R^2}{2} + \cdots \tag{6.67}$$

where k, defined in(6.63), can be expressed in terms of the wavelength $\lambda = u/f$ in the medium. We have

$$k = \frac{2\pi f}{u} = \frac{2\pi}{\lambda} \tag{6.68}$$

Thus, if

$$kR = 2\pi\frac{R}{\lambda} << 1 \tag{6.69}$$

or if the distance R is very small in comparison to the wavelength λ, $\mathrm{e}^{-\mathrm{j}kR}$ can be approximated by 1. (6.65) and(6.66) then simplify to the static expressions in(2.58) and(5.28), from which quasi-static fields can be obtained.

With the phasor solution of $\varphi(\boldsymbol{r})$ and $\boldsymbol{A}(\boldsymbol{r})$ found from(6.65) and(6.66), phasor solution of $\boldsymbol{E}$ can be found by using the phasor form of(6.20):

$$\boldsymbol{E} = -\nabla\varphi - \mathrm{j}\omega\boldsymbol{A} \tag{6.70}$$

The phasor form of **B** can be found from the curl of **A** as given in(6.17).

Example 6-8 Radiation of the Hertzian dipole(赫兹偶极子的辐射). As is shown in Figure 6-5, two small PEC spheres are connected by a short wire with length $\mathrm{d}l$. The wire carries a uniform sinusoidally varying current $I(t) = I_0\cos\omega t$. Find out the electromagnetic fields produced by the dipole.

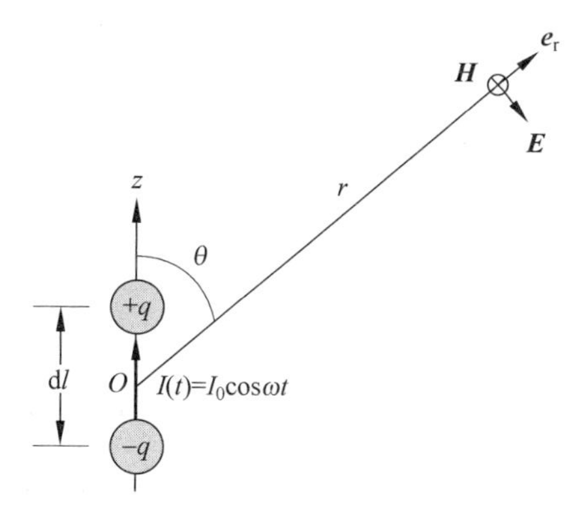

Figure 6-5 A Hertzian dipole

Solution: With the sinusoidal current $I(t)$ as is shown in Figure 6-5, the two spheres contain equal and opposite charges $\pm q(t)$ that also varies sinusoidally. And we have

$$I(t) = \frac{\mathrm{d}q(t)}{\mathrm{d}t}$$

The phasor form of this equation is easily found to be

$$I = \mathrm{j}\omega Q \quad \text{or} \quad Q = \frac{I}{\mathrm{j}\omega}$$

The positive and negative charges separated by a distance $\mathrm{d}l$ form a time-varying electric dipole with a phasor form of the electric dipole moment to be

$$\boldsymbol{p} = \boldsymbol{e}_z Q\mathrm{d}l = \boldsymbol{e}_z\left(\frac{I}{\mathrm{j}\omega}\right)\mathrm{d}l$$

Such a time-harmonic dipole is called a **Hertzian Dipole** (赫兹偶极子). It is a fundamental model of the antenna that produces electromagnetic waves that can be calculated as follows. First, by using(6.66), we find the retarded vector potential to be

$$\boldsymbol{A} = \boldsymbol{e}_z\frac{\mu_0 I\mathrm{d}l}{4\pi}\left(\frac{\mathrm{e}^{-\mathrm{j}k_0 r}}{r}\right)$$

where $k_0 = \sqrt{\varepsilon_0\mu_0}$ is the wave number in the free space. Notice that

$$\boldsymbol{e}_z = \boldsymbol{e}_r\cos\theta - \boldsymbol{e}_\theta\sin\theta$$

The vector **A** only have r and θ components as

$$A_r = \frac{\mu_0 I\mathrm{d}l}{4\pi}\left(\frac{\mathrm{e}^{-\mathrm{j}k_0 r}}{r}\right)\cos\theta$$

$$A_\theta = -\frac{\mu_0 I\mathrm{d}l}{4\pi}\left(\frac{\mathrm{e}^{-\mathrm{j}k_0 r}}{r}\right)\sin\theta$$

Therefore, from(6.17), we have

$$\boldsymbol{H}=\frac{1}{\mu_0}\nabla\times\boldsymbol{A}=\frac{1}{\mu_0}\boldsymbol{e}_\phi\left[\frac{1}{r}\frac{\partial}{\partial r}(rA_\theta)-\frac{1}{r}\frac{\partial}{\partial\theta}A_r\right]$$

$$=\boldsymbol{e}_\phi\frac{\mathrm{j}k_0I\mathrm{d}l}{4\pi}\left(1+\frac{1}{\mathrm{j}k_0r}\right)\left(\frac{\mathrm{e}^{-\mathrm{j}k_0r}}{r}\right)\sin\theta \tag{6.71}$$

By using(6.59b), we have

$$\boldsymbol{E}=\frac{1}{\mathrm{j}\omega\varepsilon_0}\nabla\times\boldsymbol{H}=\frac{1}{\mathrm{j}\omega\varepsilon_0}\left[\boldsymbol{e}_r\frac{1}{r\sin\theta}\frac{\partial}{\partial\theta}(H_\phi\sin\theta)-\boldsymbol{e}_\theta\frac{1}{r}\frac{\partial}{\partial r}(rH_\phi)\right] \tag{6.72}$$

which gives

$$E_r=2\sqrt{\frac{\mu_0}{\varepsilon_0}}\frac{I\mathrm{d}l}{4\pi}\left(\frac{1}{r}\right)\left(1+\frac{1}{\mathrm{j}k_0r}\right)\left(\frac{\mathrm{e}^{-\mathrm{j}k_0r}}{r}\right)\cos\theta \tag{6.73a}$$

$$E_\theta=\frac{\mathrm{j}k_0I\mathrm{d}l}{4\pi}\sqrt{\frac{\mu_0}{\varepsilon_0}}\left[1+\frac{1}{\mathrm{j}k_0r}-\frac{1}{k_0^2r^2}\right]\left(\frac{\mathrm{e}^{-\mathrm{j}k_0r}}{r}\right)\sin\theta \tag{6.73b}$$

(6.71) and(6.72) constitute the electromagnetic fields produced by the Hertzian dipole. Notice that when the observation point is enough far away from the dipole so that $k_0r>>1$ is satisfied, we can simply neglect the terms that decrease faster than $1/r$(that is, neglecting the $1/r^2$ and $1/r^3$ terms). In this case, (6.71) and(6.72) reduce to

$$\boldsymbol{H}=\boldsymbol{e}_\phi\frac{\mathrm{j}k_0I\mathrm{d}l}{4\pi}\left(\frac{\mathrm{e}^{-\mathrm{j}k_0r}}{r}\right)\sin\theta \tag{6.74a}$$

$$\boldsymbol{E}=\boldsymbol{e}_\theta\frac{\mathrm{j}k_0I\mathrm{d}l}{4\pi}\sqrt{\frac{\mu_0}{\varepsilon_0}}\left(\frac{\mathrm{e}^{-\mathrm{j}k_0r}}{r}\right)\sin\theta \tag{6.74b}$$

which is usually referred to as the **far-zone fields**(远区场) or **far fields**(远场) of the Hertzian dipole. Notice that

$$\frac{E_\theta}{H_\phi}=\sqrt{\frac{\mu_0}{\varepsilon_0}}\approx 120\pi(\Omega) \tag{6.75}$$

The constant $\sqrt{\mu_0/\varepsilon_0}$ is called the **intrinsic impedance**(本征阻抗) of the free space, which will be introduced in the next chapter.

6.9.4 Homogeneous Helmholtz's Equations in Source-Free Region(无源区域的齐次亥姆霍兹方程)

In a simple, non-conducting source-free medium characterized by $\rho_v=0, \boldsymbol{J}=0, \sigma=0$, the time-harmonic Maxwell's equations become

$$\nabla\times\boldsymbol{E}=-\mathrm{j}\omega\mu\boldsymbol{H} \tag{6.76a}$$

$$\nabla\times\boldsymbol{H}=\mathrm{j}\omega\varepsilon\boldsymbol{E} \tag{6.76b}$$

$$\nabla\cdot\boldsymbol{E}=0 \tag{6.76c}$$

$$\nabla\cdot\boldsymbol{H}=0 \tag{6.76d}$$

By taking the curl of(6.76a) and use(6.76b), we have

$$\nabla\times\nabla\times\boldsymbol{E}=-\mathrm{j}\omega\mu(\nabla\times\boldsymbol{H})=\omega^2\mu\varepsilon\boldsymbol{E}$$

Using the identity $\nabla\times\nabla\times\boldsymbol{E}=\nabla(\nabla\cdot\boldsymbol{E})-\nabla^2\boldsymbol{E}$ and(6.76c), we have

$$\nabla^2\boldsymbol{E}+\omega^2\mu\varepsilon\boldsymbol{E}=0 \tag{6.77}$$

Similarly, we can obtain the same equation for $\boldsymbol{H}$:

$$\nabla^2\boldsymbol{H}+\omega^2\mu\varepsilon\boldsymbol{H}=0 \tag{6.78}$$

With the definition of the wavenumber k in(6.63), (6.77) and(6.78) can be written as

$$\nabla^2\boldsymbol{E}+k^2\boldsymbol{E}=0 \tag{6.79}$$

$$\nabla^2\boldsymbol{H}+k^2\boldsymbol{H}=0 \tag{6.80}$$

which are called **homogeneous vector Helmholtz's equations**(**齐次矢量亥姆霍兹方程**). The homogeneous vector Helmholtz's equations(6.79) and(6.80) can also be derived directly from the homogeneous wave equations(6.43) and(6.44) by replacing $\partial^2/\partial t^2$ with $-\omega^2$.

In Cartesian coordinates, (6.79) and (6.80) can each be decomposed into three homogeneous scalar Helmholtz's equations. In the next chapter, we will study the plane electromagnetic waves as the solutions of homogeneous Helmholtz's equations.

6.9.5 Time-Average Power and Energy(时间平均功率与能量)

With introduction of the phasor notation, an instantaneous quantity can be expressed as the real part of the product of the phasor quantity and $\mathrm{e}^{\mathrm{j}\omega t}$, as is shown in(6.52) and(6.56). However, it is important to note that this relation between the instantaneous and phasor expressions is valid only for linear quantities with sinusoidal time dependence. For quantities involving nonlinear operations, such as a product of two sinusoidal quantities, we cannot define such phasor forms as we did for the linear quantities. For example, the instantaneous Poynting vector $\boldsymbol{S}$ is defined as the cross product of $\boldsymbol{E}$ and $\boldsymbol{H}$:

$$\boldsymbol{S}(\boldsymbol{r},t)=\boldsymbol{E}(\boldsymbol{r},t)\times\boldsymbol{H}(\boldsymbol{r},t) \tag{6.81}$$

For the time-harmonic $\boldsymbol{E}$ and $\boldsymbol{H}$, we have

$$\boldsymbol{E}(\boldsymbol{r},t)=\mathrm{Re}[\boldsymbol{E}(\boldsymbol{r})\mathrm{e}^{\mathrm{j}\omega t}]$$

$$\boldsymbol{H}(\boldsymbol{r},t)=\mathrm{Re}[\boldsymbol{H}(\boldsymbol{r})\mathrm{e}^{\mathrm{j}\omega t}]$$

where $\boldsymbol{E}(\boldsymbol{r})$ and $\boldsymbol{H}(\boldsymbol{r})$ are the phasors of $\boldsymbol{E}(\boldsymbol{r},t)$ and $\boldsymbol{H}(\boldsymbol{r},t)$. Apparently, we have

$$\boldsymbol{S}(x,y,z,t)=\mathrm{Re}[\boldsymbol{E}(x,y,z)\mathrm{e}^{\mathrm{j}\omega t}]\times\mathrm{Re}[\boldsymbol{H}(x,y,z)\mathrm{e}^{\mathrm{j}\omega t}] \tag{6.82}$$

However, there is no such phasor Poynting vector $\boldsymbol{S}(\boldsymbol{r})$ that satisfies $\boldsymbol{S}(\boldsymbol{r},t)=\mathrm{Re}[\boldsymbol{S}(\boldsymbol{r})\mathrm{e}^{\mathrm{j}\omega t}]$, and it can be easily verified that

$$\mathrm{Re}[\boldsymbol{E}(\boldsymbol{r})\mathrm{e}^{\mathrm{j}\omega t}]\times\mathrm{Re}[\boldsymbol{H}(\boldsymbol{r})\mathrm{e}^{\mathrm{j}\omega t}]\neq\mathrm{Re}[\boldsymbol{E}(\boldsymbol{r})\times\boldsymbol{H}(\boldsymbol{r})\mathrm{e}^{\mathrm{j}\omega t}] \tag{6.83}$$

The reason is that the Poynting vector $\boldsymbol{S}$, as the nonlinear product of two sinusoidal quantities, is not a simple sinusoidal quantity at the same frequency.[①] This can be seen by a close examination of the relation between instantaneous $\boldsymbol{S}$ and phasor $\boldsymbol{E}$ and $\boldsymbol{H}$ as below.

---

① 前面关于时谐场的讨论中,假设场在特定频点时谐变化,然后通过欧拉公式引入了相量表示式。然而,时谐场中不是所有的物理量都时谐变化。例如,当电场强度 $\boldsymbol{E}$ 正弦变化时,电场储能密度 $\varepsilon|\boldsymbol{E}|^2/2$ 就不是正弦或余弦变化,这是因为乘积是非线性运算,时谐函数乘积的结果不再是按照单一频率变化的时谐函数。因此储能密度、坡印廷矢量等物理量都没有直接对应的相量表达式。

We first note that, for any complex vector $\boldsymbol{F}$ and $\boldsymbol{G}$, we generally have

$$\mathrm{Re}(\boldsymbol{F})=\frac{1}{2}(\boldsymbol{F}+\boldsymbol{F}^*),\quad \mathrm{Re}(\boldsymbol{G})=\frac{1}{2}(\boldsymbol{G}+\boldsymbol{G}^*)$$

and as a result,

$$\begin{aligned}\mathrm{Re}(\boldsymbol{F})\times\mathrm{Re}(\boldsymbol{G})&=\frac{1}{2}(\boldsymbol{F}+\boldsymbol{F}^*)\times\frac{1}{2}(\boldsymbol{G}+\boldsymbol{G}^*)\\&=\frac{1}{4}[(\boldsymbol{F}\times\boldsymbol{G}^*+\boldsymbol{F}^*\times\boldsymbol{G})+(\boldsymbol{F}\times\boldsymbol{G}+\boldsymbol{F}^*\times\boldsymbol{G}^*)]\\&=\frac{1}{2}\mathrm{Re}(\boldsymbol{F}\times\boldsymbol{G}^*)+\frac{1}{2}\mathrm{Re}(\boldsymbol{F}\times\boldsymbol{G})\end{aligned}\tag{6.84}$$

Replacing $\boldsymbol{F}$ and $\boldsymbol{G}$ in (6.84) with $\boldsymbol{E}(\boldsymbol{r})\mathrm{e}^{\mathrm{j}\omega t}$ and $\boldsymbol{H}(\boldsymbol{r})\mathrm{e}^{\mathrm{j}\omega t}$ respectively, we have

$$\begin{aligned}\boldsymbol{S}(\boldsymbol{r},t)&=\mathrm{Re}[\boldsymbol{E}(\boldsymbol{r})\mathrm{e}^{\mathrm{j}\omega t}]\times\mathrm{Re}[\boldsymbol{H}(\boldsymbol{r})\mathrm{e}^{\mathrm{j}\omega t}]\\&=\frac{1}{2}\mathrm{Re}[\boldsymbol{E}(\boldsymbol{r})\times\boldsymbol{H}^*(\boldsymbol{r})]+\frac{1}{2}\mathrm{Re}[\boldsymbol{E}(\boldsymbol{r})\times\boldsymbol{H}(\boldsymbol{r})\mathrm{e}^{\mathrm{j}2\omega t}]\end{aligned}\tag{6.85}$$

(6.85) clearly explains why we have the inequality of (6.83). The right side of (6.83) has no physical meaning. But the two terms on the right side of (6.85) have clear physical meaning. Specifically, the first term is time-independent, and the second term is essentially a cosine function with frequency 2ω. Mathematically, the product of two cosine functions with the same frequency results in a function with a DC component and an AC component at doubled frequency. Therefore, the first term on the right side of (6.85) is the time-average of the $\boldsymbol{S}$ vector (denoted by $\boldsymbol{S}_{\mathrm{av}}$), and we have

$$\boldsymbol{S}_{\mathrm{av}}(\boldsymbol{r})=\frac{1}{T}\int_0^T\boldsymbol{S}(\boldsymbol{r},t)\,\mathrm{d}t=\frac{1}{2}\mathrm{Re}[\boldsymbol{E}(\boldsymbol{r})\times\boldsymbol{H}(\boldsymbol{r})^*]\tag{6.86}$$

where T is the period of the $\boldsymbol{S}$ vector. We can take T to be the shortest period of the $\boldsymbol{E}$ and $\boldsymbol{H}$ fields. However, the shortest period of $\boldsymbol{S}$ is half of the period of the $\boldsymbol{E}$ and $\boldsymbol{H}$ fields. This can also be seen by the second term in the right side of (6.85), which corresponds to the AC component at two times the frequency of the $\boldsymbol{E}$ and $\boldsymbol{H}$ fields. Note that $\boldsymbol{S}_{\mathrm{av}}(\boldsymbol{r})$ is a real vector but not dependent on time t.[①]

In a similar way, by using (6.84), we can derive expressions for time-average electric energy density w_{ave}, time-average magnetic energy density w_{avm}, and average ohmic power dissipation density $p_{\mathrm{av}\sigma}$ in time-harmonic fields as

$$w_{\mathrm{ave}}(\boldsymbol{r})=\frac{1}{T}\int_0^T\left[\frac{1}{2}\boldsymbol{E}(\boldsymbol{r},t)\cdot\boldsymbol{D}(\boldsymbol{r},t)\right]\mathrm{d}t=\frac{1}{4}\mathrm{Re}[\boldsymbol{E}(\boldsymbol{r})\cdot\boldsymbol{D}^*(\boldsymbol{r})]\tag{6.87a}$$

$$w_{\mathrm{avm}}(\boldsymbol{r})=\frac{1}{T}\int_0^T\left[\frac{1}{2}\boldsymbol{H}(\boldsymbol{r},t)\cdot\boldsymbol{B}(\boldsymbol{r},t)\right]\mathrm{d}t=\frac{1}{4}\mathrm{Re}[\boldsymbol{H}(\boldsymbol{r})\cdot\boldsymbol{B}^*(\boldsymbol{r})]\tag{6.87b}$$

---

① 时谐场的坡印廷矢量是时谐变化的电场和时谐变化的磁场的叉乘。数学上,两个同频的正/余弦函数相乘,得到的结果在一般情况下包含两项,一项是常数,另一项是两倍频的正/余弦函数,分别对应式(6.85)第二个等号右边的第一项和第二项。由于两倍频项的时间平均值为零,常数项即为乘积结果的时间平均值,也就是式(6.86)。

$$p_{\text{av}\sigma}(\boldsymbol{r}) = \frac{1}{T}\int_0^T [\boldsymbol{E}(\boldsymbol{r},t)\cdot \boldsymbol{J}(\boldsymbol{r},t)]\,\mathrm{d}t = \frac{1}{2}\mathrm{Re}[\boldsymbol{E}(\boldsymbol{r})\cdot \boldsymbol{J}^*(\boldsymbol{r})] \tag{6.87c}$$

In a simple medium with constant ε, μ and σ, (6.87a)-(6.87c) become

$$w_{\text{ave}}(\boldsymbol{r}) = \frac{1}{4}\boldsymbol{E}(\boldsymbol{r})\cdot \boldsymbol{D}(\boldsymbol{r})^* = \frac{1}{4}\varepsilon|\boldsymbol{E}(\boldsymbol{r})|^2 = \frac{1}{4\varepsilon}|\boldsymbol{D}(\boldsymbol{r})|^2 \tag{6.88a}$$

$$w_{\text{avm}}(\boldsymbol{r}) = \frac{1}{4}\boldsymbol{H}(\boldsymbol{r})\cdot \boldsymbol{B}(\boldsymbol{r})^* = \frac{1}{4}\mu|\boldsymbol{H}(\boldsymbol{r})|^2 = \frac{1}{4\mu}|\boldsymbol{B}(\boldsymbol{r})|^2 \tag{6.88b}$$

$$p_{\text{av}\sigma}(\boldsymbol{r}) = \frac{1}{2}\boldsymbol{E}(\boldsymbol{r})\cdot \boldsymbol{J}(\boldsymbol{r})^* = \frac{1}{2}\sigma|\boldsymbol{E}(\boldsymbol{r})|^2 = \frac{1}{2\sigma}|\boldsymbol{J}(\boldsymbol{r})|^2 \tag{6.88c}$$

Example 6-9 The far field expressions of the Hertzian dipole are given in (6.73) and (6.74).

(1) Write the expression for instantaneous Poynting vector in the far-zone.

(2) Find the total average power radiated by the Hertzian dipole.

Solution:

(1) The instantaneous Poynting vector is

$$\begin{aligned}\boldsymbol{S}(r,\theta,t) &= \mathrm{Re}[\boldsymbol{E}\mathrm{e}^{\mathrm{j}\omega t}]\times \mathrm{Re}[\boldsymbol{H}\mathrm{e}^{\mathrm{j}\omega t}]\\ &= \left[-\boldsymbol{e}_\theta \frac{k_0 I\mathrm{d}l}{4\pi}\sqrt{\frac{\mu_0}{\varepsilon_0}}\left(\frac{\sin\theta}{r}\right)\sin(\omega t - k_0 r)\right]\times\left[-\boldsymbol{e}_\phi \frac{k_0 I\mathrm{d}l}{4\pi}\left(\frac{\sin\theta}{r}\right)\sin(\omega t - k_0 r)\right]\\ &= \boldsymbol{e}_\mathrm{r}\left(\frac{k_0 I\mathrm{d}l}{4\pi r}\right)^2\sqrt{\frac{\mu_0}{\varepsilon_0}}\sin^2\theta\sin^2(\omega t - k_0 r)\end{aligned} \tag{6.89}$$

(2) From (a), it is obvious that the average power density vector is

$$\boldsymbol{S}_{\text{av}}(r,\theta) = \boldsymbol{e}_\mathrm{r}\frac{1}{2}\left(\frac{k_0 I\mathrm{d}l}{4\pi r}\right)^2\sqrt{\frac{\mu_0}{\varepsilon_0}}\sin^2\theta \tag{6.90}$$

which can also be obtained by using (6.86). The total average power radiated is obtained by integrating $\boldsymbol{S}_{\text{av}}$ over the surface of a sphere with radius r:

$$\begin{aligned}\text{Total } P_{\text{av}} &= \oint_S \boldsymbol{S}_{\text{av}}(r,\theta)\cdot \mathrm{d}\boldsymbol{s} = \int_0^{2\pi}\int_0^{\pi}\left[\frac{1}{2}\left(\frac{k_0 I\mathrm{d}l}{4\pi r}\right)^2\sqrt{\frac{\mu_0}{\varepsilon_0}}\sin^2\theta\right] r^2\sin\theta\mathrm{d}\theta\mathrm{d}\phi\\ &= \frac{1}{12\pi}(k_0 I\mathrm{d}l)^2\sqrt{\frac{\mu_0}{\varepsilon_0}} = 40\pi^2\left(\frac{\mathrm{d}l}{\lambda_0}\right)^2 I^2\end{aligned} \tag{6.91}$$

in which the equations (6.75) and (6.68) are used. By comparing this expression with the Ohm's law in the circuit theory, we conclude that a Hertzian dipole is equivalent to a resistor with resistance

$$R = 80\pi^2\left(\frac{\mathrm{d}l}{\lambda_0}\right)^2 \tag{6.92}$$

which is called the **radiation resistance** (**辐射电阻**) of the dipole. It is observed that the resistance is proportional to the square of the length of the dipole. For Hertzian dipole, $\mathrm{d}l \ll \lambda_0$, which makes the resistance R a very small number. Therefore, Hertzian dipole is a very poor radiator. ①

6.9.6 Phasor Form of Poynting's Theorem (相量形式的坡印廷定理)

Here, we derive the phasor form of Poynting's Theorem from (6.59), the phasor form of Maxwell's equations. We start with the vector identity

$$\nabla\cdot(\boldsymbol{E}\times\boldsymbol{H}^*)=\boldsymbol{H}^*\cdot(\nabla\times\boldsymbol{E})-\boldsymbol{E}\cdot(\nabla\times\boldsymbol{H}^*) \tag{6.93}$$

Substitute (6.59a) and conjugate of (6.59b) in (6.93), then

$$\nabla\cdot(\boldsymbol{E}\times\boldsymbol{H}^*)=\boldsymbol{H}^*\cdot(-\mathrm{j}\omega\boldsymbol{B})-\boldsymbol{E}\cdot(\boldsymbol{J}^*-\mathrm{j}\omega\boldsymbol{D}^*) \tag{6.94}$$

By integrating both sides of (6.94) over a volume V bounded by surface S and using divergence theorem, we obtain:

$$-\frac{1}{2}\oint_S(\boldsymbol{E}\times\boldsymbol{H}^*)\cdot\mathrm{d}\boldsymbol{s}=\frac{1}{2}\int_V\boldsymbol{E}\cdot\boldsymbol{J}^*\,\mathrm{d}v+2\mathrm{j}\omega\int_V\left(\frac{1}{4}\boldsymbol{H}^*\cdot\boldsymbol{B}-\frac{1}{4}\boldsymbol{E}\cdot\boldsymbol{D}^*\right)\mathrm{d}v \tag{6.95}$$

A $(-1/2)$ factor is applied to both sides of the equation to make the right-side terms consistent with power and energy expressions. (6.95) is the **phasor form of Poynting's theorem (相量形式的坡印廷定理)**. In a simple medium with constant ε, μ and σ, (6.95) becomes

$$\begin{aligned}-\frac{1}{2}\oint_S(\boldsymbol{E}\times\boldsymbol{H}^*)\cdot\mathrm{d}\boldsymbol{s}&=\frac{1}{2}\int_V\sigma|\boldsymbol{E}|^2\mathrm{d}v+2\mathrm{j}\omega\int_V\left(\frac{1}{4}\mu|\boldsymbol{H}|^2-\frac{1}{4}\varepsilon|\boldsymbol{E}|^2\right)\mathrm{d}v\\&=\int_V p_{\mathrm{av}\sigma}\mathrm{d}v+2\mathrm{j}\omega\int_V(w_{\mathrm{avm}}-w_{\mathrm{ave}})\,\mathrm{d}v\end{aligned} \tag{6.96}$$

The first term in the right side of (6.96) is a real quantity, representing the total ohmic power dissipation within the volume V. The second term in the right side of (6.96) is a purely imaginary quantity, related to the total stored magnetic and electric energy. By taking the real part of the equation (6.96) and noticing (6.86), we have

$$\mathrm{Re}\left[-\frac{1}{2}\oint_S(\boldsymbol{E}\times\boldsymbol{H}^*)\cdot\mathrm{d}\boldsymbol{s}\right]=-\oint_S\boldsymbol{S}_{\mathrm{av}}\cdot\mathrm{d}\boldsymbol{s}=\int_V p_{\mathrm{av}\sigma}\mathrm{d}v=\text{Total power loss within }V \tag{6.97}$$

which states that the time-average power flow into the volume V through surface S equals the power dissipated within V. This is consistent with the instantaneous form of Poynting's theorem because in the time-harmonic case, the time average stored electric and magnetic energy does not charge.

By taking the imaginary part of equation (6.96), we have

$$\mathrm{Im}\left[-\frac{1}{2}\oint_S(\boldsymbol{E}\times\boldsymbol{H}^*)\cdot\mathrm{d}\boldsymbol{s}\right]=2\omega\int_V(w_{\mathrm{avm}}-w_{\mathrm{ave}})\,\mathrm{d}v=\text{Reactive power entering }V \tag{6.98}$$

---

① 例6-8计算了赫兹偶极子作为电流源辐射生成的电磁波，例6-9进一步得出了该电磁波携带的功率流（坡印廷矢量）。式（6.90）表明，距离赫兹偶极子足够远时，赫兹偶极子向外辐射的功率流密度与距离平方成反比。这是合理且必须的，因为以赫兹偶极子的位置为球心的球面面积与距离的平方成正比，在该球面上对式（6.90）作通量积分得到的总功率流就是赫兹偶极子向外辐射的总功率，应该为一个常数。式（6.91）给出了该常数总功率的表达式。由于能量守恒，该辐射出去的功率必然来源于维持赫兹偶极子电流的馈源（电压源或者电流源），那么对于馈源来说，赫兹偶极子就等效于一个消耗功率的电阻，也就是式（6.92）给出的辐射电阻。

which is proportional to the difference between average stored magnetic energy and electric energy. This power is reactive, so it can be interpreted as the power flowing back and forth through the surface S to supply the instantaneous changes in net stored energy in the volume V. Particularly, if $w_{\text{avm}} = w_{\text{ave}}$, the reactive power entering V is zero and the surface integral of($\boldsymbol{E} \times \boldsymbol{H}^*$) is real, which means the fields within the volume V is at resonance(谐振状态). The volume V behaves like a resistor to the outside. If $w_{\text{avm}} > w_{\text{ave}}$, the surface integral of($\boldsymbol{E} \times \boldsymbol{H}^*$) has a positive imaginary part. The stored energy within the volume is dominant by magnetic energy. The volume V behaves like an inductor to the outside. If $w_{\text{avm}} < w_{\text{ave}}$, the surface integral of($\boldsymbol{E} \times \boldsymbol{H}^*$) has a negative imaginary part. The stored energy within the volume is primarily electric energy. The volume V behaves like a capacitor to the outside.

Summary

Concepts

Displacement current(位移电流)

Perfect electric conductor(理想导体)

Vector and scalar potential functions for time-varying fields(时变场的矢量与标量位函数)

Lorentz condition/gauge(洛伦兹规范)

Retarded scalar/vector potential(标量/矢量推迟位)

Electromagnetic wave(电磁波)

Poynting vector/power density vector(坡印廷矢量/功率密度矢量)

Time-harmonic Fields(时谐场)

Phasor(相量,复矢量).

Laws & Theorems

Faraday's law of electromagnetic induction(法拉第电磁感应定律)

Generalized Ampere's circuital Law(广义安培电路定律)

Maxwell's equations, differential form and integral form(麦克斯韦方程组,微分与积分形式)

Boundary conditions of time-varying electromagnetic fields between ①two general media; ②two perfect dielectrics; ③perfect dielectric and perfect conductor

Wave equations and their solutions(波动方程及其解)

Helmholtz's equations(亥姆霍兹方程)

Poynting's theorem, instantaneous and phasor form(坡印廷定理,瞬时形式和相量形式).

Methods

Find electric field given a time-varying magnetic field distribution, and vice versa;

Find charge and surface current distribution given electromagnetic fields on the boundary;

Convert time-harmonic instantaneous quantities to phasor forms, and vice versa;

Find time-average power and energy densities using phasor field expressions.

Review Questions

6.1 分别写出微分形式和积分形式的麦克斯韦方程组,并说明其物理意义。

6.2 麦克斯韦方程组的四个方程式是相互独立的吗?为什么?

6.3 位移电流的物理意义是什么?

6.4 时变电磁场的边界条件有哪些?其与静态场的边界条件有什么不同?

6.5 什么是PEC?PEC内部能存在电场或磁场吗?为什么?

6.6 PEC表面电位移矢量的边界条件是什么?其物理意义是什么?

6.7 PEC表面磁场强度的边界条件是什么?其物理意义是什么?

6.8 时变电磁场的位函数φ和$\boldsymbol{A}$是如何定义的?引入这两个位函数有什么作用?

6.9 写出由位函数φ和$\boldsymbol{A}$计算时变电场$\boldsymbol{E}$和$\boldsymbol{B}$的表达式。

6.10 写出洛伦兹规范表达式。引入该规范有什么作用?

6.11 什么是准静态场?准静态场是麦克斯韦方程的严格解么?

6.12 写出位函数φ和$\boldsymbol{A}$满足的达朗贝尔方程(非齐次波动方程)。

6.13 什么是标量推迟位?什么是矢量推迟位?其物理含义是什么?

6.14 如何推导得出电磁波的传播速度?

6.15 写出无源区域电场$\boldsymbol{E}$和磁场$\boldsymbol{H}$满足的波动方程。

6.16 什么是坡印廷矢量?什么是坡印廷定理?

6.17 什么是时谐场?什么是相量?引入相量有什么用处?相量是时间的函数还是频率的函数?

6.18 如何由时谐场的瞬态表达式得到相应的相量表达式?如何由相量表达式得到相应的瞬态表达式?

6.19 导电媒质中若存在时谐电磁场,同一位置的传导电流和位移电流是同相位吗?

6.20 写出简单媒质中时谐电磁场满足的相量形式的麦克斯韦方程组。

6.21 写出自由空间无源区域的时谐电场$\boldsymbol{E}$满足的相量形式的麦克斯韦方程组。

6.22 什么是波数?

6.23 写出时谐场推迟位的相量表达式。

6.24 写出以相量$\boldsymbol{E}$和$\boldsymbol{H}$表示的瞬时和时间平均坡印廷矢量表达式。

6.25 写出以相量$\boldsymbol{E}$和$\boldsymbol{H}$表示的瞬时和时间平均电、磁场储能密度表达式。

6.26 写出以相量$\boldsymbol{E}$和$\boldsymbol{J}$表示的瞬时和时间平均欧姆损耗密度表达式。

6.27 写出相量形式的坡印廷定理表达式,并阐述其物理意义。

Problems

6.1 A conducting wire loop has a rectangular shape with its four corners located at (0,0,0),(3,4,0),(3,4,4),and (0,0,4) respectively. The magnetic flux density within the loop is

$$\boldsymbol{B} = \boldsymbol{e}_x 3.4\sin 250t + \boldsymbol{e}_y 2.5\sin 250t + \boldsymbol{e}_z 0.5\cos 400t$$

Determine ①the magnetic flux linking the loop and ②the induced current in the loop if its resistance is 4Ω.

6.2 Derive the two divergence Maxwell's equations(6.10c),(6.10d) from the two curl Maxwell's equations,(6.10a),(6.10b) together with the equation of continuity(6.6).

6.3 In a source-free simple medium, if both $\boldsymbol{E}$ and $\boldsymbol{H}$ have no variation along x-direction or y-direction, prove that both $\boldsymbol{E}$ and $\boldsymbol{H}$ have no z-component(i. e. $E_z=0$ and $H_z=0$).

6.4 Prove that vector field $\boldsymbol{E}=\boldsymbol{e}_x E_0\cos(\omega t-\omega x/c)$ satisfies the wave equation in free space but does not satisfy Maxwell's equations.

6.5 Derive the boundary conditions at the interface between free space and a magnetic material with relative permeability approaching infinity.

6.6 Without using the two integral equations (6.11c), (6.11d), derive the boundary conditions(6.12c), (6.12d) by using the integral equations (6.11a), (6.11b) and the boundary conditions(6.12a),(6.12b).

6.7 Prove that the normal components of the free current density and the electric displacement across an interface between two media satisfy the following boundary condition

$$J_{n1}+\frac{\partial D_{n1}}{\partial t}=J_{n2}+\frac{\partial D_{n2}}{\partial t} \tag{6.99}$$

6.8 Prove(through direct substitution) that any twice differentiable function expressed in (6.29) is a solution to the wave equation(6.28) everywhere except at the origin.

6.9 Prove that the retarded potential in (6.38) satisfies the nonhomogeneous wave equation(6.25).

6.10 Derive the general wave equations for $\boldsymbol{E}$ and $\boldsymbol{H}$ in a nonconducting simple medium where a charge distribution ρ_v and a current distribution $\boldsymbol{J}$ exist. Write down the corresponding phasor form of the wave equations for time-harmonic fields. Write the general solutions for $\boldsymbol{E}(\boldsymbol{r},t)$ and $\boldsymbol{H}(\boldsymbol{r},t)$ in terms of ρ_v and $\boldsymbol{J}$.

6.11 An electric field in air is given as

$$\boldsymbol{E}=\boldsymbol{e}_y 0.1\sin(10\pi x)\cos(6\pi\times 10^9 t-\beta z)$$

find the constant β and determine the magnetic field $\boldsymbol{H}$.

6.12 The magnetic field intensity in a dielectric medium is

$$\boldsymbol{H}=\boldsymbol{e}_y H_0\cos(\omega t-\beta z)$$

(1) Show that the magnetic energy density is equal to the electric energy density.

(2) Evaluate the time-average Poynting vector.

6.13 An instantaneous electric field intensity in free space is

$$\boldsymbol{E}=\boldsymbol{e}_x E_0\cos[k_0(z-ct)]+\boldsymbol{e}_y\sin[k_0(z-ct)]$$

where $k_0=2\pi/\lambda_0=\omega/c$.

(1) Find the instantaneous and phasor expression of $\boldsymbol{H}$ field.

(2) Find the instantaneous Poynting vector $\boldsymbol{S}$.

(3) At $z=0$, how does the magnitude and direction of the $\boldsymbol{E}$ vector vary with time?

(4) Find the time-average electric energy density, magnetic energy density, and Poynting vector.

6.14 It is known that the electric field intensity of a spherical wave in free space is

$$\boldsymbol{E} = \boldsymbol{e}_\theta \frac{E_0}{r}\sin\theta\cos(\omega t - kr) \tag{6.100}$$

Determine ①the magnetic field intensity $\boldsymbol{H}$, ②the value of k, ③the instantaneous and time-average Poynting vector.

6.15 A lossless coaxial cable consists of an inner conductor with radius a and an outer conductor with radius b. When a DC voltage U is applied between the two conductors at one end of the cable, a current I is produced, flowing to a load resistor between the two conductors at the other end of the cable. Verify that the surface integral of the Poynting vector over the cross section of the coaxial cable equals the power UI that is transmitted to the load.

Chapter 7 Uniform Plane Waves

(均匀平面波)

7.1 Introduction(引言)

In Chapter 6, it is shown that in a source-free nonconducting simple medium, $\boldsymbol{E}$ and $\boldsymbol{H}$ fields satisfy homogeneous vector wave equations. If the medium is free space, the wave equation for $\boldsymbol{E}$ is

$$\nabla^2 \boldsymbol{E}(\boldsymbol{r},t) - \frac{1}{c^2}\frac{\partial^2 \boldsymbol{E}(\boldsymbol{r},t)}{\partial t^2} = 0 \tag{7.1}$$

where

$$c = \frac{1}{\sqrt{\mu_0 \varepsilon_0}} \approx 3\times 10^8 (\text{m/s}) \tag{7.2}$$

is the velocity of wave propagation (the speed of light) in free space. One particular solution to (7.1) is the **uniform plane wave** (均匀平面波), which will be derived in detail in the subsequent sections. Figure 7-1 illustrates a simple uniform plane wave propagating in the z-direction. The wave is composed of an $\boldsymbol{E}$ field along the x-direction and an $\boldsymbol{H}$ field along the y-direction, both having no variation along the x-direction or y-direction. Therefore, on any plane perpendicular to the z-axis, the wave is uniform. These planes are constant-phase planes, and therefore, are called **wavefronts** (波前). The main topic of this chapter is the propagation characteristics of a uniform plane wave in an unbounded space.

Strictly speaking, a uniform plane wave does not exist in practice because it carries infinitely large power, and requires its source to be infinite in extent. However, it can be proved that the fields far away from the source region always resemble uniform plane waves. Although simple, the uniform plane waves possess the essential features and typical propagation behaviors of electromagnetic waves, which will be studied in this chapter. [①]

---

① 从本章开始,重点讨论电磁波的传播问题,即时变电磁场离开源以后,如何在无源区域存在以及其传播特性等问题。在无源区域,无须借助于位函数,而是直接基于电场和磁场满足的麦克斯韦方程和波动方程研究其传播特性。数学上,均匀平面波是波动方程最基本最简单的解。物理上,均匀平面波是电磁波的一种理想的存在形式,虽然实际中并不存在,但由于其形式简单,通过均匀平面波引入和介绍电磁波相关的重要概念和性质更为便利。此外,开放空间中距离源足够远的区域,电磁波在局部表现出的特征可以近似为均匀平面波。一些实际问题中,非均匀平面波也可以分解为多个均匀平面的叠加。

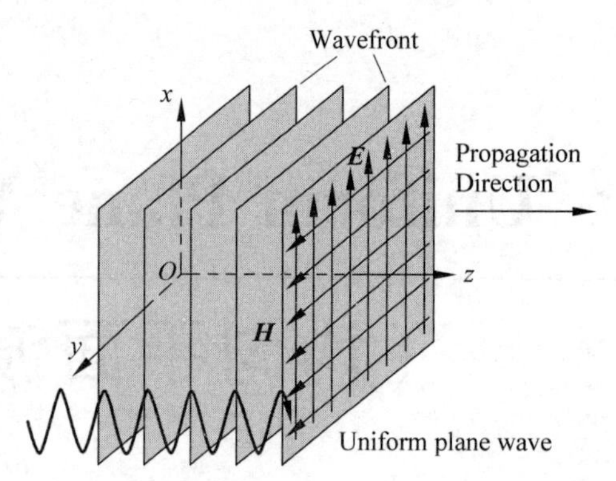

Figure 7-1 Illustration of a uniform plane wave propagating in the z-direction

7.2 Plane Waves in Lossless Medium (无损媒质中的平面波)

For time-harmonic fields, the homogeneous wave equation (7.1) in free space becomes a homogeneous vector Helmholtz's equation:

$$\nabla^2 \boldsymbol{E}(\boldsymbol{r}) + k_0^2 \boldsymbol{E}(\boldsymbol{r}) = 0 \tag{7.3}$$

where k_0 is the free-space wavenumber and

$$k_0 = \omega\sqrt{\mu_0 \varepsilon_0} = \frac{\omega}{c} \tag{7.4}$$

In Cartesian coordinates, (7.3) is equivalent to three scalar Helmholtz's equations involving E_x, E_y, and E_z respectively. Consider the component E_x, we have

$$\left(\frac{\partial^2}{\partial x^2} + \frac{\partial^2}{\partial y^2} + \frac{\partial^2}{\partial z^2} + k_0^2\right) E_x(\boldsymbol{r}) = 0 \tag{7.5}$$

Now we assume the simple case when E_x is uniform (uniform magnitude and phase) over plane surfaces perpendicular to z, i. e.,

$$\frac{\partial E_x}{\partial x} = 0 \text{ and } \frac{\partial E_x}{\partial y} = 0$$

Then $E_x(\boldsymbol{r}) = E_x(z)$ is only a function of the z coordinate. (7.5) becomes

$$\frac{\mathrm{d}^2 E_x(z)}{\mathrm{d}z^2} + k_0^2 E_x(z) = 0 \tag{7.6}$$

which is an ordinary second order differential equation. The solution of (7.6) is found to be

$$E_x(z) = E_x^+(z) + E_x^-(z) = E_0^+ \mathrm{e}^{-\mathrm{j}k_0 z} + E_0^- \mathrm{e}^{\mathrm{j}k_0 z} \tag{7.7}$$

where E_0^+ and E_0^- are arbitrary (and, in general, complex) constants that can be determined by boundary conditions in realistic applications. ①

---

① 通过若干假设得到了自由空间中最简单的一种均匀平面波的数学表达式。这些假设包括电场是时谐场，仅有 x 方向分量，且在 x 和 y 方向上无变化，从而使得三元矢量函数 $\boldsymbol{E}(\boldsymbol{r})$ 能够以一个一元标量函数 $E_x(z)$ 来描述。三维空间的亥姆霍兹方程则简化为二阶线性常微分方程，其解式(7.7) 包含的两项分别代表了沿 $+z$ 和 $-z$ 两个方向传播的均匀平面波。本节从该亥姆霍兹方程的基本解出发，阐述均匀平面波的基本概念和性质。

Now let us examine the first phasor term on the right side of(7.7). The corresponding instantaneous expression can be written as

$$E_x^+(z,t) = \mathrm{Re}[E_x^+(z)\,\mathrm{e}^{\mathrm{j}\omega t}] = \mathrm{Re}[E_0^+\mathrm{e}^{\mathrm{j}(\omega t-k_0 z)}] = |E_0^+|\cos(\omega t - k_0 z + \psi) \tag{7.8}$$

where ψ is the argument of the complex number E_0^+, or the initial phase of the field at the location $z=0$. Figure 7-2 plots E_x^+ of(7.8) as a function of z for two different values of t, in which ψ is assumed to be zero. At $t=0$,

$$E_x^+(z,0) = E_0^+\cos k_0 z$$

is a cosine function of z with an amplitude E_0^+. After a very short while at time $t=\Delta t$, the curve

$$E_x^+(z,\Delta t) = E_0^+\cos(\omega\Delta t - k_0 z) = E_0^+\cos k_0\left(z - \frac{\omega\Delta t}{k_0}\right)$$

appears to be shifted in the positive z direction for a very short distance Δz. Obviously we have

$$\frac{\Delta z}{\Delta t} = \frac{\omega}{k_0}$$

which means that the wave $E_x^+(z,t)$ is traveling in the $+z$-direction with a speed of ω/k_0. Therefore, we call this wave to be a **traveling wave**(行波).

Another way of understanding how the waves travels is by fixing observation on a particular constant phase of the wave E_x^+. Let the phase of E_x^+ in(7.8) be a constant ψ_0, then

$$\omega t - k_0 z = \psi_0$$

This means that, at different time t, the constant phase ψ_0 appears at different position z. By taking differential at both sides, we obtain

$$u_p = \frac{\mathrm{d}z}{\mathrm{d}t} = \frac{\omega}{k_0} = \frac{1}{\sqrt{\mu_0\varepsilon_0}} = c \approx 3\times 10^8(\mathrm{m/s}) \tag{7.9}$$

(7.9) indicates that the wave of(7.8) travels in free space with a velocity equal to the velocity of light. This velocity is the traveling speed of an equiphase front, and therefore, is called the **phase velocity** (**相速度**). ①

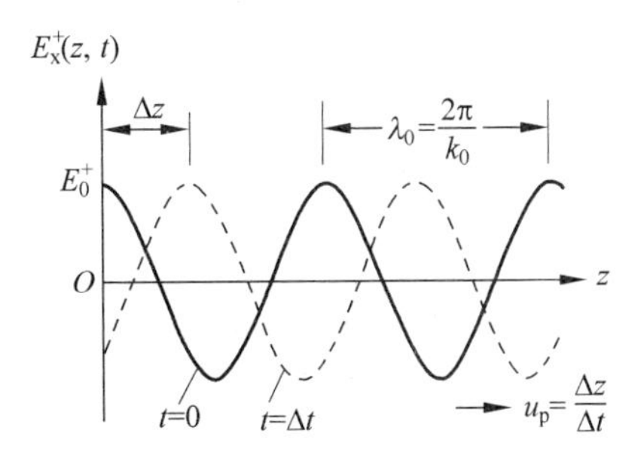

Figure 7-2 Wave traveling in positive z-direction $E_x^+(z,t)=E_0^+\cos(\omega t-k_0 z)$, for several values of t

① 之所以称为**相速度**,是因为这里的推导过程中采用了时谐场的假设,这里描述的电磁波在一个频点上以正弦/余弦波形传播,实质上就是相位随时间移动的速度。而对于更一般的情况,可能包含多个频点的电磁波,不同频点的相速度可能不同,除了相速度还可能定义其他速度(例如7.4节将要讨论的群速度)。

Notice that, as shown in Figure 7-2 the waveform $E_x^+(z,t)$ at a fixed t has a period

$$\lambda_0 = \frac{2\pi}{k_0} \tag{7.10}$$

which is the **wavelength**(波长) of the traveling wave. Hence, we have

$$k_0 = \frac{2\pi}{\lambda_0} \tag{7.11}$$

which is consistent with the expression of k_0 in(7.4) since $c=\lambda_0\omega/2\pi$. (7.11) explains why k_0 is called the **wavenumber**(波数), because it shows that k_0 is equal to the number of wave periods in 2π meters. ①

(7.10) and(7.11) are valid without the subscript "0" if the medium is a lossless material such as a perfect dielectric, instead of free space.

It is obvious that the second phasor term on the right side of(7.7), $E_0^- e^{jk_0z}$, represents a sinusoidal wave traveling in the $-z$-direction with the same velocity c. If we are concerned only with the wave traveling in the $+z$-direction, $E_0^-=0$. However, if there are discontinuities in the medium, reflected waves traveling in the opposite direction must also be considered, as we will see later in the next chapter.

The magnetic field associated with $E_x^+(z)$ component can be found from Maxwell's equation

$$-j\omega\mu_0 \boldsymbol{H} = \nabla\times\boldsymbol{E} = \begin{vmatrix} \boldsymbol{e}_x & \boldsymbol{e}_y & \boldsymbol{e}_z \\ 0 & 0 & \dfrac{\partial}{\partial z} \\ E_x^+(z) & 0 & 0 \end{vmatrix} = \boldsymbol{e}_y \frac{\partial E_x^+(z)}{\partial z}$$

which leads to

$$\boldsymbol{H} = \boldsymbol{e}_y H_y^+(z) = \boldsymbol{e}_y \frac{1}{-j\omega\mu_0}\frac{\partial}{\partial z}(E_0^+ e^{-jk_0z}) = \boldsymbol{e}_y \frac{k_0}{\omega\mu_0}(E_0^+ e^{-jk_0z}) = \boldsymbol{e}_y \frac{1}{\eta_0}E_x^+(z) \tag{7.12}$$

where

$$\eta_0 = \sqrt{\frac{\mu_0}{\varepsilon_0}} \approx 120\pi \approx 377\ (\Omega) \tag{7.13}$$

is called the **intrinsic impedance**(本征阻抗) of the free space. Thus, $\boldsymbol{H}$ has only y component and

$$H_y^+(z) = \frac{1}{\eta_0}E_x^+(z)$$

Since η_0 is a real number, $H_y^+(z)$ is in phase with $E_x^+(z)$, and we have

$$\boldsymbol{H}(z,t) = \boldsymbol{e}_y H_y^+(z,t) = \boldsymbol{e}_y \mathrm{Re}[H_y^+(z)e^{j\omega t}] = \boldsymbol{e}_y \frac{E_0^+}{\eta_0}\cos(\omega t - k_0 z) \tag{7.14}$$

Hence, for this uniform plane wave, the ratio of the magnitudes of $\boldsymbol{E}$ and $\boldsymbol{H}$ is the intrinsic

---

① 对于一个时谐均匀平面波,其在时间和空间上都是一个周期函数。时间上的周期体现在角频率 ω 上,ω 代表 2πs 时间内包含的时间周期数。而空间上的周期体现在波数 k_0 上,k_0 代表 2πm 传播距离内包含的空间周期数。

impedance of the medium. $\boldsymbol{H}$ is perpendicular to $\boldsymbol{E}$, and both fields are normal to the direction of propagation. ①

Example 7-1 A uniform plane wave with $\boldsymbol{E}=\boldsymbol{e}_x E_x$ propagates in a lossless simple medium ($\varepsilon_r=4, \mu_r=1, \sigma=0$) in the $+z$-direction. It is observed that E_x varies sinusoidally, and at $t=0$, E_x reaches a maximum value of 0.1V/m at $z=1$m and 5m consecutively.

(1) Find an expression for $\boldsymbol{E}$ as a function of t and z.

(2) Find an expression for $\boldsymbol{H}$ as a function of t and z.

Solution:

(1) Apparently, the amplitude of the sinusoidal varying E_x is 0.1V/m. Since $z=1$m and 5m are the two consecutive locations of the field with maximum value, the wavelength λ must be 4m, and

$$k=\frac{2\pi}{\lambda}=\frac{\pi}{2}(\mathrm{rad/m})$$

The phase velocity of the wave is

$$u_p=\frac{c}{\sqrt{\varepsilon_r}}=\frac{c}{2}$$

Therefore the frequency of the wave is

$$f=\frac{u_p}{\lambda}=\frac{3\times10^8}{2\times4}=3.75\times10^7(\mathrm{Hz})$$

The expression for $\boldsymbol{E}$ is then

$$\boldsymbol{E}(z,t)=\boldsymbol{e}_x E_x=\boldsymbol{e}_x 0.1\cos\left(7.5\pi\times10^7 t-\frac{\pi}{2}z+\psi\right)$$

where ψ is the initial phase of the wave at $z=0$. Since E_x reaches maximum at $t=0$ and $z=1$m, the argument of the cosine function can be considered as zero when $t=0$ and $z=1$m, i.e.,

$$7.5\pi\times10^7\times0-\frac{\pi}{2}\times1+\psi=0$$

which gives

$$\psi=\frac{\pi}{2}$$

Thus

$$\boldsymbol{E}(z,t)=\boldsymbol{e}_x 0.1\cos\left(7.5\pi\times10^7 t-\frac{\pi}{2}z+\frac{\pi}{2}\right)=-\boldsymbol{e}_x 0.1\sin\left(7.5\pi\times10^7 t-\frac{\pi}{2}z\right)$$

(2) The expression for $\boldsymbol{H}$ is

$$\boldsymbol{H}=\boldsymbol{e}_y H_y=\boldsymbol{e}_y\frac{E_x}{\eta}$$

where

---

① 这里讨论的虽然是均匀平面波的一个特殊的例子(电场仅有 x 方向分量,且沿 z 方向传播),但其结果却反映了均匀平面波的一些基本特征,包括电场、磁场、传播方向三者两两相互垂直,电场和磁场的大小的比值等于本征阻抗。

$$\eta = \sqrt{\frac{\mu}{\varepsilon}} = \frac{\eta_0}{\sqrt{\varepsilon_r}} = 60\pi\ (\Omega)$$

Hence

$$\boldsymbol{H}(z,t) = -\boldsymbol{e}_y \frac{1}{600\pi}\sin\left(7.5\pi\times10^7 t - \frac{\pi}{2}z\right)$$

With the above expressions, the $\boldsymbol{E}$ and $\boldsymbol{H}$ field distributions along the z-axis at the time $t=0$ can be illustrated by Figure 7-3.

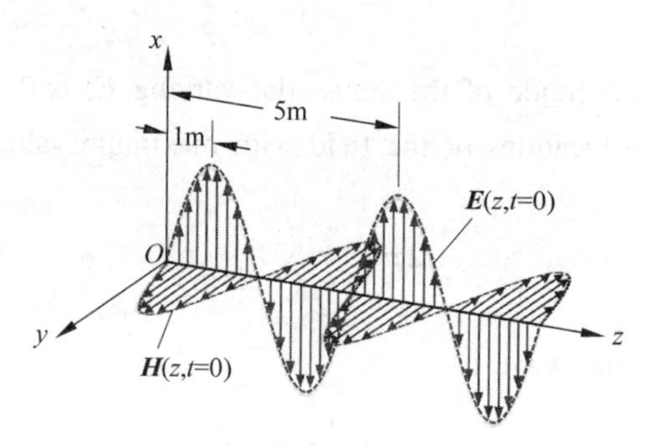

Figure 7-3 $\boldsymbol{E}$ and $\boldsymbol{H}$ fields of a uniform plane wave at $t=0$ (Example 7-1)

7.2.1 Transverse Electromagnetic Waves(横电磁波)

In the previous discussions, the simple uniform plane wave propagating along z-direction with the $\boldsymbol{E}$ field along x-direction is associated with an $\boldsymbol{H}$ field along y-direction. Thus, $\boldsymbol{E}$ and $\boldsymbol{H}$ are perpendicular to each other, and are both **transverse**(横向) to the direction of propagation. It is a particular case of a **transverse electromagnetic wave**(横电磁波) or simply called **TEM wave**(**TEM 波**). Generally, the propagation of a uniform plane wave can be along an arbitrary direction instead of being along the z-axis. Here, we prove that in a simple medium, a uniform plane wave is generally a TEM wave.

We already see that a uniform plane wave propagating in the $+z$-direction can be written as

$$\boldsymbol{E}(z) = \boldsymbol{E}_0 e^{-jkz} \tag{7.15}$$

where $\boldsymbol{E}_0$ is a constant vector, and k is the wavenumber. A more general form of (7.15) is

$$\boldsymbol{E}(x,y,z) = \boldsymbol{E}_0 e^{-jk_x x - jk_y y - jk_z z} \tag{7.16}$$

It can be easily proved by direct substitution that (7.16) is a solution of the homogeneous Helmholtz's equation, provided that

$$k_x^2 + k_y^2 + k_z^2 = k^2 \tag{7.17}$$

If we define a **wavenumber vector**(波矢量) as

$$\boldsymbol{k} = \boldsymbol{e}_x k_x + \boldsymbol{e}_y k_y + \boldsymbol{e}_z k_z = k\boldsymbol{e}_k \tag{7.18}$$

then (7.16) can be written compactly as

$$\boldsymbol{E}(\boldsymbol{r}) = \boldsymbol{E}_0 e^{-j\boldsymbol{k}\cdot\boldsymbol{r}} = \boldsymbol{E}_0 e^{-jk\boldsymbol{e}_k\cdot\boldsymbol{r}} \tag{7.19}$$

where $\boldsymbol{e}_k$ is a unit vector in the direction of propagation. From (7.18) it is clear that

$$k_x = \boldsymbol{k}\cdot\boldsymbol{e}_x = k\boldsymbol{e}_k\cdot\boldsymbol{e}_x \tag{7.20a}$$

$$k_y = \boldsymbol{k} \cdot \boldsymbol{e}_y = k\boldsymbol{e}_k \cdot \boldsymbol{e}_y \tag{7.20b}$$

$$k_z = \boldsymbol{k} \cdot \boldsymbol{e}_z = k\boldsymbol{e}_k \cdot \boldsymbol{e}_z \tag{7.20c}$$

and that $\boldsymbol{e}_k \cdot \boldsymbol{e}_x$, $\boldsymbol{e}_k \cdot \boldsymbol{e}_y$ and $\boldsymbol{e}_k \cdot \boldsymbol{e}_z$ are direction cosines of $\boldsymbol{e}_k$.

Now, we show that the expression (7.16) or (7.19) is the $\boldsymbol{E}$ field of a uniform plane wave, which is also a TEM wave. First, we use Figure 7-4 to illustrate the geometrical relations of $\boldsymbol{e}_k$, $\boldsymbol{r}$ and a plane defined by the equation

$$\boldsymbol{e}_k \cdot \boldsymbol{r} = \text{constant(length } \overline{OP}) \tag{7.21}$$

(7.21) states that all the points on the plane have a position vector $\boldsymbol{r}$ whose projection onto the direction of $\boldsymbol{e}_k$ is constant. This constant is the length of OP, with P to be a point on the ray extending from the origin and along the direction of $\boldsymbol{e}_k$. Obviously, the point P is on the plane. And for any other point Q on the plane, the line $\overline{QP}$ is perpendicular to $\boldsymbol{e}_k$. Therefore, (7.21) must define a plane containing the point P and perpendicular to $\boldsymbol{e}_k$, the direction of propagation. ①

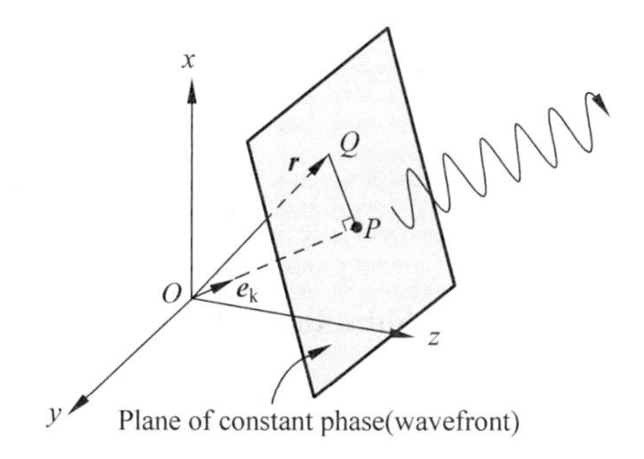

Figure 7-4 The $\boldsymbol{e}_k$, $\boldsymbol{r}$ vectors and constant-phase plane (wavefront) of a uniform plane wave

Since $\boldsymbol{e}_k \cdot \boldsymbol{r}$ determines the phase of the $\boldsymbol{E}$ field in (7.19), $\boldsymbol{e}_k \cdot \boldsymbol{r}$ = constant is a plane of constant phase of the wave. On this plane, the amplitude of $\boldsymbol{E}$ is also uniform, and therefore, (7.19) gives the $\boldsymbol{E}$ field of a uniform plane wave. Notice that, for $\boldsymbol{e}_k = \boldsymbol{e}_z$, (7.21) becomes z= constant, which denotes a plane of constant phase and uniform amplitude for the wave in (7.15). ②

In a charge-free region, the electric field given by (7.19) must satisfy $\nabla \cdot \boldsymbol{E} = 0$. As a result,

$$\boldsymbol{E}_0 \cdot \nabla(\mathrm{e}^{-jk\boldsymbol{e}_k \cdot \boldsymbol{r}}) = 0 \tag{7.22}$$

since $\boldsymbol{E}_0$ is a constant vector. Notice that

---

① 根据图 7-4,P 是坐标原点 O 沿 $\boldsymbol{e}_k$ 方向的射线上与式(7.21)描述的平面的交点。P 在该平面上,且根据式(7.21),该平面上任意一点 Q(对应于位置矢量 $\boldsymbol{r}$)与 P 的连线都垂直于 $\boldsymbol{e}_k$,因此 $\boldsymbol{e}_k$ 与该平面垂直。实际上,P 是坐标原点 O 在该平面上的投影。

② 式(7.21)描述的是式(7.19)给定的电场的等相位面,且在该等相位面上,电场幅度是均匀的。之前已经证明该等相位面垂直于 $\boldsymbol{e}_k$,因此式(7.19)描述的是一个均匀平面波的电场。接下来将要证明的是式(7.19)描述的是一个 TEM 波的电场,即电场 $\boldsymbol{E}$ 垂直于 $\boldsymbol{e}_k$,与 $\boldsymbol{E}$ 相伴的磁场 $\boldsymbol{H}$ 也垂直于 $\boldsymbol{e}_k$,且 $\boldsymbol{E}$ 和 $\boldsymbol{H}$ 也是相互垂直的。式(7.24)和式(7.28)的推导给出了该证明。

$$\nabla(e^{-jk\boldsymbol{e}_k\cdot\boldsymbol{r}}) = \left(\boldsymbol{e}_x\frac{\partial}{\partial x} + \boldsymbol{e}_y\frac{\partial}{\partial y} + \boldsymbol{e}_z\frac{\partial}{\partial z}\right)e^{-jk_xx-jk_yy-jk_zz}$$

$$= -j(\boldsymbol{e}_xk_x + \boldsymbol{e}_yk_y + \boldsymbol{e}_zk_z)e^{-jk_xx-jk_yy-jk_zz}$$

$$= -jk\boldsymbol{e}_k e^{-jk\boldsymbol{e}_k\cdot\boldsymbol{r}} \tag{7.23}$$

Hence(7.22) can be written as

$$-jk(\boldsymbol{E}_0e^{-jk\boldsymbol{e}_k\cdot\boldsymbol{r}}\cdot\boldsymbol{e}_k) = 0$$

which means

$$\boldsymbol{e}_k\cdot\boldsymbol{E} = 0 \tag{7.24}$$

Thus, the $\boldsymbol{E}$ field in(7.19) must be transverse to the direction of propagation.

The magnetic field associated with $\boldsymbol{E}$ in(7.19) can be obtained from(6.76a) as

$$\boldsymbol{H}(\boldsymbol{r}) = -\frac{1}{j\omega\mu}\nabla\times\boldsymbol{E}(\boldsymbol{r}) \tag{7.25}$$

By using(7.23), we can prove that(7.25) is equivalent to

$$\boldsymbol{H}(\boldsymbol{r}) = \frac{1}{\eta}\boldsymbol{e}_k\times\boldsymbol{E}(\boldsymbol{r}) \tag{7.26}$$

where

$$\eta = \frac{\omega\mu}{k} = \sqrt{\frac{\mu}{\varepsilon}} \tag{7.27}$$

is the intrinsic impedance of the medium. Substitution of(7.19) in(7.26) yields

$$\boldsymbol{H}(\boldsymbol{r}) = \frac{1}{\eta}(\boldsymbol{e}_k\times\boldsymbol{E}_0)e^{-jk\boldsymbol{e}_k\cdot\boldsymbol{r}} \tag{7.28}$$

It is now clear that a uniform plane wave propagating in an arbitrary direction, $\boldsymbol{e}_k$, is a TEM wave with $\boldsymbol{E}\perp\boldsymbol{H}$ and that both $\boldsymbol{E}$ and $\boldsymbol{H}$ are normal to $\boldsymbol{e}_k$.

Example 7-2 Given the magnetic field of a TEM wave to be

$$\boldsymbol{H}(\boldsymbol{r}) = \boldsymbol{H}_0e^{-jk\boldsymbol{e}_k\cdot\boldsymbol{r}} \tag{7.29}$$

Obtain an expression for $\boldsymbol{E}(\boldsymbol{r})$.

Solution: From(6.76b), we have

$$\boldsymbol{E}(\boldsymbol{r}) = \frac{1}{j\omega\varepsilon}\nabla\times\boldsymbol{H}(\boldsymbol{r}) = \frac{1}{j\omega\varepsilon}(-jk)\boldsymbol{e}_k\times\boldsymbol{H}(\boldsymbol{r})$$

which can be rewritten as

$$\boldsymbol{E}(\boldsymbol{r}) = \eta\boldsymbol{H}(\boldsymbol{r})\times\boldsymbol{e}_k \tag{7.30}$$

Alternatively, we can obtain the same result by cross-multiplying both sides of(7.26) by $\boldsymbol{e}_k$ and using the back-cab rule in(1.17).

By using(7.26) and the "back-cab" rule(1.16), we can obtain the time average Poynting vector of the uniform plane wave as

$$\boldsymbol{S}_{av} = \frac{1}{2}\text{Re}[\boldsymbol{E}\times\boldsymbol{H}^*] = \frac{1}{2}\text{Re}\left[\boldsymbol{E}\times\left(\frac{1}{\eta}\boldsymbol{e}_k\times\boldsymbol{E}^*\right)\right] = \frac{1}{2\eta}\text{Re}[\boldsymbol{e}_k(\boldsymbol{E}\cdot\boldsymbol{E}^*) - \boldsymbol{E}^*(\boldsymbol{E}\cdot\boldsymbol{e}_k)]$$

With(7.24), the above expression becomes

$$\boldsymbol{S}_{\text{av}} = \boldsymbol{e}_{\text{k}} \frac{1}{2\eta} |\boldsymbol{E}|^2 \tag{7.31}$$

Similarly, by using (7.30) we can derive another expression of the time average Poynting vector:

$$\boldsymbol{S}_{\text{av}} = \boldsymbol{e}_{\text{k}} \frac{1}{2}\eta |\boldsymbol{H}|^2 \tag{7.32}$$

Notice that (7.31) and (7.32) are based on the assumption that the characteristic impedance η is real, which is generally not true for lossy media, which will be discussed in Section 7.3.

7.2.2 Polarization of Plane Waves(平面波的极化)

In previous sections, the uniform plane wave is assumed to have electric and magnetic field vectors always oriented toward fixed directions. Specifically, the $\boldsymbol{E}$ vector of the plane wave in Example 7-1 is fixed in the x-direction, the $\boldsymbol{H}$ vector is fixed in the y-direction, and the wave is propagating along the z-direction. However, the field vectors do not necessarily remain a single orientation as the time changes. In this subsection, we examine the time-varying behavior of the field orientation of a uniform plane wave. Since the direction of the $\boldsymbol{H}$ vector has a fixed relation with that of the $\boldsymbol{E}$ vector for a uniform plane wave, it is enough to consider only the direction of the $\boldsymbol{E}$ vector. We introduce the concept of **polarization** (**极化**) as a description of how the electric field vector orientation changes with time. For example, the uniform plane wave in Example 7-1 is said to be **linearly polarized**(**线性极化**) in the x-direction, because the $\boldsymbol{E}$ vector has fixed orientation in the x-direction($\boldsymbol{E}=\boldsymbol{e}_{\text{x}}E_{\text{x}}$, where E_{x} may be positive or negative). More generally, the direction of the $\boldsymbol{E}$ vector at a given point may change with time, if the vector has two nonzero orthogonal vector components with different magnitude and phase. ①

For simplicity, consider a uniform plane wave propagating along $+z$-direction. The $\boldsymbol{E}$ vector of the wave must be perpendicular to the z-direction, and hence can only have two orthogonal components that are along x-and y-direction respectively. Assume

$$\boldsymbol{E}(z) = \boldsymbol{e}_{\text{x}}E_{\text{x}}(z) + \boldsymbol{e}_{\text{y}}E_{\text{y}}(z) = \boldsymbol{e}_{\text{x}}E_{\text{x0}}\text{e}^{-\text{j}kz}\text{e}^{\text{j}\varphi_{\text{x}}} + \boldsymbol{e}_{\text{y}}E_{\text{y0}}\text{e}^{-\text{j}kz}\text{e}^{\text{j}\varphi_{\text{y}}} \tag{7.33}$$

where E_{x0} and E_{y0} are positive real numbers denoting the amplitudes of the two linearly polarized components. φ_{x} and φ_{y} denote the phases of the two components at $z=0$. Then the instantaneous expression for $\boldsymbol{E}$ is

$$\begin{aligned}\boldsymbol{E}(z,t) &= \boldsymbol{e}_{\text{x}}E_{\text{x}}(z,t) + \boldsymbol{e}_{\text{y}}E_{\text{y}}(z,t) = \text{Re}\{[\boldsymbol{e}_{\text{x}}E_{\text{x}}(z) + \boldsymbol{e}_{\text{y}}E_{\text{y}}(z)]\,\text{e}^{\text{j}\omega t}\} \\ &= \boldsymbol{e}_{\text{x}}E_{\text{x0}}\cos(\omega t - kz + \phi_{\text{x}}) + \boldsymbol{e}_{\text{y}}E_{\text{y0}}\cos(\omega t - kz + \varphi_{\text{y}})\end{aligned}$$

Now let us examine the orientation of $\boldsymbol{E}$ at a given location as t changes. For convenience, we

---

① 对于一个沿 z 方向传播的均匀平面波，当电场 $\boldsymbol{E}$ 仅有 x 方向分量的时候，电场矢量端点始终在 x 轴上移动，我们称为沿 x 方向的线性极化。然而，在给定$+z$ 传播方向的条件下，电场还可能具有 y 方向的分量。此时，电场矢量的方向取决于 x 和 y 方向两个分量的幅度和相位，并且有可能是随时间变化的。电场矢量的方向随时间变化的特性就是所谓的电磁波的极化。更一般地，均匀平面波的电场 $\boldsymbol{E}$ 和磁场 $\boldsymbol{H}$ 在位于垂直于传播方向的平面内的方向会随着时间和位置的改变而变化，这取决于波是如何产生的或在哪类媒质中传播的。完整地描述一个电磁波不仅需要知道其波长、相速度、功率密度等参数，还需要知道场矢量的方向及其随时间的变化。

fix location to be at $z=0$, then

$$\begin{aligned} \boldsymbol{E}(0,t) &= \boldsymbol{e}_x E_x(0,t) + \boldsymbol{e}_y E_y(0,t) \\ &= \boldsymbol{e}_x E_{x0}\cos(\omega t + \varphi_x) + \boldsymbol{e}_y E_{y0}\cos(\omega t + \varphi_y) \end{aligned} \tag{7.34}$$

Obviously, both the magnitude and the direction of $\boldsymbol{E}(0,t)$ may change with time. Here, we first discuss two special cases as follows.

(1) $\varphi_x - \varphi_y = 0$ or $\pm\pi$. In this case, we have

$$\frac{E_y(0,t)}{E_x(0,t)} = \begin{cases} \dfrac{E_{y0}}{E_{x0}}, & \text{if } \varphi_x - \varphi_y = 0 \\ -\dfrac{E_{y0}}{E_{x0}}, & \text{if } \varphi_x - \varphi_y = \pm\pi \end{cases} \tag{7.35}$$

which means that the tip of the vector $\boldsymbol{E}(0,t)$ moves along a straight line that passes the origin and has a slope of E_{y0}/E_{x0} or $-E_{y0}/E_{x0}$. As is shown in Figure 7-5, the wave is linearly polarized.

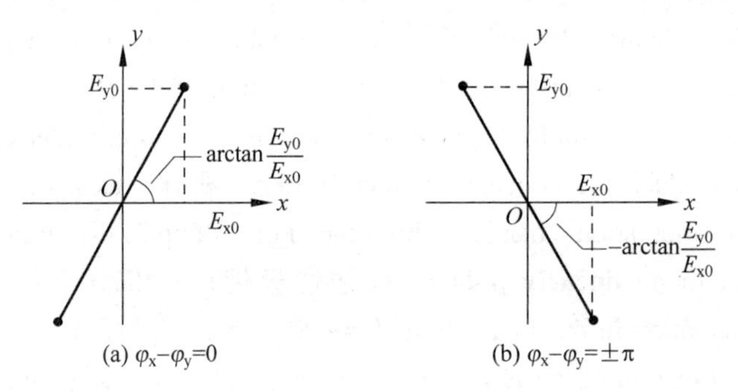

Figure 7-5 Linearly polarization diagrams of the wave with $\boldsymbol{E}(0,t)=\boldsymbol{e}_x E_{x0}\cos(\omega t+\varphi_x)+\boldsymbol{e}_y E_{y0}\cos(\omega t+\varphi_x)$

(2) $E_{x0}=E_{y0}=E_0$, and $\varphi_x-\varphi_y=\pm\pi/2$. In this case, the magnitude of $\boldsymbol{E}(0,t)$ is

$$|\boldsymbol{E}(0,t)| = \sqrt{E_0^2\cos^2\left(\omega t + \varphi_y \pm \frac{\pi}{2}\right) + E_0^2\cos^2(\omega t + \varphi_y)} = E_0^2 \tag{7.36}$$

The direction of $\boldsymbol{E}(0,t)$ can be described by the angle α between the vector and the $+x$-axis, and from (7.34) we have

$$\tan\alpha = \frac{E_y(0,t)}{E_x(0,t)} = \begin{cases} \dfrac{E_0\sin(\omega t + \varphi_x)}{E_0\cos(\omega t + \varphi_x)} = \tan(\omega t + \varphi_x), & \text{if } \varphi_x - \varphi_y = \dfrac{\pi}{2} \\ \dfrac{-E_0\sin(\omega t + \varphi_x)}{E_0\cos(\omega t + \varphi_x)} = -\tan(\omega t + \varphi_x), & \text{if } \varphi_x - \varphi_y = \dfrac{-\pi}{2} \end{cases}$$

and hence,

$$\alpha = \begin{cases} \omega t + \varphi_x, & \text{if } \varphi_x - \varphi_y = \dfrac{\pi}{2} \\ -(\omega t + \varphi_x), & \text{if } \varphi_x - \varphi_y = \dfrac{-\pi}{2} \end{cases} \tag{7.37}$$

This indicates that the tip of the vector $\boldsymbol{E}(0,t)$ moves along a circle with a radius E_0 and an angular velocity ω. Therefore, the wave in this case is called **circularly polarized** (圆极化). Particularly, if $\varphi_x-\varphi_y=\pi/2$, the vector $\boldsymbol{E}(0,t)$ rotates counterclockwise from $+x$-to $+y$-direction as is shown in Figure 7-6(a). When the fingers of the right hand follow the direction of the rotation of $\boldsymbol{E}$, the thumb points to the propagation direction($+z$-direction) of the wave. Hence the wave is called a **right-hand circularly polarized wave**(右旋圆极化波). On the other hand, if $\varphi_x-\varphi_y=-\pi/2$ as is shown in Figure 7-6(b), the vector $\boldsymbol{E}(0,t)$ rotates clockwise and the wave is called a **left-hand circularly polarized wave**(左旋圆极化波).

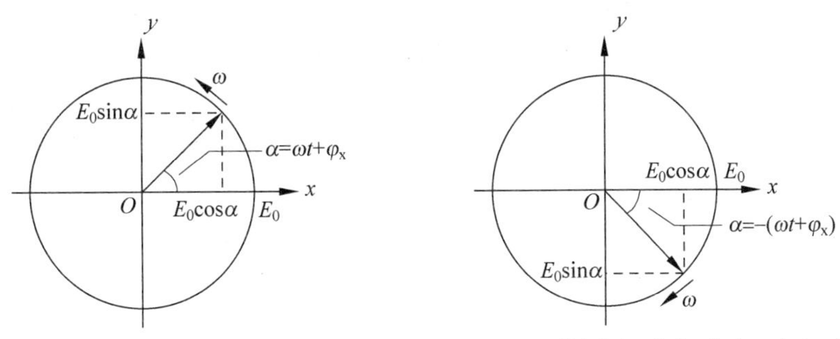

(a) $\varphi_x-\varphi_y=\pi/2$,right-hand circularly polarized (b) $\varphi_x-\varphi_y=-\pi/2$,left-hand circularly polarized

Figure 7-6 Circularly polarization diagrams of the wave with $\boldsymbol{E}(0,t)=\boldsymbol{e}_x E_{x0}\cos(\omega t+\varphi_x)+\boldsymbol{e}_y E_{y0}\cos(\omega t+\varphi_x)$

When the relations between E_{x0}, E_{y0}, φ_x and φ_y are different from the above two cases, the tip of the vector $\boldsymbol{E}(0,t)$ moves along an ellipse and the wave is called **elliptically polarized** (椭圆极化). If $0<\varphi_x-\varphi_y<\pi$, $\boldsymbol{E}(0,t)$ rotates from $+x$-to$+y$-direction, and the wave propagating along $+z$-direction is **right-hand elliptically polarized** (右旋椭圆极化). If $-\pi<\varphi_x-\varphi_y<0$, $\boldsymbol{E}(0,t)$ rotates from $+y$-to$+x$-direction, and the wave is **left-hand elliptically polarized** (左旋椭圆极化).

Generally, the $\boldsymbol{E}$ vector of a uniform plane wave rotates from the direction of the component with leading phase to the direction of the other component. Notice that, whether the polarization is right- or left-hand also depends on the propagation direction. ①

Wave polarization is a very important concept that needs to be taken into consideration in many practical applications. For example, the electromagnetic waves radiated by AM broadcast stations are linearly polarized with the $\boldsymbol{E}$-field perpendicular to the ground. Hence the antenna of an AM radio should be vertical for best reception quality. Waves from TV stations are linearly polarized in the horizontal direction, and therefore, the TV receiving antennas should be horizontal. The waves of GPS signals are right-hand circularly polarized, and as a result, GPS receiving antennas are designed to be right-hand circularly polarized.

Example 7-3 Determine the polarization of the waves represented by the following $\boldsymbol{E}$ fields, in which E_m is a positive real number.

---

① 判断均匀平面波是左旋还是右旋极化,首先应确定电场矢量的旋向。为此,找到电场矢量的相互正交的两个线性极化分量,电场矢量的旋向即为从相位领先分量的方向转向相位落后分量的方向。以平面波传播方向为拇指方向,电场矢量的旋向为四指方向,即可判断该极化方式是左旋还是右旋。

(1) $\boldsymbol{E}=\boldsymbol{e}_x E_m \sin(\omega t-kz)+\boldsymbol{e}_y E_m \cos(\omega t-kz)$

(2) $\boldsymbol{E}=\boldsymbol{e}_x E_m e^{-jkz}-\boldsymbol{e}_y jE_m e^{-jkz}$

(3) $\boldsymbol{E}=\boldsymbol{e}_x E_m \sin\left(\omega t-kz+\frac{\pi}{4}\right)+\boldsymbol{e}_y E_m \cos\left(\omega t-kz-\frac{\pi}{4}\right)$

(4) $\boldsymbol{E}=\boldsymbol{e}_x E_m \sin(\omega t-kz)+\boldsymbol{e}_y 2E_m \cos(\omega t-kz)$

Solution:

(1) The x- and y- components of the field have the same magnitude E_m, but different phases. Since $\varphi_x=-\pi/2$, $\varphi_y=0$, the $\boldsymbol{E}$ vector rotates from $+y$-direction to $+x$-direction. Notice that the wave propagates along $+z$-direction. Therefore, it is a left-hand circularly polarized wave.

(2) The x- and y- components of $\boldsymbol{E}$ have the same magnitude E_m. $\varphi_x=0$, $\varphi_y=-\pi/2$, and the wave propagates along $+z$-direction. Therefore, it is a right-hand circularly polarized wave.

(3) $\varphi_x=\varphi_y=-\pi/4$, therefore, it is a linearly polarized wave.

(4) The x and y components of $\boldsymbol{E}$ have different magnitudes. $\varphi_x=0$, $\varphi_y=-\pi/2$, and the wave propagates along $+z$-direction. Therefore, it is a left-hand elliptically polarized wave.

Example 7-4 Prove that a linearly polarized plane wave can be decomposed into a right-hand circularly polarized wave and a left-hand circularly polarized wave.

Solution: Without loss of generality, we can assume the wave is polarized in the x-direction and propagating in the $+z$-direction. Then the $\boldsymbol{E}$ field can be expressed as

$$\boldsymbol{E}(z)=\boldsymbol{e}_x E_0 e^{-jkz}$$

which can be re-written as

$$\boldsymbol{E}(z)=\boldsymbol{E}_{RC}(z)+\boldsymbol{E}_{LC}(z)$$

where

$$\boldsymbol{E}_{RC}(z)=\frac{E_0}{2}(\boldsymbol{e}_x-j\boldsymbol{e}_y)e^{-jkz}$$

is a right-hand circularly polarized wave, and

$$\boldsymbol{E}_{LC}(z)=\frac{E_0}{2}(\boldsymbol{e}_x+j\boldsymbol{e}_y)e^{-jkz}$$

is a left-hand circularly polarized wave. Hence, the statement of this problem is proved. Notice that, $\boldsymbol{E}_{RC}(z)$ and $\boldsymbol{E}_{LC}(z)$ have equal amplitude $E_0/2$. ①

7.3 Plane Waves in Lossy Medium(有损媒质中的平面波)

If a source-free medium is conducting (with a conductivity $\sigma\neq 0$), a current $\boldsymbol{J}=\sigma\boldsymbol{E}$ will flow, and the equation (6.76b) for non-conducting media should be rewritten as

$$\nabla\times\boldsymbol{H}=(\sigma+j\omega\varepsilon)\boldsymbol{E}=j\omega\left(\varepsilon+\frac{\sigma}{j\omega}\right)\boldsymbol{E}$$

① 反过来,也可以证明幅度相等、旋向相反的两个圆极化波将合成一个线性极化波。进一步,任何极化的均匀平面波都能够分解为旋向相反的圆极化波。左旋和右旋圆极化是相互正交的两种极化方式,它们的线性组合能够生成任意极化形式的均匀平面波。

$$= \mathrm{j}\omega\varepsilon_c \boldsymbol{E} \tag{7.38}$$

where

$$\varepsilon_c = \varepsilon - \mathrm{j}\frac{\sigma}{\omega} \tag{7.39}$$

is called a **complex permittivity**(复电容率/复介电常数).

The conduction of the medium causes power dissipation, and as a result, the medium is called a **lossy medium**(有损媒质). In addition to conduction, a medium can also be lossy due to polarization of the material. As is discussed in Section 2-6, an external $\boldsymbol{E}$ field causes small displacements of bound charges in dielectric medium. For time-varying fields, as the frequency increases, the inertia of the charged particles tends to cause frictional damping that leads to power loss. This effect can be effectively modeled by a complex permittivity that includes an imaginary part

$$\varepsilon_c = \varepsilon' - \mathrm{j}\varepsilon'' \tag{7.40}$$

where both ε' and ε'' may be functions of frequency. Alternatively, we may define an equivalent conductivity representing all losses and write

$$\sigma = \omega\varepsilon'' \tag{7.41}$$

In low-loss media, damping losses are very small, and we usually have $\varepsilon' \approx \varepsilon$. Combination of (7.40) and (7.41) gives (7.39).

Similarly, losses can be caused by magnetization at high frequencies, in which case, the permeability becomes complex, i. e.,

$$\mu = \mu' - \mathrm{j}\mu'' \tag{7.42}$$

In this chapter, we neglect the loss due to magnetization and assume the loss only come from conduction and polarization.

One way of characterizing the power loss of a medium is to measure the **loss tangent**(损耗角正切) which is defined as the ratio $\varepsilon''/\varepsilon'$. As indicated by its name, loss tangent is the tangent of the angle of the complex permittivity ε_c, and is denotes by

$$\tan\delta_c = \frac{\varepsilon''}{\varepsilon'} \approx \frac{\sigma}{\omega\varepsilon} \tag{7.43}$$

where δ_c is then angle of ε_c and is called the **loss angle**(损耗角). Notice that the loss tangent $\tan\delta_c$ equals the magnitude ratio of the conduction current and the displacement current for a time-harmonic field. As is demonstrated in Example 6-2 and Example 6-3, the sea water at 1MHz and copper at frequencies up to THz both have a loss tangent much larger than 1.

If the loss tangent $\tan\delta_c$ is a very large number (i. e., $\sigma \gg \omega\varepsilon$), the medium is considered as a **good conductor**(良导体). On the contrary, if $\tan\delta_c$ is much smaller than 1 (i. e., $\sigma << \omega\varepsilon$), the medium is considered as **low loss dielectric**(低损耗介质) or a **good insulator**(良绝缘体). As suggested by (7.43), whether a material is a good conductor or good insulator may depend on the frequency. A material considered as good conductor at low frequencies may be a low loss dielectric at very high frequencies. For example, a moist ground has a dielectric constant $\varepsilon_r \approx 10$ and a conductivity $\sigma \approx 10^{-2}$(S/m). $\tan\delta_c$ of the moist ground is 1.8×10^4 at 1kHz, which makes it a good conductor. However, at 10GHz, the loss tangent becomes 1.8×10^{-3}, which makes it a low-loss dielectric.

Example 7-5 **Printed circuit boards** (PCB, 印刷电路板) are laminated boards containing one or more sheet layers of copper and non-conductive substrates. It is usually desired that the substrates are low-loss dielectrics with small loss tangent, especially for high frequency applications. F4B is a PCB substrate material with a dielectric constant of typically 2.55 and a loss tangent of 0.001. If a time-harmonic $\boldsymbol{E}$ field of amplitude 250V/m and frequency 1GHz exists in the F4B material, find the average power dissipated in the medium per cubic millimeter.

Solution: First, the effective conductivity of the F4B material can be found from the loss tangent since

$$\tan\delta_c = 0.001 = \frac{\sigma}{\omega\varepsilon_0\varepsilon_r}$$

Hence

$$\sigma = 0.001(2\pi 10^9)\left(\frac{10^{-9}}{36\pi}\right)(2.55) = 1.4\times10^{-4}(\mathrm{S/m})$$

The average power dissipated per unit volume is

$$p = \frac{1}{2}\sigma E^2 = \frac{1}{2}\times(1.4\times10^{-4})\times250^2 = 4.375\ (\mathrm{W/m^3})$$

With the introduction of complex permittivity ε_c, we can derive the equation satisfied by a time-harmonic field in a source-free lossy medium by following the same procedure as in subsection 6.9.4, which also leads to the homogeneous vector Helmholtz's equation

$$\nabla^2\boldsymbol{E} + \tilde{k}^2\boldsymbol{E} = 0 \tag{7.44}$$

but with a complex wavenumber instead of the real wavenumber k in (6.79) and (6.80). The complex wavenumber is related to the complex permittivity as

$$\tilde{k} = \omega\sqrt{\mu\varepsilon_c} = \omega\sqrt{\mu\left(\varepsilon - \mathrm{j}\frac{\sigma}{\omega}\right)} \tag{7.45}$$

The derivation and discussions on the uniform plane waves in a lossless medium in Section 7.2 also applies to those in a lossy medium except that the real wave number k must be replaced with a complex $\tilde{k}$.

To conform with the conventional notation used in transmission-line theory, it is customary to define a **propagation constant**(传播常数), γ, such that

$$\gamma = \mathrm{j}\tilde{k} = \mathrm{j}\omega\sqrt{\mu\varepsilon_c} \tag{7.46}$$

Then the Helmholtz's equation (7.44) becomes

$$\nabla^2\boldsymbol{E} - \gamma^2\boldsymbol{E} = 0 \tag{7.47}$$

A typical solution of (7.47) is

$$\boldsymbol{E}(z) = \boldsymbol{e}_x E_x(z) = \boldsymbol{e}_x E_0 \mathrm{e}^{-\gamma z} \tag{7.48}$$

which represents a linearly polarized uniform plane wave propagating in the +z-direction because the field is uniform at planes z=constant.

Since γ is complex, it has a real part (denoted by α) and an imaginary part (denoted by β). Hence (7.48) can be rewritten as

$$\boldsymbol{E}(z) = \boldsymbol{e}_x E_x(z) = \boldsymbol{e}_x E_0 \mathrm{e}^{-\alpha z} \mathrm{e}^{-\mathrm{j}\beta z} \tag{7.49}$$

in which the propagation factor $\mathrm{e}^{-\gamma z}$ has been written as a product of two factors. The first factor, $\mathrm{e}^{-\alpha z}$, decreases as z increases and thus is an **attenuation factor** (衰减因子), and α is called an **attenuation constant** (衰减常数). The unit of the attenuation constant is **neper** (**Np**, 奈培) per meter (Np/m). That is, the field is attenuated by α nepers as the wave travels one meter. [①] The second factor, $\mathrm{e}^{-\mathrm{j}\beta z}$, is a **phase factor** (相位因子) as it determines the phase variation along the propagation direction. β is called a **phase constant** (相位常数) with the unit to be radians per meter (rad/m). The phase constant is equal to the amount of phase shift as the wave travels one meter. Obviously, for a lossless medium, $\sigma=0$, $\alpha=0$, and $\beta=k=\omega\sqrt{\mu\varepsilon}$.

From (7.39), we have

$$\gamma = \alpha + \mathrm{j}\beta = \mathrm{j}\omega\sqrt{\mu\varepsilon}\left(1 + \frac{\sigma}{\mathrm{j}\omega\varepsilon}\right)^{1/2} \tag{7.50}$$

from which the general expressions of α and β can be obtained as

$$\alpha = \omega\sqrt{\frac{\mu\varepsilon}{2}}\left[\sqrt{1+\left(\frac{\sigma}{\omega\varepsilon}\right)^2} - 1\right]^{\frac{1}{2}} \tag{7.51a}$$

$$\beta = \omega\sqrt{\frac{\mu\varepsilon}{2}}\left[\sqrt{1+\left(\frac{\sigma}{\omega\varepsilon}\right)^2} + 1\right]^{\frac{1}{2}} \tag{7.51b}$$

Therefore, the wavelength and the phase velocity of a uniform plane wave in the lossy medium are respectively

$$\lambda = \frac{2\pi}{\beta} = 1\Bigg/\left\{f\sqrt{\frac{\mu\varepsilon}{2}\left[\sqrt{1+\left(\frac{\sigma}{\omega\varepsilon}\right)^2} + 1\right]}\right\} \tag{7.52}$$

$$u_{\mathrm{p}} = \frac{\omega}{\beta} = 1\Bigg/\sqrt{\frac{\mu\varepsilon}{2}\left[\sqrt{1+\left(\frac{\sigma}{\omega\varepsilon}\right)^2} + 1\right]} \tag{7.53}$$

In addition to the complex wavenumber, the intrinsic impedance in a lossy medium also becomes complex:

$$\eta_{\mathrm{c}} = \sqrt{\frac{\mu}{\varepsilon_{\mathrm{c}}}} = |\eta_{\mathrm{c}}|\mathrm{e}^{\mathrm{j}\theta_\eta} \tag{7.54}$$

where θ_η is the phase angle of η_{c}. Like the case in lossless medium, the $\boldsymbol{E}$ and $\boldsymbol{H}$ fields of a TEM wave propagating along $+z$-direction in the lossy medium are related as

$$\boldsymbol{H}(\boldsymbol{r}) = \frac{1}{\eta_{\mathrm{c}}}\boldsymbol{e}_z \times \boldsymbol{E}(\boldsymbol{r}) \tag{7.55}$$

---

① 奈培是一个对数比例，用于电磁场时，1 奈培代表场的幅度的 e 倍，功率的 e^2 倍。电场强度下降 1 奈培代表该电场幅度的大小下降至原来的 1/e，功率下降 1 奈培代表其下降至原来的 $1/\mathrm{e}^2$。式(7.49)所示的电场，在其传播方向上每经过 1m，其幅度降低至原来的 $\mathrm{e}^{-\alpha}$，即下降 α 奈培。

and

$$\boldsymbol{E}(\boldsymbol{r}) = \eta_c \boldsymbol{H}(\boldsymbol{r}) \times \boldsymbol{e}_z \tag{7.56}$$

With $\boldsymbol{E}$ given by(7.48), the $\boldsymbol{H}$ in(7.55) becomes

$$\boldsymbol{H}(z) = \frac{1}{\eta_c} \boldsymbol{e}_z \times (\boldsymbol{e}_x E_0 e^{-\gamma z}) = \boldsymbol{e}_y \frac{1}{|\eta_c|} E_0 e^{-\alpha z} e^{-j(\beta z + \theta_\eta)} \tag{7.57}$$

Notice that the complex ε_c as given by(7.39) and(7.40) has a negative argument(辐角) between 0 and $-\pi/2$. Therefore, η_c given by(7.54)(assuming μ is real) has a positive argument θ_η, and $0<\theta_\eta<\pi/4$. This means that the $\boldsymbol{H}$ field of a TEM wave as given by(7.55) lags the $\boldsymbol{E}$ field by θ_η.

By using the $\boldsymbol{E}$ and $\boldsymbol{H}$ expressions(7.48) and(7.57), we can obtain the expression of time average Poynting vector:

$$\begin{aligned}
\boldsymbol{S}_{av} &= \frac{1}{2}\mathrm{Re}[\boldsymbol{E} \times \boldsymbol{H}^*] = \frac{1}{2}\mathrm{Re}\left[\boldsymbol{e}_x E_0 e^{-\alpha z} e^{-j\beta z} \times \left(\boldsymbol{e}_y \frac{1}{|\eta_c|} E_0^* e^{-\alpha z} e^{j(\beta z + \theta_\eta)}\right)\right] \\
&= \frac{1}{2}\mathrm{Re}\left[\boldsymbol{e}_z \frac{1}{|\eta_c|} |E_0|^2 e^{-2\alpha z} e^{j\theta_\eta}\right] \\
&= \boldsymbol{e}_z \frac{|E_0|^2}{2|\eta_c|} e^{-2\alpha z} \cos\theta_\eta
\end{aligned} \tag{7.58}$$

It is observed from(7.58) that, the power flow density decays exponentially along the propagation direction with an attenuation constant of 2α. Notice that, since the Poynting vector represents a power flow density, the attenuation constant of 2α means that it decays α Np per meter along the z-direction. It is also observed that the power flow density is proportional to $\cos\theta_\eta$.①

In the following two subsections, we examine two typical cases: low-loss dielectrics and good conductors, in which the constants α and β can be simplified with approximation.

7.3.1 Low Loss Dielectrics(低损耗介质)

A low-loss dielectric, or a good insulator, has a small(equivalent) conductivity that satisfies $\sigma/\omega\varepsilon \ll 1$. Under this condition, γ in(7.50) can be approximated by as:

$$\gamma = \alpha + j\beta = j\omega\sqrt{\mu\varepsilon}\left(1 + \frac{\sigma}{j\omega\varepsilon}\right)^{1/2} \approx j\omega\sqrt{\mu\varepsilon}\left[1 - j\frac{\sigma}{2\omega\varepsilon}\right]$$

which indicates that the attenuation constant

$$\alpha \approx \frac{\sigma}{2}\sqrt{\frac{\mu}{\varepsilon}} \tag{7.59}$$

and the phase constant

$$\beta \approx \omega\sqrt{\mu\varepsilon} \tag{7.60}$$

① 这里,时谐场的时间平均坡印廷矢量的大小和本征阻抗辐角的余弦成正比。本征阻抗辐角也就是电、磁场之间的相位差。对于有损媒质,该相位差在良导体中能达到最大 45°。后续章节讨论电磁波的反射以及波导的截止模式时,会遇到电、磁场某些分量之间相位差达到 90°的情形,其相应的平均坡印廷矢量为 0。

It is seen that the attenuation constant of a low-loss dielectric is approximately proportional to the conductivity (if the conductivity is not frequency dependent). The phase constant of a low-loss dielectric is approximately the same as the wavenumber of a perfect (lossless) dielectric.

The intrinsic impedance of a low-loss dielectric can be approximated as

$$\eta_c = \sqrt{\frac{\mu}{\varepsilon_c}} = \sqrt{\frac{\mu}{\varepsilon}}\left(1+\frac{\sigma}{j\omega\varepsilon}\right)^{-1/2} \approx \sqrt{\frac{\mu}{\varepsilon}}\left(1+j\frac{\sigma}{2\omega\varepsilon}\right) \tag{7.61}$$

with argument

$$\theta_\eta \approx \arctan\left(\frac{\sigma}{2\omega\varepsilon}\right) \tag{7.62}$$

which is a very small positive number (approximately half of the loss angle δ_c). This means the electric and magnetic field intensities are not in time phase, with the $\boldsymbol{E}$ field slightly leading $\boldsymbol{H}$ field.

The phase velocity u_p is still defined as ω/β, and from (7.60), we have

$$u_p = \frac{\omega}{\beta} \approx \frac{1}{\sqrt{\mu\varepsilon}} \tag{7.63}$$

which is approximately the same as that in a lossless medium. And the wavelength is

$$\lambda = \frac{2\pi}{\beta} = \frac{u_p}{f} \approx \frac{2\pi}{\omega\sqrt{\mu\varepsilon}} \tag{7.64}$$

7.3.2 Good Conductors(良导体)

A medium is considered as a good conductor if $\sigma/\omega\varepsilon \gg 1$. Under this condition, (7.50) can be approximated as

$$\gamma \approx j\omega\sqrt{\mu\varepsilon}\sqrt{\frac{\sigma}{j\omega\varepsilon}} = \frac{1+j}{\sqrt{2}}\sqrt{\omega\mu\sigma} = (1+j)\sqrt{\pi f\mu\sigma} \tag{7.65}$$

in which the relation $\omega=2\pi f$ and $\sqrt{j}=(1+j)/\sqrt{2}$ has been used. (7.65) indicates that α and β of a good conductor are approximately equal:

$$\alpha = \beta = \sqrt{\pi f\mu\sigma} \tag{7.66}$$

both proportional to $\sqrt{f}$ and $\sqrt{\sigma}$. The intrinsic impedance of a good conductor is

$$\eta_c = \sqrt{\frac{\mu}{\varepsilon_c}} \approx \sqrt{\frac{j\omega\mu}{\sigma}} = (1+j)\sqrt{\frac{\pi f\mu}{\sigma}} = (1+j)\frac{\alpha}{\sigma} \tag{7.67}$$

which has a phase angle $\theta_\eta = 45°$, which means the magnetic field lags the electric field by 45°.

The phase velocity in a good conductor is

$$u_p = \frac{\omega}{\beta} \approx \sqrt{\frac{2\omega}{\mu\sigma}} \tag{7.68}$$

which is proportional to $\sqrt{f}$ and $1/\sqrt{\sigma}$. The wavelength of a plane wave in a good conductor is

$$\lambda = \frac{2\pi}{\beta} = \frac{u_p}{f} = 2\sqrt{\frac{\pi}{f\mu\sigma}} \tag{7.69}$$

which is proportional to $1/\sqrt{f}$ and $1/\sqrt{\sigma}$.

Example 7-6 Calculate the phase velocity, wavelength, and attenuation constant of a 1MHz uniform plane wave in copper, given that $\sigma=5.8\times10^7$S/m, $\mu=4\pi\times10^{-7}$H/m.

Solution: With(7.68), the phase velocity at 1MHz is calculated to be

$$u_p \approx \sqrt{\frac{2(2\pi\times10^6)}{(4\pi\times10^{-7})(5.8\times10^7)}}=\sqrt{\frac{(\times10^6)}{(5.8)}}=415(\text{m/s})$$

With(7.69), the wavelength at 1MHz is

$$\lambda=\frac{u_p}{f}=\frac{415}{10^6}=4.15\times10^{-4}(\text{m})$$

The attenuation constant can be calculated by(7.66) which gives

$$\alpha=\sqrt{\pi(1\times10^6)(4\pi\times10^{-7})(5.8\times10^7)}\approx1.5\times10^4(\text{Np/m})$$

Example 7-6 shows that a uniform plane wave in copper is attenuated 15000Np when it travels 1m. 15000Np attenuation is equivalent to decreasing by a factor of 10^{-6514} in amplitude, which means the wave becomes negligibly small far before it can reach 1m within a copper material. Generally, high-frequency electromagnetic waves are attenuated very rapidly when propagating in a good conductor. As a result, fields and currents can be considered as confined in a very thin layer of the conductor surface which is like the skin. This is referred to as the **skin effect(趋肤效应,集肤效应)**.

To characterize the skin effect, we introduce the concept of **skin depth(趋肤深度)** or the **depth of penetration(穿透深度)**. The skin depth is defined to be the distance δ through which the amplitude of a traveling plane wave decreases by 1Np (i. e., the amplitude of the field decreases by a factor of e^{-1} or 0.368). Since the attenuation factor is $e^{-\alpha z}$, the skin depth δ is related to the attenuation coefficient α as

$$\delta=\frac{1}{\alpha}=\frac{1}{\sqrt{\pi f\mu\sigma}} \tag{7.70}$$

in which the approximate expression of α(7.66) has been used. For copper at 1MHz, this distance is $1/(1.5\times10^4)$ m, or 0.067mm. Several skin depths of typical materials are listed in Table 7-1. It is seen that, for good conductors, the skin depth usually decreases as frequency increases as is also predicted by(7.70). The skin depth is also inversely proportional to $\sqrt{\sigma}$. Hence, for PEC, $\delta\to0$ as $\sigma\to\infty$.

Table 7-1 Skin depths of typical materials①

Material	σ/(S/m)	f=60Hz	f=1MHz	f=1GHz
Silver	6.17×10^7	8.27mm	0.064mm	0.0020mm
Copper	5.80×10^7	8.53mm	0.066mm	0.0021mm

① 从表7-1中可以看到,金属材料对于一般的高频电磁场问题,都可以认为是近似于理想导体。而海水的趋肤深度虽然远大于金属,但当频率高于千赫兹量级时,都在米级及以下。因此海水中通常无法实现电磁信号的远距离传播。这也是潜水艇通常要采用声呐技术实现通信的原因。

续表

Material	σ/(S/m)	f=60Hz	f=1MHz	f=1GHz
Gold	4.10×10^7	10.14mm	0.079mm	0.0025mm
Aluminum	3.54×10^7	10.92mm	0.084mm	0.0027mm
Iron($\mu_r\approx10^3$)	1.00×10^7	0.65mm	0.005mm	0.00016mm
Sea water	4	32m	0.25m	

Since $\alpha=\beta$ for a good conductor, δ can also be written as

$$\delta = \frac{1}{\beta} = \frac{\lambda}{2\pi} \tag{7.71}$$

Substituting(7.70) into(7.67) leads to another expression of the intrinsic impedance for good conductor:

$$\eta_c \approx \frac{1+j}{\delta\sigma} = \frac{\sqrt{2}}{\delta\sigma}e^{j\pi/4} \tag{7.72}$$

Example 7-7 The electric field intensity of a uniform plane wave propagating in the $+z$-direction in the seawater is $\boldsymbol{E}=\boldsymbol{e}_x 100\cos(10^7\pi t)$ (V/m) at $z=0$. Assume the seawater has $\varepsilon_r=72$, $\mu_r=1$, and $\sigma=4$S/m. ① Determine the attenuation constant, phase constant, intrinsic impedance, phase velocity, wavelength and skin depth. ② Find the distance at which the amplitude of $\boldsymbol{E}$ becomes 1% of its value at $z=0$. ③Write the instantaneous expressions for $\boldsymbol{E}$ and $\boldsymbol{H}$ at $z=0.8$m.

Solution: From the given expression of $\boldsymbol{E}$ at $z=0$, we have $\omega=10^7\pi$(rad/s), $f=5\times10^6$Hz, and

$$\frac{\sigma}{\omega\varepsilon} = \frac{\sigma}{\omega\varepsilon_0\varepsilon_r} = \frac{4}{10^7\pi(10^{-9}/36\pi)72} = 200 \gg 1$$

Hence the seawater behaves as a good conductor.

(1) Attenuation constant:

$$\alpha = \sqrt{\pi f\mu\sigma} = \sqrt{\pi(5\times10^6)(4\pi\times10^{-7})4} = 8.89(\text{Np/m})$$

Phase constant:

$$\beta = \sqrt{\pi f\mu\sigma} = 8.89(\text{rad/m})$$

Intrinsic impedance:

$$\eta_c = (1+j)\sqrt{\frac{\pi f\mu}{\sigma}}$$

$$= (1+j)\sqrt{\frac{\pi(5\times10^6)(4\pi\times10^{-7})}{4}} = \pi e^{j\pi/4}(\Omega)$$

Phase velocity:

$$u_p = \frac{\omega}{\beta} = \frac{10^7\pi}{8.89} = 3.53\times10^6(\text{m/s})$$

Wavelength:

$$\lambda = \frac{2\pi}{\beta} = \frac{2\pi}{8.89} = 0.707(\mathrm{m})$$

Skin depth:

$$\delta = \frac{1}{\alpha} = \frac{1}{8.89} = 0.112(\mathrm{m})$$

(2) For the amplitude of wave to decrease by a factor of 1%, the distance d should satisfy

$$\mathrm{e}^{-\alpha d} = 0.01$$

$$d = -\frac{1}{\alpha}\ln 0.01 = \frac{4.605}{8.89} = 0.518(\mathrm{m})$$

(3) The phasor form of the electric field can be written as

$$\boldsymbol{E}(z) = \boldsymbol{e}_x 100\mathrm{e}^{-\alpha z}\mathrm{e}^{-\mathrm{j}\beta z}$$

By using (7.55), the phasor form of the magnetic field is

$$\boldsymbol{H}(\boldsymbol{r}) = \frac{1}{\eta_c}\boldsymbol{e}_z \times (\boldsymbol{e}_x 100\mathrm{e}^{-\alpha z}\mathrm{e}^{-\mathrm{j}\beta z}) = \boldsymbol{e}_y \frac{100}{\pi}\mathrm{e}^{-\alpha z}\mathrm{e}^{-\mathrm{j}\beta z - \mathrm{j}\pi/4}$$

The instantaneous expression for $\boldsymbol{E}$ is

$$\begin{aligned}\boldsymbol{E}(z,t) &= \mathrm{Re}[\boldsymbol{E}(z)\mathrm{e}^{\mathrm{j}\omega t}] \\ &= \mathrm{Re}[\boldsymbol{e}_x 100\mathrm{e}^{-\alpha z}\mathrm{e}^{\mathrm{j}(\omega t - \beta z)}] = \boldsymbol{e}_x 100\mathrm{e}^{-\alpha z}\cos(\omega t - \beta z)\end{aligned}$$

and the instantaneous expression for $\boldsymbol{H}$ is [①]

$$\boldsymbol{H}(z,t) = \mathrm{Re}[\boldsymbol{H}(z)\mathrm{e}^{\mathrm{j}\omega t}] = \mathrm{Re}\left[\boldsymbol{e}_y \frac{100}{\pi}\mathrm{e}^{-\alpha z}\mathrm{e}^{\mathrm{j}(\omega t - \beta z - \pi/4)}\right] = \boldsymbol{e}_y \frac{100}{\pi}\mathrm{e}^{-\alpha z}\cos(\omega t - \beta z - \pi/4)$$

At $z = 0.8\mathrm{m}$, we have

$$\boldsymbol{E}(0.8,t) = \boldsymbol{e}_x 100\mathrm{e}^{-0.8\alpha}\cos(\omega t - 0.8\beta) = \boldsymbol{e}_x 0.082\cos(10^7\pi t - 7.11)$$

$$\boldsymbol{H}(0.8,t) = \boldsymbol{e}_y \frac{100}{\pi}\mathrm{e}^{-0.8\alpha}\cos(\omega t - 0.8\beta - \pi/4) = \boldsymbol{e}_y 0.026\cos(10^7\pi t - 7.9)$$

7.3.3 Surface Resistance and Surface Impedance (表面电阻与表面阻抗)

In Chapter 4, we discussed the steady currents due to the electric field inside a conductor in the DC case. In that case, we can assume the electric field as well as the current density to be uniform throughout certain conductors such as a cylindrical wire. Based on this assumption, the DC resistance of a wire can be derived as given in (4.12). For the time-varying case, however, the electric field in a good conductor is confined in a thin layer of the surface due to the skin effect as discussed in the previous section. In this case, the resistance of a piece of wire cannot be calculated by (4.12) anymore. Instead, the resistance-behavior of a good conductor is characterized by using the concept of **surface resistance** (表面电阻,也叫 sheet resistance). In

① 对于导电媒质中沿 $+z$ 方向传播的均匀平面波,可以采用式(7.55)由已知的电场相量表达式直接得到磁场相量表达式,然后再由两者的相量表达式分别得到瞬时表达式。需要注意电场和磁场的瞬时表达式的比值不是一个固定的本征阻抗,不能由类似式(7.14)的方式由电场计算磁场。因为导电媒质的本征阻抗式复数,电、磁场的相位并不一致,两者比值不再是常数,而是随时间变化。

this subsection, we introduce the surface resistance from the point of view of power dissipation and find out the relations between the surface impedance and electromagnetic fields.

With the relations (7.70), (7.71) and (7.72), the $\boldsymbol{E}$, $\boldsymbol{H}$ and $\boldsymbol{S}_{av}$ expressions (7.48), (7.57) and (7.58) for a good conductor can be rewritten in terms of the skin depth δ as:

$$\boldsymbol{E}(z) = \boldsymbol{e}_x E_0 e^{-z/\delta} e^{-jz/\delta} \tag{7.73}$$

$$\boldsymbol{H}(z) = \boldsymbol{e}_y \frac{1}{|\eta_c|} E_0 e^{-z/\delta} e^{-jz/\delta} e^{-j\theta_\eta} = \boldsymbol{e}_y \frac{\delta\sigma}{\sqrt{2}} E_0 e^{-z/\delta} e^{-jz/\delta} e^{-j\pi/4} \tag{7.74}$$

$$\boldsymbol{S}_{av} = \boldsymbol{e}_z \frac{|E_0|^2}{2|\eta_c|} e^{-2\alpha z} \cos\theta_\eta = \boldsymbol{e}_z \frac{\delta\sigma |E_0|^2}{4} e^{-2z/\delta} \tag{7.75}$$

Suppose a plane wave enters the surface of a good conductor at $z = 0$ as is shown in Figure 7-7(a). The time-average power entering the conductor through a rectangular region ($0<x<l$, $0<y<w$) is

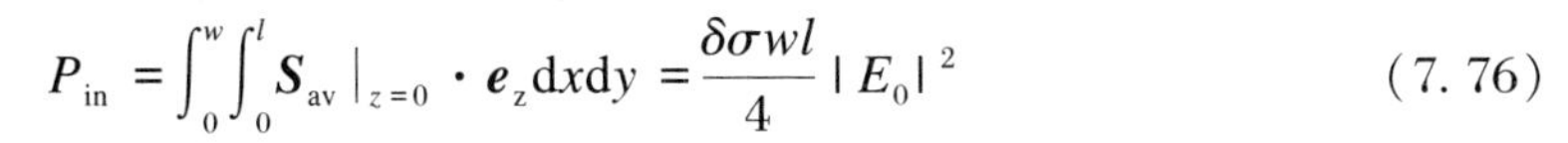

$$P_{in} = \int_0^w \int_0^l \boldsymbol{S}_{av}|_{z=0} \cdot \boldsymbol{e}_z \mathrm{d}x\mathrm{d}y = \frac{\delta\sigma wl}{4} |E_0|^2 \tag{7.76}$$

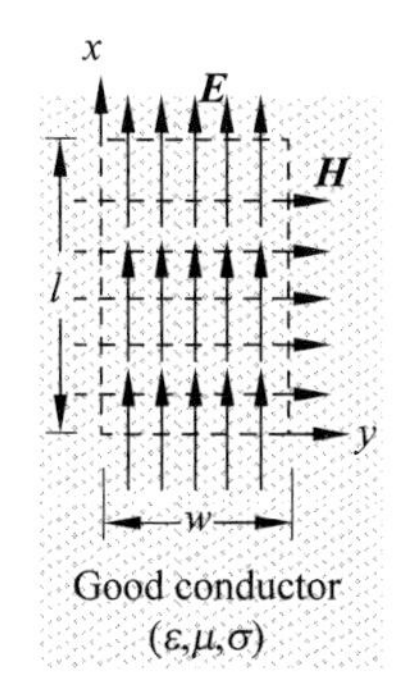

(a) tangential $\boldsymbol{E}$ and $\boldsymbol{H}$ fields on the surface of the conductor

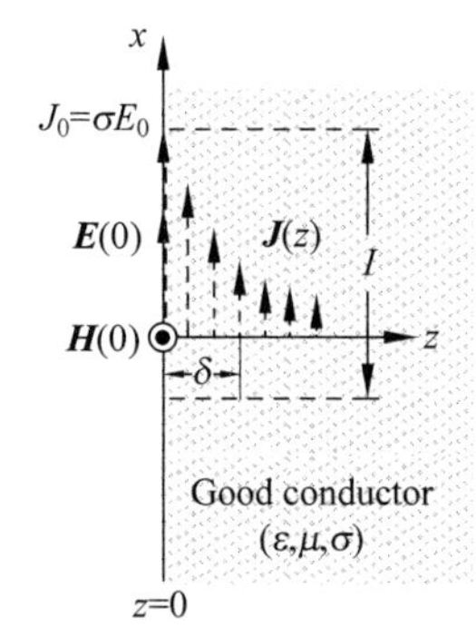

(b) current density distribution $\boldsymbol{J}(z)$ within the conductor

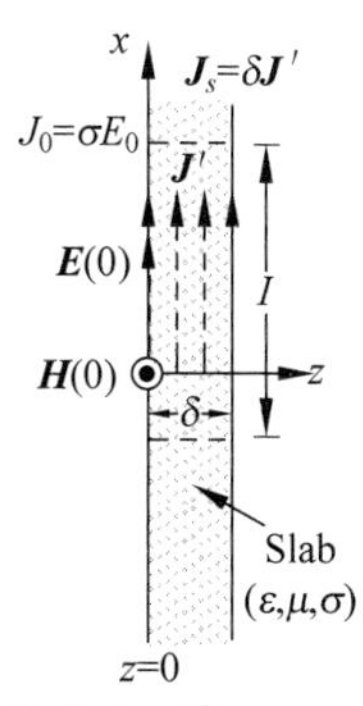

(c) equivalent uniform current density within an equivalent conductor slab with thickness δ

Figure 7-7 A plane wave enters the surface of a good conductor at $z=0$

This power is dissipated in the conductor in the region ($0<x<l$, $0<y<w$, $0<z<+\infty$) as the ohmic loss, which can be calculated as

$$P_L = \int_0^{+\infty} \int_0^w \int_0^l \frac{1}{2}\sigma |\boldsymbol{E}|^2 \mathrm{d}x\mathrm{d}y\mathrm{d}z = \frac{wl}{2}\int_0^{+\infty} \sigma |E_0|^2 e^{-2z/\delta} \mathrm{d}z = \frac{\delta\sigma wl}{4}|E_0|^2 \tag{7.77}$$

The calculated P_{in} in (7.76) is the same as P_L in (7.77), which is required by the conservation of energy.

From (7.73), we also know the current density within the conductor to be

$$\boldsymbol{J}(z) = \boldsymbol{e}_x \sigma E_0 e^{-z/\delta} e^{-jz/\delta} = \boldsymbol{e}_x J_0 e^{-z/\delta} e^{-jz/\delta} \tag{7.78}$$

where $J_0 = \sigma E_0$ is the volume current density at the surface of the conductor ($z=0$) as is shown in Figure 7-7(b). The total current (along x-direction) in the conductor in the cross section ($0<y<w$, $0<z<+\infty$) is

$$I = \int_0^{+\infty} \int_0^w J(z) \mathrm{d}y\mathrm{d}z = w\int_0^{+\infty} J_0 e^{-z/\delta} e^{-jz/\delta} \mathrm{d}z = \frac{J_0 w\delta}{1+j} \tag{7.79}$$

The current I given in(7.79) is proportional to the area $w\delta$, and therefore, can be considered as equivalent to a uniform current $\boldsymbol{J}'$ distributed within one skin depth δ into the conductor ($0<z<\delta$) as is shown in Figure 7-7(c). The equivalence requires $I=\boldsymbol{J}'w\delta$, which leads to

$$\boldsymbol{J}' = \begin{cases} \boldsymbol{e}_x \dfrac{J_0}{1+\mathrm{j}} & 0<z<\delta \\ 0 & \text{otherwise} \end{cases} \tag{7.80}$$

The time-average power dissipated by the equivalent current conductor $\boldsymbol{J}'$ in the region($0<x<l$, $0<y<w$, $0<z<\delta$) as the ohmic loss is

$$P_L' = \int_0^\delta \int_0^w \int_0^l \frac{1}{2}\frac{|\boldsymbol{J}'|^2}{\sigma}\mathrm{d}x\mathrm{d}y\mathrm{d}z = \frac{wl\delta}{2\sigma}\left|\frac{J_0}{1+\mathrm{j}}\right|^2 = \frac{wl\delta}{4\sigma}|J_0|^2 = \frac{wl\delta\sigma}{4}|E_0|^2 \tag{7.81}$$

P_L' is obviously the same as the power entering into the conductor P_{in} as given in(7.76) and the actual loss in the conductor P_L as given in(7.77). Therefore, the time-average power loss in a good conductor may be calculated by assuming that the total current is uniformly distributed in one skin depth under the surface. In other words, an infinitely thick good conductor with skin effect can be considered equivalent to a thin(thickness δ) conductor without skin effect. ①

The resistance R associated with the rectangular region($0<x<l$, $0<y<w$, $0<z<\infty$) can be determined by equating P_{in}(or P_L, P_L') with $R|I|^2/2$, which gives

$$R = \frac{2P_L'}{|I|^2} = \frac{2\dfrac{wl\delta}{4\sigma}|J_0|^2}{\left|\dfrac{J_0 w\delta}{1+\mathrm{j}}\right|^2} = \frac{l}{\sigma w\delta} \tag{7.82}$$

which is equivalent to the DC resistance of a thin slab with length l, width w and thickness δ.

(7.82) shows that the resistance of a conductor surface is proportional to the ratio of length(along the direction of current flow) and width of the surface. Therefore, we can define a **surface resistance**(**表面电阻**) as

$$R_S = \frac{R}{(l/w)} = \frac{1}{\sigma\delta} = \sqrt{\frac{\pi f\mu}{\sigma}} = \mathrm{Re}[\eta_c] \tag{7.83}$$

Then, the part of a good conductor with a surface of length l(in the direction of $\boldsymbol{E}$ field) and width w has a resistance of

$$R = R_S\left(\frac{l}{w}\right) \tag{7.84}$$

The surface resistance R_S is a property of the material. We see from(7.84) that any good

---

① 这里采用三种方法计算了良导体的($0<x<l$, $0<y<w$, $0<z<+\infty$)区域内的功率损耗：第一种方法利用式(7.76)计算由良导体表面($0<x<l$, $0<y<w$)进入该区域的时间平均坡印廷矢量的通量，即 P_{in}；第二种方法利用式(7.77)直接计算该区域内欧姆损耗 P_L；第三种方法首先通过式(7.80)将该区域 $0<z<+\infty$ 范围内指数衰减的实际电流分布 $\boldsymbol{J}$ 等效为一个只存在于 $0<z<\delta$ 表层范围的均匀电流分布 $\boldsymbol{J}'$，然后由式(7.81)计算该表层区域的欧姆损耗 P_L'。三者计算结果相等，意味着良导体的时变场功率损耗可以通过假设电流集中分布于一个趋肤深度的表层来计算。换言之，在趋肤效应之下，厚度足够(远厚于一个趋肤深度)的良导体可以等效为一个没有趋肤效应，但厚度仅为趋肤深度 δ 的导体。

conductor surface with a square shape(i. e. , with $l=w$) has a resistance $R=R_S$ independent of the length and width. Therefore, the unit of the surface resistance is ohm per square($\Omega/\square$).

(7.83) indicates that R_S is dependent on frequency. As an example, given the conductivity of copper to be 5.8×10^7 S/m, the surface resistance is $(\pi f\times4\pi\times10^{-7}/5.8\times10^7)^{1/2}=2.61\times10^{-7}f^{1/2}(\Omega)$.

The surface resistance can be used to calculate the power loss of a good conductor from the tangential $\boldsymbol{H}$ field on the surface of the conductor. From(7.74), we have

$$\boldsymbol{H}(0)=\boldsymbol{e}_y\frac{1}{1+\mathrm{j}}\delta\sigma E_0=\boldsymbol{e}_y\frac{\delta J_0}{1+\mathrm{j}} \tag{7.85}$$

From(7.81), (7.83) and(7.85), the power loss on a conductor surface with area($l\times w$) can also be expressed as

$$P_L=\frac{wl\delta}{2\sigma}\left|\frac{J_0}{1+\mathrm{j}}\right|^2=\frac{wl}{2\sigma\delta}|\boldsymbol{H}(0)|^2=\frac{1}{2}wlR_S|\boldsymbol{H}(0)|^2 \tag{7.86}$$

Notice that $\boldsymbol{H}(0)$ is the tangential magnetic field intensity on the surface of the conductor. Generally, for an arbitrary surface S of a good conductor, the power loss can be calculated from the tangential component of the $\boldsymbol{H}$ field, H_t, as

$$P_L=\frac{1}{2}\int_S R_S|H_t|^2\mathrm{d}s \tag{7.87}$$

By noticing(7.79), (7.85) can also be written as

$$\boldsymbol{H}(0)=\boldsymbol{e}_y\frac{I}{w}=\boldsymbol{e}_yJ_S=\boldsymbol{e}_z\times\boldsymbol{J}_S \tag{7.88}$$

where

$$\begin{aligned}\boldsymbol{J}_S&=\boldsymbol{e}_xJ_S=\boldsymbol{e}_x\frac{J_0\delta}{1+\mathrm{j}}=\delta\boldsymbol{J}'\\&=\boldsymbol{e}_x\frac{\sigma\delta}{1+\mathrm{j}}E_0=\frac{\sigma\delta}{1+\mathrm{j}}\boldsymbol{E}(0)\end{aligned} \tag{7.89}$$

is an equivalent surface current density if we consider all the current in the conductor are confined within the skin depth and treat it as a surface current. In this case, (7.88) essentially represents the boundary condition on the conductor surface at $z=0$. Now we can define a **surface impedance**(**表面阻抗**) Z_S to be the ratio between the electric field at the surface and the surface current:

$$Z_S=\frac{E(0)}{J_S}=\frac{1+\mathrm{j}}{\sigma\delta}=R_S+\mathrm{j}R_S \tag{7.90}$$

which is essentially the same as the intrinsic impedance as given in(7.72).

7.4 Group Velocity(群速度)

In Section 7.2, we defined the phase velocity of a uniform plane wave as the velocity of propagation of its equiphase wavefront. Given the frequency ω and the phase constant β, the phase velocity u_p is

$$u_p = \frac{\omega}{\beta} \tag{7.91}$$

In a lossless medium, $\beta=\omega\sqrt{\mu\varepsilon}$ is a linear function of ω. Consequently, the phase velocity $u_p = 1/\sqrt{\mu\varepsilon}$ is a constant independent of the frequency. However, in other cases, the phase constant may not be a linear function of ω. Such cases include wave propagation in a lossy dielectric, in which the phase constant is generally a complicated function of the frequency as indicated by (7.51b). In Chapter 9, we will also see that the wave in a waveguide has a phase constant dependent on the frequency. In these cases, waves at different frequencies travel with different velocities, which causes waveform **distortion** (失真) of signals consisting a band of frequencies. Signal distortion caused by the dependence of the phase velocity on frequency is called **dispersion**(色散). As indicated by (7.53), a lossy dielectric is apparently a **dispersive medium**(色散媒质).

In practice, an information-bearing electromagnetic wave usually has a narrow band around a high carrier frequency, which forms a wave-packet envelope. Information carried by this kind of wave is transmitted via the propagation of the envelope. The propagation velocity of the wave-packet envelope is called **group velocity**(群速度), which generally can be thought of as the signal velocity or velocity of information. In the following, we will obtain the defining equation of the group velocity from a simple example of the propagation of a two-tone wave. Then, we show that the group velocity and the phase velocity are the same in non-dispersive media but are different in dispersive media.

Consider a wave packet consisting of two traveling waves with equal amplitude but slightly different angular frequencies $\omega_0 + \Delta\omega$ and $\omega_0 - \Delta\omega$ ($\Delta\omega \ll \omega_0$). The phase constants at the two frequencies are also slightly different and can be denoted as $\beta_0 + \Delta\beta$ and $\beta_0 - \Delta\beta$ respectively. Then the instantaneous expression of the wave packet can be written as

$$\begin{aligned} E(z,t) &= E_0\cos[(\omega_0 + \Delta\omega)t - (\beta_0 + \Delta\beta)z] + E_0\cos[(\omega_0 - \Delta\omega)t - (\beta_0 - \Delta\beta)z] \\ &= 2E_0\cos(\Delta\omega t - z\Delta\beta)\cos(\omega_0 t - \beta_0 z) \end{aligned} \tag{7.92}$$

Since $\Delta\omega \ll \omega_0$, the expression in (7.92) represents a carrier wave $\cos(\omega_0 t - \beta z)$ with a slowly varying envelope $2E_0\cos(\Delta\omega_0 t - z\Delta\beta)$. Figure 7-8 illustrates a typical waveform as a function of location at a fixed time. The carrier wave propagates with a phase velocity found by letting $\omega_0 t - \beta_0 z$=constant, which is

$$u_p = \frac{\mathrm{d}z}{\mathrm{d}t} = \frac{\omega_0}{\beta_0}$$

The velocity of the envelope is determined by letting $\Delta\omega_0 t - z\Delta\beta$ = constant, which gives the group velocity

$$u_g = \frac{\mathrm{d}z}{\mathrm{d}t} = \frac{\Delta\omega}{\Delta\beta} = \frac{1}{\Delta\beta/\Delta\omega}$$

In the limit that $\Delta\omega \to 0$, we have the defining equation of the group velocity:

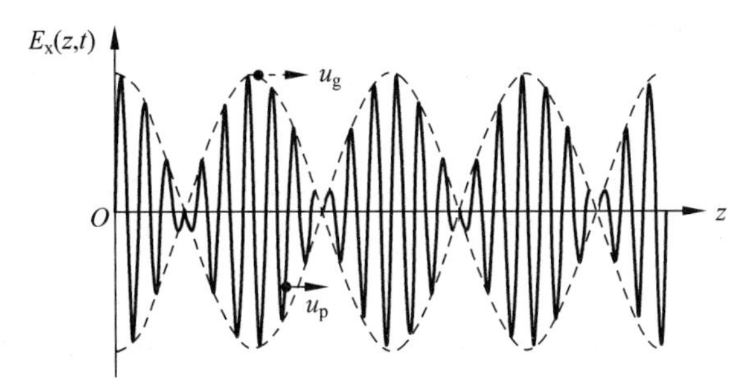

Figure 7-8 Sum of two time-harmonic traveling waves of equal amplitude and slightly different frequencies at a given t

$$u_g = \frac{1}{d\beta/d\omega} \tag{7.93}$$

A general relation between the group and phase velocities may be obtained by combining (7.91) and (7.93). From (7.91) we have

$$\frac{d\beta}{d\omega} = \frac{d}{d\omega}\left(\frac{\omega}{u_p}\right) = \frac{1}{u_p} - \frac{\omega}{u_p^2}\frac{du_p}{d\omega}$$

Substitution of the above in (7.93) yields

$$u_g = \frac{u_p}{1 - \frac{\omega}{u_p}\frac{du_p}{d\omega}} \tag{7.94}$$

From (7.94) we see three possible cases:

(1) **No dispersion(无色散)**:

$$\frac{du_p}{d\omega} = 0$$

which means that u_p is independent of ω, and β is a linear function of ω. And as a result,

$$u_g = u_p$$

(2) **Normal dispersion(正常色散)**:

$$\frac{du_p}{d\omega} < 0$$

which means that u_p decreases with ω, and as a result,

$$u_g < u_p$$

(3) **Anomalous dispersion(反常色散)**:

$$\frac{du_p}{d\omega} > 0$$

which means that u_p increases with ω, and as a result,

$$u_g > u_p$$

Example 7-8 A narrow-band signal propagates in a lossy dielectric medium which has a loss tangent 0.2 at 550kHz, the carrier frequency of the signal. The dielectric constant of the medium is 2.5.

(1) Determine α and β

(2) Determine u_p and u_g. Is the medium dispersive?

Solution:

(1) Since the loss tangent $\sigma/\omega\varepsilon = 0.2$ is much smaller than 1, the medium can be considered as a low loss medium. (7.59) and (7.60) can be used to determine α and β respectively.

$$\sigma = 0.2\omega\varepsilon = 0.2\times 2\pi\times 550\times 10^3\times 2.5\varepsilon_0 = 1.73\times 10^6\varepsilon_0\,(\text{F/m})$$

$$\alpha = \frac{\sigma}{2}\sqrt{\frac{\mu}{\varepsilon}} = \frac{1.73\times 10^6\varepsilon_0}{2}\sqrt{\frac{\mu_0}{2.5\varepsilon_0}} = \frac{1.73\times 10^6}{2}\sqrt{\frac{\varepsilon_0\mu_0}{2.5}} = 1.82\times 10^{-3}\,(\text{Np/m})$$

$$\beta = \omega\sqrt{\mu\varepsilon} = 2\pi(550\times 10^3)\frac{\sqrt{2.5}}{3\times 10^8} = 0.0182\,(\text{rad/m})$$

(2) Phase velocity (from (7.63)):

$$u_p = \frac{\omega}{\beta} = \frac{1}{\sqrt{\mu\varepsilon}} \approx \frac{3\times 10^8}{\sqrt{2.5}} = 1.897\times 10^8\,(\text{m/s})$$

Group velocity (from (7.60)):

$$u_g \approx u_p = 1.897\times 10^8\,(\text{m/s})$$

Summary

Concepts

Uniform plane wave(均匀平面波)

Wavefront(波前/波阵面)

Wavelength(波长)

Wavenumber(波数)

Phase velocity(相速度)

Intrinsic impedance(本征阻抗)

TEM wave(横电磁波)

Traveling wave(行波)

Polarization(极化)

Linearly polarized wave(线性极化波)

Circularly polarized wave(圆极化波)

left-hand/right-hand(左旋/右旋)

Complex permittivity(复介电常数)

Loss tangent, loss angle(损耗角正切,损耗角)

Complex wavenumber(复波数)

Propagation constant(传播常数)

Attenuation constant(衰减常数)

Phase constant(相位常数)

Low loss dielectric(低损耗介质/弱导电介质)

Good conductor(良导体)

Skin depth, depth of penetration(趋肤深度,穿透深度)

Dispersion, dispersive medium(色散,色散媒质)

Group velocity(群速)

Normal/anomalous dispersion(正常/反常色散)

Laws & Theorems

Relations between ***E***, ***H***, and propagation direction for TEM wave(横电磁波的场与传播方向之间的关系)

Homogeneous vector Helmholtz's equation and its solution(齐次亥姆霍兹方程及其解).

Methods

Use characteristics of TEM waves to find ***H*** field given ***E*** field expression, and vice versa;
Determine the polarization type of a TEM wave given the field expressions.

Review Questions

7.1 什么是均匀平面波?

7.2 什么是波前? 平面波的波前有什么特征?

7.3 什么是行波?

7.4 什么是波数? 波数和波长的关系是什么?

7.5 定义相速度。

7.6 什么是媒质的本征阻抗? 自由空间的本征阻抗是多少?

7.7 什么是 TEM 波? 均匀平面波是 TEM 波吗?

7.8 什么是波的极化? 线性极化和圆极化分别是什么含义?

7.9 写出沿+z 方向传播,x 方向极化的均匀平面波的 ***E*** 和 ***H*** 的相量形式表达式。

7.10 如何判定一个均匀平面波的极化形式是线性极化、左旋圆极化、右旋圆极化还是椭圆极化?

7.11 传播常数、衰减常数、相位常数分别指的是什么? 它们之间是什么关系?

7.12 什么是导体的趋肤深度? 趋肤深度与衰减常数、电导率、频率之间的分别是什么关系?

7.13 什么是色散? 色散媒质有什么特征?

7.14 什么是群速度? 正常色散和反常色散条件下,群速度和相速度之间分别是什么关系?

Problems

7.1 Derive the wave equations of the ***E*** and ***H*** fields in a source-free conducting medium with constitutive parameters ε, μ, σ.

7.2 For a uniform plane wave propagating in a source-free simple medium, with the ***E*** field expressed by (7.19), prove that the four Maxwell's equations reduce to the following equations:

$$\boldsymbol{k} \times \boldsymbol{E} = \omega\mu\boldsymbol{H}$$

$$\boldsymbol{k} \times \boldsymbol{H} = -\omega\varepsilon\boldsymbol{E}$$

$$\boldsymbol{k} \cdot \boldsymbol{E} = 0$$

$$\boldsymbol{k} \cdot \boldsymbol{H} = 0$$

7.3 The instantaneous expression for the magnetic field intensity of a uniform plane wave propagating in the air is given by

$$H = e_z 10^{-6} \cos\left(10^8 \pi t - k_0 y + \frac{\pi}{3}\right)$$

(1) Toward what direction is the wave propagating?

(2) Determine k_0, the wavelength, and the locations where H_z vanishes at $t=0$.

(3) Determine the instantaneous expression of $\boldsymbol{E}$.

7.4 The instantaneous expression for the electric field of a uniform plane wave in a nonmagnetic dielectric medium is given by

$$E(t,z) = e_x 2\cos(10^8 t - z/\sqrt{3}) - e_y \sin(10^8 t - z/\sqrt{3})$$

(1) Determine the frequency and wavelength of the wave.

(2) What is the dielectric constant of the medium?

(3) Describe the polarization of the wave.

(4) Find the corresponding $\boldsymbol{H}$-field.

7.5 Determine the polarization of the following waves:

(1) $E=200e^{-j100x} e_y + 150e^{-j100x} e_z$ V/m

(2) $E=24e^{-j\pi/4} e^{-j300z} e_x - 18e^{-j\pi/4} e^{-j300z} e_y$ V/m

(3) $E=3\cos(t-0.5y) e_x - 4\sin(t-0.5y) e_z$ V/m

7.6 Prove that, any elliptically polarized plane wave can be obtained from the superposition of a right-hand and a left-hand circularly polarized wave.

7.7 Calculate the intrinsic impedance, attenuation constant, and skin depth of the copper medium (with $\sigma=5.80\times10^7$ S/m) at the following frequencies: ①50Hz, ②1kHz, ③1MHz, and ④1GHz.

7.8 A 3GHz, x-polarized uniform plane wave propagates toward the $+z$-direction in a nonmagnetic medium with a dielectric constant 2.5 and a loss tangent 10^{-2}. The electric field at $z=0$ has an amplitude of 50V/m and a phase of $\pi/3$.

(1) Determine the distance over which the amplitude of the electric field drops by half.

(2) Determine the intrinsic impedance, wavelength, phase velocity, and group velocity of the wave.

(3) Write down the instantaneous expression for $\boldsymbol{E}$ and $\boldsymbol{H}$ for all t and z.

7.9 A 1MHz uniform plane wave is propagating toward the $+z$-direction in moist soil (with $\varepsilon_r = 15$, $\mu_r = 1$, $\sigma = 0.05$S/m). The electric field at $z = 0$ has an amplitude 110V/m. Determine ① the propagation, attenuation, and phase constants, ② the phase velocity, ③ the wavelength, ④the intrinsic impedance, ⑤the skin depth, and ⑥the time-average power density.

7.10 A 5GHz x-polarized uniform plane wave propagates toward the $+y$-direction in seawater (with $\varepsilon_r=80$, $\mu_r=1$, $\sigma=4$S/m). The magnetic field intensity of the wave at $y=0$ has an amplitude of 0.1A/m and a phase of $\pi/3$,

(1) Determine the attenuation constant, the phase constant, the intrinsic impedance, the phase velocity, the wavelength, and the skin depth.

(2) Find the location at which the amplitude of $\boldsymbol{H}$ is 0.01A/m.

(3) Write down the instantaneous expressions for $\boldsymbol{E}$ and $\boldsymbol{H}$ for all t and y.

7.11 The skin depth of graphite (石墨) is 0.16mm at 100MHz. Determine ①the conductivity of graphite, and ②the distance over which the electric field of a 1GHz uniform plane wave traveling in graphite attenuates by 30dB.

7.12 The electric field intensity in a nonmagnetic medium is given by

$$\boldsymbol{E} = \boldsymbol{e}_x 0.1 e^{-77.485y} \cos(2\pi \times 10^9 t - 302.6y)$$

Determine the dielectric constant and the conductivity of the medium. Obtain the corresponding ***H*** field. Evaluate the time-average power density.

7.13 A uniform plane wave is propagating in a good conductor. If the magnetic field intensity is given by

$$\boldsymbol{H} = \boldsymbol{e}_x 0.1 e^{-25z} \sin(2\pi \times 10^9 t - 25z)$$

Determine the conductivity and the corresponding ***E*** field. Evaluate the time-average power dissipation in a volume with unit area and a depth of penetration in the propagation direction.

7.14 A uniform plane wave at 50MHz is propagating within a nonmagnetic lossy medium. The amplitude of the electric field attenuates with a rate of 1dB per meter. The magnetic field lags the electric field by a phase of 0.2π. Find ①the attenuation and phase constants, ②the conductivity, ③the intrinsic impedance, and ④the skin depth of the medium.

7.15 Prove the following relations between group velocity u_g and phase velocity u_p in a dispersive medium:

(1) $u_g = u_p + \beta \dfrac{du_p}{d\beta}$;

(2) $u_g = u_p - \lambda \dfrac{du_p}{d\lambda}$

7.16 Show that the instantaneous Poynting vector of a circularly polarized plane wave propagating in a lossless medium is a constant, independent of time and space.

Chapter 8

Plane Wave Reflection and Transmission (平面波的反射与透射)

8.1 Introduction(引言)

In Chapter 7, we discussed the propagation of uniform plane waves in an unbounded homogeneous medium, in which single traveling wave can exist. When traveling waves reach an interface between two media with different constitutive parameters, wave reflection and transmission occur. In this chapter, we study the phenomena of reflection and transmission of plane waves when the interface is an infinitely large planar surface. As shown in Figure 8-1, if the **incident wave**(入射波) is a uniform plane wave, the **reflected wave**(反射波) will also be a uniform plane wave. Both the incident wave and reflected wave travels in the same medium (medium 1). If the medium on the other side of the interface (medium 2) is dielectric, there can also be a **transmitted wave**(透射波) that propagates in medium 2. ①

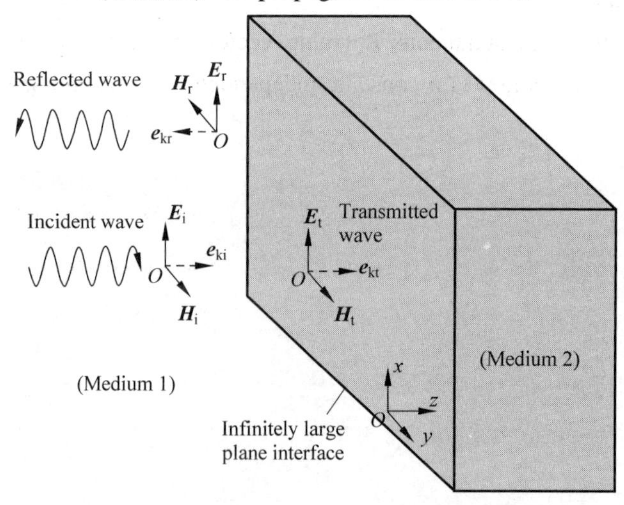

Figure 8-1 Reflection and transmission of plane waves on an interface between two different media

① 第7章讨论的均匀平面波存在于无限大均匀媒质中,当空间中存在两种不同媒质时,均匀平面波在一种媒质中传播至分界面时,一般情况下无法不受影响地穿过该分界面,而是会发生反射和透射现象。当分界面是一个无穷大的平面时,可以根据边界条件由已知的入射波求解得到反射波和透射波的解析表达式。在此基础上,本章讨论不同媒质分界面上的反射与透射规律。

In Sections 8.2 and 8.3, we examine the scenario when a uniform plane wave is incident upon a dielectric-PEC interface. Wave reflections and transmissions on dielectric-dielectric interfaces are discussed in Sections 8.4, 8.5, and 8.6.

8.2 Normal Incidence at a PEC Planar Boundary (理想导体平面的垂直入射)

Consider an incident plane wave ($\boldsymbol{E}_i, \boldsymbol{H}_i$) traveling in a lossless medium (medium 1: $\sigma_1 = 0$) along the $+z$-direction impinges upon an interface with a perfect conductor (medium 2: $\sigma_2 = \infty$). As is shown in Figure 8-2, the boundary surface is the plane $z=0$. The incident fields are assumed to be

$$\boldsymbol{E}_i(z) = \boldsymbol{e}_x E_{i0} e^{-j\beta_1 z} \tag{8.1}$$

$$\boldsymbol{H}_i(z) = \boldsymbol{e}_y \frac{E_{i0}}{\eta_1} e^{-j\beta_1 z} \tag{8.2}$$

where E_{i0} is the magnitude of $\boldsymbol{E}_i$ at $z=0$, β_1 and η_1 are respectively the phase constant and the intrinsic impedance of medium 1.

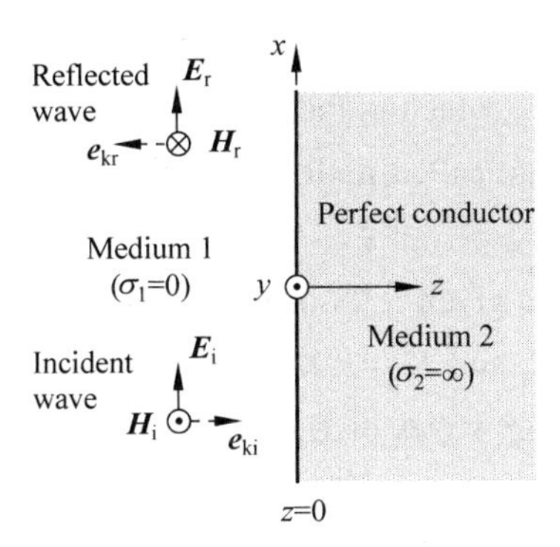

Figure 8-2 Plane wave incident normally on a planar PEC boundary

Since medium 2 is a PEC, the electric and magnetic fields in medium 2, denoted by $\boldsymbol{E}_2$ and $\boldsymbol{H}_2$, must be zero. That means no wave is transmitted across the boundary into the $z>0$ region. Physically, the incident wave is reflected, leading to a reflected wave ($\boldsymbol{E}_r, \boldsymbol{H}_r$). The reflected wave must propagate along the $-z$-direction. Hence the reflected electric field can be written as

$$\boldsymbol{E}_r(z) = \boldsymbol{e}_x E_{r0} e^{+j\beta_1 z} \tag{8.3}$$

where E_{r0} is the unknown magnitude of $\boldsymbol{E}_r$ at $z=0$. To find the unknown E_{r0}, we first write down the total electric field intensity in medium 1 as the superposition of $\boldsymbol{E}_i$ and $\boldsymbol{E}_r$:

$$\boldsymbol{E}_1(z) = \boldsymbol{E}_i(z) + \boldsymbol{E}_r(z) = \boldsymbol{e}_x (E_{i0} e^{-j\beta_1 z} + E_{r0} e^{+j\beta_1 z}) \tag{8.4}$$

The total electric field must satisfy the boundary condition of a PEC surface, which requires that

$$\boldsymbol{E}_1(0) = \boldsymbol{e}_x (E_{i0} + E_{r0}) = \boldsymbol{E}_2(0) = 0$$

Therefore, $E_{r0}=-E_{i0}$, and (8.4) becomes[①]

$$\boldsymbol{E}_1(z)=\boldsymbol{e}_x E_{i0}(\mathrm{e}^{-\mathrm{j}\beta_1 z}-\mathrm{e}^{+\mathrm{j}\beta_1 z})=-\boldsymbol{e}_x \mathrm{j}2E_{i0}\sin\beta_1 z \tag{8.5}$$

The magnetic field intensity $\boldsymbol{H}_r$ of the reflected wave is related to $\boldsymbol{E}_r$ by (7.26): [②]

$$\boldsymbol{H}_r(z)=\frac{1}{\eta_1}(-\boldsymbol{e}_z)\times\boldsymbol{E}_r(z)=\boldsymbol{e}_y\frac{E_{i0}}{\eta_1}\mathrm{e}^{+\mathrm{j}\beta_1 z} \tag{8.6}$$

The total magnetic field intensity in medium 1 is the superposition of $\boldsymbol{H}_i$ in (8.2) and $\boldsymbol{H}_r$ in (8.6):

$$\boldsymbol{H}_1(z)=\boldsymbol{H}_i(z)+\boldsymbol{H}_r(z)=\boldsymbol{e}_y 2\frac{E_{i0}}{\eta_1}\cos\beta_1 z \tag{8.7}$$

It is observed from (8.5) and (8.7) that the total fields $\boldsymbol{E}_1(z)$ and $\boldsymbol{H}_1(z)$ are 90° out of phase, and hence the time-average power associated with the total wave in medium 1 is zero.

The instantaneous expressions of the total electric and magnetic fields can be obtained from the phasor forms in (8.5) and (8.7):

$$\boldsymbol{E}_1(z,t)=\mathrm{Re}[\boldsymbol{E}_1(z)\mathrm{e}^{\mathrm{j}\omega t}]=\boldsymbol{e}_x 2E_{i0}\sin\beta_1 z\sin\omega t \tag{8.8}$$

$$\boldsymbol{H}_1(z,t)=\mathrm{Re}[\boldsymbol{H}_1(z)\mathrm{e}^{\mathrm{j}\omega t}]=\boldsymbol{e}_y 2\frac{E_{i0}}{\eta_1}\cos\beta_1 z\cos\omega t \tag{8.9}$$

which indicate that both $\boldsymbol{E}_1(z,t)$ and $\boldsymbol{H}_1(z,t)$ possess zeros and maxima at fixed locations that are independent of time t. Hence, the total wave in medium 1 is not a traveling wave. Instead, it is called a **standing wave**(**驻波**), which is the result of superposition of two waves traveling in opposite directions. These zeros are called **nodes**(**波节点**) and maxima are called **antinodes**(**波腹点**).

From (8.8) and (8.9), for a given t, both $\boldsymbol{E}_1$ and $\boldsymbol{H}_1$ vary sinusoidally with the coordinate z. Several waveforms of $\boldsymbol{E}_1=\boldsymbol{e}_x E_1$ and $\boldsymbol{H}_1=\boldsymbol{e}_y H_1$ are illustrated in Figure 8-3 at different time of ωt. It is observed that the standing waves of $\boldsymbol{E}_1$ and $\boldsymbol{H}_1$ are in **time quadrature**(with 90° phase difference) as well as in space quadrature (with a quarter wavelength shift in space). Specifically, we have the following observations: [③]

(1) Nodes of $\boldsymbol{E}_1(z,t)$ and antinodes of $\boldsymbol{H}_1(z,t)$ occur when $\beta_1 z=-n\pi$, which corresponds to locations

$$z=-\frac{n\lambda}{2},\quad n=0,1,2,\cdots$$

---

① 均匀平面波入射到 PEC 平面形成反射是因为入射波单独存在时不可能满足 PEC 表面的边界条件。垂直入射条件下，PEC 平面上入射波的电场和磁场都是均匀的，因此反射波在 PEC 平面上也必须是均匀的，这样两者之和才能满足边界条件。反射波也是均匀平面波，其传播方向也必须垂直于 PEC 平面，因此只能是沿 $-z$ 方向。

② 由于入射波和反射波都是均匀平面波，两者都各自满足式(7.26)给出的均匀平面波的基本特性，区别只是传播方向不同。而两个传播方向不同的均匀平面波叠加的总场 $\boldsymbol{E}_1$ 则不再满足式(7.26)。

③ 式(8.8)和式(8.9)表明，媒质 1 中入射波与反射波叠加形成的总的电场和总的磁场不具备行波的基本数学形式。相反，两者的零点(波节点)和最大振幅点(波腹点)在空间的位置都不随时间改变，因此被称为驻波。另外，电场和磁场在时间和空间上都正交(相差 90°)。时间上正交导致时均坡印廷矢量为 0，空间上正交意味着电场的波节点和磁场的波腹点出现在同一位置，电场的波腹点和磁场的波节点出现在同一位置，电场的两个相邻波节点的中间位置是磁场的波节点，反之亦然。

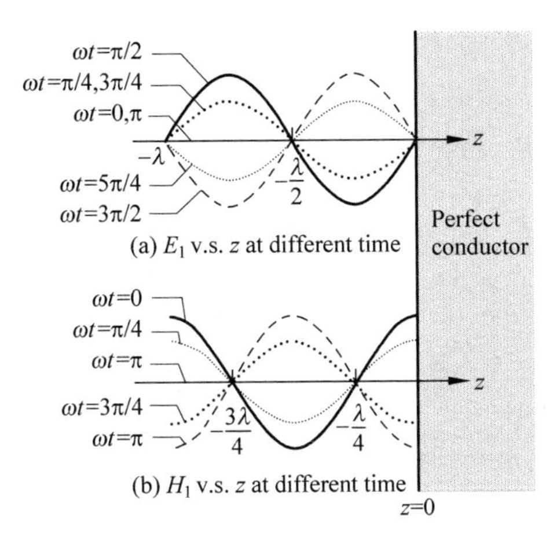

Figure 8-3 Standing waves of $\boldsymbol{E}_1=\boldsymbol{e}_x E_1$ and $\boldsymbol{H}_1=\boldsymbol{e}_y H_1$ for several values of ωt

In other words, $\boldsymbol{E}_1$ is zero whereas $\boldsymbol{H}_1$ reaches maximum at the PEC boundary as well as at locations that are multiple half-wavelengths from the PEC boundary.

(2) Antinodes of $\boldsymbol{E}_1(z,t)$ and nodes of $\boldsymbol{H}_1(z, t)$ occur when $\beta_1 z=-(2n+1)\pi/2$, corresponding to locations

$$z=-(2n+1)\frac{\lambda}{4}, \quad n=0,1,2,\cdots$$

In other words, $\boldsymbol{H}_1$ is zero whereas $\boldsymbol{E}_1$ reaches maximum at locations that are odd multiples of a quarter-wavelength from the PEC boundary.

Example 8-1 A y-polarized (with electric field linearly polarized along the y-direction) single-frequency uniform plane wave propagates in air in the $+x$-direction and impinges normally on a PEC plane at $x=0$. It is observed that $x=-0.75$m and $x=-2.25$m are two consecutive maxima locations of the total electric field, and the maximum value of the total electric field intensity is 6mV/m. Find the phasor and instantaneous expressions for ① $\boldsymbol{E}_i$ and $\boldsymbol{H}_i$ of the incident wave; ② $\boldsymbol{E}_r$ and $\boldsymbol{H}_r$ of the reflected wave; ③ $\boldsymbol{E}_1$ and $\boldsymbol{H}_1$ of the total wave in air.

Solution: In medium 1, the air, we have $\eta_1=120\pi\Omega$. The distance between two consecutive maxima locations is half wavelength. Therefore,

$$\lambda_1=2\times(2.25-0.75)=3(\text{m})$$

$$\omega=2\pi\frac{c}{\lambda_1}=2\pi\times10^8(\text{rad/s})$$

$$\beta_1=\frac{2\pi}{\lambda_1}=\frac{2\pi}{3}(\text{rad/m})$$

The maximum value of the total $\boldsymbol{E}$ is two times the amplitude of the incident field. Therefore, the amplitude of $\boldsymbol{E}_i$ is 3mV/m.

(1) The phasor expression of the incident wave is

$$\boldsymbol{E}_i(x)=\boldsymbol{e}_y 3\times10^{-3}\mathrm{e}^{-\mathrm{j}2\pi x/3}(\text{V/m})$$

$$\boldsymbol{H}_i(x)=\frac{1}{\eta_1}\boldsymbol{e}_x\times\boldsymbol{E}_i(x)=\boldsymbol{e}_z\frac{1}{4\pi}\times10^{-4}\mathrm{e}^{-\mathrm{j}2\pi x/3}(\text{A/m})$$

Here, it is assumed that the initial phase of incident fields at the interface is zero. Then, the instantaneous expressions are

$$\boldsymbol{E}_{\mathrm{i}}(x,t)=\mathrm{Re}\left[\boldsymbol{E}_{\mathrm{i}}(x)\mathrm{e}^{\mathrm{j}\omega t}\right]=\boldsymbol{e}_{y}3\times10^{-3}\cos\left(2\pi\times10^{8}t-\frac{2\pi}{3}x\right)\ (\mathrm{V/m})$$

$$\boldsymbol{H}_{\mathrm{i}}(x,t)=\boldsymbol{e}_{z}\frac{1}{4\pi}\times10^{-4}\cos\left(2\pi\times10^{8}t-\frac{2\pi}{3}x\right)\ (\mathrm{A/m})$$

(2) For the reflected wave, the phasor expressions are:

$$\boldsymbol{E}_{\mathrm{r}}(x)=-\boldsymbol{e}_{y}3\times10^{-3}\mathrm{e}^{\mathrm{j}2\pi x/3}\ (\mathrm{V/m})$$

$$\boldsymbol{H}_{\mathrm{r}}(x)=\frac{1}{\eta_1}(-\boldsymbol{e}_{x})\times\boldsymbol{E}_{\mathrm{r}}(x)=\boldsymbol{e}_{z}\frac{1}{4\pi}\times10^{-4}\mathrm{e}^{\mathrm{j}2\pi x/3}\ (\mathrm{A/m})$$

The instantaneous expressions are:

$$\boldsymbol{E}_{\mathrm{r}}(x,t)=\mathrm{Re}\left[\boldsymbol{E}_{\mathrm{r}}(x)\mathrm{e}^{\mathrm{j}\omega t}\right]=-\boldsymbol{e}_{y}3\times10^{-3}\cos\left(2\pi\times10^{8}t+\frac{2\pi}{3}x\right)\ (\mathrm{V/m})$$

$$\boldsymbol{H}_{\mathrm{r}}(x,t)=\boldsymbol{e}_{z}\frac{1}{4\pi}\times10^{-4}\cos\left(2\pi\times10^{8}t+\frac{2\pi}{3}x\right)\ (\mathrm{A/m})$$

(3) For the total wave, the phasor expressions are:

$$\boldsymbol{E}_{1}(x)=\boldsymbol{E}_{\mathrm{i}}(x)+\boldsymbol{E}_{\mathrm{r}}(x)=-\boldsymbol{e}_{y}\mathrm{j}6\times10^{-3}\sin\left(\frac{2\pi}{3}x\right)\ (\mathrm{V/m})$$

$$\boldsymbol{H}_{1}(x)=\boldsymbol{H}_{\mathrm{i}}(x)+\boldsymbol{H}_{\mathrm{r}}(x)=\boldsymbol{e}_{z}\frac{1}{2\pi}\times10^{-4}\cos\left(\frac{2\pi}{3}x\right)\ (\mathrm{A/m})$$

The instantaneous expressions are:

$$\boldsymbol{E}_{1}(x,t)=\mathrm{Re}\left[\boldsymbol{E}_{1}(x)\mathrm{e}^{\mathrm{j}\omega t}\right]=\boldsymbol{e}_{y}6\times10^{-3}\sin\left(\frac{2\pi}{3}x\right)\sin(2\pi\times10^{8}t)\ (\mathrm{V/m})$$

$$\boldsymbol{H}_{1}(x,t)=\boldsymbol{e}_{z}\frac{1}{2\pi}\times10^{-4}\cos\left(\frac{2\pi}{3}x\right)\cos(2\pi\times10^{8}t)\ (\mathrm{A/m})$$

8.3 Oblique Incidence at a PEC Planar Boundary (理想导体平面的斜入射)

Now we consider an incident plane wave impinges upon a PEC plane obliquely from a lossless medium. In this case, we first define a **plane of incidence** (入射面) to be the one determined by the propagation direction of the incident wave and the normal direction of the interface plane. Unlike the case of normal incidence, the reflected wave depends on the polarization of the incident wave. Here we study the reflection of an incident wave for two different linear polarizations: the **perpendicular polarization** (垂直极化) with the incident electric field perpendicular to the plane of incidence, and the **parallel polarization** (平行极化) with the incident electric field parallel to the plane of incidence. An incident wave with arbitrary polarization can always be decomposed into two components: one with perpendicular polarization and the other with parallel polarization. ①

8.3.1 Perpendicular Polarization(垂直极化)

In the case of perpendicular polarization, $\boldsymbol{E}_{\mathrm{i}}$ is perpendicular to the plane of incidence and parallel to the PEC plane as illustrated in Figure 8-4. The propagation direction of the incident wave can be written as

$$\boldsymbol{e}_{\mathrm{ki}}=\boldsymbol{e}_{\mathrm{x}}\sin\theta_{\mathrm{i}}+\boldsymbol{e}_{\mathrm{z}}\cos\theta_{\mathrm{i}} \tag{8.10}$$

where θ_{i} is the **angle of incidence**(入射角) between the normal direction of the boundary surface and the propagation direction of the incident field. Then, from(7.19) and(7.26), the incident fields can be written as

$$\boldsymbol{E}_{\mathrm{i}}(x,z)=\boldsymbol{e}_{\mathrm{y}}E_{\mathrm{i0}}\mathrm{e}^{-\mathrm{j}\beta_1\boldsymbol{e}_{\mathrm{ki}}\cdot\boldsymbol{r}}=\boldsymbol{e}_{\mathrm{y}}E_{\mathrm{i0}}\mathrm{e}^{-\mathrm{j}\beta_1(x\sin\theta_{\mathrm{i}}+z\cos\theta_{\mathrm{i}})} \tag{8.11}$$

$$\begin{aligned}\boldsymbol{H}_{\mathrm{i}}(x,z)&=\frac{1}{\eta_1}[\boldsymbol{e}_{\mathrm{ki}}\times\boldsymbol{E}_{\mathrm{i}}(x,z)]\\&=\frac{E_{\mathrm{i0}}}{\eta_1}(-\boldsymbol{e}_{\mathrm{x}}\cos\theta_{\mathrm{i}}+\boldsymbol{e}_{\mathrm{z}}\sin\theta_{\mathrm{i}})\,\mathrm{e}^{-\mathrm{j}\beta_1(x\sin\theta_{\mathrm{i}}+z\cos\theta_{\mathrm{i}})}\end{aligned} \tag{8.12}$$

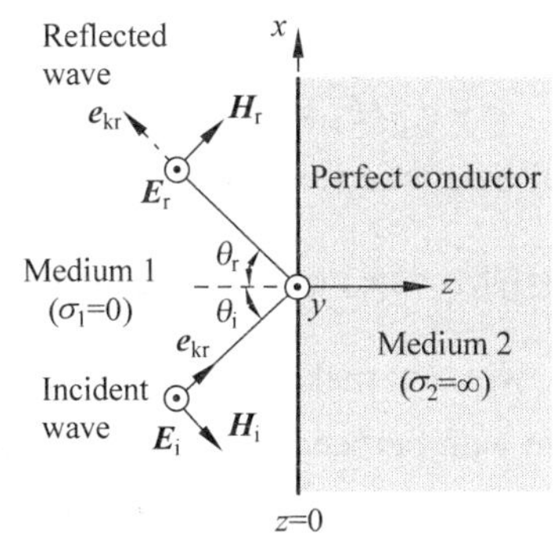

Figure 8-4 Plane wave incident obliquely on a PEC planar boundary(perpendicular polarization)

The propagation direction of the reflected wave is

$$\boldsymbol{e}_{\mathrm{kr}}=\boldsymbol{e}_{\mathrm{x}}\sin\theta_{\mathrm{r}}-\boldsymbol{e}_{\mathrm{z}}\cos\theta_{\mathrm{r}} \tag{8.13}$$

where θ_{r} is the **angle of reflection**(反射角) between the normal direction of the boundary surface and the propagation direction of the reflected wave. We have

$$\boldsymbol{E}_{\mathrm{r}}(x,z)=\boldsymbol{e}_{\mathrm{y}}E_{\mathrm{r0}}\mathrm{e}^{-\mathrm{j}\beta_1(x\sin\theta_{\mathrm{r}}-z\cos\theta_{\mathrm{r}})} \tag{8.14}$$

At the boundary surface, $z=0$, the boundary condition requires that the total electric field vanish. Thus,

$$\begin{aligned}\boldsymbol{E}_1(x,0)&=\boldsymbol{E}_{\mathrm{i}}(x,0)+\boldsymbol{E}_{\mathrm{r}}(x,0)\\&=\boldsymbol{e}_{\mathrm{y}}(E_{\mathrm{i0}}\mathrm{e}^{-\mathrm{j}\beta_1x\sin\theta_{\mathrm{i}}}+E_{\mathrm{r0}}\mathrm{e}^{-\mathrm{j}\beta_1x\sin\theta_{\mathrm{r}}})=0\end{aligned}$$

---

① 垂直入射时,无论入射波是什么极化,入射电场和磁场都平行于 PEC 表面,PCE 平面的反射特性不变。然而,斜入射条件下,PEC 平面的反射特性与入射波的极化形式有关。由于任何极化形式的均匀平面波都能分解为相互正交的两个线性极化波,我们只需讨论两种极化的入射波即可:一种是垂直极化,其电场垂直于入射面,平行于 PEC 平面,磁场平行于入射面;另一种是平行极化,其电场平行于入射面,磁场平行于 PEC 平面,垂直于入射面。

This relation must hold for all values of x, which is possible only if $E_{r0}=-E_{i0}$ and $\theta_r=\theta_i$. $\theta_r=\theta_i$ means that the angle of reflection equals the angle of incidence, which is referred to as **Snell's law of reflection**(斯涅尔反射定律). $E_{r0}=-E_{i0}$ means that the reflected fields have the same magnitude as the incident fields, and (8.14) becomes

$$\boldsymbol{E}_r(x,z)=-\boldsymbol{e}_y E_{i0}\mathrm{e}^{-\mathrm{j}\beta_1(x\sin\theta_i-z\cos\theta_i)} \tag{8.15}$$

The corresponding $H_r(x,z)$ is

$$\begin{aligned}\boldsymbol{H}_r(x,z)&=\frac{1}{\eta_1}[\boldsymbol{e}_{kr}\times\boldsymbol{E}_r(x,z)]\\&=\frac{E_{i0}}{\eta_1}(-\boldsymbol{e}_x\cos\theta_i-\boldsymbol{e}_z\sin\theta_i)\,\mathrm{e}^{-\mathrm{j}\beta_1(x\sin\theta_i-z\cos\theta_i)}\end{aligned} \tag{8.16}$$

Notice that the actual directions of $\boldsymbol{E}_r$, and $\boldsymbol{H}_r$ is opposite to the reference directions in Figure 8-4 due to the negative sign in (8.15).

The total field is obtained by superposing the incident and reflected fields. From (8.11) and (8.15),

$$\begin{aligned}\boldsymbol{E}_1(x,z)&=\boldsymbol{E}_i(x,z)+\boldsymbol{E}_r(x,z)\\&=\boldsymbol{e}_y E_{i0}(\mathrm{e}^{-\mathrm{j}\beta_1 z\cos\theta_i}-\mathrm{e}^{\mathrm{j}\beta_1 z\cos\theta_i})\,\mathrm{e}^{-\mathrm{j}\beta_1 x\sin\theta_i}\\&=-\boldsymbol{e}_y\mathrm{j}2E_{i0}\sin(\beta_1 z\cos\theta_i)\,\mathrm{e}^{-\mathrm{j}\beta_1 x\sin\theta_i}\end{aligned} \tag{8.17}$$

With (8.12) and (8.16), the total magnetic field is

$$\boldsymbol{H}_1(x,z)=-2\frac{E_{i0}}{\eta_1}[\boldsymbol{e}_x\cos\theta_i\cos(\beta_1 z\cos\theta_i)\,\mathrm{e}^{-\mathrm{j}\beta_1 x\sin\theta_i}+\boldsymbol{e}_z\mathrm{j}\sin\theta_i\sin(\beta_1 z\cos\theta_i)\,\mathrm{e}^{-\mathrm{j}\beta_1 x\sin\theta_i}] \tag{8.18}$$

With (8.17) and (8.18), we can make the following observations about the oblique incidence of a uniform plane wave with perpendicular polarization on a PEC planar boundary: ①

(1) Along the normal direction (z-direction) of the PEC boundary, the transverse field components E_{1y} and H_{1x} maintain standing-wave patterns according to $\sin\beta_{1z}z$ and $\cos\beta_{1z}z$, respectively, where $\beta_{1z}=\beta_1\cos\theta$. No average power is propagated in the z-direction since E_{1y} and H_{1x} are 90° out of phase. ②

(2) Along the parallel direction (x-direction) of the PEC boundary, the transverse field components E_{1y} and H_{1z} are in phase and propagate along the parallel direction with a phase velocity③

$$u_{1x}=\frac{\omega}{\beta_{1x}}=\frac{\omega}{\beta_1\sin\theta_i}=\frac{u_1}{\sin\theta_i} \tag{8.19}$$

---

① 按照式(8.17)和式(8.18)的表达式,垂直极化条件下总电场只有 y 方向一个分量,而总磁场则有 x 和 z 两个方向的分量。相应的坡印廷矢量也有 x 和 z 两个方向的分量。

② 坡印廷矢量 z 方向的分量来源于 y 方向的电场分量和 x 方向的磁场分量,由于 y 方向的电场分量和 x 方向的磁场分量相位相差 90°,坡印廷矢量 z 方向的分量时均值为 0,z 方向形成驻波,没有传播。

③ 坡印廷矢量 x 方向的分量来源于 y 方向的电场分量和 z 方向的磁场分量,由于 y 方向的电场分量和 z 方向的磁场分量相位相同,坡印廷矢量 x 方向的分量时均值不为 0,沿 x 方向存在行波。

The wavelength in this direction is

$$\lambda_{1x}=\frac{2\pi}{\beta_{1x}}=\frac{\lambda_1}{\sin\theta_i} \tag{8.20}$$

(3) The wave propagating in the x-direction is a **nonuniform plane wave(非均匀平面波)** because its is amplitude varies along the direction z. The wave is also called a **transverse electric(TE) wave(横电波)**, because the electric field of the wave has no component along the propagation direction($E_{1x}=0$).

(4) The total electric field $E_1=0$ when $\sin(\beta_1 z\cos\theta_i)=0$ or when

$$\beta_1 z\cos\theta_i=\frac{2\pi}{\lambda_1}z\cos\theta_i=-m\pi,\quad m=1,2,3,\cdots$$

Therefore, a PEC plate could be inserted at

$$z=-\frac{m\lambda_1}{2\cos\theta_i},\quad m=1,2,3,\cdots \tag{8.21}$$

without changing the field distributions between the PEC plate and the boundary at $z=0$. That makes a parallel-plate waveguide that guides the wave to propagate along its axis direction with the phase factor $e^{-j\beta_1 x\sin\theta_i}$.

On the PEC boundary($z=0$), the total magnetic field $\boldsymbol{H}_1(x,0)$ can be obtained from(8.18):

$$\boldsymbol{H}_1(x,0)=-\boldsymbol{e}_x\frac{2E_{i0}}{\eta_1}\cos\theta_i e^{-j\beta_1 x\sin\theta_i} \tag{8.22}$$

According to the boundary condition(6.14b), we have

$$\boldsymbol{J}_S(x)=\boldsymbol{e}_n\times\boldsymbol{H}_1(x,0)=(-\boldsymbol{e}_z)\times(-\boldsymbol{e}_x)\frac{2E_{i0}}{\eta_1}\cos\theta_i e^{-j\beta_1 x\sin\theta_i}$$

$$=\boldsymbol{e}_y\frac{2E_{i0}}{\eta_1}\cos\theta_i e^{-j\beta_1 x\sin\theta_i} \tag{8.23}$$

The time average Poynting vector in medium 1 is found by using(8.17) and(8.18) in (6.86). Since E_{1y} and H_{1x} are in time quadrature, $\boldsymbol{S}_{av}$ does not have a z component, but has an x-component arising from E_{1y} and H_{1z}:

$$\boldsymbol{S}_{av}=\frac{1}{2}\mathrm{Re}[\boldsymbol{E}_1(x,z)\times\boldsymbol{H}_1^*(x,z)]=\boldsymbol{e}_x 2\frac{E_{i0}^2}{\eta_1}\sin\theta_i\sin^2\beta_{1z}z \tag{8.24}$$

where $\beta_{1z}=\beta_1\cos\theta_i$. The time-average Poynting vector in medium 2(PEC) is zero.

8.3.2 Parallel Polarization(平行极化)

For a parallel polarization, as depicted in Figure 8-5, $\boldsymbol{E}_i$ is parallel to the plane of incidence. Then $\boldsymbol{H}_i$ is perpendicular to the plane of incidence. The propagation direction of the incident wave $\boldsymbol{e}_{ki}$ has the same expression as in(8.10), and we have

$$\boldsymbol{E}_i(x,z)=E_{i0}(\boldsymbol{e}_x\cos\theta_i-\boldsymbol{e}_z\sin\theta_i)e^{-j\beta_1(x\sin\theta_i+z\cos\theta_i)} \tag{8.25}$$

$$\boldsymbol{H}_i(x,z)=\boldsymbol{e}_y\frac{E_{i0}}{\eta_1}e^{-j\beta_1(x\sin\theta_i+z\cos\theta_i)} \tag{8.26}$$

The reflected field $\boldsymbol{E}_r$ is also expected to be parallel to the plane of incidence, and $\boldsymbol{H}_r$ is perpendicular to the plane of incidence. The propagation direction of the reflected wave $\boldsymbol{e}_{kr}$, has the same expression as in(8.13). We make the reference direction of $\boldsymbol{H}_r$ to be along the $-y$-direction. Then

$$\boldsymbol{E}_r(x,z) = E_{r0}(\boldsymbol{e}_x\cos\theta_r + \boldsymbol{e}_z\sin\theta_r)\,\mathrm{e}^{-\mathrm{j}\beta_1(x\sin\theta_r - z\cos\theta_r)} \tag{8.27}$$

$$\boldsymbol{H}_r(x,z) = -\boldsymbol{e}_y\frac{E_{r0}}{\eta_1}\mathrm{e}^{-\mathrm{j}\beta_1(x\sin\theta_r - z\cos\theta_r)} \tag{8.28}$$

At the surface of the PEC medium, $z=0$, the tangential component (the x-component) of the total electric field intensity must vanish, i. e. , $E_{ix}(x,0)+E_{rx}(x,0)=0$ for all x. From(8.25) and (8.27) we have

$$E_{i0}\cos\theta_i\mathrm{e}^{-\mathrm{j}\beta_1 x\sin\theta_i} + E_{r0}\cos\theta_r\mathrm{e}^{-\mathrm{j}\beta_1 x\sin\theta_r} = 0$$

for all x. This boundary condition can be satisfied only if $E_{r0}=-E_{i0}$ and $\theta_r=\theta_i$, which is the same as the case for perpendicular polarization. Again, the wave reflection satisfies Snell's law of reflection, which can be expressed as

$$\boldsymbol{E}_r(x,z) = -E_{i0}(\boldsymbol{e}_x\cos\theta_i + \boldsymbol{e}_z\sin\theta_i)\,\mathrm{e}^{-\mathrm{j}\beta_1(x\sin\theta_i - z\cos\theta_i)} \tag{8.29}$$

$$\boldsymbol{H}_r(x,z) = \boldsymbol{e}_y\frac{E_{i0}}{\eta_1}\mathrm{e}^{-\mathrm{j}\beta_1(x\sin\theta_i - z\cos\theta_i)} \tag{8.30}$$

Notice that, the actual directions of $\boldsymbol{E}_r$ and $\boldsymbol{H}_r$ at the interface $z=0$ are opposite to the reference directions in Figure 8-5.

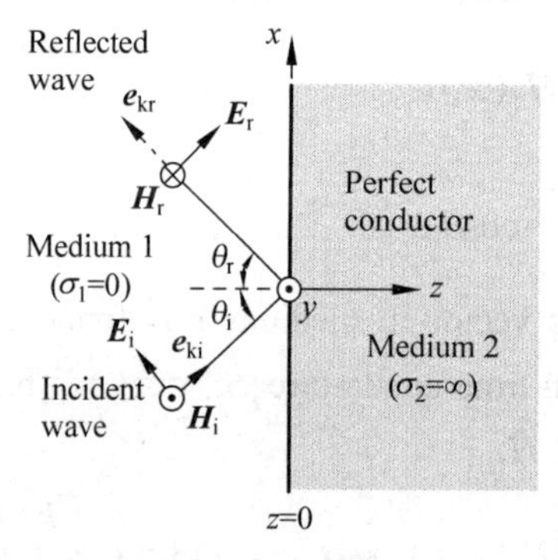

Figure 8-5 Plane wave incident obliquely on a PEC planar boundary (parallel polarization)

The total fields in medium 1 are the sum of the incident and reflected fields given in (8.25)-(8.28):

$$\begin{aligned}\boldsymbol{E}_1(x,z) &= \boldsymbol{E}_i(x,z) + \boldsymbol{E}_r(x,z)\\ &= -2E_{i0}\left[\boldsymbol{e}_x\mathrm{j}\cos\theta_i\sin(\beta_1 z\cos\theta_i) + \boldsymbol{e}_z\sin\theta_i\cos(\beta_1 z\cos\theta_i)\right]\mathrm{e}^{-\mathrm{j}\beta_1 x\sin\theta_i}\end{aligned} \tag{8.31}$$

$$\boldsymbol{H}_1(x,z) = \boldsymbol{H}_i(x,z) + \boldsymbol{H}_r(x,z) = \boldsymbol{e}_y 2\frac{E_{i0}}{\eta_1}\cos(\beta_1 z\cos\theta_i)\,\mathrm{e}^{-\mathrm{j}\beta_1 x\sin\theta_i} \tag{8.32}$$

With (8.31) and (8.32), we can make the following observations about the oblique incidence of a uniform plane wave with parallel polarization on a PEC planar boundary: ①

(1) Along the normal direction (z-direction) of the PEC boundary, the transverse field components E_{1x} and H_{1y} maintain standing-wave patterns according to $\sin\beta_{1z}z$ and $\cos\beta_{1z}z$ respectively, where $\beta_{1z}=\beta_1\cos\theta_i$. No average power is propagated in the z-direction since E_{1x} and H_{1y} are 90° out of time phase. ②

(2) Along the parallel direction (x-direction) of the PEC boundary, the transverse field components E_{1z} and H_{1y} are in phase and propagate along the parallel direction with a phase velocity u_{1x} and a wavelength λ_{1x} given by (8.19) and (8.20) respectively. ③

(3) As in the case of perpendicular polarization, the wave propagating in the x-direction is a nonuniform plane wave. The wave is also called a **transverse magnetic (TM) wave (横磁波)**, because the magnetic field of the wave has no component along the propagation direction ($H_{1x}=0$).

(4) The x-component of the total electric field $E_{1x}=0$ when $\sin(\beta_1 z\cos\theta_i)=0$. As in the case of perpendicular polarization, the insertion of a PEC plate at locations with coordinate z given by (8.21) would not affect the field distributions between the PEC plate and the boundary. That makes a parallel-plate waveguide that guides the wave to propagate along the x-direction with the exponential term $e^{-j\beta_1 x\sin\theta_i}$.

8.4 Normal Incidence at a Dielectric Planar Boundary (理想介质平面的垂直入射)

When an electromagnetic wave is incident from one dielectric medium upon the surface of another dielectric medium, part of the incident power is reflected, and part is transmitted. In this section, we first consider the case of the normal incidence. Both media are assumed to be lossless ($\sigma_1=\sigma_2=0$).

Consider the situation in Figure 8-6, where the incident wave travels in the $+z$-direction and the boundary surface is the plane $z=0$. The incident electric and magnetic field intensity phasors can be expressed as

$$\boldsymbol{E}_i(z)=\boldsymbol{e}_x E_{i0}e^{-j\beta_1 z} \tag{8.33}$$

$$\boldsymbol{H}_i(z)=\boldsymbol{e}_y \frac{E_{i0}}{\eta_1}e^{-j\beta_1 z} \tag{8.34}$$

which are the same as (8.1) and (8.2). The reflected fields can be written as

---

① 按照式(8.31)和式(8.32)的表达式,平行极化条件下总磁场只有 y 方向一个分量,而总电场则有 x 和 z 两个方向的分量。相应的坡印廷矢量也有 x 和 z 两个方向的分量。

② 坡印廷矢量 z 方向的分量来源于 y 方向的磁场分量和 x 方向的电场分量,由于 y 方向的磁场分量和 x 方向的电场分量相位相差 90°,坡印廷矢量 z 方向的分量时均值为 0,z 方向形成驻波,没有传播。

③ 坡印廷矢量 x 方向的分量来源于 y 方向的磁场分量和 z 方向的电场分量,由于 y 方向的磁场分量和 z 方向的电场分量相位相差 180°,坡印廷矢量 x 方向的分量时均值不为零,沿 x 方向存在行波。

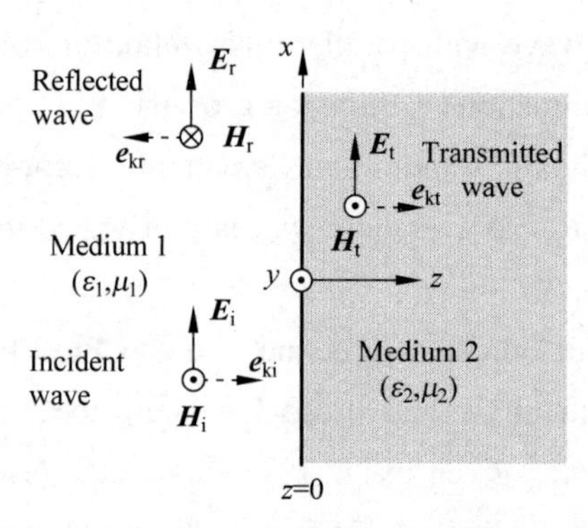

Figure 8-6 Plane wave incident normally on a dielectric planar boundary

$$\boldsymbol{E}_{\rm r}(z)=\boldsymbol{e}_{\rm x}E_{\rm r0}{\rm e}^{{\rm j}\beta_1 z} \tag{8.35}$$

$$\boldsymbol{H}_{\rm r}(z)=(-\boldsymbol{e}_{\rm z})\times\frac{1}{\eta_1}\boldsymbol{E}_{\rm r}(z)=-\boldsymbol{e}_{\rm y}\frac{E_{\rm r0}}{\eta_1}{\rm e}^{{\rm j}\beta_1 z} \tag{8.36}$$

And the transmitted fields are

$$\boldsymbol{E}_{\rm t}(z)=\boldsymbol{e}_{\rm x}E_{\rm t0}{\rm e}^{-{\rm j}\beta_2 z} \tag{8.37}$$

$$\boldsymbol{H}_{\rm t}(z)=\boldsymbol{e}_{\rm z}\times\frac{1}{\eta_2}\boldsymbol{E}_{\rm t}(z)=\boldsymbol{e}_{\rm y}\frac{E_{\rm t0}}{\eta_2}{\rm e}^{-{\rm j}\beta_2 z} \tag{8.38}$$

where $E_{\rm t0}$ is the magnitude of $\boldsymbol{E}_{\rm t}$ at $z=0$, and β_2 and η_2 are the phase constant and the intrinsic impedance of medium 2. ①

In the above expressions(8.33)-(8.38), the incident fields are assumed to be known, and the only two unknowns are the magnitudes $E_{\rm r0}$ and $E_{\rm t0}$. To determine the two unknowns, two equations are needed, which come from the boundary conditions over the interface. At the dielectric interface $z=0$, the tangential components of the electric and magnetic field intensities must be continuous, i. e. ,

$$\boldsymbol{E}_{\rm i}(0)+\boldsymbol{E}_{\rm r}(0)=\boldsymbol{E}_{\rm t}(0)\quad\text{and}\quad\boldsymbol{H}_{\rm i}(0)+\boldsymbol{H}_{\rm r}(0)=\boldsymbol{H}_{\rm t}(0) \tag{8.39}$$

With(8.33)-(8.38), the two equations in(8.39) become

$$E_{\rm i0}+E_{\rm r0}=E_{\rm t0}\quad\text{and}\quad\frac{1}{\eta_1}(E_{\rm i0}-E_{\rm r0})=\frac{E_{\rm t0}}{\eta_2} \tag{8.40}$$

which leads to the solution

$$E_{\rm r0}=\frac{\eta_2-\eta_1}{\eta_2+\eta_1}E_{\rm i0} \tag{8.41}$$

$$E_{\rm t0}=\frac{2\eta_2}{\eta_2+\eta_1}E_{\rm i0} \tag{8.42}$$

The ratios $E_{\rm r0}/E_{\rm i0}$ and $E_{\rm t0}/E_{\rm i0}$ are called **reflection coefficient（反射系数）** and

① 与 PEC 不同,介质内部可以存在电磁场。均匀平面波入射到两种介质分界面时,部分被反射,部分穿过分界面形成透射。与 PEC 平面反射类似,沿+z 方向垂直入射的条件下,理想介质平面的反射也是沿-z 方向,而越过分界面的透射波则会在另一种介质内沿+z 方向继续传播。这里在入射波已知的条件下,根据介质分界面电场和磁场分别满足的边界条件,可以求解得到反射波和透射波。

transmission coefficient（**透射系数**）,and are respectively denoted by

$$\Gamma = \frac{E_{r0}}{E_{i0}} = \frac{\eta_2 - \eta_1}{\eta_2 + \eta_1} \tag{8.43}$$

and

$$\tau = \frac{E_{t0}}{E_{i0}} = \frac{2\eta_2}{\eta_2 + \eta_1} \tag{8.44}$$

From the above expressions, it is obvious that①

$$1 + \Gamma = \tau \tag{8.45}$$

If the two media are perfect dielectrics, η_1 and η_2 will be both positive real numbers. Depending on whether η_2 is larger or smaller than η_1, the reflection coefficient Γ in(8.43) can be positive or negative, but $|\Gamma|$ cannot be larger than 1. The transmission coefficient τ, however, is always positive, and its value must be smaller than 2. In other words, we have $\Gamma \in [-1,1]$ and $\tau \in [0,2]$.

The expressions for Γ and τ in(8.43) and(8.44) are still valid if the media are lossy, in which case, η_1 and/or η_2 are complex, and Γ and/or τ may also be complex. A complex Γ or τ means that a phase shift is introduced at the interface upon reflection or transmission.

Notice that, if medium 2 is a perfect conductor, $\eta_2 = 0$, (8.43) and(8.44) yield $\Gamma = -1$ and $\tau = 0$. Consequently, $E_{r0} = -E_{i0}$ and $E_{t0} = 0$. The incident wave will be totally reflected, and a standing wave will be produced in medium 1 as discussed in Section 8.2.

As long as medium 2 is not a perfect conductor, partial reflection will result. The total electric field in medium 1 can be written as

$$\begin{aligned} \boldsymbol{E}_1(z) &= \boldsymbol{E}_i(z) + \boldsymbol{E}_r(z) = \boldsymbol{e}_x E_{i0}(\mathrm{e}^{-\mathrm{j}\beta_1 z} + \Gamma \mathrm{e}^{\mathrm{j}\beta_1 z}) \\ &= \boldsymbol{e}_x E_{i0}[(1 + \Gamma)\mathrm{e}^{-\mathrm{j}\beta_1 z} + \Gamma(\mathrm{e}^{\mathrm{j}\beta_1 z} - \mathrm{e}^{-\mathrm{j}\beta_1 z})] \\ &= \boldsymbol{e}_x E_{i0}[\tau \mathrm{e}^{-\mathrm{j}\beta_1 z} + \Gamma(\mathrm{j}2\sin\beta_1 z)] \end{aligned} \tag{8.46}$$

in which(8.45) is applied. (8.46) shows that, the total field $\boldsymbol{E}_1(z)$ can be decomposed into two parts: a traveling wave with an amplitude τE_{i0} and a standing wave with an amplitude $2\Gamma E_{i0}$.② The standing wave part makes the amplitude of the total fields non-uniform along the propagation direction (z-direction). The total fields have fixed locations of maximum and minimum values. To see that, we rewrite $\boldsymbol{E}_1(z)$ as

$$\boldsymbol{E}_1(z) = \boldsymbol{e}_x E_{i0}\mathrm{e}^{-\mathrm{j}\beta_1 z}(1 + \Gamma \mathrm{e}^{\mathrm{j}2\beta_1 z}) \tag{8.47}$$

From (8.47), it is easy to figure out that the maximum and minimum amplitudes are respectively

$$|\boldsymbol{E}_1(z)|_{\max} = |E_{i0}|(1 + |\Gamma|) \tag{8.48}$$

---

① 根据式(8.43)和式(8.44)的定义,反射系数 Γ 是分界面上反射电场与入射电场的比值,透射系数 τ 是分界面上透射电场与入射电场的比值。如果分界面上入射电场是 1,那么反射电场就是 Γ,透射电场就是 τ。分界面上电场连续,因此就有 $1+\Gamma=\tau$,即式(8.45)。

② 式(8.46)表明,由于部分反射形成的反射波振幅小于入射波振幅,反射波与部分入射波叠加形成了驻波,而入射波的其余部分仍然为行波。这种行波与驻波共存的情形,也称为**行驻波**。

$$|\boldsymbol{E}_1(z)|_{\min} = |E_{i0}|(1 - |\Gamma|) \tag{8.49}$$

For lossless media, depending on whether Γ is positive or negative, we have the following two cases.

(1) $\Gamma>0(\eta_2>\eta_1)$:

The maximum value of $|\boldsymbol{E}_1(z)|$ is $E_{i0}(1+\Gamma)$, which occurs when $2\beta_1 z=-2\pi n$, or when

$$z = -\frac{n\pi}{\beta_1} = -\frac{n\lambda_1}{2}, \quad n = 0,1,2,\cdots \tag{8.50}$$

The minimum value of $|\boldsymbol{E}_1(z)|$ is $E_{i0}(1-\Gamma)$, which occurs when $2\beta_1 z=-(2n+1)\pi$, or when

$$z = -\frac{(2n+1)\pi}{2\beta_1} = -\frac{(2n+1)\lambda_1}{4}, \quad n = 0,1,2,\cdots \tag{8.51}$$

In other words, $\boldsymbol{E}$ reaches maximum at the interface ($z=0$) and at all the planes that are multiple half wavelengths away from the interface. And in the midpoint between two adjacent maxima, $\boldsymbol{E}$ reaches minimum.

(2) $\Gamma<0(\eta_2<\eta_1)$:

$|\boldsymbol{E}_1(z)|$ reaches its maximum value of $E_{i0}(1-\Gamma)$ at locations with z given by (8.51). $|\boldsymbol{E}_1(z)|$ reaches its minimum value of $E_{i0}(1+\Gamma)$ at locations with z given by (8.50). In other words, $\boldsymbol{E}$ reaches minimum at the interface ($z=0$) and at all the planes that are multiple half wavelengths away from the interface. In the midpoint between two adjacent minima, $\boldsymbol{E}$ reaches maximum. Hence, the locations for $|E_1(z)|_{\max}$ and $|E_1(z)|_{\min}$ when $\Gamma>0$ and when $\Gamma<0$ are interchanged.

The ratio of the maximum value to the minimum value of the electric field intensity of a standing wave is called the **standing-wave ratio**(**SWR**,驻波比),

$$S = \frac{|E|_{\max}}{|E|_{\min}} = \frac{1+|\Gamma|}{1-|\Gamma|} \tag{8.52}$$

An inverse relation of (8.52) is

$$|\Gamma| = \frac{S-1}{S+1} \tag{8.53}$$

While $|\Gamma|$ grows from 0 to 1, the value of S grows from 1 to $+\infty$. Apparently, the larger the reflection is, the larger the standing wave ratio is.

The total magnetic field intensity in medium 1 is the sum of $\boldsymbol{H}_i(z)$ and $\boldsymbol{H}_r(z)$. With (8.34) and (8.36),

$$\boldsymbol{H}_1(z) = \boldsymbol{e}_y \frac{E_{i0}}{\eta_1}(\mathrm{e}^{-\mathrm{j}\beta_1 z} - \Gamma \mathrm{e}^{\mathrm{j}\beta_1 z}) = \boldsymbol{e}_y \frac{E_{i0}}{\eta_1}\mathrm{e}^{-\mathrm{j}\beta_1 z}(1 - \Gamma \mathrm{e}^{\mathrm{j}2\beta_1 z}) \tag{8.54}$$

Compared with $\boldsymbol{E}_1(z)$ in (8.47), $|\boldsymbol{H}_1(z)|$ will be a minimum at locations where $|\boldsymbol{E}_1(z)|$ is a maximum, and vice versa.

The transmitted wave ($\boldsymbol{E}_t, \boldsymbol{H}_t$) in medium 2 can be found from (8.37), (8.38) and (8.44):

$$\boldsymbol{E}_t(z)=\boldsymbol{e}_x\tau E_{i0}e^{-j\beta_2 z}=\boldsymbol{e}_x\left(\frac{2\eta_2}{\eta_2+\eta_1}\right)E_{i0}e^{-j\beta_2 z}\tag{8.55}$$

$$\boldsymbol{H}_t(z)=\boldsymbol{e}_y\frac{\tau}{\eta_2}E_{i0}e^{-j\beta_2 z}=\boldsymbol{e}_y\left(\frac{2}{\eta_2+\eta_1}\right)E_{i0}e^{-j\beta_2 z}\tag{8.56}$$

Now we can find the time-average power densities in both media using(6.86). In medium 1, we use(8.47) and(8.54):

$$\boldsymbol{S}_{av1}=\frac{1}{2}\mathrm{Re}[\boldsymbol{E}_1(z)\times\boldsymbol{H}_1^*(z)]=\boldsymbol{e}_z\frac{|E_{i0}|^2}{2\eta_1}\mathrm{Re}[(1+\Gamma e^{j2\beta_1 z})(1-\Gamma e^{-j2\beta_1 z})]$$

$$=\boldsymbol{e}_z\frac{|E_{i0}|^2}{2\eta_1}\mathrm{Re}[(1-\Gamma^2)+j2\Gamma\sin 2\beta_1 z]=\boldsymbol{e}_z\frac{|E_{i0}|^2}{2\eta_1}(1-\Gamma^2)\tag{8.57}$$

In medium 2, we use(8.55) and(8.56) to obtain

$$\boldsymbol{S}_{av2}=\frac{1}{2}\mathrm{Re}[\boldsymbol{E}_t(z)\times\boldsymbol{H}_t^*(z)]=\boldsymbol{e}_z\frac{E_{i0}^2}{2\eta_2}\tau^2\tag{8.58}$$

where τ is a real number because media are lossless.

From the principle of conservation of energy, we must have $\boldsymbol{S}_{av1}=\boldsymbol{S}_{av2}$, leading to

$$1-\Gamma^2=\frac{\eta_1}{\eta_2}\tau^2\tag{8.59}$$

which can be verified by using(8.43) and(8.44).

8.5 Normal Incidence at Multiple Dielectric Interfaces (多层介质分界面的垂直入射)

We now consider the wave reflection and transmission problem with multiple dielectric interfaces. As depicted in Figure 8-7, an x-polarized incident wave($\boldsymbol{E}_i$, $\boldsymbol{H}_i$) travels in the $+z$-direction in medium 1(ε_1, μ_1) and impinges normally upon the interface with medium 2(ε_2, μ_2) at $z=0$. Medium 2 has a finite thickness d, and in the $z>d$ region is medium 3(ε_3, μ_3). Multiple reflection and transmission occur at both $z=0$ and $z=d$, producing multiple forward-propagating(propagating along $+z$-direction) and backward-propagating(propagating along $-z$-direction) waves.

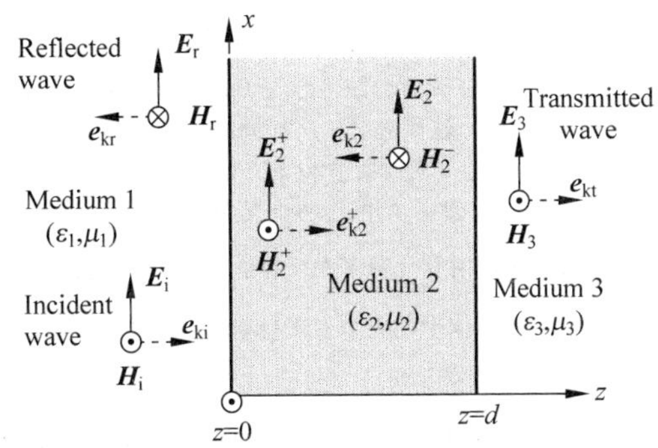

Figure 8-7 Normal incidence at multiple dielectric interfaces

The total electric field in medium 1 is the superposition of the incident wave($\boldsymbol{E}_i, \boldsymbol{H}_i$) and the reflected wave($\boldsymbol{E}_r, \boldsymbol{H}_r$), which can be written as

$$\boldsymbol{E}_1 = \boldsymbol{E}_i + \boldsymbol{E}_r = \boldsymbol{e}_x (E_{i0} e^{-j\beta_1 z} + E_{r0} e^{j\beta_1 z}) \tag{8.60}$$

where E_{i0} and E_{r0} are respectively the phasors of $\boldsymbol{E}_i$ and $\boldsymbol{E}_r$ at $z=0$. The corresponding magnetic field $\boldsymbol{H}_1$ in medium 1 is then

$$\boldsymbol{H}_1 = \boldsymbol{H}_i + \boldsymbol{H}_r = \boldsymbol{e}_y \frac{1}{\eta_1} (E_{i0} e^{-j\beta_1 z} - E_{r0} e^{j\beta_1 z}) \tag{8.61}$$

Within medium 2, multiple reflection and transmission on the boundaries can be combined into two waves propagating in opposite directions. The total fields can be written as

$$\boldsymbol{E}_2 = \boldsymbol{e}_x (E_2^+ e^{-j\beta_2 z} + E_2^- e^{j\beta_2 z}) \tag{8.62}$$

$$\boldsymbol{H}_2 = \boldsymbol{e}_y \frac{1}{\eta_2} (E_2^+ e^{-j\beta_2 z} - E_2^- e^{j\beta_2 z}) \tag{8.63}$$

where E_2^+ and E_2^- are respectively the phasors of the forward-propagating electric field $\boldsymbol{E}_2^+$ and the backward-propagating electric field $\boldsymbol{E}_2^-$ at $z=0$. ①

In medium 3, only a transmitted wave traveling in $+z$-direction exists, Hence, the fields in medium 3 can be written as②

$$\boldsymbol{E}_3 = \boldsymbol{e}_x E_{t0} e^{-j\beta_3 (z-d)} \tag{8.64}$$

$$\boldsymbol{H}_3 = \boldsymbol{e}_y \frac{E_{t0}}{\eta_3} e^{-j\beta_3 (z-d)} \tag{8.65}$$

where E_{t0} is the phasor of $\boldsymbol{E}_3$ at $z=d$.

In the field expressions (8.60)-(8.65), there are altogether four unknowns: E_{r0}, E_2^+, E_2^- and E_{t0} (E_{i0} is assumed to be known). They can be determined by solving the four boundary-condition equations required by the continuity of the tangential components of the electric and magnetic fields as follows.

(1) At $z=0$, $\boldsymbol{E}_1(0)=\boldsymbol{E}_2(0)$ and $\boldsymbol{H}_1(0)=\boldsymbol{H}_2(0)$, leading to

$$E_{i0} + E_{r0} = E_2^+ + E_2^- \tag{8.66}$$

$$\frac{1}{\eta_1}(E_{i0} - E_{r0}) = \frac{1}{\eta_2}(E_2^+ - E_2^-) \tag{8.67}$$

---

① 理论上,入射波首先在 $z=0$ 的分界面上发生一次反射和透射,透射进入媒质 2 形成的沿 $+z$ 方向传播的波到达 $z=d$ 的分界面时再次发生反射和透射,二次反射波沿 $-z$ 方向回传至 $z=0$ 的分界面时发生第三次反射和透射,其中的反射波沿 $+z$ 方向传播,直至遇到分界面时发生第四次反射和透射。这样继续下去,在两个分界面上会发生无穷多次反射和透射。媒质 1 中的反射波包含了 $z=0$ 的分界面上的一次反射波和无穷多次透射波,但由于所有这些反射、透射波都是沿 $-z$ 方向传播且极化相同的均匀平面波,它们叠加也是一个沿 $-z$ 方向传播的均匀平面波,其电场由式(8.60)中的第二项 $\boldsymbol{e}_x E_{r0} e^{j\beta_1 z}$ 代表。同理,$z=0$ 的分界面上的一次透射和无穷多次反射形成媒质 2 中沿 $+z$ 方向传播的波,其电场叠加结果由式(8.62)中第一项 $\boldsymbol{e}_x E_2^+ e^{-j\beta_2 z}$ 代表。$z=d$ 的分界面上的无穷多次反射形成媒质 2 中沿 $-z$ 方向传播的波,其电场叠加结果由式(8.62)中第二项 $\boldsymbol{e}_x E_2^- e^{j\beta_2 z}$ 代表。

② 媒质 3 中包含 $z=d$ 的分界面上无穷多次透射形成中沿 $+z$ 方向传播的波,它们的极化相同,因此叠加形成一个总的透射波,可以由式(8.64)和式(8.65)代表。

(2) At $z=d$, $\boldsymbol{E}_2(d)=\boldsymbol{E}_3(d)$ and $\boldsymbol{H}_2(d)=\boldsymbol{H}_3(d)$, leading to

$$E_2^+ e^{-j\beta_2 d} + E_2^- e^{j\beta_2 d} = E_{t0} \tag{8.68}$$

$$\frac{1}{\eta_2}(E_2^+ e^{-j\beta_2 d} - E_2^- e^{j\beta_2 d}) = \frac{E_{t0}}{\eta_3} \tag{8.69}$$

From (8.68) and (8.69), we have

$$E_2^- = \left(\frac{\eta_3 - \eta_2}{\eta_2 + \eta_3}\right) E_2^+ e^{-2j\beta_2 d} = \Gamma_{23} E_2^+ e^{-2j\beta_2 d} \tag{8.70}$$

where

$$\Gamma_{23} = \frac{\eta_3 - \eta_2}{\eta_2 + \eta_3} \tag{8.71}$$

is the reflection coefficient on the interface between medium 2 and medium 3 when a forward-propagating wave come from medium 2. ①Then, by substituting (8.70) into (8.66) and (8.67), we can solve E_{r0} and find the reflection coefficient at $z=0$ to be ②

$$\Gamma = \frac{E_{r0}}{E_{i0}} = \frac{\eta_{eff} - \eta_1}{\eta_{eff} + \eta_1} \tag{8.72}$$

where

$$\eta_{eff} = \eta_2 \frac{\eta_3 \cos\beta_2 d + j\eta_2 \sin\beta_2 d}{\eta_2 \cos\beta_2 d + j\eta_3 \sin\beta_2 d} \tag{8.73}$$

can be considered as the effective impedance seen by the incident wave $(\boldsymbol{E}_i, \boldsymbol{H}_i)$. Substitute $E_{r0} = \Gamma E_{i0}$ and (8.70) into (8.66), we have

$$E_2^+ = \frac{1+\Gamma}{1+\Gamma_{23} e^{-2j\beta_2 d}} E_{i0} \tag{8.74}$$

And as a result,

$$E_2^- = \frac{1+\Gamma}{1+\Gamma_{23} e^{-2j\beta_2 d}} \Gamma_{23} E_{i0} e^{2j\beta_2 d} \tag{8.75}$$

$$E_{t0} = \frac{(1+\Gamma)(1+\Gamma_{23})}{1+\Gamma_{23} e^{-2j\beta_2 d}} E_{i0} e^{-j\beta_2 d} \tag{8.76}$$

A better understanding of the expressions of Γ in (8.72) and the effective impedance η_{eff} in (8.73) can be provided by using the concept of wave impedance, which is introduced in the following subsections.

---

① 在 $z=d$ 的分界面上，$+z$ 方向传播的电场为 $\boldsymbol{e}_x E_2^+ e^{-j\beta_2 d}$，可以看作分界面上的入射电场，$-z$ 方向传播的电场为 $\boldsymbol{e}_x E_2^- e^{j\beta_2 d}$，可以看作分界面上的反射电场，媒质 3 中仅有的沿 $+z$ 方向传播的波为透射场，因此电场 $E_2^- e^{j\beta_2 d}$ 与电场 $E_2^+ e^{-j\beta_2 d}$ 的比值应等于式(8.43) 给出的反射系数，也即式(8.70)、式(8.71)。

② 由于媒质 3 的存在(导致媒质 2 的两个边界上的无穷多次反射和透射)，媒质 2 对来自于媒质 1 的入射波的反射系数不再满足式(8.43)，但可以表示为式(8.72)。式(8.72)从另一个角度表明了媒质 3 对媒质 1、2 分界面上反射特性的影响：媒质 1 中的入射波到达与媒质 2 的分界面时，感受到的媒质 2 的阻抗不是 η_2，而是一个等效阻抗 η_{eff}。η_{eff} 与 η_2、η_3 以及媒质 2 的厚度 d 均有关，由式(8.73)确定。

8.5.1 Wave Impedance and Impedance Transformation (波阻抗与阻抗变换)

We introduce the **wave impedance**(波阻抗) of an electromagnetic wave(which can be a traveling wave or a standing wave) as the ratio of the tangential component of the $\boldsymbol{E}$ field to the tangential component of the $\boldsymbol{H}$ field on a given plane. To specify a wave impedance, we first need to specify the plane and the tangential directions of $\boldsymbol{E}$ and $\boldsymbol{H}$. The plane is usually selected to be the one that is perpendicular to the propagation direction or parallel to the boundary plane. For example, the wave impedance on a PEC boundary is zero. The wave impedance on the dielectric boundary in the problem of Figure 8-6 is η_2. ①

For the problem of Figure 8-7, we define the wave impedance at a plane parallel to the interfaces as

$$Z(z)=\frac{\text{Total } E_x(z)}{\text{Total } H_y(z)}(\Omega) \tag{8.77}$$

Notice that, for an x-polarized traveling wave propagating in the $+z$-direction in an unbounded medium(without reflection), the wave impedance Z equals the intrinsic impedance η of the medium for all z. For an x-polarized traveling wave propagating in the $-z$-direction, the wave impedance $Z=-\eta$. However, with reflections, the wave impedance Z in the example of Figure 8-7 becomes a function of z.

To find an expression for the wave impedance in Figure 8-7, we first find a general relation between the wave impedances at two different planes separated by a distance l. As shown in Figure 8-8, two traveling waves ($\boldsymbol{E}^+$, $\boldsymbol{H}^+$) and ($\boldsymbol{E}^-$, $\boldsymbol{H}^-$) propagate along $+z$ and $-z$-directions respectively within a simple medium with phase constant β and intrinsic impedance η. The fields of the two waves can be written as

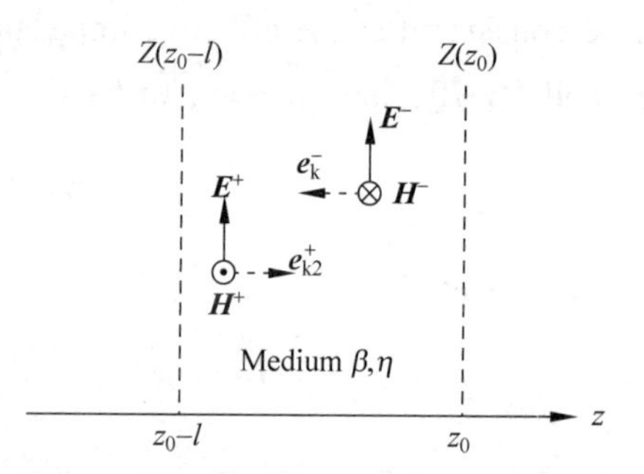

Figure 8-8 Wave impedance transformation

$$\boldsymbol{E}^+(z)=\boldsymbol{e}_x E_0^+ \mathrm{e}^{-\mathrm{j}\beta z},\quad \boldsymbol{H}^+(z)=\boldsymbol{e}_y H_0^+ \mathrm{e}^{-\mathrm{j}\beta z}=\boldsymbol{e}_y\frac{E_0^+}{\eta}\mathrm{e}^{-\mathrm{j}\beta z}$$

$$\boldsymbol{E}^-(z)=\boldsymbol{e}_x E_0^- \mathrm{e}^{-\mathrm{j}\beta z},\quad \boldsymbol{H}^-(z)=\boldsymbol{e}_y H_0^- \mathrm{e}^{-\mathrm{j}\beta z}=-\boldsymbol{e}_y\frac{E_0^-}{\eta}\mathrm{e}^{-\mathrm{j}\beta z}$$

① “波阻抗”名称上就意味着该阻抗取决于与其相关的电磁波，是该电磁波(可能是行波，也可能是驻波或行驻波)的电场或电场的某个指定分量与磁场或磁场的某个指定分量之间的比值，而不是媒质所固有的特性。同一种媒质内，不同形式的电磁波，或者多个电磁波(由于反射等各种原因)的叠加会呈现出不同的波阻抗。此外，波阻抗取决于观察者感兴趣的场的分量，同一个电磁波取不同的场分量，以及在不同位置观察到的波阻抗都有可能不同。因此谈到波阻抗，需要结合具体的电磁波给出明确的定义。这里讨论均匀平面波的反射和透射问题，将波阻抗定义为平行于无限大媒质分界面的平面上的切向电场与切向磁场的比值。均匀平面波垂直入射的条件下，上述切向电场和切向磁场就是总的电场和总的磁场本身。

Then the wave impedance at a plane $z=z_0$, as defined by(8.77), is

$$Z(z_0)=\frac{E_0^+ e^{-j\beta z_0}+E_0^- e^{j\beta z_0}}{H_0^+ e^{-j\beta z_0}+H_0^- e^{j\beta z_0}}=\eta\frac{E_0^+ e^{-j\beta z_0}+E_0^- e^{j\beta z_0}}{E_0^+ e^{-j\beta z_0}-E_0^- e^{j\beta z_0}}$$

from which, we can easily find that

$$\frac{E_0^- e^{j\beta z_0}}{E_0^+ e^{-j\beta z_0}}=\frac{Z(z_0)-\eta}{Z(z_0)+\eta} \tag{8.78}$$

Now we write down the wave impedance at the plane $z=z_0-l$ as

$$Z(z_0-l)=\eta\frac{E_0^+ e^{-j\beta(z_0-l)}+E_0^- e^{j\beta(z_0-l)}}{E_0^+ e^{-j\beta(z_0-l)}-E_0^- e^{j\beta(z_0-l)}}=\eta\frac{e^{j\beta l}+\left(\dfrac{E_0^- e^{j\beta z_0}}{E_0^+ e^{-j\beta z_0}}\right)e^{-j\beta l}}{e^{j\beta l}-\left(\dfrac{E_0^- e^{j\beta z_0}}{E_0^+ e^{-j\beta z_0}}\right)e^{-j\beta l}} \tag{8.79}$$

Substituting(8.78) into(8.79), we have①

$$Z(z_0-l)=\eta\frac{Z(z_0)\cos\beta l+j\eta\sin\beta l}{\eta\cos\beta l+jZ(z_0)\sin\beta l} \tag{8.80}$$

Now, we apply the relation(8.80) to the problem of Figure 8-7 by letting $z_0=l=d$. First, we realize that the wave impedance in medium 3 is constant η_3 since there is only a single transmitted wave traveling along the $+z$-direction. In other words, $Z(z_0)=Z(d)=\eta_3$. Let $\beta=\beta_2$, $\eta=\eta_2$, then(8.80) becomes

$$Z(0)=\eta_2\frac{\eta_3\cos\beta_2 d+j\eta_2\sin\beta_2 d}{\eta_2\cos\beta_2 d+j\eta_3\sin\beta_2 d} \tag{8.81}$$

which is the same expression of η_{eff} given in(8.73). Therefore, the effective impedance seen by the incident wave is essentially the wave impedance at the interface at $z=0$. The two-layer dielectric(medium 2 medium 3) can be replaced by an equivalent single-layer dielectric with an intrinsic impedance $Z(0)=\eta_{\text{eff}}$ without affecting the waves in medium 1. This is because, in both cases, the ratio between the total E_x and H_y at the interface at $z=0$ is enforced to be $Z(0)=\eta_{\text{eff}}$. In other words, the insertion of a dielectric medium 2 between the medium 1 and medium 3 has the effect of transforming the impedance seen by the incident wave from η_3 to $Z(0)$. Consequently, the reflection coefficient at $z=0$ for the incident wave in medium 1 is

$$\Gamma=\frac{E_{r0}}{E_{i0}}=-\frac{H_{r0}}{H_{i0}}=\frac{Z(0)-\eta_1}{Z(0)+\eta_1} \tag{8.82}$$

which is the same as(8.72). Once Γ has been found from(8.82), E_{r0} of the reflected wave in medium 1 can be calculated as $E_{r0}=\Gamma E_{i0}$, which solves the problem of Figure 8-7. ②Given η_1

---

① 式(8.80)给出了 $z=z_0$ 和 $z=z_0-l$ 两个平面上波阻抗的关系。该关系仅取决于这两个平面之间的距离以及其间的媒质特性(即 l、β 和 η),与 $z=z_0$ 右边以及 $z=z_0-l$ 左边是什么媒质都没有关系。

② 根据定义,$Z(0)$ 给定了 $z=0$ 处切向电场与切向磁场的比值,对于媒质 1 中沿 $+z$ 方向传播的入射波而言,其效果相当于将 $z>0$ 的整个半空间区域换成特性阻抗为 $Z(0)$ 的媒质。因此 $Z(0)$ 相当于式(8.73)中的 η_{eff}, $z=0$ 平面上的反射系数可由式(8.82)给出。

and η_3, Γ can be adjusted by suitable choices of η_2 and d, which leads to the applications as introduced in the next subsection.

(8.81), or more generally, (8.80) is a very useful impedance-transformation formula in many engineering design problems. For example, we can apply (8.80) to the problem of Figure 8-2, with $z_0 = 0$, $\beta = \beta_1$ and $\eta = \eta_1$. The PEC boundary has an impedance $Z(0) = 0$, and (8.80) becomes

$$Z(-l) = \mathrm{j}\eta_1 \tan\beta_1 l \tag{8.83}$$

which is the same as the ratio between E_1 in (8.5) and H_1 in (8.7) with $z=-l$. This means, by designing the length l, we can have an equivalent medium with any reactive intrinsic impedance.

8.5.2 Half-wave Dielectric Window and Quarter-wave Impedance transformer(半波长介质窗与 1/4 波长阻抗变换器)

Consider again the example as illustrated in Figure 8-7. In some engineering applications, it is desired that no reflection occurs when a uniform plane wave in medium 1 impinges normally on the interface with medium 2. In this case, we need to design d and η_2 such that $\Gamma=0$ at the interface $z=0$. In other words, the design must make $Z_2(0)=\eta_1$. To realize that, from (8.81), we need to let

$$\eta_2(\eta_3\cos\beta_2 d + \mathrm{j}\eta_2\sin\beta_2 d) = \eta_1(\eta_2\cos\beta_2 d + \mathrm{j}\eta_3\sin\beta_2 d) \tag{8.84}$$

Equating the real and imaginary part on both sides of (8.84), we have

$$\eta_3\cos\beta_2 d = \eta_1\cos\beta_2 d \tag{8.85}$$

$$\eta_2^2\sin\beta_2 d = \eta_1\eta_3\sin\beta_2 d \tag{8.86}$$

(8.85) is satisfied if either

$$\eta_3 = \eta_1 \tag{8.87}$$

or

$$\cos\beta_2 d = 0 \tag{8.88}$$

Therefore, we have then two possibilities for the condition of no reflection, which is discussed as follows.

(1) The condition (8.87) holds ($\eta_1=\eta_3$). We only need to consider the case $\eta_3=\eta_1\neq\eta_2$, because otherwise, we will essentially have only one medium with no discontinuities. In this case, (8.86) can be satisfied only if $\sin\beta_2 d=0$, or equivalently,

$$d = n\frac{\lambda_2}{2}, \quad (n = 0,1,2,\cdots) \tag{8.89}$$

where $\lambda_2 = 2\pi/\beta_2$ is the wavelength of the uniform plane wave in medium 2. That is, the thickness of the dielectric layer be a multiple of a half-wavelength in the dielectric at the operating frequency. Such a dielectric layer is referred to as a **half-wave dielectric window**(半波长介质窗). A typical application of the half-wave dielectric window is the **radome**(雷达天线罩). A radome is a dome-shaped enclosure installed around a radar antenna to protect it from inclement weather. In this application, $\eta_3=\eta_1=\eta_0$. Therefore, the radome should be a multiple of the half-wavelength thick at the operating frequency of the radar, so that electromagnetic waves can propagation through it with as little reflection as possible.

(2) The condition(8.87) does not hold($\eta_1\neq\eta_3$). Then the condition(8.88) must hold, which leads to

$$d=(2n+1)\frac{\lambda_2}{4}, \quad (n=0,1,2,\cdots) \tag{8.90a}$$

and(8.86) can be satisfied only if $\eta_2^2=\eta_1\eta_3$, i.e.,

$$\eta_2=\sqrt{\eta_1\eta_3} \tag{8.90b}$$

In other words, when medium 1 and medium 3 are different, η_2 should be the geometric mean of η_1 and η_3, and d should be an odd multiple of a quarter wavelength in the dielectric layer at the operating frequency to eliminate reflection. Under these conditions the dielectric layer(medium 2) is called a **quarter-wave impedance transformer**(1/4 波长阻抗变换器). It can be used to design anti-reflection coatings applied to optical surfaces in order to reduce the reflectivity and improve transmission of the light. This concept of impedance transformation is also very useful in transmission-line problems for impedance matching. ①

8.6 Oblique Incidence at a Dielectric Planar Boundary (介质平面上的斜入射)

We now consider the case in which a uniform plane wave is incident upon an interface between two dielectric media obliquely. The two media are assumed to be lossless and simple with constitutive parameters(ε_1,μ_1) and(ε_2,μ_2) respectively. Similar to the case of normal incidence, part of the incident wave is reflected back into medium 1, and part is transmitted into medium 2. The reflected and transmitted waves depend on the polarization of the incident wave. We discuss the two polarizations, perpendicular polarization and parallel polarization, separately as follows.

8.6.1 Perpendicular Polarization(垂直极化)

For the perpendicular polarization, as illustrated in Figure 8-9, the incident wave propagates in medium 1 along the direction of

$$\boldsymbol{e}_{\mathrm{ki}}=\boldsymbol{e}_{\mathrm{x}}\sin\theta_{\mathrm{i}}+\boldsymbol{e}_{\mathrm{z}}\cos\theta_{\mathrm{i}} \tag{8.91}$$

where θ_{i} is the incident angle. The incident field phasors can be written as

$$\boldsymbol{E}_{\mathrm{i}}(x,z)=\boldsymbol{e}_{\mathrm{y}}E_{\mathrm{i0}}\mathrm{e}^{-\mathrm{j}\beta_1\boldsymbol{e}_{\mathrm{ki}}\cdot\boldsymbol{r}}=\boldsymbol{e}_{\mathrm{y}}E_{\mathrm{i0}}\mathrm{e}^{-\mathrm{j}\beta_1(x\sin\theta_{\mathrm{i}}+z\cos\theta_{\mathrm{i}})} \tag{8.92}$$

$$\boldsymbol{H}_{\mathrm{i}}(x,z)=\frac{1}{\eta_1}[\boldsymbol{e}_{\mathrm{ki}}\times\boldsymbol{E}_{\mathrm{i}}(x,z)]=\frac{E_{\mathrm{i0}}}{\eta_1}(-\boldsymbol{e}_{\mathrm{x}}\cos\theta_{\mathrm{i}}+\boldsymbol{e}_{\mathrm{z}}\sin\theta_{\mathrm{i}})\mathrm{e}^{-\mathrm{j}\beta_1(x\sin\theta_{\mathrm{i}}+z\cos\theta_{\mathrm{i}})} \tag{8.93}$$

---

① 需要注意的是,半波长介质窗和 1/4 波长阻抗变换器都是针对特定频率(波长)的电磁波设计的。对于偏离设计频率的电磁波,发射系数将不再为零。因此,半波长介质窗和 1/4 波长阻抗变换器均呈现出某种带通滤波器的特性。

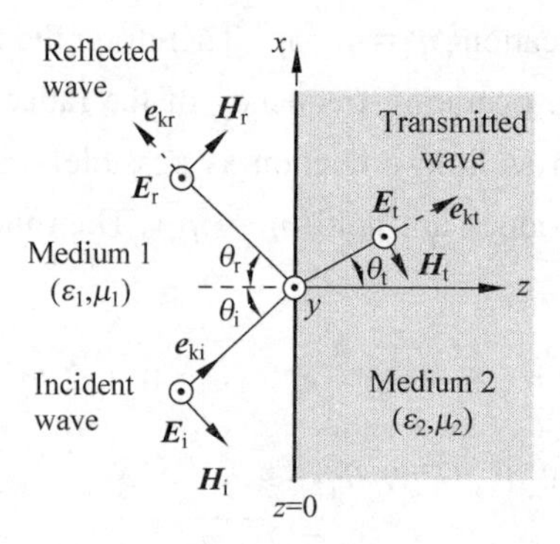

Figure 8-9 Plane wave incident obliquely on a dielectric planar boundary (perpendicular polarization)

The reflected wave propagates along the direction of

$$\boldsymbol{e}_{kr} = \boldsymbol{e}_x \sin\theta_r - \boldsymbol{e}_z \cos\theta_r \tag{8.94}$$

where θ_r is the angle of reflection. The reflected fields can be written in the same form as those given by (8.14) and (8.16)

$$\boldsymbol{E}_r(x,z) = \boldsymbol{e}_y E_{r0} e^{-j\beta_1(x\sin\theta_r - z\cos\theta_r)} \tag{8.95}$$

$$\boldsymbol{H}_r(x,z) = \frac{E_{r0}}{\eta_1}(\boldsymbol{e}_x \cos\theta_r + \boldsymbol{e}_z \sin\theta_r) e^{-j\beta_1(x\sin\theta_r - z\cos\theta_r)} \tag{8.96}$$

However, notice that E_{r0} is no longer equal to $-E_{i0}$ as in the case of reflection at a PEC plane.

In medium 2, the transmitted wave propagates along the direction of $\boldsymbol{e}_{kt}$

$$\boldsymbol{e}_{kt} = \boldsymbol{e}_x \sin\theta_t + \boldsymbol{e}_z \cos\theta_t \tag{8.97}$$

where θ_t is the **angle of transmission** (**透射角**) between the normal direction of the boundary surface and the propagation direction $\boldsymbol{e}_{kt}$. Then the transmitted field phasors can be written as

$$\boldsymbol{E}_t(x,z) = \boldsymbol{e}_y E_{t0} e^{-j\beta_2(x\sin\theta_t + z\cos\theta_t)} \tag{8.98}$$

$$\boldsymbol{H}_t(x,z) = \frac{E_{t0}}{\eta_2}(-\boldsymbol{e}_x \cos\theta_t + \boldsymbol{e}_z \sin\theta_t) e^{-j\beta_2(x\sin\theta_t + z\cos\theta_t)} \tag{8.99}$$

Given the incident wave, i. e., given E_{i0}, there are four unknown quantities in the reflected and transmitted field expressions: E_{r0}, E_{t0}, θ_r and θ_t, which can be determined by the boundary conditions at $z=0$. Continuity of the tangential $\boldsymbol{E}$ and $\boldsymbol{H}$ fields respectively requires

$$\boldsymbol{E}_i(x,0) + \boldsymbol{E}_r(x,0) = \boldsymbol{E}_t(x,0) \quad \text{and} \quad \boldsymbol{e}_z \times [\boldsymbol{H}_i(x,0) + \boldsymbol{H}_r(x,0)] = \boldsymbol{e}_z \times \boldsymbol{H}_t(x,0) \tag{8.100}$$

With the expressions (8.92)-(8.99), the two equations in (8.100) become

$$E_{i0} e^{-j\beta_1 x\sin\theta_i} + E_{r0} e^{-j\beta_1 x\sin\theta_r} = E_{t0} e^{-j\beta_2 x\sin\theta_t} \tag{8.101}$$

$$\frac{1}{\eta_1}(-E_{i0}\cos\theta_i e^{-j\beta_1 x\sin\theta_i} + E_{r0}\cos\theta_r e^{-j\beta_1 x\sin\theta_r}) = -\frac{E_{t0}}{\eta_2}\cos\theta_t e^{-j\beta_2 x\sin\theta_t} \tag{8.102}$$

Since (8.101) and (8.102) have to be satisfied for all x, all three exponential factors must be identical, which is called the **phase-matching** (**相位匹配**) condition. Thus,

$$\beta_1 x\sin\theta_i = \beta_1 x\sin\theta_r = \beta_2 x\sin\theta_t \quad \text{for all } x$$

which leads to

$$\theta_r = \theta_i \tag{8.103}$$

$$\frac{\sin\theta_t}{\sin\theta_i} = \frac{\beta_1}{\beta_2} = \frac{u_{p2}}{u_{p1}} = \frac{n_1}{n_2} \tag{8.104}$$

where n_1 and n_2 are **the indices of refraction**(折射率) of medium 1 and 2, respectively. The index of refraction of a medium is defined as the ratio of the speed of light in free space to that in the medium. That is, $n_1 = c/u_{p1}$ and $n_2 = c/u_{p2}$. (8.103) and (8.104) are respectively called **Snell's law of reflection**(斯涅尔反射定律) and **Snell's law of refraction**(斯涅尔折射定律).①

With Snell's laws, (8.101) and (8.102) can now be simplified to be

$$E_{i0} + E_{r0} = E_{t0} \tag{8.105}$$

$$\frac{1}{\eta_1}(E_{i0} - E_{r0})\cos\theta_i = \frac{E_{t0}}{\eta_2}\cos\theta_t \tag{8.106}$$

from which E_{r0} and E_{t0} can be found in terms of E_{i0}, and the reflection and transmission coefficients are

$$\Gamma_{\perp} = \frac{E_{r0}}{E_{i0}} = \frac{\eta_2\cos\theta_i - \eta_1\cos\theta_t}{\eta_2\cos\theta_i + \eta_1\cos\theta_t} = \frac{(\eta_2/\cos\theta_t) - (\eta_1/\cos\theta_i)}{(\eta_2/\cos\theta_t) + (\eta_1/\cos\theta_i)} \tag{8.107}$$

$$\tau_{\perp} = \frac{E_{t0}}{E_{i0}} = \frac{2\eta_2\cos\theta_i}{\eta_2\cos\theta_i + \eta_1\cos\theta_t} = \frac{2(\eta_2/\cos\theta_t)}{(\eta_2/\cos\theta_t) + (\eta_1/\cos\theta_i)} \tag{8.108}$$

which are the same as the formulas (8.43) and (8.44) for Γ and τ of normal incidence with η_1 and η_2 replaced by ($\eta_1/\cos\theta_i$) and ($\eta_2/\cos\theta_t$) respectively. Notice that ($\eta_1/\cos\theta_i$) and ($\eta_2/\cos\theta_t$) are essentially the wave impedances of the incident and transmitted waves at the boundary plane, which reduce to η_1 and η_2 respectively in the case of normal incidence.②

Also notice that, depending on whether $\Gamma_{\perp}$ is positive or negative, the actual directions of $\boldsymbol{E}_r$ and $\boldsymbol{H}_r$ may or may not be the same as the reference directions as shown Figure 8-9.

(8.107) and (8.108) shows that $\Gamma_{\perp}$ and $\tau_{\perp}$ are related in the following way:

$$1 + \Gamma_{\perp} = \tau_{\perp} \tag{8.109}$$

which is the same as (8.45) for normal incidence.

If medium 2 is a perfect conductor, $\eta_2 = 0$, we have $\Gamma_{\perp} = -1$ (as $E_{r0} = -E_{i0}$) and $\tau_{\perp} = 0$ (as $E_{t0} = 0$). The tangential $\boldsymbol{E}$ field on the surface of the conductor vanishes. No power or wave is

---

① 从这里的推导过程可以看出,斯涅尔反射和折射定律本质上是电磁场边界条件的要求,即切向电场和切向磁场的连续性要求反射波和透射波(折射波)的传播方向要能够使得分界面上总场的相位分布满足匹配条件。分界面上电磁场的相位由相位常数和传播方向共同确定。由于入射波和反射波位于同一种媒质,相位常数相同,因此入射角等于反射角。然而,透射波位于另一种不同的媒质,其相位常数与入射波的相位常数不同,因此透射角通常与入射角不同,也就是发生了所谓的折射。

② 式(8.107)中的($\eta_1/\cos\theta_i$)和($\eta_2/\cos\theta_t$)分别可以看成是入射波和透射波在媒质分界面上的波阻抗(注意:分界面上的波阻抗的定义是分界面上切向电场与切向磁场的比值)。因此,式(8.107)实际上给出了均匀平面波反射的一个基本规律,即分界面上反射电场与入射电场的切向分量之比等于透射波波阻抗与入射波波阻抗之差与两者之和的比值。垂直极化条件下,入射和反射电场都只有切向分量,因此上述计算所得的比值就是反射系数 $\Gamma_{\perp}$。同样,式(8.108)给出了均匀平面波透射的一个基本规律,即分界面上透射电场与入射电场的切向分量之比等于2倍的透射波波阻抗比上入射波与透射波的波阻抗之和。该比值在垂直极化条件下就是透射系数。

transmitted across the PEC boundary, and the reflected fields become the same as those given by (8.15) and (8.16).

8.6.2 Parallel Polarization(平行极化)

Referring to Figure 8-10 for the case of parallel polarization, the propagation directions of the incident, reflected, and transmitted waves are the same as in the case of perpendicular polarization, which are given by (8.91), (8.94) and (8.97) respectively. Then the incident and reflected fields in medium 1 are

$$\boldsymbol{E}_{\mathrm{i}}(x,z)=E_{\mathrm{i0}}(\boldsymbol{e}_{\mathrm{x}}\cos\theta_{\mathrm{i}}-\boldsymbol{e}_{\mathrm{z}}\sin\theta_{\mathrm{i}})\,\mathrm{e}^{-\mathrm{j}\beta_1(x\sin\theta_{\mathrm{i}}+z\cos\theta_{\mathrm{i}})} \tag{8.110}$$

$$\boldsymbol{H}_{\mathrm{i}}(x,z)=\boldsymbol{e}_{\mathrm{y}}\frac{E_{\mathrm{i0}}}{\eta_1}\mathrm{e}^{-\mathrm{j}\beta_1(x\sin\theta_{\mathrm{i}}+z\cos\theta_{\mathrm{i}})} \tag{8.111}$$

$$\boldsymbol{E}_{\mathrm{r}}(x,z)=E_{\mathrm{r0}}(\boldsymbol{e}_{\mathrm{x}}\cos\theta_{\mathrm{r}}+\boldsymbol{e}_{\mathrm{z}}\sin\theta_{\mathrm{r}})\,\mathrm{e}^{-\mathrm{j}\beta_1(x\sin\theta_{\mathrm{r}}-z\cos\theta_{\mathrm{r}})} \tag{8.112}$$

$$\boldsymbol{H}_{\mathrm{r}}(x,z)=-\boldsymbol{e}_{\mathrm{y}}\frac{E_{\mathrm{r0}}}{\eta_1}\mathrm{e}^{-\mathrm{j}\beta_1(x\sin\theta_{\mathrm{r}}-z\cos\theta_{\mathrm{r}})} \tag{8.113}$$

which are the same as those given by (8.25)-(8.28). The transmitted field phasors in medium 2 are

$$\boldsymbol{E}_{\mathrm{t}}(x,z)=E_{\mathrm{t0}}(\boldsymbol{e}_{\mathrm{x}}\cos\theta_{\mathrm{t}}-\boldsymbol{e}_{\mathrm{z}}\sin\theta_{\mathrm{t}})\,\mathrm{e}^{-\mathrm{j}\beta_2(x\sin\theta_{\mathrm{t}}+z\cos\theta_{\mathrm{t}})} \tag{8.114}$$

$$\boldsymbol{H}_{\mathrm{t}}(x,z)=\boldsymbol{e}_{\mathrm{y}}\frac{E_{\mathrm{t0}}}{\eta_2}\mathrm{e}^{-\mathrm{j}\beta_2(x\sin\theta_{\mathrm{t}}+z\cos\theta_{\mathrm{t}})} \tag{8.115}$$

Again, the continuity conditions for the tangential components of $\boldsymbol{E}$ and $\boldsymbol{H}$ at the boundary $z=0$ lead to Snell's laws of reflection and refraction, as well as to the following two equations:

$$(E_{\mathrm{i0}}+E_{\mathrm{r0}})\cos\theta_{\mathrm{i}}=E_{\mathrm{t0}}\cos\theta_{\mathrm{t}} \tag{8.116}$$

$$\frac{1}{\eta_1}(E_{\mathrm{i0}}-E_{\mathrm{r0}})=\frac{1}{\eta_2}E_{\mathrm{t0}} \tag{8.117}$$

Solving the two equations for E_{r0} and E_{t0} in terms of E_{i0}, we obtain

$$\Gamma_{\parallel}=\frac{E_{\mathrm{r0}}}{E_{\mathrm{i0}}}=\frac{\eta_2\cos\theta_{\mathrm{t}}-\eta_1\cos\theta_{\mathrm{i}}}{\eta_2\cos\theta_{\mathrm{t}}+\eta_1\cos\theta_{\mathrm{i}}} \tag{8.118}$$

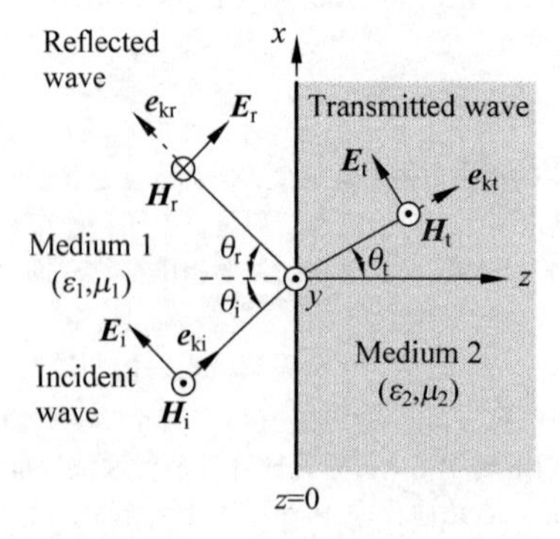

Figure 8-10 Plane wave incident obliquely on a dielectric planar boundary (parallel polarization)

and

$$\tau_{\parallel}=\frac{E_{t0}}{E_{i0}}=\frac{2\eta_2\cos\theta_i}{\eta_2\cos\theta_t+\eta_1\cos\theta_i}=\frac{2\eta_2\cos\theta_t}{\eta_2\cos\theta_t+\eta_1\cos\theta_i}\left(\frac{\cos\theta_i}{\cos\theta_t}\right) \tag{8.119}$$

The expression for $\Gamma_{\parallel}$ in(8.118) can be obtained from that for Γ in(8.43) by replacing η_1 and η_2 with ($\eta_1\cos\theta_i$) and ($\eta_2\cos\theta_t$) respectively. Notice that ($\eta_1\cos\theta_i$) and ($\eta_2\cos\theta_t$) are respectively the wave impedances of the incident and transmitted waves at the boundary plane for the parallel polarization, which are different from those for the perpendicular polarization as introduced in Subsection 8.6.1. The expression for $\tau_{\parallel}$ in(8.119) can also be obtained from that for τ in(8.44), but in two steps. First, replacing η_1 and η_2 with($\eta_1\cos\theta_i$) and($\eta_2\cos\theta_t$) respectively in(8.44) gives the ratio of the tangential component of $\boldsymbol{E}_t$ to that of $\boldsymbol{E}_i$, which then multiplied by($\cos\theta_i/\cos\theta_t$), leads to(8.119), the ratio of the total magnitude of $\boldsymbol{E}_t$ to that of $\boldsymbol{E}_i$.①

Also notice that, the expression of $\Gamma_{\parallel}$ in(8.118) is specific for the reference directions of the reflected fields indicated in Figure 8-10. If the opposite reference directions are used(i.e., if the reference direction of $\boldsymbol{H}_r$ is selected to be $+y$ direction), there should be an additional negative sign on the right side of the expression(8.118). However, the ultimate expressions for the reflected fields should be the same no matter which reference directions are used. As is the same with the case of perpendicular polarization, the actual directions of $\boldsymbol{E}_r$ and $\boldsymbol{H}_r$ may or may not be the same as the reference directions, depending on whether $\Gamma_{\parallel}$ is positive or negative.

From(8.118) and(8.119), it is easy to verify that

$$1+\Gamma_{\parallel}=\tau_{\parallel}\left(\frac{\cos\theta_t}{\cos\theta_i}\right) \tag{8.120}$$

Notice that(8.120) is different from the relations(8.45) and(8.109).

If medium 2 is a perfect conductor($\eta_2=0$), (8.118) and(8.119) is simplified to $\Gamma_{\parallel}=-1$ and $\tau_{\parallel}=0$ respectively. The tangential $\boldsymbol{E}$ field on the surface of the conductor vanishes. No power or wave is transmitted into medium 2, and the reflected fields become the same as those given by(8.29) and(8.30).

For the planar interface between two dielectric media, we can verify that $|\Gamma_{\perp}|^2>|\Gamma_{\parallel}|^2$ for all θ_i larger than 0. This means that the reflection of an unpolarized wave (a wave whose polarization direction changes rapidly and randomly, such as the sunlight) usually contain more

---

① 平行极化条件下,($\eta_1\cos\theta_i$)和($\eta_2\cos\theta_t$)分别为入射波和透射波在媒质分界面上的波阻抗,因此式(8.118)与式(8.107)一样,也是均匀平面波反射的一个基本规律,即分界面上反射电场与入射电场的切向分量之比等于透射波波阻抗与入射波波阻抗之差与两者之和的比值。对于平行极化,分界面上电场除了由切向分量,还有法向分量。但由于反射角等于入射角,反射电场与入射电场的切向分量之比就等于反射电场与入射电场整体的比值,也就是反射系数。同样地,根据均匀平面波透射的一个基本性质,可以得到分界面上透射电场与入射电场的切向分量之比等于2倍的透射波波阻抗比上入射波与透射波的波阻抗之和。然后根据斯涅尔折射定律,透射电场与入射电场的切向分量之比与($\cos\theta_i/\cos\theta_t$)的乘积就是透射电场与入射电场整体的比值,也就是式(8.119)给出的透射系数。

power in the wave with perpendicular polarization than that with parallel polarization.[①] This phenomenon is utilized in the invention of **polarized lenses**(偏振镜片) to reduce **sun glare**(眩光). Since the reflected wave of strong sunlight from a planar surface is mostly perpendicularly polarized, the glare can be significantly reduced by the polarized lenses that blocks the light with perpendicular polarization (i. e., the lenses are able to filter out the waves with electric field parallel to the surface of reflection). Therefore, we see polarized lenses commonly used in sunglasses, binoculars, telescopes, and cameras.

8.6.3 Total Reflection(全反射)

As is seen in previous subsections, the Snell's law of refraction given in (8.104) is independent of the polarization of the wave, but dependent on the property of the media. It states that at an interface between two dielectric media, the ratio of the sine of the angle of transmission to the sine of the angle of incidence is equal to the inverse ratio of indices of refraction n_1/n_2.

For nonmagnetic media, $\mu_1=\mu_2=\mu_0$, (8.104) becomes

$$\frac{\sin\theta_t}{\sin\theta_i}=\sqrt{\frac{\varepsilon_1}{\varepsilon_2}}=\sqrt{\frac{\varepsilon_{r1}}{\varepsilon_{r2}}}=\frac{n_1}{n_2}=\frac{\eta_2}{\eta_1} \tag{8.121}$$

where η_1 and η_2 are the intrinsic impedances of the media. Furthermore, if medium 1 is free space such that $\varepsilon_{ri}=1$ and $n_1=1$, (8.121) reduces to

$$\frac{\sin\theta_t}{\sin\theta_i}=\frac{1}{\sqrt{\varepsilon_{r2}}}=\frac{1}{n_2}=\frac{\eta_2}{120\pi} \tag{8.122}$$

Since $n_2\geqslant 1$ (the relative permittivity of a natural medium is usually larger than 1), if a plane wave is incident obliquely upon the surface of a dielectric medium from free space, the transmitted wave in the dielectric will be bent toward the normal. Generally, when $n_2>n_1$, we call medium 2 is a **denser medium** (光密媒质) as compared to medium 1, which is a **rarer medium**(光疏媒质). In this case, $\theta_t<\theta_i$.

On the other hand, if medium 1 is a denser medium whereas medium 2 is a rarer medium, i. e., $n_1>n_2(\varepsilon_1>\varepsilon_2)$, then according to (8.121), we must have $\theta_t>\theta_i$. Let θ_i increase from 0 to $\pi/2$, then θ_t will reach $\pi/2$ before θ_i increases to $\pi/2$. In this case, the transmitted wave will propagate along the interface. The angle of incidence at which θ_t reaches $\pi/2$ is called the **critical angle**(临界角), and is denoted by θ_c. This situation is illustrated in Figure 8-11, where $\boldsymbol{e}_{ki}$ $\boldsymbol{e}_{kr}$ and $\boldsymbol{e}_{kt}$ are unit vectors denoting the directions of propagation of the incident, reflected, and transmitted waves, respectively.

By setting $\theta_t=\pi/2$ in (8.121), we have

---

① 许多电磁辐射,例如阳光、火焰,是众多随机极化(偏振)方向的波的叠加。这样形成的电磁波称为**非极化波**(unpolarized wave)。这些非极化波经过多次反射和散射后,会成为部分极化波,其中一个原理就是垂直极化波的反射系数大于平行极化波。

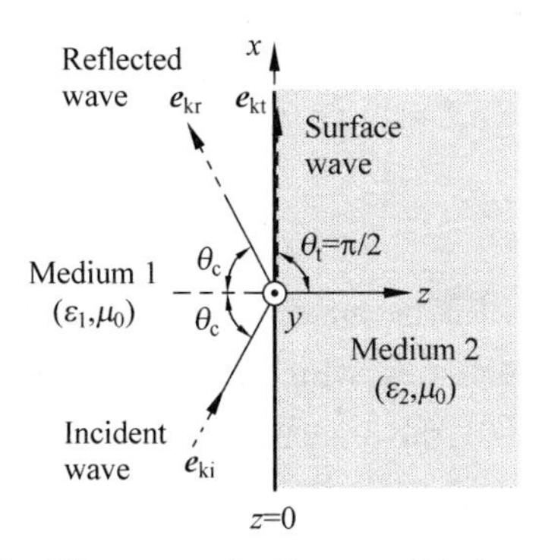

Figure 8-11 Plane wave incident at critical angle, $\varepsilon_1 > \varepsilon_2$

$$\sin\theta_c = \sqrt{\frac{\varepsilon_2}{\varepsilon_1}} \tag{8.123}$$

Hence,

$$\theta_c = \sin^{-1}\sqrt{\frac{\varepsilon_2}{\varepsilon_1}} = \sin^{-1}\left(\frac{n_2}{n_1}\right) \tag{8.124}$$

Now the question is, what will happen if θ_i is further increased. If we let $\theta_i > \theta_c$ in (8.121), we have

$$\sin\theta_t = \sqrt{\frac{\varepsilon_1}{\varepsilon_2}}\sin\theta_i > 1 \tag{8.125}$$

and mathematically, we have

$$\cos\theta_t = \sqrt{1-\sin^2\theta_t} = \pm j\sqrt{\frac{\varepsilon_1}{\varepsilon_2}\sin^2\theta_i - 1} \tag{8.126}$$

which becomes purely imaginary. (8.125) and (8.126) mean that θ_t is not a realistic angle when $\theta_i > \theta_c$. In other words, when the incident angle is larger than the critical angle, the transmitted wave is no longer a uniform plane wave propagating along the direction of $\boldsymbol{e}_{kt}$ as given in (8.97). Nevertheless, the expressions (8.98), (8.99), (8.114) and (8.115) are still valid due to the requirement of phase matching and the homogeneous Helmholtz's equation that the transmitted fields must satisfy. Therefore, $\boldsymbol{E}_t$ and $\boldsymbol{H}_t$ vary spatially with the same factor①

$$e^{-j\beta_2(x\sin\theta_t + z\cos\theta_t)} = e^{-j\beta_2 x\sin\theta_t}e^{-j\beta_2 z\cos\theta_t}$$

which, with (8.125) and (8.126), becomes

$$e^{-j\beta_{2x}x}e^{-\alpha_2 z} \tag{8.127}$$

---

① 式(8.125)和式(8.126)表明,当入射角大于临界角时,θ_t 不再是一个有几何意义的透射角,$\boldsymbol{e}_{kt}$ 不再是代表某个实际方向的单位矢量,媒质2中的透射场也不再是一个沿 $\boldsymbol{e}_{kt}$ 代表的方向的均匀平面波。然而垂直极化和平行极化条件下透射场的表达式(8.98)~式(8.99)和式(8.114)~式(8.115)在数学上仍然是成立的,因为它们满足相位匹配(边界条件)和媒质2中的亥姆霍兹方程。透射场必须沿分界面满足相位匹配条件,这就决定了其沿分界面的切向必然呈现出以 $\beta_2\sin\theta_t(=\beta_1\sin\theta_i)$ 为相位常数的传播特性。在此基础上,由于透射场还要满足媒质2中的亥姆霍兹方程,这就决定了其沿分界面的法向只能是指数衰减的,其衰减常数为 $\beta_2\cos\theta_t$。

where

$$\alpha_2 = \beta_2 \sqrt{\varepsilon_1/\varepsilon_2 \sin^2\theta_i - 1}$$

and

$$\beta_{2x} = \beta_2 \sqrt{\varepsilon_1/\varepsilon_2} \sin\theta_i$$

The plus sign in(8.126) is discarded because it leads to an impossible result of a transmitted field growing exponentially as z increases, which is not practical.

(8.127) shows that, for $\theta_i > \theta_c$, the transmitted wave is an **evanescent wave**(渐逝波), which propagates along the tangential direction(in the x-direction) of the interface but decays exponentially along the normal direction(z-direction). Therefore, this wave is tightly bound to the interface and is also called a **surface wave**(表面波). Obviously, it is a nonuniform plane wave.

With(8.125) and(8.126), it can be proved that the reflection coefficients $\Gamma\perp$ and $\Gamma_{\parallel}$ both have a unit magnitude, and the time-average power flow along the normal direction of the interface is zero. Therefore, the incident wave is totally reflected when $\theta_i > \theta_c$, and the critical angle θ_c is the threshold of **total reflection**(全反射). ①

Example 8-2 An isotropic light source is at a distance 1m under water. Assume the relative permitivity of the water is 1.75 at optical frequencies. Find the illuminated area seen from outside the water.

Solution: The index of refraction of water is $n = \sqrt{1.75} = 1.32$. So the critical angle is

$$\theta_c = \arcsin\left(\frac{1}{1.32}\right) = 49.2°$$

As is shown in Figure 8-12, the illuminated area must be a circle right above the light source, with its center at point O'. Within the circle, the light arrives at the water surface with an incident angle smaller than the critical angle. Partial light is transmitted into the air above. On the boundary of the circle, point P for instance, the incident angle is equal to the critical angle. Outside the circle, the incident angle is larger than the critical angle, and the light from the source is totally reflected. Hence the radius of the illuminated circle is

$$\overline{O'P} = \overline{OO'}\tan\theta_c = (1\text{m})\tan 49.2° = 1.16\text{m}$$

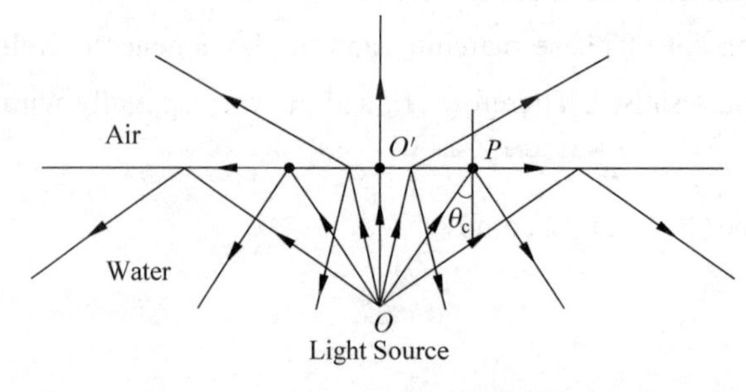

Figure 8-12 An underwater light source(Example 8-2)

① 全反射条件下,两种极化的反射系数 $\Gamma\perp$ 和 $\Gamma_{\parallel}$ 仍然可以分别通过式(8.107)和式(8.118)计算。由于 $\cos\theta_i$ 是实数,而 $\cos\theta_t$ 是纯虚数,两个反射系数的计算结果都是幅度为 1 的复数,因此,反射波的功率密度和入射波相同,所有入射功率都被反射回媒质 1。

And the area of the circle is $\pi(1.16^2)=4.2(\mathrm{m}^2)$.

Example 8-3 An important application of total reflection is the using of a dielectric rod to guide the transmission of an electromagnetic wave over long distances with very little loss. This is basically how optical fiber (光纤) works. Suppose an electromagnetic wave enters the dielectric rod from the left end and leaves from the right end as is shown in Figure 8-13. Determine the minimum dielectric constant of the dielectric rod so that the wave is confined within the rod.

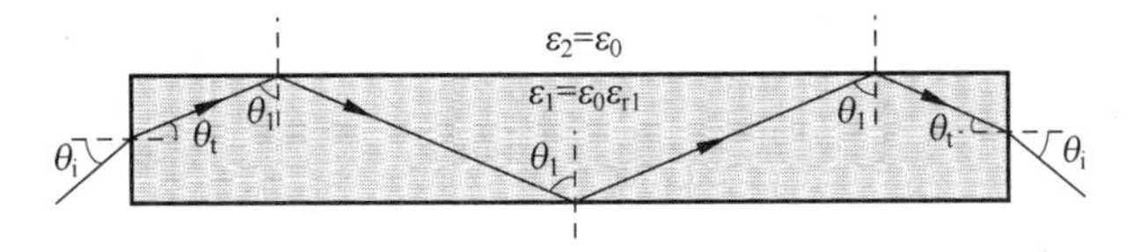

Figure 8-13 Dielectric rod or fiber guiding electromagnetic wave by total internal reflection

Solution: For the total internal reflection, the incident angle θ_1 must be larger than the critical angle θ_c of the dielectric medium. Hence we must let $\sin\theta_1>\sin\theta_c$. Notice that the angle of transmission at the left end of the dielectric is $\theta_t=(\pi/2)-\theta_1$. $\sin\theta_1=\cos\theta_t$, and therefore we have

$$\cos\theta_t > \sin\theta_c \tag{8.128}$$

From Snell's law of refraction,

$$\sin\theta_t=\frac{1}{\sqrt{\varepsilon_{r1}}}\sin\theta_i$$

Therefore, we have

$$\cos\theta_t=\sqrt{1-\frac{1}{\varepsilon_{r1}}\sin^2\theta_i} \tag{8.129}$$

Substituting(8.129) and(8.123) into(8.128), we obtain

$$\sqrt{1-\frac{1}{\varepsilon_{r1}}\sin^2\theta_i} > \frac{1}{\sqrt{\varepsilon_{r1}}}$$

which requires

$$\varepsilon_{r1} \geqslant 1+\sin^2\theta_i \tag{8.130}$$

(8.130) can be satisfied for any incident angle θ_i if ε_{r1} is selected to be larger than 2. Hence the minimum dielectric constant of the rod is 2, corresponding to a minimum refraction index $n_1=\sqrt{2}$.

8.6.4 Brewster Angle and Total Transmission (布儒斯特角与全透射)

In addition to the total reflection discussed in the previous subsection, **total transmission** (全透射) or equivalently, **zero reflection** (零反射) is also possible in the case of oblique incidence. We will see that total transmission could occur only when the incident angle takes a particular value θ_B called **Brewster angle** (布儒斯特角). In other words, when the incident angle $\theta_i=\theta_B$, the reflection coefficient is zero. As previously discussed, the reflection coefficient is different for the two types of polarizations. Therefore, the condition for total transmission (zero reflection) is different for different polarizations.

We start with the parallel polarization first. Denote the Brewster angle for the case of parallel polarization by $\theta_{B\parallel}$. When $\theta_i=\theta_{B\parallel}$, the reflection coefficient $\Gamma_{\parallel}$ must vanish. From (8.118), that means

$$\eta_2\cos\theta_t=\eta_1\cos\theta_{B\parallel} \tag{8.131}$$

Using Snell's law of refraction, we have

$$\cos\theta_t=\sqrt{1-\sin^2\theta_t}=\sqrt{1-\frac{n_1^2}{n_2^2}\sin^2\theta_i}=\sqrt{1-\frac{n_1^2}{n_2^2}\sin^2\theta_{B\parallel}} \tag{8.132}$$

Substituting (8.132) into (8.131), we have

$$\sin^2\theta_{B\parallel}=\frac{1-\mu_2\varepsilon_1/\mu_1\varepsilon_2}{1-(\varepsilon_1/\varepsilon_2)^2} \tag{8.133}$$

Obviously, the Brewster angle $\theta_{B\parallel}$ exists only if $\varepsilon_1\neq\varepsilon_2$. Usually, for two nonmagnetic media, we have $\mu_1=\mu_2=\mu_0$, and

$$\sin\theta_{B\parallel}=\frac{1}{\sqrt{1+(\varepsilon_1/\varepsilon_2)}}\quad(\text{if }\mu_1=\mu_2) \tag{8.134}$$

An alternative form for (8.134) is

$$\theta_{B\parallel}=\arctan\sqrt{\frac{\varepsilon_2}{\varepsilon_1}}=\arctan\left(\frac{n_2}{n_1}\right)\quad(\text{if }\mu_1=\mu_2) \tag{8.135}$$

Similarly, the Brewster angle for perpendicular polarization, denoted by $\theta_{B\perp}$, can be derived by letting the reflection coefficient in (8.107) vanish. That is,

$$\eta_2\cos\theta_{B\perp}=\eta_1\cos\theta_t \tag{8.136}$$

from which we obtain

$$\sin^2\theta_{B\perp}=\frac{1-\mu_1\varepsilon_2/\mu_2\varepsilon_1}{1-(\mu_1/\mu_2)^2} \tag{8.137}$$

For nonmagnetic media, $\mu_1=\mu_2=\mu_0$. The right side of (8.137) becomes infinite, and $\theta_{B\perp}$ does not exist. Therefore, total transmission of perpendicularly polarized wave can exist only if at least one of the media is made of magnetic materials. In the case of $\varepsilon_1=\varepsilon_2$ and $\mu_1\neq\mu_2$, (8.137) reduces to

$$\sin\theta_{B\perp}=\frac{1}{\sqrt{1+(\mu_1/\mu_2)}} \tag{8.138}$$

And therefore, the Brewster angle under perpendicular polarization is

$$\theta_{B\perp}=\arctan\sqrt{\frac{\mu_2}{\mu_1}}\quad(\text{if }\varepsilon_1=\varepsilon_2) \tag{8.139}$$

Notice that the Brewster angle is different for different polarization of the incident wave. For interface between non-magnetic media, Brewster angle only exists under parallel polarization, whereas for interface between magnetic media with equal permittivity (which is a much rarer case) Brewster angle only exists under perpendicular polarization. Hence Brewster angle is also called **polarizing angle**(**极化角**). These phenomena can be used to separate the two types of polarization from an unpolarized wave. When an unpolarized wave such as the sun light is incident upon a non-magnetic material at the Brewster angle $\theta_{B\parallel}$, only the component with perpendicular polarization will be reflected.

Example 8-4 A perpendicularly polarized wave is incident from the air upon a non-magnetic dielectric medium at the incident angle $\theta_i = \theta_{B\parallel}$. Given that the permittivity of the medium is $80\varepsilon_0$, determine the Brewster angle $\theta_{B\parallel}$, the reflection and transmission coefficients.

Solution: From(8.134),

$$\theta_{B\parallel} = \arcsin \frac{1}{\sqrt{1 + (1/80)}} = 81.0°$$

From(8.122), the angle of transmission is

$$\theta_t = \arcsin\left(\frac{\sin\theta_{B\parallel}}{\sqrt{80}}\right) = \arcsin\left(\frac{1}{\sqrt{81}}\right) = 6.38°$$

By letting $\eta_1 = 377(\Omega)$, $\eta_2 = 377/\sqrt{80} = 40.1(\Omega)$, $\theta_i = 81.0°$ and $\theta_t = 6.38°$ in(8.107) and(8.108), we find

$$\Gamma_\perp = -0.967 \quad \text{and} \quad \tau_\perp = 0.033$$

This result means about 93.5% of the energy carried by the wave components with perpendicular polarization is reflected, whereas no energy carried by those with parallel polarization is reflected since the incident angle is the Brewster angle $\theta_{B\parallel}$.

Summary

Concepts

Incident/reflected/transmitted wave(入射/反射/透射波)

Normal/oblique incidence(垂直/斜入射)

Standing wave(驻波)

Plane of incidence(入射面)

Angle of incidence/ reflection/transmission(入射/反射/透射角)

Perpendicular polarization(垂直极化)

Parallel polarization(平行极化)

Nonuniform plane wave(非均匀平面波)

Transverse electric(TE) wave(横电波)

Transverse magnetic(TM) wave(横磁波)

Reflection coefficient(反射系数)

Transmission coefficient(透射系数)

Standing wave ratio(驻波比)

Wave impedance(波阻抗)

Impedance transformation(阻抗变换)

Half-wave dielectric window(半波长介质窗)

Quarter-wave impedance transformer(四分之一波长阻抗变换器)

Phase matching(相位匹配)

Total reflection(全反射)

Critical angle(临界角)

Total transmission(全透射)

Brewster Angle(布儒斯特角)

Evanescent wave(渐逝波)

Surface wave(表面波)

Laws & Theorems
Snell's law of reflection/transmission(斯涅尔反射/折射定律)

Methods
Determine the reflected and transmitted waves given an incident plane wave using boundary conditions on an infinitely large plane interface.

Review Questions

8.1 什么是驻波?驻波与行波的区别有哪些?

8.2 给定一个垂直入射波,如何求得 PEC 表面的反射波?

8.3 什么是入射面,什么是入射角?

8.4 垂直极化和平行极化的定义是什么?

8.5 反射系数和透射系数是如何定义的?两者之间是什么关系?

8.6 PEC 表面的反射系数和透射系数分别是多少?

8.7 在什么条件下,反射系数是正实数?什么条件下反射系数是负实数?

8.8 在什么条件下,透射系数是正实数?什么条件下透射系数是负实数?

8.9 两种不同的理想介质的分界面上,电场强度达到最大值和最小值的条件分别是什么?

8.10 两种不同的理想介质的分界面上,磁场强度达到最大值和最小值的条件分别是什么?

8.11 什么是驻波比?驻波比与反射系数的关系是什么?

8.12 如何定义波阻抗?波阻抗与本征阻抗有什么区别?在什么条件下两者相等?

8.13 什么是四分之一波长阻抗变换器?其有什么应用?

8.14 什么是半波长介质窗?其有什么应用?

8.15 简述斯涅尔反射定律和折射定律。

8.16 什么是临界角?什么条件下存在临界角?

8.17 两种介质分界面上发生全反射的条件是什么?发生全反射时,是否还存在透射波?如果存在,透射波的特征是什么?

8.18 什么是布儒斯特角?布儒斯特角存在的条件是什么?

8.19 为什么布儒斯特角也叫极化角?

Problems

8.1 A uniform plane wave with electric field given by

$$\boldsymbol{E}_{\mathrm{i}}(z)=E_0(\boldsymbol{e}_{\mathrm{x}}-\mathrm{j}\boldsymbol{e}_{\mathrm{y}})\,\mathrm{e}^{-\mathrm{j}\beta z}$$

is normally incident from air onto a PEC plane at $z=0$.

(1) Determine the polarization of the incident and reflected waves.

(2) Obtain the instantaneous expression of the incident, reflected and total magnetic fields.

(3) Find the instantaneous and phasor expressions of the induced current on the PEC plane.

8.2 Repeat Problem 8.1 for oblique incidence of a uniform plane wave with incident angle θ_i

$$\boldsymbol{E}_i(x,z) = E_0(\boldsymbol{e}_x\cos\theta_i + j\boldsymbol{e}_y - \boldsymbol{e}_z\sin\theta_i)\,e^{-j\beta(x\sin\theta_i + z\cos\theta_i)}$$

8.3 A uniform plane wave with a magnetic field intensity given by

$$\boldsymbol{H}_i(x,z) = \boldsymbol{e}_y 0.1e^{-j(3\pi x+4\pi z)}$$

is incident from air onto a PEC plane at $z=0$.

(1) Find the frequency, wavelength and the angle of incidence of the wave.

(2) Write down the phasor expression of the incident electric field $\boldsymbol{E}_i(x,z)$.

(3) Find $\boldsymbol{E}_r(x,z)$ and $\boldsymbol{H}_r(x,z)$ of the reflected wave.

(4) Find the total fields $\boldsymbol{E}(x,z)$ and $\boldsymbol{H}(x,z)$ in the air.

8.4 Repeat Problem 8.3 for $\boldsymbol{E}_i(x,z)=\boldsymbol{e}_y 120\pi e^{-j(3\pi x+4\pi z)}$.

8.5 A uniform plane wave is normally incident upon a planar interface between two lossless dielectric media. If the magnitude of the reflection coefficient equals the magnitude of the transmission coefficient, determine the reflection coefficient and the standing-wave ratio.

8.6 A z-poloarized uniform plane wave is normally incident from air onto a dielectric medium with $\varepsilon_r=2.5$ with the interface at $x=0$. If the maximum magnitude of the electric field intensity observed at the interface is 500mV/m, determine ①the reflection coefficient, ②the transmission coefficient, and ③the time-average power densities of the incident, and transmitted waves.

8.7 An x polarized uniform plane wave is normally incident from air onto a good conductor with conductivity $\sigma \gg \omega\varepsilon_0$. The interface between air and the good conductor is a plane at $z=0$.

(1) Find the reflection coefficient.

(2) Find the ratio between the transmitted power density and the incident power density.

8.8 A uniform plane wave is normally incident from air upon a lossy medium with the planar interface at $z=0$. The medium has a dielectric constant 2.5 and a loss tangent 0.5. Assume the incident wave has an electric field $\boldsymbol{E}_i(z)=\boldsymbol{e}_x 10e^{-j6\pi z}$(V/m). Find

(1) The reflected fields $\boldsymbol{E}_r(z)$ and $\boldsymbol{H}_r(z)$.

(2) The time-average Poynting vectors in air and in the lossy medium.

8.9 An anti-reflection coating (AR coating, 防反射膜,增透膜) is a dielectric thin-film deposited on optical surfaces in order to reduce the optical reflectivity and enhance the transmission. For an AR coating on the surface of glass ($\varepsilon_r=4, \mu_r=1$) for red light ($\lambda_0=0.75\mu m$), determine

(1) The dielectric constant and thickness of the coating,

(2) The reflection coefficient of the coated glass for violet light ($\lambda_0=0.42\mu m$).

Note: normal incidence is assumed in this problem.

8.10 Consider a radome made of a dielectric slab with $\varepsilon_r=2.8$.

(1) Determine the thickness of the radome operating at f=3GHz.

(2) For the above radome design, find its reflection coefficient at 3.1GHz and 2.9GHz.

8.11 A perpendicularly polarized plane wave is incident upon a planar interface at $z=0$ with $\theta_i>\theta_c$, as shown in Figure 8-9. With the incident ($\boldsymbol{E}_i$, $\boldsymbol{H}_i$) fields given by (8.92) and (8.93),

(1) Write down expressions of the transmitted fields($\boldsymbol{E}_t$, $\boldsymbol{H}_t$).

(2) Verify that the time-average power transmitted into medium 2 is zero.

8.12 An infinitely large dielectric plate with thickness d and refraction index n_1 is inserted into a dielectric medium with refraction index n_0 ($n_0>n_1$). A uniform plane wave is incident from one side of the dielectric medium onto the dielectric plate with an incident angle larger than the critical angle. Evaluate the reflection coefficient as a function of the thickness d. What is the fraction of the incident power transmitted to the other side of the dielectric medium?

8.13 A circularly polarized wave is incident from a perfect dielectric (ε_1, μ_1) onto the planar surface of another perfect dielectric(ε_2, μ_2).

(1) Is total transmission possible? If yes, under what condition does it happen?

(2) Is total reflection possible? If yes, under what condition does it happen?

8.14 A plane wave in free space is incident at an angle θ onto a conducting half-space. Show that, for large $\sigma/\omega\varepsilon_0$, the transmitted angle is

$$\theta_t \approx \frac{\sqrt{2\omega\varepsilon_0}}{\sigma}\sin\theta$$

Chapter 9 Waveguides and Cavity Resonators(波导与谐振腔)

9.1 Introduction(引言)

In previous chapters, we studied the characteristics of plane waves propagating in unbounded medium and the reflection and transmission phenomena at infinitely large plane interfaces between different media. In many practical electrical/electronic systems, however, the electromagnetic waves are supposed to travel along specified direction within a bounded region. In other words, we have a **guided electromagnetic wave**(导行电磁波). In this chapter, we study the wave propagation along uniform guiding structures. Wave-guiding structures are called **waveguides**(波导), in which electromagnetic fields are confined within the cross-sectional area and propagate in the longitudinal direction.①

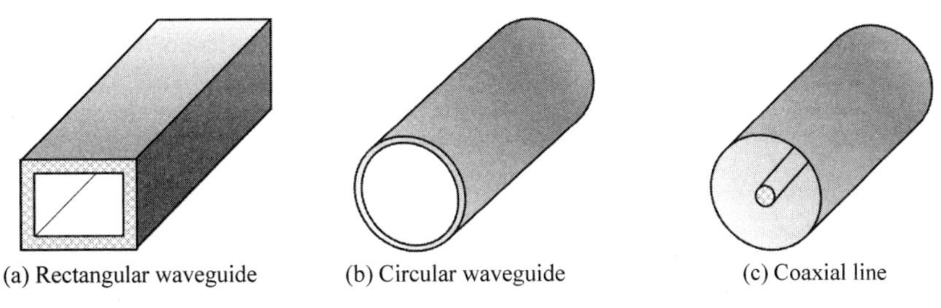

Figure 9-1 Common types of waveguides

A waveguide may consist of a single conductor or multiple conductors. The waveguides studied in this chapter include rectangular waveguide, circular waveguides, and coaxial lines. As shown in Figure 9-1, rectangular and circular waveguides are hollow or dielectric-filled metal pipes. Coaxial lines are the most seen two-conductor transmission lines. This chapter starts with a general analysis of the characteristics of the waves propagating inside waveguides with uniform cross section. Then we will study each of the waveguides in detail. Based on the analysis

---

① 前两章讨论的均匀平面波是无界空间中电磁波的基本解。实际的许多应用要求将电磁波限制在一个有限区域中，沿特定方向传播。为此需要专门设计的波导结构。波导可以由导线、导体管或介质板实现。常见的电路、传输线都可以视为波导。本章讨论典型的金属边界管状波导，给出其中能够传播的电磁波的形态和特性。在此基础上，进一步讨论完全封闭空间及谐振腔内的电磁场。

methods used for waveguide structures, we will discuss cavity resonators that can be considered as fully-enclosed waveguides.

9.2 Wave Propagation along Uniform Guiding Structures (均匀波导结构内波的传播)

In this chapter, we consider the electromagnetic wave propagation along a straight waveguide with a uniform cross section, as is shown in Figure 9-2. That means, we assume that the cross section of the waveguide does not vary along the **longitudinal direction** (纵向方向), although it may be of an arbitrary shape. For simplicity, the longitudinal direction (the direction of the axis of the waveguide) is assumed to be the z-direction, and the electromagnetic wave is assumed to propagate in the $+z$-direction. Then, for a time-harmonic wave, the variation of all field components along the z-direction can be described by the exponential propagation factor

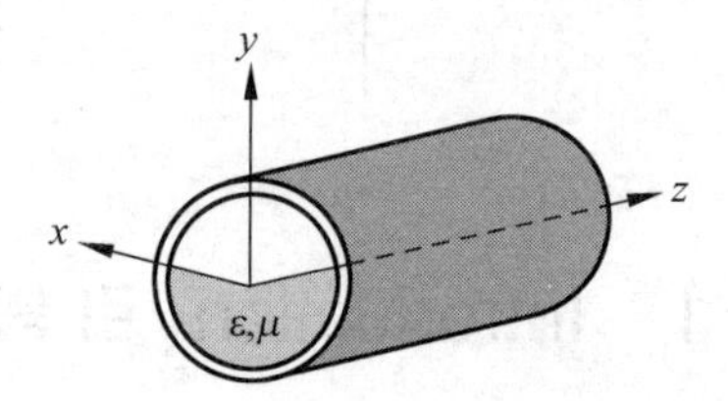

Figure 9-2 Uniform waveguide with arbitrary cross section

$$e^{-\gamma z} = e^{-\alpha z} e^{-j\beta z} \tag{9.1}$$

where $\gamma = \alpha + j\beta$ is the propagation constant that is to be determined.

The propagation factor of (9.1) is used in Section 7-3 to express the uniform plane wave in an unbounded medium. In those cases, the fields have no variation along x- and y-directions. However, the fields in a waveguide are usually nonuniform along **transverse directions** (横向方向). Hence the phasor expressions for the $\boldsymbol{E}$ and $\boldsymbol{H}$ fields in Cartesian coordinates should be written as

$$\boldsymbol{E}(x,y,z) = \boldsymbol{E}^0(x,y)e^{-\gamma z} = [\boldsymbol{e}_x E_x^0(x,y) + \boldsymbol{e}_y E_y^0(x,y) + \boldsymbol{e}_z E_z^0(x,y)]e^{-\gamma z} \tag{9.2a}$$

$$\boldsymbol{H}(x,y,z) = \boldsymbol{H}^0(x,y)e^{-\gamma z} = [\boldsymbol{e}_x H_x^0(x,y) + \boldsymbol{e}_y H_y^0(x,y) + \boldsymbol{e}_z H_z^0(x,y)]e^{-\gamma z} \tag{9.2b}$$

where $\boldsymbol{E}^0(x,y) = \boldsymbol{e}_x E_x^0(x,y) + \boldsymbol{e}_y E_y^0(x,y) + \boldsymbol{e}_z E_z^0(x,y)$ and $\boldsymbol{H}^0(x,y) = \boldsymbol{e}_x H_x^0(x,y) + \boldsymbol{e}_y H_y^0(x,y) + \boldsymbol{e}_z H_z^0(x,y)$ are complex vector fields that depends only on the cross-sectional coordinates (x and y). In fact, $\boldsymbol{E}^0(x,y)$ and $\boldsymbol{H}^0(x,y)$ are exactly the fields in the waveguide at $z=0$. Similarly, under cylindrical coordinates, the phasor expression for the $\boldsymbol{E}$ and $\boldsymbol{H}$ fields can be written as

$$\boldsymbol{E}(\rho,\phi,z) = \boldsymbol{E}^0(\rho,\phi)e^{-\gamma z} = [\boldsymbol{e}_x E_\rho^0(\rho,\phi) + \boldsymbol{e}_\phi E_\phi^0(\rho,\phi) + \boldsymbol{e}_z E_z^0(\rho,\phi)]e^{-\gamma z} \tag{9.3a}$$

$$\boldsymbol{H}(\rho,\phi,z) = \boldsymbol{H}^0(\rho,\phi)e^{-\gamma z} = [\boldsymbol{e}_x H_\rho^0(\rho,\phi) + \boldsymbol{e}_\phi H_\phi^0(\rho,\phi) + \boldsymbol{e}_z H_z^0(\rho,\phi)]e^{-\gamma z} \tag{9.3b}$$

With the fields expressed by (9.2) and (9.3), the variable z is separated, and the partial derivative with respect to z is equivalent to products with $-\gamma$, i.e., [①]

$$\frac{\partial}{\partial z} \xleftrightarrow{\text{Equivalent}} -\gamma \tag{9.4}$$

---

① 波导中,我们感兴趣的是沿着轴向传播的电磁波,通过假设其沿 z 方向的变化为传播因子 $e^{-\gamma z}$,得到电磁场表达式(9.2)和式(9.3),从而实现了横向和纵向的变量分离。只要确定传播常数 γ,场的纵向变化就能够确定,这样,波导中电场 $\boldsymbol{E}(x,y,z)$ 和磁场 $\boldsymbol{H}(x,y,z)$ 的三元矢量函数求解问题就简化为求解二元矢量函数 $\boldsymbol{E}^0(x,y)$ 和 $\boldsymbol{H}^0(x,y)$ 的问题。

In the source-free region of the waveguide (i. e. , the air or dielectric region), the **E** and **H** fields must satisfy the two source-free curl equations given by (6.76a) and (6.76b).① With the **E** and **H** field expressions given by (9.2), these two vector equations can be expanded under Cartesian coordinates as

From $\nabla\times\boldsymbol{E}=-\mathrm{j}\omega\mu\boldsymbol{H}$		From $\nabla\times\boldsymbol{H}=\mathrm{j}\omega\varepsilon\boldsymbol{E}$	
$\dfrac{\partial E_z^0}{\partial y}+\gamma E_y^0=-\mathrm{j}\omega\mu H_x^0$	(9.5a)	$\dfrac{\partial H_z^0}{\partial y}+\gamma H_y^0=\mathrm{j}\omega\varepsilon E_x^0$	(9.6a)
$-\gamma E_x^0-\dfrac{\partial E_z^0}{\partial x}=-\mathrm{j}\omega\mu H_y^0$	(9.5b)	$-\gamma H_x^0-\dfrac{\partial H_z^0}{\partial x}=\mathrm{j}\omega\varepsilon E_y^0$	(9.6b)
$\dfrac{\partial E_y^0}{\partial x}-\dfrac{\partial E_x^0}{\partial y}=-\mathrm{j}\omega\mu H_z^0$	(9.5c)	$\dfrac{\partial H_y^0}{\partial x}-\dfrac{\partial H_x^0}{\partial y}=\mathrm{j}\omega\varepsilon E_z^0$	(9.6c)

In equations (9.5) and (9.6), the common factor $e^{-\gamma z}$ in each component of the fields has been dropped, and the partial derivatives with respect to z have been replaced by $(-\gamma)$. All the component field quantities in the equations above are complex scalar fields that depend only on x and y. Similarly, the two source-free curl equations vector equations can be expanded under cylindrical coordinates as

From $\nabla\times\boldsymbol{E}=-\mathrm{j}\omega\mu\boldsymbol{H}$		From $\nabla\times\boldsymbol{H}=\mathrm{j}\omega\varepsilon\boldsymbol{E}$	
$\dfrac{1}{\rho}\dfrac{\partial E_z^0}{\partial\phi}+\gamma E_\varphi^0=-\mathrm{j}\omega\mu H_\rho^0$	(9.7a)	$\dfrac{1}{\rho}\dfrac{\partial H_z^0}{\partial\phi}+\gamma H_\phi^0=\mathrm{j}\omega\varepsilon E_\rho^0$	(9.8a)
$-\gamma E_\rho^0-\dfrac{\partial E_z^0}{\partial\rho}=-\mathrm{j}\omega\mu H_\phi^0$	(9.7b)	$-\gamma H_\rho^0-\dfrac{\partial H_z^0}{\partial\rho}=\mathrm{j}\omega\varepsilon E_\phi^0$	(9.8b)
$\dfrac{1}{\rho}\dfrac{\partial(\rho E_\phi^0)}{\partial\rho}-\dfrac{1}{\rho}\dfrac{\partial E_\rho^0}{\partial\phi}=-\mathrm{j}\omega\mu H_z^0$	(9.7c)	$\dfrac{1}{\rho}\dfrac{\partial(\rho H_\phi^0)}{\partial\rho}-\dfrac{1}{\rho}\dfrac{\partial H_\rho^0}{\partial\phi}=\mathrm{j}\omega\varepsilon H_z^0$	(9.8c)

With the assumption that all the fields vary in the z-direction as the expression in (9.1), the wave propagation problem in a waveguide is reduced to finding solutions for 6 two-variable scalar fields that satisfy (9.5)-(9.6), or (9.7)-(9.8), with a propagation constant γ to be determined. To further simplify the analysis, we classify the solution of the propagating waves into the following three types:

(1) **Transverse electromagnetic (TEM 横电磁波)** waves. These are the waves that contain neither E_z nor H_z. In other words, both the **E** and **H** fields are transverse to the longitudinal direction.

(2) **Transverse magnetic (TM 横磁波)** waves. These are the waves that contain no H_z but have a nonzero E_z. In other words, only the **H** field is transverse to the longitudinal direction.

(3) **Transverse electric (TE 横电波)** waves. These are the waves that contain no E_z but

---

① 这里需要强调,无源区域指的是自由电荷和自由电流为零的区域,但不能认为无源区域中的场是无源场。事实上,无源区域时变场的旋度通常并不为零,时变电场是磁场的漩涡源,时变磁场是电场的漩涡源。

has a nonzero H_z. In other words, only the $\boldsymbol{E}$ field is transverse to the longitudinal direction.

In the following sections, the characteristics of the three types of waves are studied separately.[①]

9.3 TEM Waves and Two-Conductor Transmission Lines(TEM 波与双导体传输线)

The concept of TEM wave was first introduced in Chapter 7, in which it is concluded that uniform plane waves are TEM waves. In a waveguide, however, uniform plane waves are generally not supported, but TEM waves may exist under certain conditions which is discussed in this section.

9.3.1 General Characteristics of TEM Waves (TEM 波的一般特征)

For the TEM wave within a waveguide, $E_z = 0$ and $H_z = 0$. Then, (9.5b) and (9.6a) respectively give

$$\frac{E_x^0(x,y)}{H_y^0(x,y)} = \frac{\mathrm{j}\omega\mu}{\gamma} \quad \text{and} \quad \frac{E_x^0(x,y)}{H_y^0(x,y)} = \frac{\gamma}{\mathrm{j}\omega\varepsilon} \tag{9.9a}$$

(9.5a) and (9.6b) respectively lead to

$$-\frac{E_y^0(x,y)}{H_x^0(x,y)} = \frac{\mathrm{j}\omega\mu}{\gamma} \quad \text{and} \quad -\frac{E_y^0(x,y)}{H_x^0(x,y)} = \frac{\gamma}{\mathrm{j}\omega\varepsilon} \tag{9.9b}$$

Both results require that

$$\frac{\mathrm{j}\omega\mu}{\gamma} = \frac{\gamma}{\mathrm{j}\omega\varepsilon}$$

which means

$$\gamma = \mathrm{j}\omega\sqrt{\mu\varepsilon} = \mathrm{j}k \tag{9.10}$$

where

$$k = \omega\sqrt{\mu\varepsilon} \tag{9.11}$$

is the wavenumber in the source-free region in the waveguide. Since (9.10) is the same expression of the propagation constant of a uniform plane wave in an unbounded medium, the velocity of propagation (phase velocity) for TEM waves is the same as that of the uniform plane wave, i.e.,

$$u_\mathrm{p} = \frac{\omega}{k} = \frac{1}{\sqrt{\mu\varepsilon}} \tag{9.12}$$

(9.12) indicates that the phase velocity of the TEM waves is independent of the frequency. In

---

① 在横纵变量分离的基础上,由无源区域麦克斯韦方程组的两个旋度方程得出6个场分量之间的关系式(9.5)~式(9.8)。以下将基于这些关系式,结合TEM、TE、TM三种基本波型对场分量做进一步的简化和分离,推导得到各个场分量的横向分布和传播特性。

other words, TEM waves are nondispersive if the medium within the waveguide is lossless. ①

By substituting (9.10) back to (9.9), we can obtain the ratio between the transverse components of the $\boldsymbol{E}$ and $\boldsymbol{H}$ fields, which is the **wave impedance** (**波阻抗**). Specifically, we have

$$Z_{\mathrm{TEM}} = \frac{E_x^0}{H_y^0} = -\frac{E_y^0}{H_x^0} = \sqrt{\frac{\mu}{\varepsilon}} = \eta \tag{9.13}$$

which indicates that the wave impedance of a TEM wave is the same as the intrinsic impedance of the medium as given in (7.13). Since a TEM wave has no field components along z-direction, (9.13) can be formulated in the vector form as

$$\boldsymbol{H} = \frac{1}{Z_{\mathrm{TEM}}} \boldsymbol{e}_z \times \boldsymbol{E} = \frac{1}{\eta} \boldsymbol{e}_z \times \boldsymbol{E} \tag{9.14a}$$

$$\boldsymbol{E} = Z_{\mathrm{TEM}} \boldsymbol{H} \times \boldsymbol{e}_z = \eta \boldsymbol{H} \times \boldsymbol{e}_z \tag{9.14b}$$

which are similar to the relations between $\boldsymbol{E}$ and $\boldsymbol{H}$ fields of a uniform plane wave propagating along the $+z$-direction in an unbounded medium. ②

It is also observed that, for TEM waves, equations (9.5c) and (9.6c) become

$$\frac{\partial E_y^0(x,y)}{\partial x} - \frac{\partial E_x^0(x,y)}{\partial y} = 0 \tag{9.15a}$$

$$\frac{\partial H_y^0(x,y)}{\partial x} - \frac{\partial H_x^0(x,y)}{\partial y} = 0 \tag{9.15b}$$

which are respectively equivalent to the following vector equations:

$$\nabla_{\mathrm{T}} \times \boldsymbol{E} = 0 \tag{9.16a}$$

$$\nabla_{\mathrm{T}} \times \boldsymbol{H} = 0 \tag{9.16b}$$

where ($\nabla_{\mathrm{T}} \times$) is a two-dimensional curl operator in the transverse plane. Under Cartesian coordinates,

$$\nabla_{\mathrm{T}} \equiv \boldsymbol{e}_x \frac{\partial}{\partial x} + \boldsymbol{e}_y \frac{\partial}{\partial y} \tag{9.17}$$

It is worth pointing out that, although the above discussions are based on Cartesian coordinates, the conclusions, including (9.10)-(9.14) and (9.16), are all valid under cylindrical coordinates, except that ∇_{T} becomes a transverse operator involving partial derivatives along the ρ-direction and ϕ-direction, i. e.,

$$\nabla_{\mathrm{T}} \equiv \boldsymbol{e}_\rho \frac{1}{\rho} \frac{\partial}{\partial \rho} \rho + \boldsymbol{e}_y \frac{\partial}{\rho \partial \phi} \tag{9.18}$$

Notice that, the two equations in (9.16) do not mean the $\boldsymbol{E}$ and $\boldsymbol{H}$ fields are curl-free

---

① TEM 波(横电磁波)的纵向电场和磁场分量均为零,使得式(9.5a),式(9.5b)和式(9.6a),式(9.6b)简化为式(9.9a),式(9.9b),从而得出 TEM 波的第一个特征:TEM 波的传播常数 $\gamma = \mathrm{j}k$ 与均匀平面波的传播常数相同,相速度也与均匀平面波的相速度相同。当媒质无损时,该相速度与频率无关,就等于该媒质中的光速。

② TEM 波的第二个特征:波阻抗等于无源区域媒质的本征阻抗。电场、磁场、传播方向两两相互垂直,满足右手关系。

because the operator($\nabla_T \times$) does not include differentiation along the z-direction. In fact, from Maxell equations, neither $\boldsymbol{E}$ nor $\boldsymbol{H}$ is curl-free due to their time-varying nature. Nevertheless, (9.16) indicates that the vortex source of the $\boldsymbol{E}$ and $\boldsymbol{H}$ fields has no z component within the source-free region(i. e., in the air or the dielectric region). Particularly for the $\boldsymbol{E}$ field, the vortex source has no z component everywhere in the waveguide including the region inside the conductor. This means that circulation of the $\boldsymbol{E}$ field along any closed path within the cross-sectional plane of the wave guide is zero, and therefore, we have the following conclusions: ①

(1) The $\boldsymbol{E}$ field lines cannot form closed paths, and therefore, must originate from positive charges and terminate at negative charges on the surface of conductors.

(2) Within any cross-sectional plane, the line integral of the $\boldsymbol{E}$ field is independent of the path. Therefore, the $\boldsymbol{E}$ field can be represented as the negative transverse gradient of a scalar potential field($-\nabla_T \varphi$), based on which an electric voltage associated with the TEM wave can be defined without ambiguity.

(3) One necessary condition for a uniform waveguide to support TEM waves is that an electrostatic field with no component along the z-direction can be established on the cross section of the waveguide.

The $\boldsymbol{H}$ field, however, must form closed paths along which the circulation is nonzero. Since the $\boldsymbol{H}$ field lies in cross-sectional planes, the closed paths with nonzero circulation must also be within cross-sectional planes. The nonzero circulation of the $\boldsymbol{H}$ field must be equal to the magnitude of the vortex source along the z-direction. The vortex source cannot be a displacement current as $E_z = 0$, and hence, must be the conduction current carried by conductors. Therefore, we have the following conclusions: ②

(1) The $\boldsymbol{H}$ field lines form closed paths around conductors.

(2) The circulation of the $\boldsymbol{H}$ field over any closed path around a conductor must be equal to the current carried by the conductor. Based on that, an electric current associated with the TEM wave can be defined without ambiguity.

(3) One condition for a uniform waveguide to support TEM waves is that a static magnetic field with no component along the z-direction can be established on the cross section of the waveguide.

From the above discussions, we conclude that TEM waves cannot exist within a single-conductor hollow or dielectric-filled waveguide, such as the rectangular or circular waveguides

---

① TEM 波满足的式(9.5c)和式(9.6c)简化为式(9.15)和式(9.16),从而得出其第三个特征：波导无源区域中,TEM 波的电场和磁场均没有纵向的漩涡源。对于电场而言,其漩涡源是变化的磁场。由于构成波导的金属导体(近似为理想导体)中不存在磁场,整个波导系统中电场不存在纵向漩涡源,因此 TEM 波电场的电力线不可能形成闭合回路,在任意横截面上呈现出静电场的特性。

② 任何实际存在的电磁场的磁力线都是闭合的,闭合磁力线的磁场环量非零,对应的漩涡源非零。磁场的漩涡源有两种,一种是自由电流(包括传导电流),另一种是位移电流。TEM 波的磁力线只能在波导的横截面上形成闭合回路,因此必然存在纵向的漩涡源。式(9.15b)和式(9.16b)表明波导无源区域(空气或填充介质)中,TEM 波的磁场没有纵向的漩涡源,因此磁场的纵向漩涡源只能是导体中的电流,TEM 波必然含有包围导体的闭合磁力线。在波导横截面上,TEM 波的磁场呈现出恒定磁场的特性。这也表明,空心或者填充介质的单导体波导内部不存在 TEM 波。

as shown in Figure 9-1. This is because no electrostatic field can be established without free charges inside these waveguides, and there are no inner conductors that carry the necessary longitudinal conduction current to support the $\boldsymbol{H}$ field lines within the cross-section of these waveguides. On the other hand, two-conductor transmission lines, such as the coaxial lines and the two-wire lines can support TEM waves.

9.3.2 TEM Waves in Coaxial Lines(同轴线中的 TEM 波)

The coaxial line(or coaxial cable) is one of the most commonly used transmission lines. It can operate from DC to very high frequencies. Figure 9-3 illustrates the basic structure of the coaxial lines, which consists of an inner conductor, an outer conducting shield, and an insulating layer in between. Assume the inner and outer conductors are made of PEC. Then the electromagnetic fields are confined within the dielectric between the conductors. We have already studied the capacitance and inductance per unit length of the coaxial lines in Chapter 2 and Chapter 5. In this section, we focus on the TEM waves that can propagate within coaxial lines.

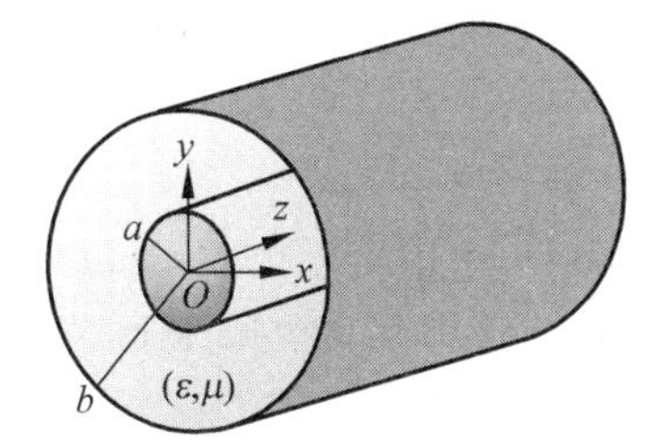

Figure 9-3 A coaxial line

A coaxial line supports TEM waves because it is a two-conductor system that can establish two-dimensional static electric and magnetic fields within the cross-sectional plane as is seen in Chapter 2 and Chapter 5 respectively.① Specifically, the static electric field lines originate normally from the positive charges on the surface of one conductor and terminate normally on the surface of the other conductor. Therefore, it is expected that the $\boldsymbol{E}$ field of the TEM wave only has a ρ component as well. On the other hand, the $\boldsymbol{H}$ field must form closed paths around the inner conductor as is the case with steady magnetic fields. Therefore, the $\boldsymbol{H}$ field is expected to have only a ϕ component. With these considerations, we can write down the expressions for $\boldsymbol{E}$ and $\boldsymbol{H}$ as

$$\boldsymbol{E}(\rho,\phi,z)=\boldsymbol{e}_\rho E_\rho^0(\rho,\phi)\mathrm{e}^{-jkz} \tag{9.19}$$

$$\boldsymbol{H}(\rho,\phi,z)=\boldsymbol{e}_\phi H_\phi^0(\rho,\phi)\mathrm{e}^{-jkz} \tag{9.20}$$

where the propagation constant of the TEM wave γ has been replaced by jk as required by (9.10).

With the expressions in(9.19) and(9.20), the equation(9.7c) is reduced to

$$-\frac{1}{\rho}\frac{\partial E_\rho^0(\rho,\phi)}{\partial\phi}=0$$

which means that E_ρ^0 does not depend on the ϕ coordinate. Therefore, the $\boldsymbol{E}$ field given by(9.19) is not a function of ϕ. From the relation(9.14a), the $\boldsymbol{H}$ field is not a function of ϕ either. This is

① 同轴线是针对高频电磁信号最常用的一种传输线,作为双导体波导,能够支持 TEM 波。事实上,第 2 章讨论过同轴线内的静电场问题,第 5 章讨论过同轴线内的恒定电场问题,因此同轴线满足 9.3.1 节中阐述的 TEM 波存在的基本条件。参照静电场和恒定磁场在横截面上的分布,可以预期电力线沿 ρ 方向,磁力线沿 ϕ 方向。

as expected because of the symmetry of the coaxial structure. Now from the equation (9.8c), which is reduced to

$$\frac{1}{\rho}\frac{\partial(\rho H_{\phi}^{0})}{\partial\rho}=0$$

we know that ρH_{ϕ}^{0} is not a function of ρ. We have already known that ρH_{ϕ}^{0} is not a function of ϕ or z. So, it must be a constant. As a result, H_{ϕ}^{0} must be a constant divided by ρ, and we can write

$$H_{\phi}^{0}=H_{\phi}^{0}(\rho)=\frac{a}{\rho}H_{\mathrm{m}} \tag{9.21}$$

where H_{m} is a constant representing the magnitude of the $\boldsymbol{H}$ field on the surface of the inner conductor ($\rho=a$). By using (9.14b), we obtain the solution of $E_{\rho}(\rho)$ as

$$E_{\rho}(\rho)=\eta H_{\phi}(\rho)=\frac{a\eta H_{\mathrm{m}}}{\rho}=\frac{aE_{\mathrm{m}}}{\rho} \tag{9.22}$$

where $E_{\mathrm{m}}=\eta H_{\mathrm{m}}$ is the magnitude of the $\boldsymbol{E}$ field on the surface of the inner conductor. Substitute (9.21) and (9.22) back into (9.19) and (9.20), we obtain the solution of the TEM wave in a coaxial line as

$$\boldsymbol{E}(\rho,z)=\boldsymbol{e}_{\rho}E_{\mathrm{m}}\left(\frac{a}{\rho}\right)\mathrm{e}^{-\mathrm{j}kz} \tag{9.23}$$

$$\boldsymbol{H}(\rho,z)=\boldsymbol{e}_{\phi}\frac{E_{\mathrm{m}}}{\eta}\left(\frac{a}{\rho}\right)\mathrm{e}^{-\mathrm{j}kz} \tag{9.24}$$

Figure 9-4 illustrates the $\boldsymbol{E}$ and $\boldsymbol{H}$ field lines of a TEM wave within the coaxial transmission line. The transmitted power can be derived to be

$$P=\frac{1}{2}\mathrm{Re}\left[\int_{S}(\boldsymbol{E}\times\boldsymbol{H}^{*})\cdot\mathrm{d}\boldsymbol{s}\right]=\frac{\pi a^{2}}{\eta}|E_{\mathrm{m}}|^{2}\ln\frac{b}{a} \tag{9.25}$$

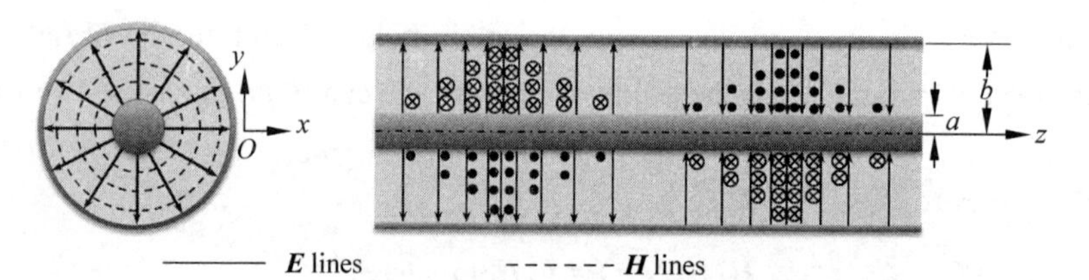

Figure 9-4 TEM fields in coaxial lines

9.4 TM and TE Waves(横磁波与横电波)

As noted in the previous section, TEM waves cannot exist within a single-conductor hollow or dielectric-filled metal pipe such as the rectangular or circular waveguide. Only TM and TE waves can exist in such waveguides. In these cases, either the **longitudinal field component**(纵向场分量) E_{z}^{0} or H_{z}^{0} is nonzero. To analyze TE and TM waves, we start by expressing the **transverse field components**(横向场分量) H_{x}^{0} H_{y}^{0}, E_{x}^{0} and E_{y}^{0} in terms of the two longitudinal components E_{z}^{0} and H_{z}^{0}. For instance, (9.5a) and (9.6b) can be combined to eliminate E_{y}^{0} and obtain an expression of H_x^0 in terms of E_z^0 and H_z^0. Similarly, we can have the expressions of all the transverse field components in terms of the longitudinal components as

$$H_x^0 = -\frac{1}{k_c^2}\left(\gamma\frac{\partial H_z^0}{\partial x} - j\omega\varepsilon\frac{\partial E_z^0}{\partial y}\right) \tag{9.26a}$$

$$H_y^0 = -\frac{1}{k_c^2}\left(\gamma\frac{\partial H_z^0}{\partial y} + j\omega\varepsilon\frac{\partial E_z^0}{\partial x}\right) \tag{9.26b}$$

$$E_x^0 = -\frac{1}{k_c^2}\left(\gamma\frac{\partial E_z^0}{\partial x} + j\omega\mu\frac{\partial H_z^0}{\partial y}\right) \tag{9.26c}$$

$$E_y^0 = -\frac{1}{k_c^2}\left(\gamma\frac{\partial E_z^0}{\partial y} - j\omega\mu\frac{\partial H_z^0}{\partial x}\right) \tag{9.26d}$$

where

$$k_c^2 = \gamma^2 + k^2 \tag{9.27}$$

with k to be the wavenumber as given by(9.11). k_c is called the **cutoff wavenumber**(**截止波数**) whose significance will be discussed later.

Equations(9.26a) ~ (9.26d) give the general **transverse-longitudinal relationship**(**横纵关系**) of the electromagnetic fields of a non-TEM wave that propagates along the z-direction with a propagation constant γ. They indicate that the six components of $\boldsymbol{E}$ and $\boldsymbol{H}$ are not all independent, and we can consider E_z^0 and H_z^0 as the only independent field components. In other words, the solution of TM or TE waves in the waveguide can be obtained by first finding the longitudinal components E_z^0 and H_z^0, and then using(9.26a) ~ (9.26d) to determine all the other components. ①

9.4.1 General Characteristics of TM Waves (横磁波的一般特征)

For TM waves, the magnetic field does not have the longitudinal component, i. e., $H_z = 0$. Then the equations in(9.26) become

$$H_x^0 = \frac{j\omega\varepsilon}{k_c^2}\frac{\partial E_z^0}{\partial y} \tag{9.28a}$$

$$H_y^0 = -\frac{j\omega\varepsilon}{k_c^2}\frac{\partial E_z^0}{\partial x} \tag{9.28b}$$

$$E_x^0 = -\frac{\gamma}{k_c^2}\frac{\partial E_z^0}{\partial x} \tag{9.28c}$$

---

① 对于波导中的非 TEM 波，E_z 和 H_z 两个分量至少有一个非零。此时，由式(9.5)和式(9.6)得到横纵场分量之间的关系式(9.26a) ~ (9.26d)。该横纵关系表明电、磁场的 6 个分量不是完全独立的，横向场分量可以由纵向场分量表出。因此波导中非 TEM 波的传播问题可以简化为 E_z 和 H_z 两个标量场的求解问题。进一步，对于 TE 和 TM 波，只需要求解两个标量场其中之一即可。

$$E_y^0 = -\frac{\gamma}{k_c^2}\frac{\partial E_z^0}{\partial y} \tag{9.28d}$$

(9.28c) and (9.28d) can be combined to form a transverse vector field as

$$\boldsymbol{E}_T^0 = \boldsymbol{e}_x E_x^0 + \boldsymbol{e}_y E_y^0 = -\frac{\gamma}{k_c^2}\nabla_T E_z^0 \tag{9.29}$$

where

$$\nabla_T E_z^0 = \left(\boldsymbol{e}_x\frac{\partial}{\partial x} + \boldsymbol{e}_y\frac{\partial}{\partial y}\right)E_z^0 \tag{9.30}$$

denotes the transverse gradient of E_z^0 in the cross-sectional plane. (9.29) states that the transverse vector component of the **E** field of a TM wave is proportional to the transverse gradient of the longitudinal component of the **E** field. ①

From (9.28b) and (9.28c), it is easy to find that the ratio between E_x^0 and H_y^0 is a constant $\gamma/(j\omega\varepsilon)$. From (9.28a) and (9.28d), the ratio between $(-E_y^0)$ and H_x^0 is the same constant $\gamma/(j\omega\varepsilon)$. This ratio is defined as the **wave impedance** for the TM wave, denoted by Z_{TM}. So, we have

$$Z_{TM} = \frac{E_x^0}{H_y^0} = -\frac{E_y^0}{H_x^0} = \frac{\gamma}{j\omega\varepsilon} \tag{9.31}$$

Notice that Z_{TM} is not equal to $j\omega\mu/\gamma$, because we will see that γ for TM waves is not equal to $j\omega\sqrt{\mu\varepsilon}$. From (9.31), we have the following relations between the transverse components of the **E** and **H** fields for TM waves:

$$\boldsymbol{H}_T^0 = \boldsymbol{e}_x H_x^0 + \boldsymbol{e}_y H_y^0 = \frac{1}{Z_{TM}}\boldsymbol{e}_z \times \boldsymbol{E}_T^0 \tag{9.32}$$

(9.32) appears to be a similar form as (9.14a) for TEM waves. It indicates that the ratio between the transverse components of the **E** and **H** fields is Z_{TM}. The direction of the transverse component of **E**, the transverse component of **H**, and the longitudinal direction $\boldsymbol{e}_z$ follow the right-hand rule. ②

As a summary, for TM waves, we only need to find a solution of E_z^0, and then the other components of the **E** and **H** fields can be readily determined by using (9.29) and (9.32).

To find the solution of E_z^0, we start with the homogeneous Helmholtz's equation that the **E** field must satisfy in the source-free region of the waveguide. As derived in Section 6.9.4,

$$\nabla^2\boldsymbol{E} + k^2\boldsymbol{E} = 0 \tag{9.33}$$

Under Cartesian coordinates, the three-dimensional Laplacian operator ∇^2 may be broken into

① 式(9.29)所示的TM波的横纵关系表明，得到标量场 E_z 的解之后，只需要对 E_z 作横向梯度运算，再乘以常数 $-\gamma/k_c^2$ 即可得到电场横向矢量的分量，从而得到电场完整的解。换言之，TM波电场的沿某个横向方向分量的大小正比于 E_z 在该方向上的方向导数。从式(9.28c)，式(9.28d)也可以看出该结论。

② TM波中横向电、磁场和纵向传播方向两两垂直，满足右手关系。类似于均匀平面波中电、磁场与传播方向之间的关系。与均匀平面波相比不同之处在于：横向电、磁场的比值，即式(9.31)所示的波阻抗不等于本征阻抗；磁场只有横向分量，但电场既有横向分量，也有纵向分量，总的电场与传播方向不是相互垂直。

two parts: ∇_{xy}^2 for the cross-sectional coordinates and $\partial^2/\partial z^2$ for the longitudinal coordinate. By noticing(9.4), we have

$$\nabla^2 \boldsymbol{E} = \left(\nabla_{xy}^2 + \frac{\partial^2}{\partial z^2}\right)\boldsymbol{E} = \nabla_{xy}^2\boldsymbol{E} + \gamma^2\boldsymbol{E} = \left(\frac{\partial^2}{\partial x^2} + \frac{\partial^2}{\partial y^2}\right)\boldsymbol{E} + \gamma^2\boldsymbol{E} \tag{9.34}$$

Combination of(9.33) and(9.34) gives

$$\nabla_{xy}^2\boldsymbol{E} + (\gamma^2 + k^2)\boldsymbol{E} = 0 \tag{9.35}$$

By substituting(9.2a) into(9.35) and extracting the longitudinal component, we have the following scalar Helmholtz's equation

$$\nabla_{xy}^2 E_z^0(x,y) + k_c^2 E_z^0(x,y) = 0 \tag{9.36}$$

where the common factor $e^{-\gamma z}$ is dropped, and (γ^2+k^2) is replaced by k_c^2 as defined in(9.27).

The specific solution of(9.36) will be derived later for specific waveguides. Here, we notice that, mathematically, k_c^2 is the **characteristic value**(**特征值**) or **eigenvalue**(**本征值**) of the Laplacian operator on the scalar function E_z^0. For single-conductor waveguide problems, E_z^0 is distributed within a finite cross-sectional area with the boundary value to be fixed zero, which mathematically decides that k_c^2 can take only **discrete values**(**离散值**). These discrete values are related to the boundary conditions, involving the geometrical shape and size of the waveguide. Although k_c is introduced by(9.27), its possible value is independent of k or γ. Instead, given the frequency of the wave, γ is determined by the value of k_c. Each discrete value of k_c corresponds to a specific TM mode, which determines the characteristic properties the TM wave of the given waveguide. ①

From(9.27), we have

$$\gamma = \sqrt{k_c^2 - k^2} = \sqrt{k_c^2 - \omega^2\mu\varepsilon} \tag{9.37}$$

Given a specific k_c, two distinct ranges of γ are resulted depending on ω. The dividing point is when $\omega = \omega_c$, the **cutoff angular frequency**(**截止角频率**) that makes $\gamma = 0$. That is, $k_c^2 = \omega_c^2\mu\varepsilon$, or②

$$\omega_c = \frac{k_c}{\sqrt{\mu\varepsilon}} = uk_c \tag{9.38}$$

where

$$u = \frac{1}{\sqrt{\mu\varepsilon}} \tag{9.39}$$

---

① TM 波的传播常数 γ 并不是可以任意选取的,这是由于 k_c^2,也就是 γ^2+k^2,是各个场分量满足的横向二维亥姆霍兹方程的系数,数学上是二维拉普拉斯线性算子的特征值。在横向尺寸受限的条件下,k_c^2 只能取特定的离散值,且这些离散值取决于波导横向结构尺寸以及边界条件,与电磁波的频率无关。而 k 取决于频率和媒质,因此不同频率和媒质特性条件下,γ 的取值也不同。

② 从式(9.37)可以看出 k_c 被称为截止波数的物理意义。波导中的 TM 波如果要能够沿纵向传播,其传播常数 γ 必须是纯虚数(或者在考虑损耗的条件下,γ 有非零的虚部)。根据式(9.37),k 必须大于 k_c 才能形成传播的 TM 波。当 k 小于 k_c 时,γ 变为实数,沿纵向呈现出衰减特性,称之为截止状态,因此 k_c 是波导中 TM 波的传播和截止两种状态的临界点,称为截止波数。

is the speed of light in the unbounded medium(ε,μ). Now, we introduce the concepts of **cutoff frequency**(截止频率) f_c and **cutoff wavelength**(截止波长) λ_c associated with ω_c and k_c as [①]

$$f_c = \frac{\omega_c}{2\pi} = \frac{uk_c}{2\pi} = \frac{k_c}{2\pi\sqrt{\mu\varepsilon}} \tag{9.40}$$

$$\lambda_c = \frac{2\pi}{k_c} = \frac{u}{f_c} = \frac{2\pi}{\omega_c\sqrt{\mu\varepsilon}} \tag{9.41}$$

Using(9.40) and(9.41), we can write(9.37) as

$$\gamma = jk\sqrt{1-\left(\frac{f_c}{f}\right)^2} = jk\sqrt{1-\left(\frac{\lambda}{\lambda_c}\right)^2} \tag{9.42}$$

where

$$\lambda = \frac{2\pi}{k} = \frac{1}{f\sqrt{\mu\varepsilon}} = \frac{u}{f} \tag{9.43}$$

is the wavelength of a uniform plane wave at the frequency f in the unbounded medium(ε,μ). For ease of discrimination, the frequency f of the wave is particularly referred as **operating frequency**(工作频率), $\omega=2\pi f$ is referred as **operating angular frequency**(工作角频率) and the wavelength λ is referred as **operating wavelength**(工作波长).

The two distinct ranges of γ are defined depending on whether the operating frequency f is higher than the cutoff frequency f_c(or equivalently, whether the operating wavelength λ is lower than the cutoff wavelength λ_c).

(1) $f > f_c$ or $\lambda<\lambda_c$. In this range, $k^2>k_c^2$, and γ is imaginary. According to(9.1), it is a propagating mode and from(9.42), the phase constant β is

$$\beta = k\sqrt{1-\left(\frac{f_c}{f}\right)^2} = k\sqrt{1-\left(\frac{\lambda}{\lambda_c}\right)^2} \tag{9.44}$$

The corresponding **wavelength in the guide**(波导波长) or **guide wavelength** is

$$\lambda_g = \frac{2\pi}{\beta} = \frac{\lambda}{\sqrt{1-\left(\frac{f_c}{f}\right)^2}} = \frac{\lambda}{\sqrt{1-\left(\frac{\lambda}{\lambda_c}\right)^2}} > \lambda \tag{9.45}$$

Equation (9.45) can be rearranged to give a simple relation among the operating wavelength λ, the guide wavelength λ_g and the cutoff wavelength λ_c:

$$\frac{1}{\lambda^2} = \frac{1}{\lambda_g^2} + \frac{1}{\lambda_c^2} \tag{9.46}$$

The phase velocity of the propagating wave in the guide is

$$u_p = \frac{\omega}{\beta} = \frac{u}{\sqrt{1-\left(\frac{f_c}{f}\right)^2}} = \frac{u}{\sqrt{1-\left(\frac{\lambda}{\lambda_c}\right)^2}} = \frac{\lambda_g}{\lambda}u > u \tag{9.47}$$

---

① 截止频率 f_c 和截止波长 λ_c 都是相对于截止波数 k_c 而言的，即波导填充媒质中波数为 k_c 的均匀平面波的频率与波长。因此 λ_c 和 f_c 均与波导中电磁波的实际频率(即工作频率)无关。

(9.47) shows that the phase velocity within a waveguide is higher than the velocity of light u in an unbounded medium and is frequency dependent. Hence single-conductor waveguides are dispersive systems. The group velocity for a propagating wave can be determined by using(9.44): ①

$$u_g = \frac{1}{\mathrm{d}\beta/\mathrm{d}\omega} = \frac{1}{\frac{\mathrm{d}}{\mathrm{d}\omega}\left[\frac{\omega}{u}\sqrt{1-\left(\frac{\omega_c}{\omega}\right)^2}\right]} = u\sqrt{1-\left(\frac{f_c}{f}\right)^2} = u\sqrt{1-\left(\frac{\lambda}{\lambda_c}\right)^2}$$

$$= \frac{\lambda}{\lambda_g}u < u \qquad (9.48)$$

Hence the group velocity of a TM wave is lower than the velocity of light u. From(9.47) and(9.48), we also have ②

$$u_g u_p = u^2 \qquad (9.49)$$

For air dielectric, $u=c$, (9.49) becomes $u_g u_p = c^2$. Substitution of(9.42) in(9.31) yields

$$Z_{TM} = \eta\sqrt{1-\left(\frac{f_c}{f}\right)^2} = \eta\sqrt{1-\left(\frac{\lambda}{\lambda_c}\right)^2} \qquad (9.50)$$

Therefore, the wave impedance of propagating TM waves in a waveguide with a lossless dielectric is purely resistive and is less than the intrinsic impedance of the dielectric medium. ③

(2) $f<f_c$ or $\lambda>\lambda_c$. In this range, $k^2<k_c^2$, and hence γ is real. From(9.42), we have

$$\gamma = \alpha = k\sqrt{\left(\frac{\lambda}{\lambda_c}\right)^2 - 1} = k\sqrt{\left(\frac{f_c}{f}\right)^2 - 1} \qquad (9.51)$$

which is an attenuation constant. Since all field components contain the propagation factor $e^{-\gamma z} = e^{-\alpha z}$, the wave decays exponentially with z and is said to be evanescent(渐逝的). Therefore, a waveguide exhibits the property of a high-pass filter. For a given mode, only waves with a frequency higher than the cutoff frequency of the mode can propagate in the guide. Substitution of(9.51) in(9.31) gives the wave impedance of TM waves for $f<f_c$.

$$Z_{TM} = -\mathrm{j}\eta\sqrt{\left(\frac{\lambda}{\lambda_c}\right)^2 - 1} = -\mathrm{j}\eta\sqrt{\left(\frac{f_c}{f}\right)^2 - 1} \qquad (9.52)$$

Thus, the wave impedance of evanescent TM waves at frequencies below cutoff is purely reactive, indicating that there is no power flow along the longitudinal direction of the waveguide. Notice that the imaginary part of Z_{TM} is negative. This means that the waveguide operating in the evanescent TM mode is equivalent to a capacitor. ④

---

① 式(9.47)表明 TM 波的相速度与频率有关,后面也会看到 TE 波的相速度也与频率有关。单导体波导不能支持 TEM 波,只能传播 TE 和 TM 波,因此是色散系统。

② 式(9.47)~式(9.49)表明,TM 波的相速度大于光速,而群速度小于光速。

③ 式(9.50)表明,在传播条件下,TM 波的波阻抗为实数,且小于媒质的本征阻抗。

④ 式(9.52)表明,在截止条件下,TM 波的波阻抗为纯虚数,且呈现为容抗(相当于电容),纵向坡印廷矢量时均值为零,没有功率沿纵向方向流动。因此,波导对于给定截止波数 k_c 的 TM 波而言是一个高通滤波器。

9.4.2 General Characteristics of TE Waves(横电波的一般特征)

For TE waves, $E_z=0$, then the equations in(9.26) become

$$H_x^0=-\frac{\gamma}{k_c^2}\frac{\partial H_z^0}{\partial x} \tag{9.53a}$$

$$H_y^0=-\frac{\gamma}{k_c^2}\frac{\partial H_z^0}{\partial y} \tag{9.53b}$$

$$E_x^0=-\frac{\mathrm{j}\omega\mu}{k_c^2}\frac{\partial H_z^0}{\partial y} \tag{9.53c}$$

$$E_y^0=\frac{\mathrm{j}\omega\mu}{k_c^2}\frac{\partial H_z^0}{\partial x} \tag{9.53d}$$

(9.53a) and(9.53b) can be combined to form a transverse vector field as

$$\boldsymbol{H}_T^0=\boldsymbol{e}_x H_x^0+\boldsymbol{e}_y H_y^0=-\frac{\gamma}{k_c^2}\nabla_T H_z^0 \tag{9.54}$$

where

$$\nabla_T H_z^0=\left(\boldsymbol{e}_x\frac{\partial}{\partial x}+\boldsymbol{e}_y\frac{\partial}{\partial y}\right)H_z^0 \tag{9.55}$$

denotes the gradient of H_z^0 in the transverse plane. Therefore, the transverse vector component of the $\boldsymbol{H}$ field of a TE wave is proportional to the transverse gradient of the longitudinal component of the $\boldsymbol{H}$ field. ①

The transverse components of the $\boldsymbol{E}$ and $\boldsymbol{H}$ fields are related through the **wave impedance** of TE waves. We have, from the equations in(9.53),

$$Z_{TE}=\frac{E_x^0}{H_y^0}=-\frac{E_y^0}{H_x^0}=\frac{\mathrm{j}\omega\mu}{\gamma} \tag{9.56}$$

Note that Z_{TE} in(9.56) is different from Z_{TM} in(9.31) because γ for TE waves is not equal to $\mathrm{j}\omega\sqrt{\mu\varepsilon}$. From(9.56), we have

$$\boldsymbol{E}_T^0=\boldsymbol{e}_x E_x^0+\boldsymbol{e}_y E_y^0=Z_{TE}\boldsymbol{H}_T^0\times\boldsymbol{e}_z \tag{9.57}$$

It indicates that the ratio between the transverse components of the $\boldsymbol{E}$ and $\boldsymbol{H}$ fields is Z_{TE}. The transverse vector component of $\boldsymbol{E}$, the transverse vector component of $\boldsymbol{H}$, and the wave propagation direction vector $\boldsymbol{e}_z$ follow the right-hand rule. ②

As a summary, for TE waves, we only need to find a solution of H_z^0, and then the other components of the $\boldsymbol{E}$ and $\boldsymbol{H}$ fields can be easily determined by using(9.54) and(9.57). To find

---

① 式(9.54)给出的TE波的横纵关系表明，磁场的横向矢量分量等于纵向分量在横向平面的梯度与常数$-\gamma/k_c^2$的乘积。换言之，TE波磁场的沿某个横向方向分量的大小正比于H_z在该方向上的方向导数。

② TE波中横向电、磁场和纵向传播方向两两垂直，满足右手关系。类似于均匀平面波中电、磁场与传播方向之间的关系。与均匀平面波相比不同之处在于：横向电、磁场的比值，即波阻抗不等于本征阻抗，也不等于TM波的波阻抗；电场只有横向分量，但磁场既有横向分量也有纵向分量，总的磁场与传播方向不是相互垂直。

the solution of H_z^0, we start with the homogeneous Helmholtz's equation that the $\boldsymbol{H}$ field must satisfy in the source-free region of the waveguide. As derived in Section 6.9.4,

$$\nabla^2 \boldsymbol{H} + k^2 \boldsymbol{H} = 0 \tag{9.58}$$

Like the derivation of the equation(9.36) for E_z^0, the scalar Helmholtz's equation satisfied by H_z^0 can be derived from(9.58), which leads to

$$\nabla_{xy}^2 H_z^0(x,y) + k_c^2 H_z^0(x,y) = 0 \tag{9.59}$$

Again, it will be demonstrated that solutions of(9.59) are possible only for discrete values of k_c that depend on boundary conditions of the specific waveguides. Equations(9.37)-(9.49), involving the propagation/phase constants, cutoff frequency/wavelength, guide wavelength, and phase/group velocities for TM waves, also apply to TE waves. ①

There are also two distinct ranges of γ, depending on whether the operating frequency f is higher than the cutoff frequency f_c given in(9.40)(or equivalently, whether the operating wavelength λ is lower than the cutoff wavelength λ_c).

(1) $f>f_c$ or $\lambda<\lambda_c$. In this range, $k^2>k_c^2$, and γ is imaginary. The expressions for β, λ_g, u_p and u_g are the same as those for TE waves as given by(9.44)-(9.49). Substituting(9.42) into (9.56), we obtain

$$Z_{TE} = \frac{\eta}{\sqrt{1-\left(\frac{f_c}{f}\right)^2}} = \frac{\eta}{\sqrt{1-\left(\frac{\lambda}{\lambda_c}\right)^2}} \tag{9.60}$$

which is obviously different from the expression for Z_{TM} in(9.50). (9.60) indicates that the wave impedance of propagating TE waves in a waveguide with a lossless dielectric is purely resistive and is always larger than the intrinsic impedance of the dielectric medium. ②

(2) $f<f_c$ or $\lambda>\lambda_c$. In this range, $k^2<k_c^2$, and γ is real and we have an evanescent or non-propagating wave, with the attenuation constant also given by(9.51). Substitution of(9.51) into(9.56) gives the wave impedance of TE waves for $f<f_c$:

$$Z_{TE} = \frac{j\omega\mu}{\gamma} = \frac{j\omega\mu}{k\sqrt{\left(\frac{f_c}{f}\right)^2 - 1}} = j\frac{\eta}{\sqrt{\left(\frac{f_c}{f}\right)^2 - 1}} \tag{9.61}$$

which is purely reactive, indicating that there is no power flow for evanescent waves. Notice that the imaginary part of Z_{TE} is positive. This means that the waveguide operating in the evanescent TE mode is equivalent to an inductor. ③

Example 9-1 Consider a waveguide made of metal and filled with air. A wave in the

---

① 与TM波一样,TE波的截止波数平方k_c^2也只能取特定的离散值,且这些离散值取决于波导横向结构尺寸以及边界条件,与电磁波的频率无关。TE波的传播(相位常数)、截止频率、截止波长、波导波长相速、群速等,与截止波数的关系都与TM波无异。

② 式(9.60)表明,在传播条件下,TE波的波阻抗为实数,且大于媒质的本征阻抗。

③ 式(9.61)表明,在截止条件下,TE波的波阻抗为纯虚数,且呈现为感抗(相当于电感),纵向坡印廷矢量的时均值为零,没有功率沿纵向方向流动。因此,波导对于给定截止波数k_c的TE波而言是一个高通滤波器。

guide operates at 2GHz. Determine its guide wavelength, phase velocity and wave impedance if the wave is ①a TE/TM wave with 1GHz cutoff frequency, ②a TE/TM wave with 4GHz cutoff frequency, ③a TEM wave.

Solution: The medium is air with the speed of the light $c=3\times10^8$m/s and the intrinsic impedance $\eta=377\Omega$. The operating frequency of is $f=2$GHz. Hence the operating wavelength is $\lambda=15$cm.

(1) For $f_c=1$GHz, the wave is propagating.

$$\sqrt{1-\left(\frac{f_c}{f}\right)^2}=\sqrt{1-\left(\frac{1}{2}\right)^2}=0.866$$

The guide wavelength and phase velocity are the same for TE and TM waves, and can be obtained from (9.45) and (9.47) as

$$\lambda_g=\frac{\lambda}{\sqrt{1-\left(\frac{f_c}{f}\right)^2}}=1.155\lambda=17.325(\text{cm})$$

$$u_p=\frac{c}{\sqrt{1-\left(\frac{f_c}{f}\right)^2}}=1.155c=3.465\times10^8(\text{m/s})$$

The wave impedance for TE wave is obtained from (9.60):

$$Z_{TE}=\frac{\eta}{\sqrt{1-\left(\frac{f_c}{f}\right)^2}}=1.155\eta=435.4(\Omega)$$

The wave impedance for TM wave is obtained from (9.50):

$$Z_{TM}=\eta\sqrt{1-\left(\frac{f_c}{f}\right)^2}=0.866\eta=326.5(\Omega)$$

(2) For $f_c=4$GHz, the wave is cutoff.

$$\sqrt{\left(\frac{f_c}{f}\right)^2-1}=\sqrt{2^2-1}=1.732$$

The guide wavelength and phase velocity have no significance. The wave impedance for TE and TM wave can be obtained from (9.52) and (9.61) respectively:

$$Z_{TE}=\text{j}\frac{\eta}{\sqrt{\left(\frac{f_c}{f}\right)^2-1}}=\text{j}0.577\eta$$

$$Z_{TM}=-\text{j}\eta\sqrt{\left(\frac{f_c}{f}\right)^2-1}=-\text{j}1.732\eta$$

(3) For the TEM wave, the guide wavelength is the same as operating wavelength, phase velocity is the same as the speed of light, and the wave impedance is the same as the intrinsic impedance. i. e.,

$$u_{TEM}=c=3\times10^8(\text{m/s})$$

$$\lambda_{\text{TEM}} = 15(\text{cm})$$

$$Z_{\text{TEM}} = \eta = 377(\Omega)$$

For propagating modes, $\gamma = j\beta$ and the variation of β versus frequency determines the characteristics of a wave along a guide. Figure 9-5 plots a typical ω-β diagram. For the TEM mode, β is linearly proportional to ω, which is represented by the dashed line in the figure. The dashed line has a constant slope $\omega/\beta = u = 1/\sqrt{\mu\varepsilon}$, which is equal to the speed of light in the dielectric medium with constitutive parameters μ and ε. For the TE and TM modes, the relation between ω and β is already given in(9.44), which can be rewritten as

$$\omega = \sqrt{\beta^2 u^2 + \omega_c^2} \tag{9.62}$$

The relation(9.62) is illustrated in Figure 9-5 by the solid curve that intersects the ω-axis at $\omega = \omega_c$. At an arbitrary point P on the curve, the slope of the line OP is equal to the phase velocity u_p, whereas the slope of the line tangential to the curve at P is the group velocity u_g. The curve clearly shows that $u_p > u$ and $u_g < u$ for the propagating TE and TM waves. As the operating frequency increases, both u_p and u_g approach u asymptotically. ①

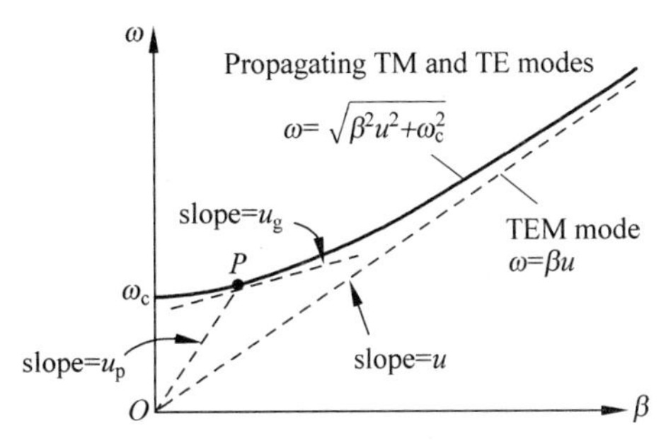

Figure 9-5 ω-β diagram for waveguide

9.5 Rectangular Waveguides(矩形波导)

In this section, we analyze the wave propagation in rectangular waveguides. The cross-section of a rectangular waveguide is a rectangle with size $a \times b$ as illustrated in Figure 9-6. Conventionally, the longer side of the rectangle with length a is aligned along the x-direction, and the shorter side of the rectangle with height b is aligned along the y-direction. The enclosed dielectric medium is assumed to have a permittivity ε and permeability μ. As concluded previously, the rectangular waveguide can only support TM and TE waves, which are discussed separately in this section.

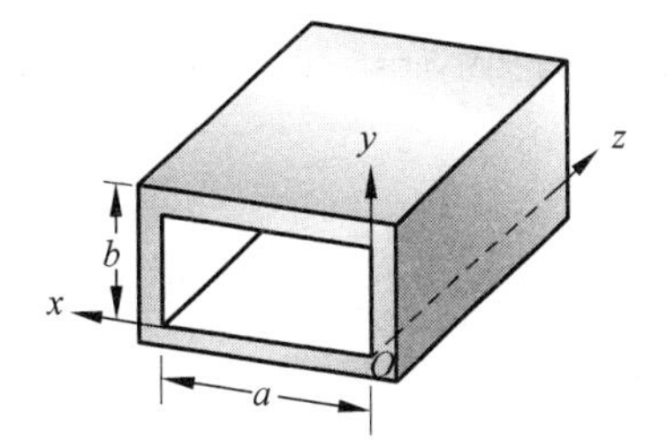

Figure 9-6 Rectangular waveguide

① 当β越小，ω越接近于ω_c时，TE/TM波的ω-β曲线受截止频率影响越大，偏离过原点直线（TEM波的ω-β线）越远，相速度越大，群速度越小。

9.5.1 TM Waves in Rectangular Waveguides (矩形波导内的 TM 波)

For transverse magnetic waves, $H_z=0$, and $E_z=E_z^0(x,y)\mathrm{e}^{-\gamma z}$ is to be solved from (9.36), which is rewritten as below:

$$\left(\frac{\partial^2}{\partial x^2}+\frac{\partial^2}{\partial y^2}+k_c^2\right)E_z^0(x,y)=0 \tag{9.63}$$

To solve for E_z^0, we use the method of separation of variables that is introduced in Section 3.5.[①] Let

$$E_z^0(x,y)=X(x)Y(y) \tag{9.64}$$

Substitute (9.64) in (9.63) and divide the resulting equation by $X(x)Y(y)$. Then we have

$$\frac{1}{X(x)}\frac{\mathrm{d}^2X(x)}{\mathrm{d}x^2}+\frac{1}{Y(y)}\frac{\mathrm{d}^2Y(x)}{\mathrm{d}y^2}+k_c^2=0 \tag{9.65}$$

which can be separated into two ordinary differential equations:

$$\frac{\mathrm{d}^2X(x)}{\mathrm{d}x^2}+k_x^2X(x)=0 \tag{9.66}$$

$$\frac{\mathrm{d}^2Y(y)}{\mathrm{d}y^2}+k_y^2Y(y)=0 \tag{9.67}$$

where k_x and k_y are two separation constants satisfying

$$k_x^2+k_y^2=k_c^2 \tag{9.68}$$

The possible solutions of (9.66) and (9.67) are listed in Table 3-1 in Section 3.5. The appropriate forms to be chosen must satisfy the following boundary conditions.

(1) In the x-direction:

$$E_z^0(0,y)=0 \tag{9.69}$$

$$E_z^0(a,y)=0 \tag{9.70}$$

(2) In the y-direction:

$$E_z^0(x,0)=0 \tag{9.71}$$

$$E_z^0(x,b)=0 \tag{9.72}$$

Obviously, to satisfy the above boundary conditions, $X(x)$ must be in the form of $\sin k_x x$, with

$$k_x=\frac{m\pi}{a},\quad m=1,2,3,\cdots$$

$Y(y)$ must be in the form of $\sin k_y y$, with

$$k_y=\frac{n\pi}{b},\quad n=1,2,3,\cdots$$

---

① 分离变量法是求解二阶线性偏微分方程的一种重要方法。在 3.5 节中，分离变量法被用于求解满足拉普拉斯方程的静电场边界值问题，这里被用于求解 TM 波的纵向分量 E_z^0 满足的亥姆霍兹方程。两者的区别在于，静电场边界值问题中，作用于电位的拉普拉斯算子的本征值为固定值零，而这里矩形波导 TM 波的问题中，作用于 E_z^0 的拉普拉斯算子的本征值 k_c^2 需要由边界条件确定。

Hence the solution for $E_z^0(x,y)$ must take the form of

$$E_z^0(x,y) = E_0 \sin\left(\frac{m\pi}{a}x\right) \sin\left(\frac{n\pi}{b}y\right) \tag{9.73}$$

From(9.68), we have

$$k_c = \sqrt{\left(\frac{m\pi}{a}\right)^2 + \left(\frac{n\pi}{b}\right)^2} \tag{9.74}$$

which means that the cutoff wave number can only take discrete values that depend on the size of the waveguide.①

The transverse field components are obtained from(9.28a)-(9.28d) as

$$E_x^0(x,y) = -\frac{\gamma}{k_c^2}\left(\frac{m\pi}{a}\right) E_0 \cos\left(\frac{m\pi}{a}x\right) \sin\left(\frac{n\pi}{b}y\right) \tag{9.75}$$

$$E_y^0(x,y) = -\frac{\gamma}{k_c^2}\left(\frac{n\pi}{b}\right) E_0 \sin\left(\frac{m\pi}{a}x\right) \cos\left(\frac{n\pi}{b}y\right) \tag{9.76}$$

$$H_x^0(x,y) = \frac{j\omega\varepsilon}{k_c^2}\left(\frac{n\pi}{b}\right) E_0 \sin\left(\frac{m\pi}{a}x\right) \cos\left(\frac{n\pi}{b}y\right) \tag{9.77}$$

$$H_y^0(x,y) = -\frac{j\omega\varepsilon}{k_c^2}\left(\frac{m\pi}{a}\right) E_0 \cos\left(\frac{m\pi}{a}x\right) \sin\left(\frac{n\pi}{b}y\right) \tag{9.78}$$

where

$$\gamma = j\beta = j\sqrt{\omega^2\mu\varepsilon - \left(\frac{m\pi}{a}\right)^2 - \left(\frac{n\pi}{b}\right)^2} \tag{9.79}$$

Each combination of the integers m and n corresponds to a mode that is referred to as a TM_{mn} mode. Here, the subscripts m and n represent the number of half-cycle variations of the fields in the x- and y-direction respectively.

The cutoff frequency of the TM_{mn} mode can be found by using(9.40) or by making γ in (9.79) be zero, which leads to

$$(f_c)_{mn} = \frac{1}{2\sqrt{\mu\varepsilon}}\sqrt{\left(\frac{m}{a}\right)^2 + \left(\frac{n}{b}\right)^2} = \frac{u}{2}\sqrt{\left(\frac{m}{a}\right)^2 + \left(\frac{n}{b}\right)^2} \tag{9.80}$$

Then, from(9.41), the cutoff wavelength is expressed as

$$(\lambda_c)_{mn} = \frac{2}{\sqrt{\left(\frac{m}{a}\right)^2 + \left(\frac{n}{b}\right)^2}} \tag{9.81}$$

The expressions for the phase constant β and the wave impedance Z_{TM} of the propagating TM_{mn} modes are the same as given in(9.44) and(9.50) respectively, with f_c given by(9.80) and λ_c by(9.81).

Notice that, for TM modes in rectangular waveguides, neither m nor n can be zero. Hence,

---

① 矩形波导 TM 波的截止波数 k_c 由波导横截面的尺寸确定,可以取无穷多个离散值,每个取值对应一个 TM_{mn} 模式。相应地,截止频率 f_c,截止波长 λ_c 也只能取离散值。

the TM_{11} mode has the lowest cutoff frequency and largest cutoff wavelength of all TM modes. The phasor expressions for the fields of a TM_{11} wave can be obtained by letting $m=n=1$ in(9.73) and(9.75)-(9.78). We can write down the corresponding instantaneous expressions as①

$$E_x(x,y,z,t)=\frac{\beta}{k_c^2}\left(\frac{\pi}{a}\right)E_0\cos\left(\frac{\pi}{a}x\right)\sin\left(\frac{\pi}{b}y\right)\sin(\omega t-\beta z) \tag{9.82}$$

$$E_y(x,y,z,t)=\frac{\beta}{k_c^2}\left(\frac{\pi}{b}\right)E_0\sin\left(\frac{\pi}{a}x\right)\cos\left(\frac{\pi}{b}y\right)\sin(\omega t-\beta z) \tag{9.83}$$

$$E_z(x,y,z,t)=E_0\sin\left(\frac{\pi}{a}x\right)\sin\left(\frac{\pi}{b}y\right)\cos(\omega t-\beta z) \tag{9.84}$$

$$H_x(x,y,z,t)=-\frac{\omega\varepsilon}{k_c^2}\left(\frac{\pi}{b}\right)E_0\sin\left(\frac{\pi}{a}x\right)\cos\left(\frac{\pi}{b}y\right)\sin(\omega t-\beta z) \tag{9.85}$$

$$H_y(x,y,z,t)=\frac{\omega\varepsilon}{k_c^2}\left(\frac{\pi}{a}\right)E_0\cos\left(\frac{\pi}{a}x\right)\sin\left(\frac{\pi}{b}y\right)\sin(\omega t-\beta z) \tag{9.86}$$

$$H_z(x,y,z,t)=0 \tag{9.87}$$

where

$$k_c=\sqrt{\left(\frac{\pi}{a}\right)^2+\left(\frac{\pi}{b}\right)^2} \tag{9.88}$$

$$\beta=\sqrt{k^2-k_c^2}=\sqrt{\omega^2\mu\varepsilon-\left(\frac{\pi}{a}\right)^2-\left(\frac{\pi}{b}\right)^2} \tag{9.89}$$

Figure 9-7 sketches of ***E*** and ***H*** field lines in *xy*- and *yz*-planes. Notice that the ***E*** lines are normal and ***H*** lines are parallel to conducting guide walls. ***E*** and ***H*** lines are everywhere perpendicular to each another. ***H*** lines form closed loops in the *xy*-plane as expected.

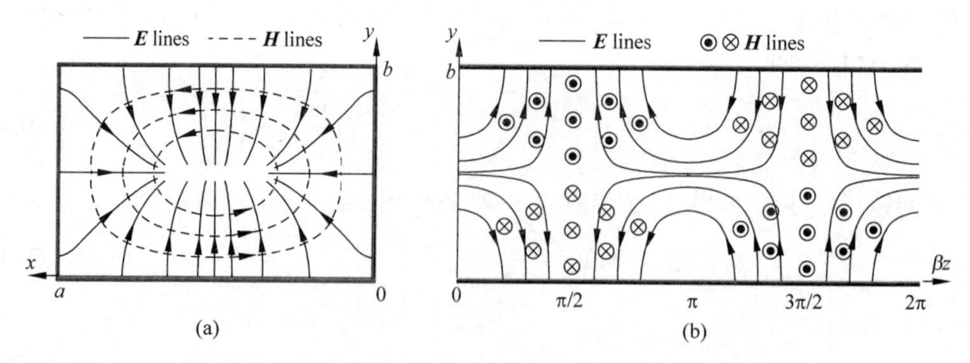

Figure 9-7 Field lines for TM_{11} mode in rectangular waveguide

① 矩形波导 TM 波截止频率最低的模式为 TM_{11} 模。该模式的场在 x 和 y 两个方向上都呈现出半个空间周期的变化。磁场在横截面上形成一个闭合回路,4 个波导壁上的电流均沿 z 方向流动。

9.5.2 TE Waves in Rectangular Waveguides (矩形波导内的 TE 波)

For transverse electric waves, $E_z=0$, and $H_z=H_z^0(x,y)\mathrm{e}^{-\gamma z}$, where $H_z^0(x, y)$ satisfies the second-order partial differential equation(9.59) which is rewritten as below

$$\left(\frac{\partial^2}{\partial x^2}+\frac{\partial^2}{\partial y^2}+k_c^2\right)H_z^0(x,y)=0 \tag{9.90}$$

The boundary conditions satisfied by $H_z^0(x,y)$ can be derived by using(9.53c) and (9.53d). Specifically,①

(1) In the x-direction:

$$E_y(0,y)=0 \Rightarrow \left.\frac{\partial H_z^0}{\partial x}\right|_{x=0}=0 \tag{9.91}$$

$$E_y(a,y)=0 \Rightarrow \left.\frac{\partial H_z^0}{\partial x}\right|_{x=a}=0 \tag{9.92}$$

(2) In the y-direction:

$$E_x(x,0)=0 \Rightarrow \left.\frac{\partial H_z^0}{\partial y}\right|_{y=0}=0 \tag{9.93}$$

$$E_x(x,b)=0 \Rightarrow \left.\frac{\partial H_z^0}{\partial y}\right|_{y=b}=0 \tag{9.94}$$

By using the method of separation of variables, we find that the cutoff wavenumber k_c for TE waves also can only take discrete values given by(9.74), and the solution for $H_z^0(x,y)$ can be expressed as

$$H_z^0(x,y)=H_0\cos\left(\frac{m\pi}{a}x\right)\cos\left(\frac{n\pi}{b}y\right) \tag{9.95}$$

where m and n are integers. Each combination of m and n corresponds to a mode that is referred to as a TE_{mn} mode. Notice that either m or n(but not both) can be zero for the TE_{mn} mode.

The transverse field components are obtained from(9.53a-d):

$$E_x^0(x,y)=\frac{\mathrm{j}\omega\mu}{k_c^2}\left(\frac{n\pi}{b}\right)H_0\cos\left(\frac{m\pi}{a}x\right)\sin\left(\frac{n\pi}{b}y\right) \tag{9.96}$$

$$E_y^0(x,y)=-\frac{\mathrm{j}\omega\mu}{k_c^2}\left(\frac{m\pi}{a}\right)H_0\sin\left(\frac{m\pi}{a}x\right)\cos\left(\frac{n\pi}{b}y\right) \tag{9.97}$$

$$H_x^0(x,y)=\frac{\gamma}{k_c^2}\left(\frac{m\pi}{a}\right)H_0\sin\left(\frac{m\pi}{a}x\right)\cos\left(\frac{n\pi}{b}y\right) \tag{9.98}$$

---

① H_z 在波导壁上是磁场切向分量,其边界值等于波导壁上的垂直于 z 方向的表面电流分量。由于该表面电流未知,因此无法通过 H_z 的边界值求解。然而根据 9.4.2 节中讨论的 TE 波的横纵关系,即式(9.53),H_z 沿 x 方向的方向导数正比于 H_x,H_x 又正比于 E_y,在 $x=0$ 和 $x=a$ 的边界上 H_x 和 E_y 均为零,因此 H_z 沿 x 方向的方向导数在这两个边界上应为零,即式(9.91)和式(9.92)给出的两个边界条件。同样可以得到式(9.93)和式(9.94)两个边界条件。

$$H_y^0(x,y) = \frac{\gamma}{k_c^2}\left(\frac{n\pi}{b}\right) H_0\cos\left(\frac{m\pi}{a}x\right)\sin\left(\frac{n\pi}{b}y\right) \tag{9.99}$$

where γ is the same as that for TM modes as given by(9.79).

The cutoff frequency f_c and cutoff wavelength λ_c of the TE_{mn} mode can also be expressed by(9.80) and(9.81) respectively. Consequently, the TE_{mn} mode has the same cutoff frequency as TM_{mn} mode has. Two modes having the same cutoff frequency but different field distributions are called **degenerate modes**(简并模式). Hence the TE_{mn} mode and TM_{mn} mode are degenerate modes.

Since $a>b$, the cutoff frequency is the lowest when $m=1$ and $n=0$:

$$(f_c)_{TE10} = \frac{1}{2a\sqrt{\mu\varepsilon}} = \frac{u}{2a} \tag{9.100}$$

And the corresponding cutoff wavelength is

$$(\lambda_c)_{TE10} = 2a \tag{9.101}$$

Notice that the cutoff frequency of the TE_{10} mode given by(9.100) is the lowest among all the TE and TM modes. Therefore, we consider the TE_{10} mode as the **dominant mode**(主模) of the rectangular waveguide. The phasor expressions for the fields of a TE_{10} wave can be obtained by letting $m=1$ and $n=0$ in(9.95)-(9.99). The corresponding instantaneous expressions can be written as①

$$E_x(x,y,z,t) = 0 \tag{9.102}$$

$$E_y(x,y,z,t) = \frac{\omega\mu}{k_c^2}\left(\frac{\pi}{a}\right) H_0\sin\left(\frac{\pi}{a}x\right)\sin(\omega t - \beta z) \tag{9.103}$$

$$E_z(x,y,z,t) = 0 \tag{9.104}$$

$$H_x(x,y,z,t) = -\frac{\beta}{k_c^2}\left(\frac{\pi}{a}\right) H_0\sin\left(\frac{\pi}{a}x\right)\sin(\omega t - \beta z) \tag{9.105}$$

$$H_y(x,y,z,t) = 0 \tag{9.106}$$

$$H_z(x,y,z,t) = H_0\cos\left(\frac{\pi}{a}x\right)\cos(\omega t - \beta z) \tag{9.107}$$

where

$$k_c = \frac{\pi}{a} \tag{9.108}$$

and

$$\beta = \sqrt{k^2 - k_c^2} = \sqrt{\omega^2\mu\varepsilon - \left(\frac{\pi}{a}\right)^2} \tag{9.109}$$

It is seen that the TE_{10} mode has only three nonzero field components namely, E_y, H_x and H_z. In the

① 矩形波导 TE 波截止频率最低的模式为 TM_{10} 模，由于其截止频率也低于所有 TM 模式的截止频率，故称为矩形波导的**主模**。TE_{10} 模的场在 x 方向上呈现半个空间周期的变化，在 y 方向上没有变化。由于 H_y 和 E_x 与 H_z 的 y 方向的方向导数成正比，故 H_y 和 E_x 均为零。

xy-plane, when $\sin(\omega t-\beta z)=1$, both E_y and H_x vary as $\sin(\pi x/a)$ as shown in Figure 9-8(a). In the yz-plane at $x=a/2$ and $t=0$, we observe E_y and H_x vary with($-\sin\beta z$) as is seen in Figure 9-8(b). The magnetic field lines form closed paths in the xz-plane as is shown in Figure 9-8(c).

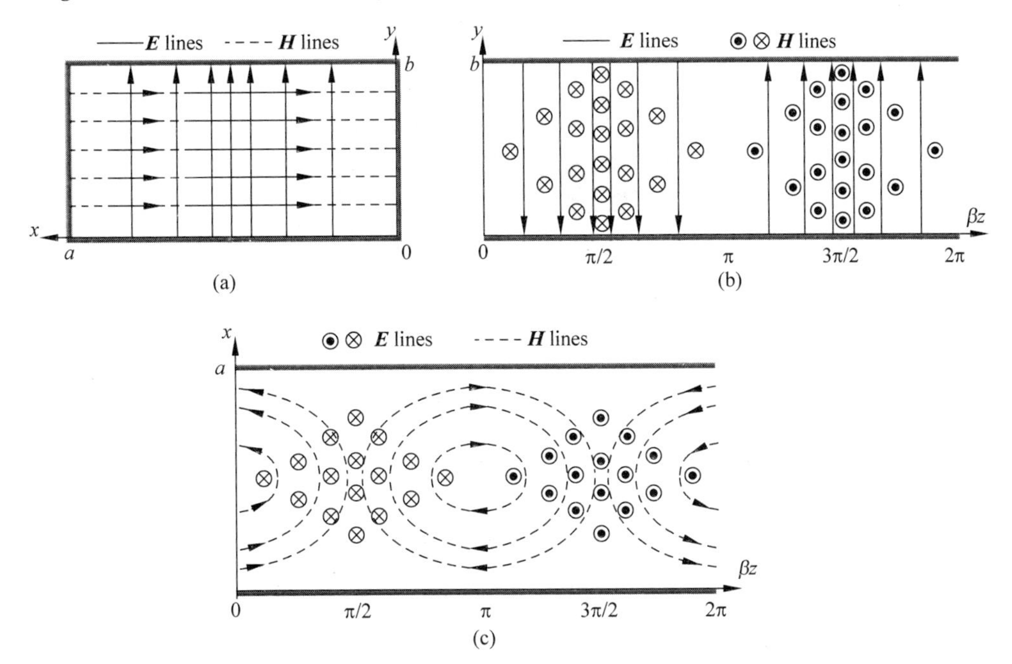

Figure 9-8 Field lines for TE_{10} mode in rectangular waveguide

With the expressions of the magnetic fields(9.105) and(9.107), the surface current J_s on the four PEC walls of the waveguide can be derived by applying the boundary condition(6.14b):

$$\boldsymbol{J}_s=\boldsymbol{e}_n\times\boldsymbol{H} \tag{9.110}$$

where $\boldsymbol{e}_n$ is the outward normal from the wall surface. Specifically, at $t=0$, the surface current on the inside wall $x=0$ is

$$\boldsymbol{J}_s(x=0)=\boldsymbol{e}_x\times[\boldsymbol{e}_xH_z(0,y,z;\ 0)+\boldsymbol{e}_zH_z(0,y,z;\ 0)]=-\boldsymbol{e}_yH_0\cos\beta z \tag{9.111}$$

where $\boldsymbol{e}_n=\boldsymbol{e}_x$. On the inside wall $x=a$, $\boldsymbol{e}_n=-\boldsymbol{e}_x$, and

$$\begin{aligned}\boldsymbol{J}_s(x=a)&=(-\boldsymbol{e}_x)\times[\boldsymbol{e}_xH_z(a,y,z;\ 0)+\boldsymbol{e}_zH_z(a,y,z;\ 0)]=-\boldsymbol{e}_yH_0\cos\beta z\\&=\boldsymbol{J}_s(x=0)\end{aligned} \tag{9.112}$$

On the inside wall $y=0$, $\boldsymbol{e}_n=\boldsymbol{e}_y$, and

$$\begin{aligned}\boldsymbol{J}_s(y=0)&=\boldsymbol{e}_y\times[\boldsymbol{e}_xH_x(x,0,z;0)+\boldsymbol{e}_zH_z(x,0,z;0)]\\&=\boldsymbol{e}_xH_0\cos\left(\frac{\pi}{a}x\right)\cos\beta z-\boldsymbol{e}_z\frac{\beta}{k_c^2}\left(\frac{\pi}{a}\right)H_0\sin\left(\frac{\pi}{a}x\right)\sin\beta z\end{aligned} \tag{9.113}$$

On the inside wall $y=b$, $\boldsymbol{e}_n=-\boldsymbol{e}_y$, and

$$\begin{aligned}\boldsymbol{J}_s(y=b)&=(-\boldsymbol{e}_y)\times[\boldsymbol{e}_xH_x(x,b,z;\ 0)+\boldsymbol{e}_zH_z(x,b,z;\ 0)]\\&=-\boldsymbol{e}_xH_0\cos\left(\frac{\pi}{a}x\right)\cos\beta z+\boldsymbol{e}_z\frac{\beta}{k_c^2}\left(\frac{\pi}{a}\right)H_0\sin\left(\frac{\pi}{a}x\right)\sin\beta z\\&=-\boldsymbol{J}_s(y=0)\end{aligned} \tag{9.114}$$

The surface currents on the inside walls at $x=0$ and at $y=b$ are sketched in Figure 9-9. Notice that the surface current on the wall at $x=a$ is the same as that on the wall $x=0$, whereas the surface current on the wall at $y=b$ is the opposite to that on the wall $y=0$.

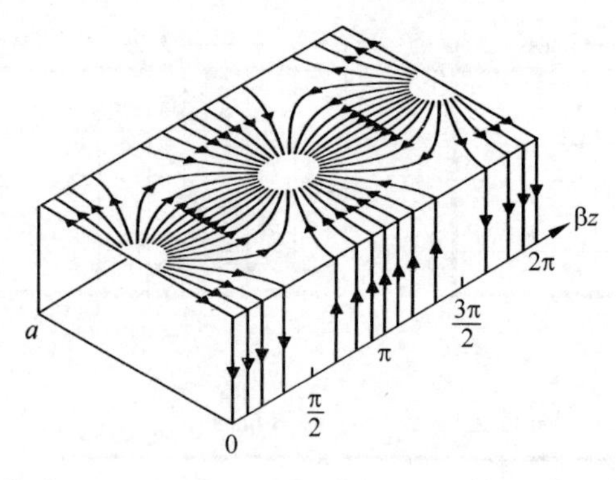

Figure 9-9 Surface currents on rectangular waveguide walls with TE_{10} mode

Example 9-2 An air-filled waveguide has dimensions $a=2.29$cm and $b=1.02$cm. The waveguide is designed to operate only in the dominant mode, with the operating frequency at least 25% above the cutoff frequency of the dominant mode but no higher than 95% of the next higher cutoff frequency. What is the operating-frequency range of the waveguide?

Solution The dominant mode of the waveguide is TE_{10} mode. For $a=2.29\times10^{-2}$m and $b=1.02\times10^{-2}$m, the mode with the next higher cutoff frequency is TE_{20} mode. Using (9.80), we find

$$(f_c)_{10}=\frac{c}{2a}=\frac{3\times10^8}{2\times2.29\times10^{-2}}=6.55\times10^9(\text{Hz})$$

$$(f_c)_{20}=\frac{c}{a}=13.10\times10^9(\text{Hz})$$

Thus, the operating-frequency range is

$$1.25(f_c)_{TE10}\leqslant f\leqslant 0.95(f_c)_{TE20}$$

i. e. ,

$$8.19\text{GHz}\leqslant f\leqslant 12.45\text{GHz}$$

9.5.3 Attenuation in Rectangular Waveguides(矩形波导的衰减)

In a realistic waveguide, there will be unavoidable attenuation in the propagating waves due to the losses in the dielectric and the imperfectly conducting walls. The attenuation causes both the $\boldsymbol{E}$ and $\boldsymbol{H}$ field decay exponentially along the propagation direction. Usually these losses are very small, and it can be assumed that the field patterns within the guide are not appreciably affected by the losses. Then the general expressions of $\boldsymbol{E}$ and $\boldsymbol{H}$ in (9.2) can be modified as ①

① 波导系统的损耗主要来源于金属和填充介质材料的非理想特性,即金属不是 PEC,填充材料不是理想介质。有损媒质和非理想导体中电磁波传播的分析会变得很复杂。但实际的波导系统损耗通常很小,小到足以假设其中的电磁场横向分布不受损耗的影响,即式(9.115)中的$\boldsymbol{E}^0(x,y)$和$\boldsymbol{H}^0(x,y)$可以直接采用前面无损波导系统分析的结果,损耗的影响则由 z 方向上的衰减常数 α 表征。

$$\boldsymbol{E}(x,y,z)=\boldsymbol{E}^0(x,y)\mathrm{e}^{-\alpha z}\mathrm{e}^{-\mathrm{j}\beta z} \tag{9.115a}$$

$$\boldsymbol{H}(x,y,z)=\boldsymbol{H}^0(x,y)\mathrm{e}^{-\alpha z}\mathrm{e}^{-\mathrm{j}\beta z} \tag{9.115b}$$

where α is the attenuation constant due to the losses. Let $\alpha=\alpha_d+\alpha_c$, where α_d and α_c represent the attenuation constants due to the dielectric loss and conductor loss respectively. α_d can be obtained by substituting $\varepsilon_c=\varepsilon+(\sigma/\mathrm{j}\omega)$ for ε in(9.37), which leads to

$$\alpha_d=\frac{\sigma\eta}{2\sqrt{1-(f_c/f)^2}} \tag{9.116}$$

where σ is the equivalent conductivity of the dielectric medium, and f_c is the cutoff frequency given by(9.80). It is seen from(9.116) that, as the operating frequency increases from f_c, α_d decreases **monotonically**(单调递减) from infinite to $\sigma\eta/2$.

To determine the attenuation constant α_c due to wall losses, let $P(z)$ be the time-average power flowing through a cross section of the waveguide. Then, by using(9.115), we have

$$P(z)=\int_0^a\int_0^b(E\times H^*)\cdot\boldsymbol{e}_z\mathrm{d}x\mathrm{d}y=\left[\int_0^a\int_0^b(\boldsymbol{E}^0\times\boldsymbol{H}^{0*})\cdot\boldsymbol{e}_z\mathrm{d}x\mathrm{d}y\right]\mathrm{e}^{-2\alpha z}$$

and therefore,

$$\frac{\mathrm{d}P(z)}{\mathrm{d}z}=-2\alpha_c P(z) \tag{9.117}$$

As a result, α_c can be calculated as

$$\alpha_c=\frac{-\mathrm{d}P(z)/\mathrm{d}z}{2P(z)} \tag{9.118}$$

where the numerator, $-\mathrm{d}P(z)/\mathrm{d}z$, is the time-average power loss per unit length.

For the dominant TE_{10} mode, $P(z)$ is the surface integral of the Poynting vector contributed by E_y and H_x over the waveguide cross section. By letting $m=1$, $n=0$, and $k_c=(\pi/a)$ in(9.97) and(5.94), we have

$$P(z)=\mathrm{e}^{-2\alpha z}\int_0^b\int_0^a-\frac{1}{2}(E_y^0)(H_x^0)^*\mathrm{d}x\mathrm{d}y=\omega\mu\beta ab\left(\frac{aH_0}{2\pi}\right)^2\mathrm{e}^{-2\alpha z} \tag{9.119}$$

The time-average power lost per unit length in the conducting walls can be calculated by using (7.87):

$$-\frac{\mathrm{d}P}{\mathrm{d}z}=\underbrace{2\mathrm{e}^{-2\alpha z}\int_0^a\frac{1}{2}[|H_z^0|_{y=0}^2+|H_x^0|_{y=0}^2]R_S\mathrm{d}x}_{\text{Power loss on the walls at } y=0 \text{ and } y=b}+\underbrace{2\mathrm{e}^{-2\alpha z}\int_0^b\frac{1}{2}|H_z^0|_{x=0}^2R_S\mathrm{d}y}_{\text{Power loss on the walls } x=0 \text{ and } x=a} \tag{9.120}$$

From(9.95) and(5.94) we have

$$|H_z^0|_{x=0}=|H_0| \tag{9.121}$$

$$|H_z^0|_{y=0}=\left|H_0\cos\left(\frac{\pi}{a}x\right)\right| \tag{9.122}$$

$$|H_x^0|_{y=0}=\left|\frac{\beta a}{\pi}H_0\sin\left(\frac{\pi}{a}x\right)\right| \tag{9.123}$$

Substitute(9.121)-(9.123) into(9.120), we have

$$-\frac{\mathrm{d}P}{\mathrm{d}z}=\left\{b+\frac{a}{2}\left[1+\left(\frac{\beta a}{\pi}\right)^2\right]\right\}H_0^2R_s\mathrm{e}^{-2\alpha z}=\left[b+\frac{a}{2}\left(\frac{f}{f_c}\right)^2\right]H_0^2R_s\mathrm{e}^{-2\alpha z} \tag{9.124}$$

The last expression is the result of recognizing that

$$\beta=\sqrt{\omega^2\mu\varepsilon-\left(\frac{\pi}{a}\right)^2}=\omega\sqrt{\mu\varepsilon}\sqrt{1-\left(\frac{f_c}{f}\right)^2} \tag{9.125}$$

Substituting (5.99) and (9.124) in (9.118), we obtain

$$(\alpha_c)_{TE10}=\frac{R_s[1+(2b/a)(f_c/f)^2]}{\eta b\sqrt{1-(f_c/f)^2}}$$

$$=\frac{1}{\eta b}\sqrt{\frac{\pi f\mu_c}{\sigma_c[1-(f_c/f)^2]}}\left[1+\frac{2b}{a}\left(\frac{f_c}{f}\right)^2\right] \tag{9.126}$$

By following a similar procedure, we can derive the attenuation constant due to wall losses for TM modes. For the TM_{11} mode, we obtain

$$(\alpha_c)_{TM11}=\frac{2R_s(b/a^2+a/b^2)}{\eta ab\sqrt{1-(f_c/f)^2}(1/a^2+1/b^2)} \tag{9.127}$$

Figure 9-10 plots the calculated $(\alpha_c)_{TE10}$ and $(\alpha_c)_{TM11}$ as a function of frequency for a standard air-filled rectangular copper waveguide with $a=2.29$cm and $b=1.02$cm. From (9.80), we find $(f_c)_{10}=6.55$GHz and $(f_c)_{11}=16.10$GHz. The curves show that the attenuation constant increases rapidly toward infinity as the operating frequency approaches the cutoff frequency. The attenuation constant of the TE_{10} mode is everywhere lower than that of the TM_{11} mode.

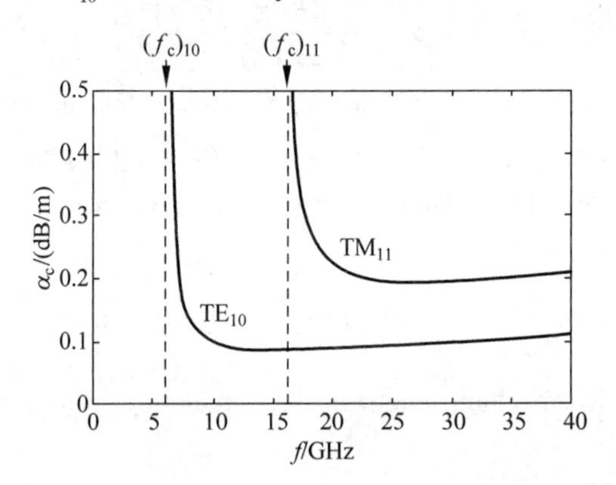

Figure 9-10 Attenuation constants due to wall losses in a rectangular copper waveguide with $a=2.29$cm, $b=1.02$cm

Example 9-3 A TE_{10} wave propagates in a copper rectangular waveguide filled with a dielectric ($\varepsilon_r=2.25$, $\mu_r=1$, $\tan\delta_c=10^{-4}$). The cross-sectional size of the waveguide is $a=1.5$cm and $b=0.6$cm and the operating frequency is 10GHz. Determine the phase constant, guide wavelength, phase velocity, wave impedance, attenuation constant due to loss in the dielectric and in the metal walls.

Solution: The speed of light in the dielectric filling the waveguide is

$$u=\frac{c}{\sqrt{\varepsilon_r}}=2\times10^8(\text{m/s})$$

With the operating frequency $f=10^{10}$Hz, the operating wavelength is

$$\lambda = \frac{u}{f} = \frac{2 \times 10^8}{10^{10}} = 0.02(\text{m})$$

The cutoff frequency for the TE_{10} mode is, from (9.100),

$$f_c = \frac{u}{2a} = \frac{2 \times 10^8}{2 \times (1.5 \times 10^{-2})} = 0.667 \times 10^{10}(\text{Hz})$$

From (5.103), the phase constant is

$$\beta = \frac{\omega}{u}\sqrt{1 - \left(\frac{f_c}{f}\right)^2} = \frac{2\pi \times 10^{10}}{2 \times 10^8}\sqrt{1 - 0.667^2} = 74.5\pi = 234(\text{rad/m})$$

From (9.45), the guide wavelength is,

$$\lambda_g = \frac{\lambda}{\sqrt{1 - (f_c/f)^2}} = \frac{0.02}{0.745} = 0.0268(\text{m})$$

From (9.47), the phase velocity is

$$u_p = \frac{u}{\sqrt{1 - (f_c/f)^2}} = \frac{2 \times 10^8}{0.745} = 2.68 \times 10^8(\text{m/s})$$

From (9.60), the wave impedance is

$$(Z_{TE})_{10} = \frac{\eta}{\sqrt{1 - (f_c/f)^2}} = \frac{377/\sqrt{2.25}}{0.745} = 337.4(\Omega)$$

The effective conductivity for the dielectric at 10GHz can be determined from the loss tangent by using (7.43), which gives

$$\sigma = \omega\varepsilon\tan\delta_c = (2\pi \times 10^{10}) \times \left(2.25 \times \frac{1}{36\pi} \times 10^{-9}\right) \times 10^{-4} = 1.25 \times 10^{-4}(\text{S/m})$$

Then the attenuation constant due to dielectric loss is obtained from (9.116) to be

$$\alpha_d = \frac{\sigma}{2} Z_{TE} = \frac{1.25 \times 10^{-4}}{2} \times 337.4 = 0.021(\text{Np/m})$$
$$= 0.18(\text{dB/m})$$

The attenuation constant due to loss in the guide walls is found from (9.126). We have, from (7.83),

$$R_s = \sqrt{\frac{\pi f \mu_0}{\sigma_c}} = \sqrt{\frac{\pi \times 10^{10} \times 4\pi \times 10^{-7}}{5.8 \times 10^7}} = 0.026(\Omega)$$

$$\alpha_c = \frac{R_s[1 + (2b/a)(f_c/f)^2]}{\eta b\sqrt{1 - (f_c/f)^2}} = \frac{0.026[1 + (1.2/1.5)(0.667)^2]}{251 \times 0.006 \times 0.745}$$
$$= 0.031(\text{Np/m}) = 0.273(\text{dB/m})$$

9.6 Circular Waveguides(圆波导)

In this section, we study wave behaviors in circular waveguides whose cross section is circle. Like rectangular waveguides, circular waveguides do not support TEM waves. Only TE and TM waves can exist in the circular waveguides.

The solution of TM and TE waves can be obtained by solving the same equations(9.36) and(9.59) for E_z^0 and H_z^0 respectively, except that the transverse coordinates x and y need to be replaced with cylindrical coordinates ρ and ϕ. Specifically,

$$E_z(\rho,\phi,z) = E_z^0(\rho,\phi)\mathrm{e}^{-\gamma z} \tag{9.128}$$

$$H_z(\rho,\phi,z) = H_z^0(\rho,\phi)\mathrm{e}^{-\gamma z} \tag{9.129}$$

and the equations(9.36) and(9.59) become

$$\nabla_{\rho\phi}^2 E_z^0(\rho,\phi) + k_c^2 E_z^0(\rho,\phi) = 0 \tag{9.130}$$

$$\nabla_{\rho\phi}^2 H_z^0(\rho,\phi) + k_c^2 H_z^0(\rho,\phi) = 0 \tag{9.131}$$

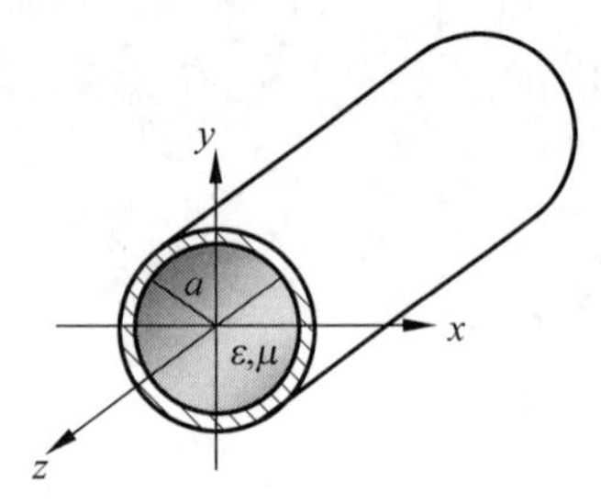

Figure 9-11 Circular waveguide

With the solution of E_z^0 and H_z^0, the complete $\boldsymbol{E}$ and $\boldsymbol{H}$ fields can be found by using the relations between longitudinal and transverse components that will be given in the following Subsections for TM and TE modes separately.

In the following discussions, we suppose the circular waveguide has a radius a and consists of a metal pipe centered at the z-axis, as is shown in Figure 9-11. The enclosed dielectric medium is assumed to have constitutive parameters ε and μ.

9.6.1 TM Waves in Circular Waveguides(圆波导内的 TM 波)

For TM waves, $H_z = 0$. E_z is to be solved through equation(9.130). Expansion of(9.130) gives

$$\frac{1}{\rho}\frac{\partial}{\partial\rho}\left(\rho\frac{\partial E_z^0}{\partial\rho}\right) + \frac{1}{\rho^2}\frac{\partial^2 E_z^0}{\partial\phi^2} + k_c^2 E_z^0 = 0 \tag{9.132}$$

Using the method of separation of variables, we let

$$E_z^0(\rho,\phi) = R(\rho)\Phi(\phi) \tag{9.133}$$

where R is a single-variable function of ρ, and $\Phi(\phi)$ is a single-variable function of ϕ. Substituting(9.133) in(9.132) and dividing both sides by the product $R(\rho)\Phi(\phi)$, we obtain

$$\frac{\rho}{R(\rho)}\frac{\mathrm{d}}{\mathrm{d}\rho}\left[\rho\frac{\mathrm{d}R(\rho)}{\mathrm{d}\rho}\right] + k_c^2\rho^2 = -\frac{1}{\Phi(\phi)}\frac{\mathrm{d}^2\Phi(\phi)}{\mathrm{d}\phi^2} \tag{9.134}$$

The left side of(9.134) is a function of ρ only, whereas the right side is a function of ϕ only. That means both sides must be equal to the same constant. Let this constant be n^2, which separates(9.134) into two ordinary differential equations:

$$\frac{\mathrm{d}^2\Phi(\phi)}{\mathrm{d}\phi^2} + n^2\Phi(\phi) = 0 \tag{9.135}$$

$$\frac{\rho}{R(\rho)}\frac{\mathrm{d}}{\mathrm{d}\rho}\left[\rho\frac{\mathrm{d}R(\rho)}{\mathrm{d}\rho}\right] + k_c^2\rho^2 = n^2 \tag{9.136}$$

Notice that, $\Phi(\phi)$ must be periodic with a period of 2π, which requires the separation constant n to be an integer. The resulting solutions for (9.135) can be $\sin n\phi$, $\cos n\phi$, or generally, a linear combination of the two (see Table 3-1), i. e.,

$$\Phi(\phi)=A\sin n\phi+B\cos n\phi=C\cos[n(\phi-\phi_0)] \tag{9.137}$$

where ϕ_0 is a constant angle. However, whether $\sin n\phi$, $\cos n\phi$ or $\cos[n(\phi-\phi_0)]$ is selected, the resulting solution is the same except that the location of the reference $\phi=0$ angle is different. Therefore, in the following discussion, we select $\cos n\phi$ as the solution of $\Phi(\phi)$.

Equation (9.136) can be rewritten as

$$\frac{d^2R(\rho)}{d\rho^2}+\frac{1}{\rho}\frac{dR(\rho)}{d\rho}+\left(k_c^2-\frac{n^2}{\rho^2}\right)R(\rho)=0 \tag{9.138}$$

which is known as **Bessel's differential equation** (**贝塞尔微分方程**). The general solution of this equation is

$$R(\rho)=C_nJ_n(k_c\rho)+D_nN_n(k_c\rho) \tag{9.139}$$

where C_n and D_n are arbitrary constants, and

$$J_n(k_c\rho)=\sum_{m=0}^{\infty}\frac{(-1)^m(k_c\rho)^{n+2m}}{m!\,(n+m)!\,2^{n+2m}} \tag{9.140}$$

is called the **Bessel function of the first kind** (**第一类贝塞尔函数**) of nth order with an argument $k_c\rho$.

$$N_n(k_c\rho)=\frac{(\cos n\pi)J_n(k_c\rho)-J_{-n}(k_c\rho)}{\sin n\pi} \tag{9.141}$$

is called **Bessel function of the second kind** (**第二类贝塞尔函数**) or **Neumann function** (**纽曼函数**). However, the function $N_n(k_c\rho)$ goes to infinity when the argument $k_c\rho$ approaches zero, and therefore cannot appear in the solution of the field. In other words, the coefficient D_n in (9.139) must be zero for all n, and therefore, the solution of $E_z^0(\rho,\phi)$ can be written as

$$E_z^0=C_nJ_n(k_c\rho)\cos n\phi \tag{9.142}$$

in which $\cos n\phi$ is selected to be the solution of $\Phi(\phi)$.

The cutoff wavenumber k_c can be determined by the boundary condition, which requires E_z^0 to be zero at $\rho=a$, i. e.,

$$J_n(k_ca)=0 \tag{9.143}$$

That means k_ca must be a zero of the Bessel function $J_n(x)$. There are infinitely many zeros of $J_n(x)$. Figure 9-12 illustrates the function $J_n(x)$ versus x for the first few orders, in which the pth zero of $J_n(x)$ is denoted by x_{np}. Then, k_c for TM modes can only take discrete values as

$$k_c=\frac{x_{np}}{a} \tag{9.144}$$

Each combination of n and p corresponds to a TM_{np} mode. Table 9-1 lists the first a few zeros, x_{np} of $J_n(x)$. Notice that the listed values of x_{np} do not include the point $x=0$, although $J_n(0)=0$ for all n except when $n=0$. This is because k_c cannot be zero for the TM mode.

The transverse components E_ρ^0 and E_ϕ^0 can be found using the same relation (9.29) for rectangular waveguides, but under the cylindrical coordinates, i. e.,

$$(\boldsymbol{E}_T^0)_{TM}=\boldsymbol{e}_\rho E_\rho^0+\boldsymbol{e}_\phi E_\phi^0=-\frac{\gamma}{k_c^2}\nabla_T E_z^0=-\frac{\gamma}{k_c^2}\left(\boldsymbol{e}_\rho\frac{\partial}{\partial\rho}+\boldsymbol{e}_\phi\frac{\partial}{\rho\partial\phi}\right)E_z^0 \tag{9.145}$$

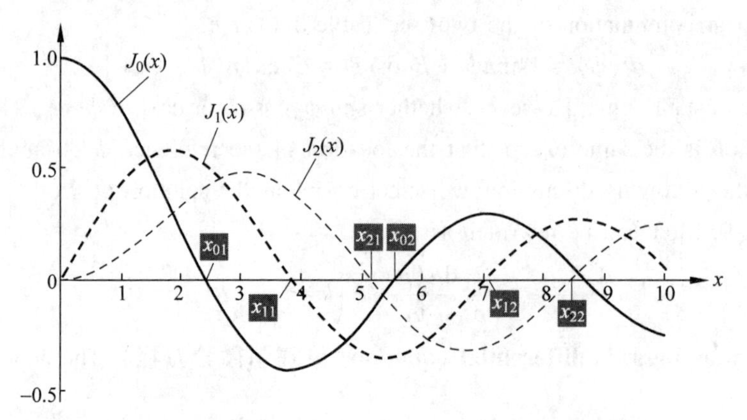

Figure 9-12 Bessel functions of the first kind and their zeros

where

$$\nabla_T E_z^0 = \left(\boldsymbol{e}_\rho \frac{\partial}{\partial \rho} + \boldsymbol{e}_\phi \frac{\partial}{\rho \partial \phi} \right) E_z^0 \tag{9.146}$$

Table 9-1 Zeros of $J_n(x)$, x_{np}

n	0	1	2
x_{n1}	2.405	3.832	5.136
x_{n2}	5.520	7.016	8.417

The magnetic field components can then be obtained by using (9.32). Specifically,

$$(\boldsymbol{H}^0)_{TM} = \boldsymbol{e}_\rho H_\rho^0 + \boldsymbol{e}_\phi H_\phi^0 = \frac{1}{Z_{TM}} (\boldsymbol{e}_z \times E_T^0) = - \boldsymbol{e}_\rho \left(\frac{E_\phi^0}{Z_{TM}} \right) + \boldsymbol{e}_\phi \left(\frac{E_\rho^0}{Z_{TM}} \right) \tag{9.147}$$

Therefore, in addition to E_z^0 in (9.142), we have,

$$E_\rho^0 = - \frac{j\beta}{k_c} C_n J_n'(k_c \rho) \cos n\phi \tag{9.148}$$

$$E_\phi^0 = \frac{j\beta n}{k_c^2 \rho} C_n J_n(k_c \rho) \sin n\phi \tag{9.149}$$

$$H_\rho^0 = - \frac{j\omega\varepsilon n}{k_c^2 \rho} C_n J_n(k_c \rho) \sin n\phi \tag{9.150}$$

$$H_\phi^0 = - \frac{j\omega\varepsilon}{k_c} C_n J_n'(k_c \rho) \cos n\phi \tag{9.151}$$

$$H_z^0 = 0 \tag{9.152}$$

where γ has been replaced by $j\beta$, and $J_n'(x)$ is the derivative of the Bessel function $J_n(x)$.

The first zero point of the Bessel function $J_n(x)$, $x_{01} = 2.405$, is the lowest among all the zeros. So

$$(k_c)_{TM01} = \frac{x_{01}}{a} = \frac{2.405}{a} \tag{9.153}$$

is the smallest wavenumber, and the TM_{01} mode has the lowest cutoff frequency among all the

TM modes, which according to (9.40), is

$$(f_c)_{TM01} = \frac{(k_c)_{TM01}}{2\pi\sqrt{\mu\varepsilon}} = \frac{0.383}{a\sqrt{\mu\varepsilon}} \tag{9.154}$$

The phase constant β and the guide wavelength λ_g can be found from (9.44) and (9.45). For the TM_{01} mode, the only nonzero field components are E_z^0, E_ρ^0 and H_ϕ^0. A sketch of the electric and magnetic field lines in a typical transverse plane is given in Figure 9-13.

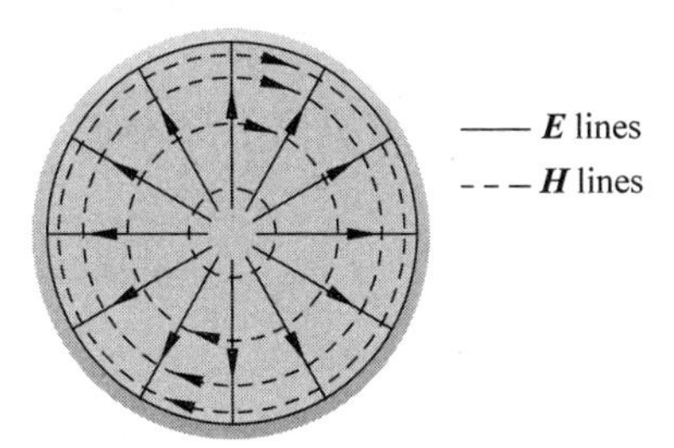

Figure 9-13 Field lines of TM_{01} mode in a transverse plane of a circular waveguide

9.6.2 TE Waves in Circular Waveguides(圆波导内的 TE 波)

For TE modes, $E_z = 0$, H_z is to be solved with equation (9.131). Using the method of separation of variables leads to a solution with the same form as (9.142), which is written as

$$H_z^0 = C_n' J_n(k_c\rho)\cos n\phi \tag{9.155}$$

The boundary condition for TE waves requires that derivative of H_z^0 with respect to ρ is zero at $\rho = a$. Therefore, from (9.155), we must have

$$J_n'(k_c a) = 0 \tag{9.156}$$

which means $k_c a$ must be a zero of the function $J_n'(x)$. The zeros of the function $J_n'(x)$ correspond to the local minimums and/or maximums of the $J_n(x)$ curves. As illustrated in Figure 9-14, the pth zero of $J_n'(x)$ is denoted by x_{np}'. Therefore, k_c for TE modes can only take discrete values as

$$k_c = \frac{x_{np}'}{a} \tag{9.157}$$

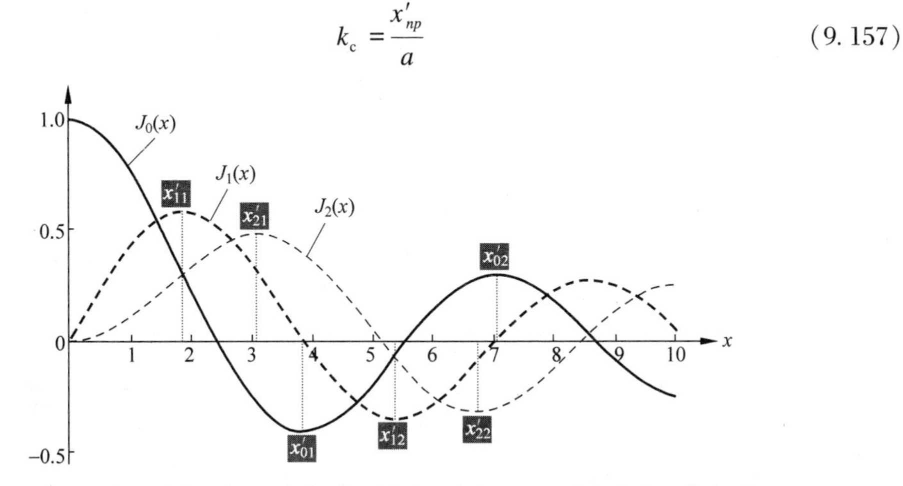

Figure 9-14 Bessel functions of the first kind and the zeros of their first derivatives

Each combination of n and p corresponds to a TE_{np} mode.

From H_z^0, we find the transverse field components by using (9.54) and (9.57) under cylindrical coordinates:

$$(\boldsymbol{H}_T^0)_{TE} = \boldsymbol{e}_\rho H_\rho^0 + \boldsymbol{e}_\phi H_\phi^0 = -\frac{\gamma}{k_c^2}\nabla_T H_z^0 = -\frac{\gamma}{k_c^2}\left(\boldsymbol{e}_\rho \frac{\partial}{\partial\rho} + \boldsymbol{e}_\phi \frac{\partial}{\rho\partial\phi}\right)H_z^0 \quad (9.158)$$

$$(\boldsymbol{E}^0)_{TE} = \boldsymbol{e}_\rho E_\rho^0 + \boldsymbol{e}_\phi E_\phi^0 = -Z_{TE}(\boldsymbol{e}_z \times \boldsymbol{H}^0) = \boldsymbol{e}_\rho Z_{TE} H_\phi^0 - \boldsymbol{e}_\phi Z_{TE} H_\rho^0 \quad (9.159)$$

With (9.155), we have

$$H_\rho^0 = -\frac{\mathrm{j}\beta}{k_c} C'_n J'_n(k_c\rho)\cos n\phi \quad (9.160)$$

$$H_\phi^0 = \frac{\mathrm{j}\beta n}{k_c^2\rho} C'_n J_n(k_c\rho)\sin n\phi \quad (9.161)$$

$$E_\rho^0 = \frac{\mathrm{j}\omega\mu n}{k_c^2\rho} C'_n J_n(k_c\rho)\sin n\phi \quad (9.162)$$

$$E_\phi^0 = \frac{\mathrm{j}\omega\mu}{k_c} C'_n J'_n(k_c\rho)\cos n\phi \quad (9.163)$$

$$E_z^0 = 0 \quad (9.164)$$

From Table 9-2, we see that the smallest x'_{np} is $x'_{11} = 1.841$. This corresponds to the smallest cutoff wavenumber

$$(k_c)_{TE11} = \frac{1.841}{a} \quad (9.165)$$

and the lowest cutoff frequency

$$(f_c)_{TE11} = \frac{(k_c)_{TE11}}{2\pi\sqrt{\mu\varepsilon}} = \frac{0.293}{a\sqrt{\mu\varepsilon}} \quad (9.166)$$

which is even lower than $(f_c)_{TM01}$ given in (9.153). Hence TE_{11} is the dominant mode in a circular waveguide. In an air-filled circular waveguide of radius a, the cutoff wavelength of the TE_{11} mode is

$$(\lambda_c)_{TE11} = \frac{a}{0.293} = 3.41a \quad (9.167)$$

A sketch of the electric and magnetic field lines for the TE_{11} mode in a typical transverse plane is shown in Figure 9-15. ①

Table 9-2 Zeros of $J'_n(x)$, x'_{np}

n	0	1	2
x'_{n1}	3.832	1.841	3.054
x'_{n2}	7.016	5.331	6.706

In the above discussions, $\cos n\phi$ has been selected to be the solution of $\Phi(\phi)$ for E_z^0 in

① 圆柱形波导的主模是 TE_{11} 模，其截止波长为 $3.41a$。

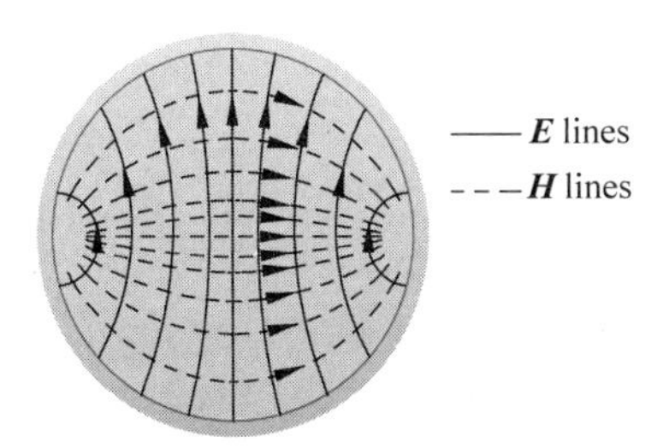

Figure 9-15 Field lines of TE_{11} mode in a transverse plane of a circular waveguide

(9.142) and H_z^0 in (9.155). However, as previously mentioned, the solution of $\Phi(\phi)$ is generally a linear combination of $\sin n\phi$ and $\cos n\phi$, or equivalently, $\cos[n(\phi-\phi_0)]$ as given by (9.137). Replacing ϕ with $(\phi-\phi_0)$ in the above field expressions also leads to valid solutions of the guided wave with different polarization, but still the same cutoff frequencies. Typically, $\sin n\phi$ and $\cos n\phi$ functions ($\phi_0 = 0$ and $\pi/2$, respectively) correspond to two orthogonal polarizations. Since they have the same cutoff frequencies, the two polarizations are degenerate, and this phenomenon is called **polarization degeneracy**(**极化简并**).

In additional to polarization degeneracy, mode degeneracy also exists as different modes in a circular waveguide can have the same cutoff frequency. For instance, $(f_c)_{TE0n}=(f_c)_{TM1n}$ for all n, because the zeros of functions $J_n'(x)$ and $J_n(x)$ have the relation $x_{0n}'=x_{1n}$ for all n. Therefore, TE_{0n} and TM_{1n} are degenerate modes. This phenomena are also called **E-H degeneracy**(**E-H 简并**).①

The attenuation constant due to losses in the imperfectly conducting wall of a circular waveguide can be calculated by following the same procedure used in Subsection 9.5.3 for a rectangular waveguide. The attenuation constants of the dominant-mode propagating waves in circular and rectangular waveguides having comparable dimensions are of the same order of magnitude.

Example 9-4 A hollow circular waveguide is designed to operate at 10GHz. It is required that the lowest cutoff frequency is 20% below operating frequency. ① Determine the inside diameter of the waveguide. ②If the signal frequency within the same waveguide is increased to 15GHz, what are the modes that can propagate?

Solution:

(1) The TE_{11} mode has the lowest cutoff frequency, which, according to the example problem, must be $(f_c)_{TE11} = 10\times(1-20\%) = 8(\text{GHz})$. From (9.166), the radius of a circular waveguide can be found as

$$a = \frac{0.293c}{(f_c)_{TE11}} = \frac{0.879\times10^8}{8\times10^9} \approx 0.011(\text{m})$$

That is, the required inside diameter is 2.2(cm).

(2) From Table 9-1 and Table 9-2, the cutoff frequencies of the first a few modes of the

① 圆柱形波导中存在的双重简并：模式简并(E-H 简并)和极化简并。

circular waveguide with radius $a=0.011\text{m}$ are

$$(f_c)_{\text{TE11}} = 8(\text{GHz})$$

$$(f_c)_{\text{TM01}} = \left(\frac{x_{01}}{a}\right)\left(\frac{c}{2\pi}\right) = \frac{2.405}{a}\left(\frac{c}{2\pi}\right) \approx 10.44(\text{GHz})$$

$$(f_c)_{\text{TE21}} = \left(\frac{x_{21}}{a}\right)\left(\frac{c}{2\pi}\right) = \frac{3.054}{a}\left(\frac{c}{2\pi}\right) \approx 13.26(\text{GHz})$$

The f_c of all other modes are higher than 15GHz. Hence only TE_{11}, TM_{01} and TE_{21} modes can propagate in the pipe.

9.6.3 Non-TEM Waves in Coaxial Lines(同轴线中的非 TEM 波)

As a two-conductor waveguide, the coaxial line can support TEM waves as is studied in Section 9. 3. 2. Nevertheless, a coaxial line can also support TE and TM waves when the operating frequency is high enough. The solution of these non-TEM waves can be derived in a similar way as for the circular waveguide. The difference is that the coefficient D_n in(9. 139), the expression of the function $R(\rho)$ is generally not zero due to the existence of the inner conductor. Bessel functions of the second kind $N_n(x)$ is needed to satisfy the boundary conditions on both surfaces of the inner and outer conductors. Of particular interests to us is the cutoff wavelengths, the longest of which is also associated with the TE_{11} mode and found to be

$$(\lambda_c)_{\text{TE11}} = \pi(a+b) \tag{9.168}$$

To restrain the non-TEM wave, the operating wavelength λ must be larger than the longest cutoff wavelength of these waves, i. e. ,

$$\lambda > \pi(a+b) \tag{9.169}$$

Therefore, given the operating wavelength λ or operating frequency f, the dimension of the coaxial line should satisfy

$$a+b < \frac{\lambda}{\pi} = \frac{1}{\pi f\sqrt{\varepsilon\mu}} \tag{9.170}$$

so that higher order modes are eliminated and only the TEM mode can propagate. Hence, in order to eliminate the higher order modes in a coaxial line, the dimensions have to be decreased as the frequency increases.

9.7 Cavity Resonators(谐振腔)

At microwave frequencies or above (larger than 300MHz), high-performance lumped-circuit elements such as R, L, and C become difficult to produce, as a lot of parasitic and distributed effects are not ignorable. Moreover, in GHz frequencies and above, the size of the circuits may become comparable to the operating wavelength, leading to radiations that may cause interfere with other circuits and systems. All these effects make it difficult to realize the traditional *L*-*C* resonating circuits with a high **quality factor** (品质因数). To provide high-Q **resonant circuits**(谐振电路) at microwave frequencies, metal cavities with conducting walls can be employed. A cavity resonator confines electromagnetic fields inside and provides large areas for current flow, which can effectively eliminate radiation and reduce loss. In this section, we study the properties (resonant frequencies and quality factors) of rectangular and circular cylindrical cavity resonators. ①

9.7.1 Rectangular Cavity Resonators(矩形谐振腔)

A rectangular cavity resonator can be considered as a rectangular waveguide with two ends closed by a conducting wall. The interior size of the cavity is $a \times b \times d$ as shown in Figure 9-16, with a assumed to be the longest side (i. e., $a>b$ and $a>d$). To be consistent with the analysis in previous sections, we select the z-axis as the reference longitudinal direction, along which the size is d. And the side with length a is arranged along the x-direction. Then the field patterns in the cavity can also be classified into TM and TE modes, and these resonant modes can be considered as the results of multiple reflections of the TM and TE waves of an $a \times b$ rectangular waveguide on the two end walls at $z=0$ and $z=d$. In the following, we analyze the TM and TE resonant modes separately. ②

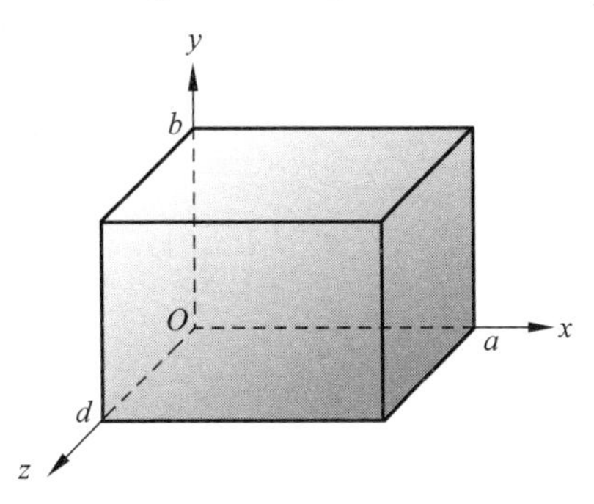

Figure 9-16 Rectangular cavity resonator

TM resonant modes

For TM resonant modes, we have $H_z=0$ and E_z is the only longitudinal field component. Solutions of E_z and the other transverse field components can be obtained by solving the Helmholtz's equation with boundary conditions on the six conducting walls. To do that, the method of separation of variables is applied. The solving process is similar but more complex than that in analyzing the TM waves in rectangular waveguides, because the simple transverse-longitudinal relations given by (9.28) does not apply here. In other words, E_z cannot be solved alone, and instead, needs to be solved together with E_x, E_y, H_x and H_y at the same time. However, since the TM resonant modes can be considered as the results of the multiple reflections of the guided TM waves, it is reasonable to expect that ①the transverse variation of the fields follows the same pattern as that of the TM waves of the rectangular waveguide, and ② the longitudinal variation of the fields must be in the form of a linear combination of sine and cosine functions. This can somehow simply the solving process.

Particularly, by referring to (9.76), we expect E_y in the rectangular cavity can be written as

---

① 随着频率的增高,电路的集总参数元件寄生效应不可忽略,且当电磁波的波长接近元件尺寸时,由集总参数元件组成的振荡回路容易产生辐射,损耗增大,因此微波频段的谐振电路通常采用谐振腔实现。谐振腔将电磁波包围在空腔内部,避免了辐射。同时,大面积的腔体表面能够显著降低电流流经的电阻,减小欧姆损耗,提高 Q 值。

② 谐振腔的工作原理是电磁波在腔体中来回反射形成振荡,因此,可以将谐振腔看作一段两端短路的波导,利用波导的场分布导出谐振腔的场分布以及相应参数。

$$E_y(x,y,z) = \sin\left(\frac{m\pi}{a}x\right)\cos\left(\frac{n\pi}{b}y\right)(A_1\cos k_z z + B_1\sin k_z z) \tag{9.171}$$

where A_1 and B_1 are two coefficients to be determined. k_z is a constant that must satisfy

$$\left(\frac{m\pi}{a}\right)^2 + \left(\frac{n\pi}{b}\right)^2 + k_z^2 = k^2 \tag{9.172}$$

because the E_y given by(9.171) must satisfy the scalar Helmholtz's equation. Notice that E_y must be zero at $z=0$ and $z=d$ according to the boundary conditions on a PEC surface. This means $A_1=0$ and $k_z=p\pi/d$. Therefore,(9.171) and(9.172) reduces to

$$E_y(x,y,z) = B_1\sin\left(\frac{m\pi}{a}x\right)\cos\left(\frac{n\pi}{b}y\right)\sin\left(\frac{p\pi}{d}z\right) \tag{9.173}$$

$$\left(\frac{m\pi}{a}\right)^2 + \left(\frac{n\pi}{b}\right)^2 + \left(\frac{p\pi}{d}\right)^2 = k^2 = \omega^2\varepsilon\mu \tag{9.174}$$

Similarly, by referring to(9.77), the solution of H_x can be written as

$$H_x(x,y,z) = \sin\left(\frac{m\pi}{a}x\right)\cos\left(\frac{n\pi}{b}y\right)\left[A_2\cos\left(\frac{p\pi}{d}z\right) + B_2\sin\left(\frac{p\pi}{d}z\right)\right] \tag{9.175}$$

where the same $k_z=p\pi/d$ is used because(9.172) and(9.174) must be satisfied for all the field components. To determine the coefficients A_2 and B_2, we use Maxell's equations:

$$\frac{\partial}{\partial z}H_x = \mathrm{j}\omega\varepsilon E_y \tag{9.176}$$

in which the fact that $H_z=0$ for TM modes has been used. By substituting(9.173) and(9.175) into(9.176), we have

$$\left(\frac{p\pi}{d}\right)\left[-A_2\sin\left(\frac{p\pi}{d}z\right) + B_2\cos\left(\frac{p\pi}{d}z\right)\right] = \mathrm{j}\omega\varepsilon B_1\sin\left(\frac{p\pi}{d}z\right)$$

which can be satisfied for all z only if $B_2=0$ and $A_2=-\mathrm{j}\omega\varepsilon B_1(d/p\pi)$. Consequently, (9.175) reduces to

$$H_x(x,y,z) = -\mathrm{j}\omega\varepsilon B_1\left(\frac{d}{p\pi}\right)\sin\left(\frac{m\pi}{a}x\right)\cos\left(\frac{n\pi}{b}y\right)\cos\left(\frac{p\pi}{\mathrm{d}}z\right) \tag{9.177}$$

By referring to(9.73), the solutions of E_z can be written as

$$E_z(x,y,z) = \sin\left(\frac{m\pi}{a}x\right)\sin\left(\frac{n\pi}{b}y\right)\left[A_3\cos\left(\frac{p\pi}{d}z\right) + B_3\sin\left(\frac{p\pi}{d}z\right)\right] \tag{9.178}$$

From Maxwell's equation,

$$\frac{\partial E_z}{\partial y} - \frac{\partial E_y}{\partial z} = -\mathrm{j}\omega\mu H_x \tag{9.179}$$

By substituting(9.173),(9.177) and(9.178) into(9.179), we have

$$\left(\frac{n\pi}{b}\right)\left[A_3\cos\left(\frac{p\pi}{d}z\right) + B_3\sin\left(\frac{p\pi}{d}z\right)\right] - \left(\frac{p\pi}{d}\right)E_{y0}\cos\left(\frac{p\pi}{d}z\right) = -k^2B_1\left(\frac{d}{p\pi}\right)\cos\left(\frac{p\pi}{d}z\right)$$

which can be satisfied for all z only if $B_3=0$ and

$$A_3 = \left(\frac{b}{n\pi}\right)\left(\frac{d}{p\pi}\right)\left[\left(\frac{p\pi}{d}\right)^2 - k^2\right]B_1 = -\left(\frac{b}{n\pi}\right)\left(\frac{d}{p\pi}\right)k_c^2B_1$$

where k_c is the same constant as given in(9.74), but not in the sense of cutoff wavenumber. Consequently, the expression of E_z of(9.178) is reduced to

$$E_z(x,y,z) = -\left(\frac{b}{n\pi}\right)\left(\frac{d}{p\pi}\right)k_c^2 B_1 \sin\left(\frac{m\pi}{a}x\right)\sin\left(\frac{n\pi}{b}y\right)\cos\left(\frac{p\pi}{d}z\right) \tag{9.180}$$

For better formulation, we let

$$E_z(x,y,z) = E_0 \sin\left(\frac{m\pi}{a}x\right)\sin\left(\frac{n\pi}{b}y\right)\cos\left(\frac{p\pi}{d}z\right) \tag{9.181}$$

where E_0 denotes the magnitude of the field E_z. Then the coefficient B_1 in(9.173) and(9.177) can be replaced by

$$B_1 = -\frac{1}{k_c^2}\left(\frac{n\pi}{b}\right)\left(\frac{p\pi}{d}\right)E_{z0}$$

which leads to

$$E_y(x,y,z) = -\frac{1}{k_c^2}\left(\frac{n\pi}{b}\right)\left(\frac{p\pi}{d}\right)E_0 \sin\left(\frac{m\pi}{a}x\right)\cos\left(\frac{n\pi}{b}y\right)\sin\left(\frac{p\pi}{d}z\right) \tag{9.182}$$

$$H_x(x,y,z) = \frac{j\omega\varepsilon}{k_c^2}\left(\frac{n\pi}{b}\right)E_0 \sin\left(\frac{m\pi}{a}x\right)\cos\left(\frac{n\pi}{b}y\right)\cos\left(\frac{p\pi}{d}z\right) \tag{9.183}$$

By following the similar procedures, we can obtain the expressions for E_x and H_y as

$$E_x(x,y,z) = -\frac{1}{k_c^2}\left(\frac{m\pi}{a}\right)\left(\frac{p\pi}{d}\right)E_0 \cos\left(\frac{m\pi}{a}x\right)\sin\left(\frac{n\pi}{b}y\right)\sin\left(\frac{p\pi}{d}z\right) \tag{9.184}$$

$$H_y(x,y,z) = -\frac{j\omega\varepsilon}{k_c^2}\left(\frac{m\pi}{a}\right)E_0 \cos\left(\frac{m\pi}{a}x\right)\sin\left(\frac{n\pi}{b}y\right)\cos\left(\frac{p\pi}{d}z\right) \tag{9.185}$$

In all the above expressions, the integers m, n and p denote the number of half-wave variations in the x-, y-, and z-direction, respectively. Each combination of the integers m, n and p corresponds to a resonant mode that is referred to as a TM_{mnp} mode. Equations(9.181)-(9.185) give the complete formulation of the fields of the resonant TM_{mnp} mode.

It is important to realize that, from(9.174), the **resonant frequency**(**谐振频率**) of the rectangular cavity can take only discrete values:

$$\omega_{mnp} = \frac{1}{\sqrt{\mu\varepsilon}}\sqrt{\left(\frac{m\pi}{a}\right)^2 + \left(\frac{n\pi}{b}\right)^2 + \left(\frac{p\pi}{d}\right)^2}$$

or

$$f_{mnp} = \frac{u}{2}\sqrt{\left(\frac{m}{a}\right)^2 + \left(\frac{n}{b}\right)^2 + \left(\frac{p}{d}\right)^2} \tag{9.186}$$

(9.186) states that the resonant frequency increases as the order of a mode becomes higher. It is seen from(9.181)-(9.185) that neither m nor n can be zero, but p can be zero. Therefore, TM_{110} mode has the lowest resonant frequency among all the TM modes.

TE resonant modes

For the TE modes, $E_z=0$. The phasor expressions for H_z, E_x, E_y, H_x and H_y can be derived in the similar way as is used for the TM modes. For example, the transverse variation of H_z must follow the same pattern as given in(9.95). The longitudinal variation of H_z must in the form of $\sin(p\pi z/d)$ as the normal magnetic field must vanish $z=0$ and $z=d$. Hence H_z can be expressed as,

$$H_z(x,y,z)=H_0\cos\left(\frac{m\pi}{a}x\right)\cos\left(\frac{n\pi}{b}y\right)\sin\left(\frac{p\pi}{d}z\right) \tag{9.187}$$

The other field components can be derived subsequently according to Maxwell's equations and the corresponding boundary conditions, leading to

$$E_x(x,y,z)=\frac{j\omega\mu}{k_c^2}\left(\frac{n\pi}{b}\right)H_0\cos\left(\frac{m\pi}{a}x\right)\sin\left(\frac{n\pi}{b}y\right)\sin\left(\frac{p\pi}{d}z\right) \tag{9.188}$$

$$E_y(x,y,z)=-\frac{j\omega\mu}{k_c^2}\left(\frac{m\pi}{a}\right)H_0\sin\left(\frac{m\pi}{a}x\right)\cos\left(\frac{n\pi}{b}y\right)\sin\left(\frac{p\pi}{d}z\right) \tag{9.189}$$

$$H_x(x,y,z)=-\frac{1}{k_c^2}\left(\frac{m\pi}{a}\right)\left(\frac{p\pi}{d}\right)H_0\sin\left(\frac{m\pi}{a}x\right)\cos\left(\frac{n\pi}{b}y\right)\cos\left(\frac{p\pi}{d}z\right) \tag{9.190}$$

$$H_y(x,y,z)=-\frac{1}{k_c^2}\left(\frac{n\pi}{b}\right)\left(\frac{p\pi}{d}\right)H_0\cos\left(\frac{m\pi}{a}x\right)\sin\left(\frac{n\pi}{b}y\right)\cos\left(\frac{p\pi}{d}z\right) \tag{9.191}$$

where k_c is the same constant as given in(9.74). Each combination of the integers m, n and p corresponds to a resonant mode that is referred to as a TE_{mnp} mode.

The expression for resonant frequency, f_{mnp}, remains the same as that for TM_{mnp} modes in (9.186). It is seen from(9.187)-(9.191) that either m or n(but not both m and n) can be zero, but that p cannot be zero. Therefore, the TE mode with the lowest resonant frequency could be TE_{101}(if $a>b$) or TE_{011}(if $b>a$).

For cavity resonators, different modes having the same resonant frequency are called **degenerate modes**(简并模式). Thus TM_{mnp} and TE_{mnp} modes are always degenerate if none of the mode indices is zero. The mode with the lowest resonant frequency for a given cavity size is referred to as the **dominant mode**(主模).

Example 9-5 Determine the dominant modes and their frequencies in an air filled rectangular cavity resonator for ①$a>b>d$, ②$a>d>b$, and ③$a=b=d$, where a, b, and d are the dimensions in the x-, y-, and z-direction, respectively.

Solution: The modes of the lowest orders are TM_{110}, TE_{011} and TE_{101}, with the resonant frequencies given by(9.186).

(1) For $a>b>d$, the lowest resonant frequency is

$$f_{110}=\frac{c}{2}\sqrt{\frac{1}{a^2}+\frac{1}{b^2}} \tag{9.192}$$

where c is the velocity of light in air. Therefore TM_{110} is the dominant mode.

(2) For $a>d>b$, the lowest resonant frequency is

$$f_{101}=\frac{c}{2}\sqrt{\frac{1}{a^2}+\frac{1}{d^2}} \tag{9.193}$$

Therefore, TE_{101} is the dominant mode.

(3) For $a=b=d$, all three of the lowest-order modes (namely, TM_{110}, TE_{011}, and TE_{101}) have the same field patterns. The resonant frequency of these degenerate modes is

$$f_{110} = \frac{c}{\sqrt{2}a} \tag{9.194}$$

Excitation of a rectangular cavity resonator

A resonant mode in a cavity (or a propagating mode in a waveguide) can be excited by using a coaxial line with its tip of the inner conductor protruding into the cavity in the form of a probe or loop. To excite a certain resonant mode, the source frequency from the coaxial line must be the same as the resonant frequency of the desired mode. In Figure 9-17 (a), the excitation is realized with the inner conductor of a coaxial line as a probe. The probe is placed in a location where the electric field of the desired mode is a maximum along the orientation of the probe. In this location, the probe can most effectively couple electromagnetic energy into the resonator (or waveguide). Alternatively, Figure 9-17 (b) illustrates excitation of a cavity resonator by using the inner conductor of a coaxial line as a small loop. The loop is placed in a position such that the magnetic flux density of the desired mode linking the loop is a maximum.

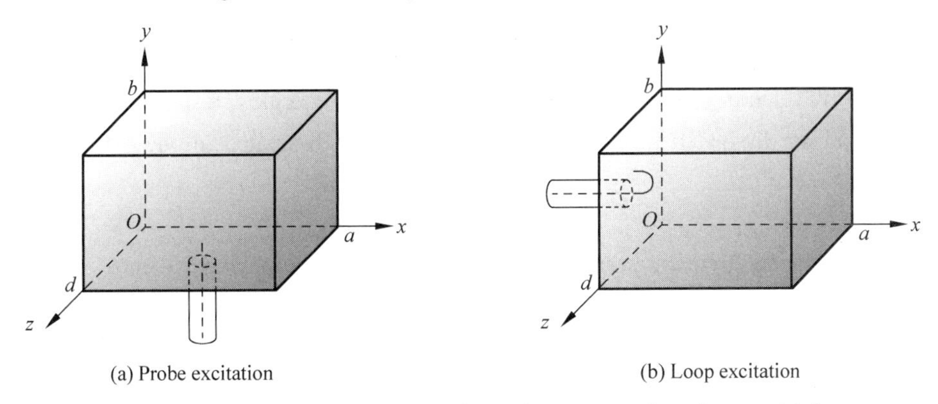

Figure 9-17 Excitation of a rectangular cavity resonator by using coaxial line

As an example, for the TE_{101} mode in an $a \times b \times d$ rectangular cavity, there are only three nonzero field components:

$$E_y = -\frac{j\omega\mu a}{\pi} H_0 \sin\left(\frac{\pi}{a}x\right) \sin\left(\frac{\pi}{d}z\right) \tag{9.195}$$

$$H_x = -\frac{a}{d} H_0 \sin\left(\frac{\pi}{a}x\right) \cos\left(\frac{\pi}{d}z\right) \tag{9.196}$$

$$H_z = H_0 \cos\left(\frac{\pi}{a}x\right) \sin\left(\frac{\pi}{d}z\right) \tag{9.197}$$

This mode may be excited by a probe inserted in the center region of the top or bottom face where E_y is maximum, as shown in Figure 9-17 (a), or by a loop to couple a maximum H_z placed inside the left or right face, as shown in Figure 9-17 (b).

From the field expressions (9.181)-(9.185) and (9.187)-(9.191), it is observed that, the electric field components are all in time quadrature with the magnetic field components, whether in TM or TE modes. As a result, the time-average Poynting vector is zero everywhere inside the cavity, which is as expected because the cavity is assumed to be lossless. In other words, the waves in the cavity must be standing waves. This contrasts with the field expressions in a waveguide, in which the transverse electric field components are in time phase with the transverse magnetic field components in a propagating mode, with the wave traveling along the longitudinal direction.

9.7.2 Quality Factor of Cavity Resonator(谐振腔的品质因数)

In practical cavities, the finite conductivity of the metal walls inevitably causes power loss, causing a dissipation of the stored energy in the electric and magnetic fields. Like in the resonant circuit theory, we define the quality factor, or Q, of a resonator as

$$Q = 2\pi \frac{\text{Time-average energy stored at a resonant frequency}}{\text{Energy dissipated in one period of this frequency}} \tag{9.198}$$

Apparently, higher Q factor indicates lower rate of energy loss, and narrower bandwidth can be achieved in realistic applications.

Let W be the total time-average energy in a cavity resonator. We can write

$$W = W_e + W_m \tag{9.199}$$

where W_e and W_m denote the stored electric and magnetic energies, respectively. If P_L is the time-average power dissipated in the cavity, then the energy dissipated in one period T is

$$Q = 2\pi \frac{W}{TP_L} = 2\pi \frac{fW}{P_L} = \frac{\omega W}{P_L} \tag{9.200}$$

where f is the resonant frequency and ω is the corresponding angular frequency. Therefore, to determine the Q of a cavity at a resonant frequency, we can first calculate W_e, W_m and P_L of the given resonant mode. To simplify the calculation, it is assumed that the loss is small enough so that the field patterns remain the same as if there is no loss. Hence the expressions (9.181)-(9.185) for TM modes and (9.187)-(9.191) for TE modes can be used directly.

As an example, consider calculation of Q of an $a\times b\times d$ cavity resonant at the TE_{101} mode. The fields of the TE_{101} mode have only three nonzero components as given in (9.195)-(9.197). The time average stored electric energy is

$$\begin{aligned} W_e &= \frac{\varepsilon_0}{4}\int |E_y|^2 \mathrm{d}v = \frac{\varepsilon_0\omega^2\mu_0^2\pi^2}{4k_c^4a^2}H_0^2\int_0^d\int_0^b\int_0^a \sin^2\left(\frac{\pi}{a}x\right)\sin^2\left(\frac{\pi}{d}z\right)\mathrm{d}x\mathrm{d}y\mathrm{d}z \\ &= \frac{\varepsilon_0\omega_{101}^2\mu_0^2a^2}{4\pi^2}H_0^2\left(\frac{a}{2}\right)b\left(\frac{d}{2}\right) \\ &= \frac{1}{4}\varepsilon_0\mu_0^2a^3bdf_{101}^2H_0^2 \end{aligned} \tag{9.201}$$

in which the fact that $k_c = \pi/a$ for the TE_{101} mode has been used. The total time-average stored magnetic energy is

$$W_m = \frac{\mu_0}{4}\int(|H_x|^2 + |H_z|^2)\,\mathrm{d}v$$

$$=\frac{\mu_0}{4}H_0^2\int_0^d\int_0^b\int_0^a\left[\frac{\pi^4}{k_c^4a^2d^2}\sin^2\left(\frac{\pi}{a}x\right)\cos^2\left(\frac{\pi}{d}z\right)+\cos^2\left(\frac{\pi}{a}x\right)\sin^2\left(\frac{\pi}{d}z\right)\right]\mathrm{d}x\mathrm{d}y\mathrm{d}z$$

$$=\frac{\mu_0}{4}H_0^2\left[\frac{a^2}{d^2}\left(\frac{a}{2}\right)b\left(\frac{d}{2}\right)+\left(\frac{a}{2}\right)b\left(\frac{d}{2}\right)\right]$$

$$=\frac{\mu_0}{16}abd\left(\frac{a^2}{d^2}+1\right)H_0^2 \tag{9.202}$$

Substituting(9.193) into(9.201), we can verify that $W_e=W_m$ at the resonant frequency f_{101} as expected. Hence, we have

$$W=2W_e=2W_m=\frac{\mu_0H_0^2}{8}abd\left(\frac{a^2}{d^2}+1\right) \tag{9.203}$$

To find P_L, we make use of(7.87), in which $|H_t|$ denotes the magnitude of the tangential component of the magnetic field and $R_S=\sqrt{\pi f_{101}\mu_0/\sigma}$ is the surface resistance of the conducting walls. Notice that the power loss in the $z=0$ wall is equal to that in the $z=d$ wall, the power loss in the $x=0$ wall is equal to that in the $x=a$ wall, and the power loss in the $y=0$ wall is equal to that in the $y=b$ wall. We have

$$P_L=2\times\left[\underbrace{\frac{1}{2}R_S\int_0^b\int_0^a|H_x|_{z=0}^2\mathrm{d}x\mathrm{d}y}_{\text{Power loss in the } z=0 \text{ wall}}+\underbrace{\frac{1}{2}R_S\int_0^d\int_0^b|H_z|_{x=0}^2\mathrm{d}y\mathrm{d}z}_{\text{Power loss in the } x=0 \text{ wall}}+\right.$$

$$\left.\underbrace{\frac{1}{2}R_S\int_0^d\int_0^a(|H_x|_{y=0}^2+|H_z|_{y=0}^2)\,\mathrm{d}x\mathrm{d}z}_{\text{Power loss in the } y=0 \text{ wall}}\right]$$

$$=\frac{R_SH_0^2a}{2}\left\{\frac{a^2}{d}\left(\frac{b}{d}+\frac{1}{2}\right)+d\left(\frac{b}{a}+\frac{1}{2}\right)\right\} \tag{9.204}$$

Using(9.203) and(9.204) in(9.200), we obtain

$$Q_{101}=\frac{\pi f_{101}\mu_0abd(a^2+d^2)}{R_S[2b(a^3+d^3)+ad(a^2+d^2)]} \tag{9.205}$$

where f_{101} has been given in (9.193).

Example 9-6 A cubic cavity made of copper has the dominant resonant frequency to be 10GHz. ① What is the size of the cavity? ② What is the Q of the cavity at the dominant resonant frequency?

Solution:

(1) From Example 9-5, TM_{110}, TE_{011} and TE_{101} are degenerate dominant modes with the resonant frequency to be

$$f_{101}=\frac{3\times10^8}{\sqrt{2}a}=10^{10}(\mathrm{Hz})$$

Therefore

$$a=\frac{3\times10^8}{\sqrt{2}\times10^{10}}=2.12\times10^{-2}(\mathrm{m})$$

$$=21.2(\mathrm{mm})$$

(2) By letting $b=d=a$ in(9.205), the Q of a cubic cavity with side length a is

$$Q=\frac{\pi f_{101}\mu_0 a}{3R_S}=\frac{a}{3}\sqrt{\pi f_{101}\mu_0\sigma} \tag{9.206}$$

For copper, $\sigma=5.80\times10^7$ S/m, we have

$$Q_{101}=\left(\frac{2.12}{3}\times10^{-2}\right)\sqrt{\pi10^{10}(4\pi\times10^{-7})(5.80\times10^{7})}=10700$$

9.7.3 Circular Cavity Resonator(圆形谐振腔)

A circular cavity resonator can be made from a circular waveguide with two ends enclosed by conducting walls. The analysis method is similar to that used for rectangular cavity resonators. Here we only discuss the TM_{010} resonant mode that has the simplest field distribution. The TM_{010} resonant mode can be considered as the TM_{01} mode in a circular waveguide operating at the cutoff frequency with no variation in the z-direction ($\beta=0$). That means the resonant frequency $(f)_{TM010}$ of a circular cavity is exactly the cutoff frequency of the TM_{01} mode of a circular waveguide with the same radius a. From(9.153), we have

$$(f)_{TM010}=\frac{c}{2\pi}(k_c)_{TM01}=\frac{1}{2\pi\sqrt{\mu_0\varepsilon_0}}\left(\frac{2.405}{a}\right)=\frac{0.115}{a}(\text{GHz}) \tag{9.207}$$

By letting $n=0,\beta=0$ and substituting(9.153) in the field component expressions in(9.142) and(9.148)-(9.152), we found that only two field components inside the cavity are nonzero, which are

$$E_z=C_0J_0(k_c\rho)=C_0J_0\left(\frac{2.405}{a}\rho\right) \tag{9.208}$$

$$H_\phi=-\frac{jC_0}{\eta_0}J_0'(k_c\rho)=\frac{jC_0}{\eta_0}J_1\left(\frac{2.405}{a}\rho\right) \tag{9.209}$$

where the relation $J_0'(k_c\rho)=-J_1(k_c\rho)$ has been used. The electric and magnetic field patterns for the TM_{010} mode are sketched in Figure 9-18. Notice that the electric and magnetic fields are in time quadrature, resulting in no time-average power flow in the cavity walls.

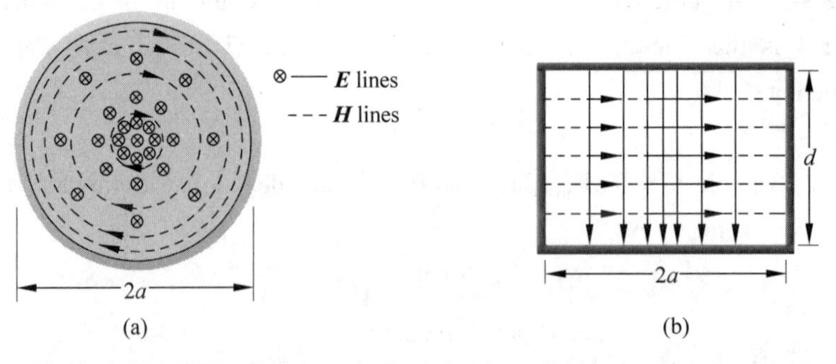

Figure 9-18 TM_{010} field patterns in a circular cylindrical cavity resonator

The Q of a circular cavity can be calculated by follow the same procedure as that used in for the rectangular resonator. As an example, we find the Q of a circular cylindrical cavity of radius a and length d for the TM_{010} mode as follows. From (9.208) and (9.209). The time-average stored energy is

$$W = 2W_e = \frac{\varepsilon_0}{2}\int_V |E_z|^2 \mathrm{d}v$$
$$= \frac{\varepsilon_0 C_0^2}{2}(2\pi d)\int_0^a J_0^2\left(\frac{2.405}{a}\rho\right)\rho \mathrm{d}\rho$$
$$= (\pi\varepsilon_0 d)C_0^2\left[\frac{a^2}{2}J_1^2(2.405)\right] \tag{9.210}$$

The time-average power loss is calculated by (7.87), in which $H_t = H_\phi$ on the cylindrical side walls as well as the two end walls. Therefore,

$$P_L = \underbrace{\frac{R_S}{2}(2\pi ad)\,|H_\phi|^2_{\rho=a}}_{\text{Power loss in the side wall}} + \underbrace{2\left(\frac{R_S}{2}\int_0^a |H_\phi|^2_{z=0}2\pi\rho \mathrm{d}\rho\right)}_{\text{Power loss in the } z=0 \text{ wall}}$$
$$= \frac{\pi R_S C_0^2}{\eta_0^2}\left[(ad)J_1^2(2.405) + 2\int_0^a J_1^2\left(\frac{2.405}{a}\rho\right)\rho \mathrm{d}\rho\right]$$
$$= \frac{\pi a R_S C_0^2}{\eta_0^2}(a+d)J_1^2(2.405) \tag{9.211}$$

Substituting (9.210) and (9.211) in (9.200), we obtain

$$(Q)_{\mathrm{TM010}} = \left(\frac{\eta_0}{R_S}\right)\frac{2.405}{2(1+a/d)} \tag{9.212}$$

where $R_S = \sqrt{\pi (f)_{\mathrm{TM010}}\mu_0/\sigma}$.

Example 9-7 A circular cavity resonator made of copper has the resonant frequency of TM_{010} mode to be 10GHz. The cavity has a length d equal to its diameter $2a$. ①Determine the sizes a and d. ②Find the Q of the cavity at the TM_{010} resonant mode.

Solution:

(1) From (9.207) we have

$$\frac{0.115}{a} = 10$$

Hence

$$a = 1.15\times10^{-2}(\mathrm{m}) = 1.15(\mathrm{cm})$$

and

$$d = 2a = 2.30(\mathrm{cm})$$

(2) First we calculate the surface resistance

$$R_S = \sqrt{\frac{\pi f\mu_0}{\sigma}} = \sqrt{\frac{\pi\times10^{10}\times(4\pi\times10^{-7})}{5.80\times10^7}} = 2.61\times10^{-2}(\Omega)$$

From (9.212) we obtain

$$Q = \left(\frac{377}{2.61\times10^{-2}}\right)\frac{2.405}{2(1+1/2)} = 11580$$

Summary

Concepts

Uniform waveguide(均匀波导)
TEM wave(横电磁波)
TE wave(横电波)
TM wave(横磁波)
Propagation/evanescent mode(传输/截止模式)
Cutoff wavenumber(截止波数)
Cutoff frequency(截止频率)
Cutoff wavelength(截止波长)
Guide wavelength(波导波长)
Wave impedance(波阻抗)
Rectangular waveguide(矩形波导)
Dominant mode(主模)
Circular waveguide(圆波导)
Cavity resonator(谐振腔)
Resonant frequency(谐振频率)
TM/TE resonant modes(横电/横磁谐振模式)
Degenerate modes(模式简并)
Quality factor(品质因数)

Laws & Theorems

Single-conductor hollow or dielectric-filled waveguides can support TE and TM waves.
Single-conductor hollow or dielectric-filled waveguides cannot support TEM waves.
The relations between transverse and longitudinal fields in a uniform wave guide.

Methods

Find the field patterns in a waveguide by using the method of separation of variables;
Determine the propagating TE/TM modes in a waveguide given its size;
Calculate the attenuation constant of the waveguide for specific propagating modes;
Determine the resonant modes and quality factor of a cavity resonator given its size.

Review Questions

9.1 什么是均匀波导? 波导中能传播均匀平面波吗?
9.2 在均匀波导内的介质或空气区域,电场 $\boldsymbol{E}$ 和磁场 $\boldsymbol{H}$ 满足什么控制方程?
9.3 均匀波导内传播的波有哪三种基本的类型?
9.4 均匀波导内 TEM 波的相速度和波阻抗分别是多少?
9.5 解释为什么单导体空腔或者填充介质波导内不能传播 TEM 波?
9.6 写出同轴波导内 TEM 波的电场 $\boldsymbol{E}$ 和磁场 $\boldsymbol{H}$ 的相量表达式。
9.7 波导内 TM 波的横纵场关系是什么?
9.8 什么是波导的截止频率和截止波长?
9.9 什么是截止模式?
9.10 波导中的波导波长与无界空间中的工作波长相比,哪个更长?
9.11 波导内 TE 波的横纵场关系是什么?
9.12 TM 和 TE 波的波阻抗与频率分别是什么关系?

9.13 波导内电磁波的波阻抗在什么情况下变为纯虚数？纯虚数的波阻抗的物理意义是什么？

9.14 如何从给定传播模式的 ω-β 图中确定其相速度和群速度？

9.15 矩形波导中 TM 波 E_z 满足的边界条件是哪些？

9.16 矩形波导中截止频率最低的模式是什么？

9.17 矩形波导中 TE 波 H_z 满足的边界条件是哪些？

9.18 波导的主模是如何定义的？矩形波导的主模是哪个模式？

9.19 矩形波导 TE_{10} 模式的截止波长是多少？

9.20 矩形波导 TE_{10} 模式的电磁波有哪些非零场分量？

9.21 描述矩形波导 TE_{10} 模式下波导壁上的表面电流流向。

9.22 波导内哪些因素会导致传输模式的波发生衰减？

9.23 矩形波导横截面尺寸的选择应考虑哪些因素？

9.24 圆波导的主模是什么模式？

9.25 什么是谐振腔？谐振腔最重要的性质是什么？

9.26 谐振腔中的电磁场是以行波还是驻波的形式存在？谐振腔中的电磁场与波导中的电磁场有什么异同？

9.27 $a\times b\times d$ 的矩形谐振腔 TM_{mnp} 模式和 TE_{mnp} 模式的谐振频率分别是多少？

9.28 什么是模式简并？

9.29 矩形谐振腔的最低谐振模式是什么模式？

9.30 品质因数是如何定义的？如何计算一个谐振腔的品质因数？

Problems

9.1 Write down the expressions of the surface current on the walls of a rectangular waveguide operating at the TM_{11} and TE_{11} mode respectively. Sketch the surface currents on the bottom and side walls.

9.2 A standard air-filled S-band rectangular waveguide has dimensions $a=7.21$cm and $b=3.40$cm. What mode types can be used to transmit electromagnetic waves having the following wavelengths?

(1) $\lambda=10$cm.

(2) $\lambda=5$cm.

9.3 In a hollow rectangular waveguide, the phase constant of a TE_{10} wave at 12GHz is 150rad/m. Given the shorter side length $b=1$cm, what is the value of the longer side length a?

9.4 An air-filled $a\times b$ (with $b<a<2b$) rectangular waveguide is designed to operate at 3GHz. It is required that the operating frequency is at least 20% higher than the cutoff frequency of the dominant mode and at least 20% lower than the cutoff frequency of the next higher-order mode.

(1) Determine the range of the dimensions a and b.

(2) Select a typical design of a and b, and find β, u_p, λ_g and Z_{TE10} at the operating

frequency.

9.5 Calculate the values of β, u_p, u_g, λ_g and Z_{TE10} for a 2.5cm × 1.5cm rectangular waveguide operating at 7.5GHz

(1) If the waveguide is hollow,

(2) If the waveguide is filled with a dielectric medium characterized by $\varepsilon_r = 2$, $\mu_r = 1$, and $\sigma = 0$.

9.6 An air-filled (10mm×5mm) rectangular waveguide operates in the dominant mode with a propagation constant of 200rad/m. The maximum amplitude of the electric field intensity is 250V/m. Find ①the operating frequency, ②the expressions of the fields, ③ the time-average power that is transmitted in the waveguide.

9.7 An air-filled rectangular waveguide made of copper with transverse dimensions $a=7.20$cm and $b=3.40$cm operates at 3GHz in the dominant mode. Find f_c, λ_g, α_c and the distance over which the magnitude of the fields is attenuated by 50%.

9.8 A 4cm×1cm copper rectangular waveguide is filled with a dielectric with loss tangent of 10^{-4}. For the operating frequency of 12GHz, determine the attenuation constant associated with the dominant mode.

9.9 Given an air-filled lossless rectangular cavity resonator with dimensions 8cm×6cm×5cm, find the twelve lowest-order modes and their resonant frequencies.

9.10 An air-filled rectangular cavity with copper walls has the dimensions 4cm × 3cm×5cm.

(1) Determine the resonant frequency and Q of the cavity associated with the dominant mode.

(2) Calculate the time-average stored electric and magnetic energies, when the electric field has a maximum magnitude of 1V/m.

9.11 Repeat solving Problem 9.10 if the rectangular cavity is filled with a lossless dielectric material with $\varepsilon_r = 2.5$.

9.12 An air-filled rectangular cavity resonator of length d is constructed by using a piece of a×b rectangular waveguide made of copper. The cavity operates in the TE_{101} mode.

(1) For fixed b, determine the relative magnitudes of a and d such that the cavity Q is maximized.

(2) Obtain an expression for Q as a function of a/b under the above conditions.

9.13 For an air-filled circular cylindrical cavity resonator of radius a and length d:

(1) Write the general expressions of the resonant frequencies and the corresponding wavelengths for TM_{mnp} and TE_{mnp} modes.

(2) For $d=a$, list the first seven modes that have the lowest resonant frequencies.

图 书 资 源 支 持

感谢您一直以来对清华大学出版社图书的支持和爱护。为了配合本书的使用，本书提供配套的资源，有需求的读者请扫描下方的“书圈”微信公众号二维码，在图书专区下载，也可以拨打电话或发送电子邮件咨询。

如果您在使用本书的过程中遇到了什么问题，或者有相关图书出版计划，也请您发邮件告诉我们，以便我们更好地为您服务。

我们的联系方式：

地　　址：北京市海淀区双清路学研大厦 A 座 714

邮　　编：100084

电　　话：010-83470236　010-83470237

资源下载：http://www.tup.com.cn

客服邮箱：tupjsj@vip.163.com

QQ：2301891038（请写明您的单位和姓名）

教学资源・教学样书・新书信息

人工智能科学与技术
人工智能|电子通信|自动控制

资料下载・样书申请

书圈

用微信扫一扫右边的二维码，即可关注清华大学出版社公众号。